Principles and Practice of Image-Guided Abdominal Radiation Therapy

Online at: https://doi.org/10.1088/978-0-7503-2468-7

About the Series

The series in Physics and Engineering in Medicine and Biology will allow the Institute of Physics and Engineering in Medicine (IPEM) to enhance its mission to 'advance physics and engineering applied to medicine and biology for the public good'.

It is focused on key areas including, but not limited to:
- clinical engineering
- diagnostic radiology
- informatics and computing
- magnetic resonance imaging
- nuclear medicine
- physiological measurement
- radiation protection
- radiotherapy
- rehabilitation engineering
- ultrasound and non-ionising radiation.

A number of IPEM–IOP titles are being published as part of the EUTEMPE Network Series for Medical Physics Experts.

A full list of titles published in this series can be found here: https://iopscience.iop.org/bookListInfo/physics-engineering-medicine-biology-series.

Principles and Practice of Image-Guided Abdominal Radiation Therapy

Edited by

Edited by
Yu Kuang
University of Nevada, Las Vegas, Las Vegas, NV, USA

IOP Publishing, Bristol, UK

ISBN   978-0-7503-2468-7 (ebook)
ISBN   978-0-7503-2466-3 (print)
ISBN   978-0-7503-2469-4 (myPrint)
ISBN   978-0-7503-2467-0 (mobi)

DOI   10.1088/978-0-7503-2468-7

Version: 20230201

IOP ebooks

British Library Cataloguing-in-Publication Data: A catalogue record for this book is available from the British Library.

Published by IOP Publishing, wholly owned by The Institute of Physics, London

IOP Publishing, No.2 The Distillery, Glassfields, Avon Street, Bristol, BS2 0GR, UK

US Office: IOP Publishing, Inc., 190 North Independence Mall West, Suite 601, Philadelphia, PA 19106, USA

Contents

Part II Principles of image-guided radiation therapy for abdominal cancer

3 Imaging simulation for abdominal cancer image-guided radiation therapy   **3-1**
Emily Hirata and Manju Sharma

4 Treatment planning for abdominal cancer   **4-1**
Bingqi Guo and Peng Qi

Part IV Advances in image-guided radiation therapy for abdominal cancer

15 Advances in simulation imaging for abdominal radiotherapy   15-1
Chia-ho Hua, Hui Wang and Yu Kuang

23 Synthetic computed tomography for abdominal image-guided radiation therapy

Richard L J Qiu, Yang Lei, Tonghe Wang, Walter J Curran, Tian Liu and Xiaofeng Yang

Editor biography

Yu Kuang

Yu Kuang is an associate professor of medical physics and director of the Commission on Accreditation of Medical Physics Education Programs–accredited professional doctorate in medical physics degree program at University of Nevada, Las Vegas. He is also an American Board of Radiology–certified therapeutic medical physicist. His research comprises the development and clinical integration of novel medical imaging devices with medical linear accelerator and proton therapy devices, real-time image-guided and adaptive radiation therapy, combining biological and imaging biomarkers for early detection of cancers and cancer interventions, and biological imaging for radiation biology and clinical applications.

List of contributors

Ergun Ahunbay
Medical College of Wisconsin, Milwaukee, WI, USA

Ronik Bhangoo
Mayo Clinic in Arizona, Phoenix, AZ, USA

Joy Carter
Alliance Oncology, Scottsdale, AZ, USA

David Chamberlain
University of Nevada, Las Vegas, Alliance Oncology, Reno, NV, USA

Maria F Chan
Department of Radiation Oncology, Memorial Sloan Kettering Cancer Center, New York, NY, USA

Ting Chen
Department of Radiation Oncology, New York University, NY, USA

Xinfeng Chen
Medical College of Wisconsin, Milwaukee, WI, USA

Taoran Cui
Rutgers, Cancer Institute of New Jersey, New Brunswick, NJ, USA

Walter J Curran
Emory University, Atlanta, GA, USA

Bingqi Guo
Cleveland Clinic, Cleveland, OH, USA

Wendy Harris
Duke University, Durham, NC, USA

Emily Hirata
UCSF Radiation Oncology, University of California San Francisco, San Francisco, CA, USA

Dimitre Hristov
Stanford University, Stanford, CA, USA

Yanle Hu
Mayo Clinic in Arizona, Phoenix, AZ, USA

Chia-ho Hua
St. Jude Children's Research Hospital, Memphis, TN, USA

Xun Jia
Johns Hopkins University, Baltimore, MD, USA

Natalia Kovachuk
Radiation Oncology Department, Stanford University, Stanford, CA, USA

Yu Kuang
University of Nevada, Las Vegas, Las Vegas, NV, USA

Yang Lei
Emory University, Atlanta, GA, USA

X Allen Li
Medical College of Wisconsin, Milwaukee, WI, USA

Xiadong Li
Department of Radiation Oncology, Affiliated Hangzhou Cancer Hospital Zhejiang University School of Medicine, Hangzhou First People's Hospital Group, Hangzhou, Zhejiang, China

Seng Boh Lim
Department of Medical Physics, Memorial Sloan Kettering Cancer Center, New York, USA

Mu-Han Lin
University of Texas Southwestern Medical Center, Dallas, TX, USA

Bo Liu
Rutgers, Cancer Institute of New Jersey, New Brunswick, NJ, USA

Tian Liu
Emory University, Atlanta, GA, USA

Wei Liu
Mayo Clinic in Arizona, Phoenix, AZ, USA

Chi Ma
Rutgers, Cancer Institute of New Jersey, New Brunswick, NJ, USA

Shenglin Ma
Department of Radiation Oncology, Affiliated Hangzhou Cancer Hospital Zhejiang University School of Medicine, Hangzhou First People's Hospital Group, Hangzhou, Zhejiang, China

Tianjun Ma
Department of Radiation Oncology, Virginia Commonwealth University, Richmond, VA, USA

Boris Mueller
Department of Radiation Oncology, Memorial Sloan Kettering Cancer Center, New York, NY, USA

Dan Nguyen
University of Texas Southwestern Medical Center, Dallas, TX, USA

Ke Nie
Rutgers, Cancer Institute of New Jersey, New Brunswick, NJ, USA

Adam Paxton
University of Utah, Salt Lake City, UT, USA

Eric Paulson
Medical College of Wisconsin, Milwaukee, WI, USA

Gregory P Penoncello
Mayo Clinic in Arizona, Phoenix, AZ, USA

Erqi Pollom
Radiation Oncology Department, Stanford University, Stanford, CA, USA

Peng Qi
Cleveland Clinic, Cleveland, OH, USA

Richard L J Qiu
Emory University, Atlanta, GA, USA

Lei Ren
Duke University Medical Center, Durham, NC, USA

Meral Reyhan
Rutgers, Cancer Institute of New Jersey, New Brunswick, NJ, USA

Daniel Robertson
Mayo Clinic in Arizona, Phoenix, AZ, USA

William Rule
Mayo Clinic in Arizona, Phoenix, AZ, USA

Manju Sharma
UCSF Radiation Oncology, University of California San Francisco, San Francisco, CA, USA

Yulin Song
Department of Radiation Oncology, Memorial Sloan Kettering Cancer Center, New York, NY, USA

Grace Tang
Department of Medical Physics, Memorial Sloan Kettering Cancer Center, New York, USA

Carlos Vargas
Mayo Clinic in Arizona, Phoenix, AZ, USA

Irina Vergalasova
Rutgers, Cancer Institute of New Jersey, New Brunswick, NJ, USA

Hesheng Wang
Department of Radiation Oncology, New York University, NY, USA

Hui Wang
University of Nevada, Las Vegas, Las Vegas, NV, USA

Jing Wang
University of Texas Southwestern Medical Center, Dallas, TX, USA

Lei Wang
Radiation Oncology Department, Stanford University, Stanford, CA, USA

Tonghe Wang
Emory University, Atlanta, GA, USA

Xiao Wang
Rutgers, Cancer Institute of New Jersey, New Brunswick, NJ, USA

Lei Xing
Stanford University, Stanford, CA, USA

Xudong Xue
Department of Radiation Oncology, Hubei Cancer Hospital, Tongji Medical College, Huazhong University of Science and Technology, Wuhan, China

Xiaofeng Yang
Emory University, Atlanta, GA, USA

Yidong Yang
University of Science and Technology of China, Hefei, China

Yunze Yang
Mayo Clinic in Arizona, Phoenix, AZ, USA

Hao Yao
Washington University in St. Louis, St. Louis, MO, USA

Amy Yu
Stanford University, Stanford, CA, USA

Naichang Yu
Radiation Oncology, Cleveland Clinic, Cleveland, OH, USA

Ning Yue
Rutgers, Cancer Institute of New Jersey, New Brunswick, NJ, USA

Yin Zhang
Rutgers, Cancer Institute of New Jersey, New Brunswick, NJ, USA

You Zhang
University of Texas Southwestern Medical Center, Dallas, TX, USA

Hui Zhao
University of Utah, Salt Lake City, UT, USA

Tianyu Zhao
Washington University in St. Louis, St. Louis, MO, USA

Wei Zhao
Stanford University, Palo Alto, CA, USA

Part I

Introduction

IOP Publishing

Principles and Practice of Image-Guided Abdominal
Radiation Therapy

Yu Kuang

Chapter 1

Overview of image-guided radiation therapy

Irina Vergalasova, Ning J Yue and Meral Reyhan

This chapter is dedicated to presenting a general overview of image-guided radiation therapy (IGRT). We describe the evolution of various imaging modalities within the field of radiation oncology by first starting with two-dimensional (2D) methods of imaging such as kilovoltage (kV) and megavoltage (MV) projection images, ultrasound imaging and magnetic resonance imaging (MRI). This is then further expanded upon by emphasizing the critical role three-dimensional (3D) imaging technologies play in support of achieving accurate radiation treatments. The volumetric imaging techniques discussed include megavoltage cone-beam CT (MV-CBCT), kilovoltage cone-beam CT (kV-CBCT), CT systems that have the ability to translate around a stationary patient (CT-on-rails) and, lastly, MRI.

1.1 Evolution of IGRT

Imaging has always been and will continue to be the foundation of precisely and accurately delivering radiation treatment. It plays a critical role across all stages of radiotherapy, from initial diagnosis to simulation and daily verification prior to— and often during—treatment delivery, all the way to follow-up and surveillance of treatment response. Over the past few decades, the term IGRT has significantly evolved along with the advancements of multiple modalities that are now accessible immediately prior to and/or during radiation treatment delivery. Since the early days of radiation therapy with Cobalt-60 machines, the field has significantly progressed to linear accelerators with 3D conformal radiation therapy at first and now with intensity-modulated radiation therapy. This progression in linear accelerator hardware and delivery techniques has largely been supported by developments made in the imaging arena. Their evolution together has enabled radiation oncologists to make significant strides in treatment outcomes and reduced toxicities by way of hypofractionated treatment schema and tighter margins on target volumes. The following sections will summarize the history and explain the technological progression of different imaging modalities that play a role in modern-day IGRT in the context of using image guidance to monitor inter- and intrafractional changes in patient setup and anatomy during the course of radiotherapy.

1.2 2D imaging technology

1.2.1 MV

The importance of imaging in guiding radiotherapy treatment was realized not long after its inception. The first example of IGRT on a linear accelerator involved the use of the treatment MV beam itself to create a 2D projection of the anatomy onto a portal film—with the shape of the treatment field overlaying the anatomy—to ensure treatment accuracy in terms of anatomic location and size. Since the photon energies were in the MV range, the film required the use of metallic screens and specific exposure techniques to result in optimal image resolution [1]. With the development of the digital age and all of the associated difficulties of using film (i.e. tedious procedure of processing, delayed viewing due to additional time required for processing, poor image resolution for energies above 6 MV, etc), the use of films was replaced with electronic portal imaging detectors (EPIDs). Throughout its history, detector technology has been employed in several different iterations (video cameras, liquid ion chambers, etc) to achieve online digital imaging [2, 3]. Today, the most modern form of an EPID consists of an amorphous silicon flat panel array of solid state detectors mounted onto a retractable arm that is attached to the linac, directly facing—with the panel plane orthogonal to—the treatment beamline and directly underneath the couch to collect the exit photons.

Historically, the MV portal film that was acquired on the day of treatment was compared to a 2D kV film acquired at simulation to initiate treatment. These two films (kV vs MV) with entirely different resolution and contrast were compared and deemed acceptable if they were within ±5 mm. The MV films were only acquired on a weekly basis for treatment verification [4]. Today, MV portal images are still largely used for treatment field verification, but they are immensely improved in terms of contrast and resolution (since the days of MV films) and are registered online with digitally reconstructed radiographs (DRRs) generated from the 3D CT scan acquired during simulation and used for treatment planning. Although this still involves matching radiographs with different image contrasts (MV to kV), sub-millimeter accuracy can be achieved when aligning the patient at the time of delivery to the setup acquired during simulation, upon which the entire treatment regimen is solely based. Currently, 2D MV portal imaging is primarily used for the setup and beams-eye-view verification of 3D conformal radiation therapy fields immediately prior to treatment delivery.

1.2.2 kV

Similar to MV imaging, 2D kV imaging also first originated with the use of radiographic film (even before the development of MV film imaging). There are multiple advantages of imaging anatomy with kV energies to that of MV energies, with the main advantages being markedly improved contrast and resolution due to the dominance of photoelectric effect in the MV energy range versus that of the Compton effect for the MV energy range. kV imaging also delivers much less of a radiation dose to the patient than that of MV. Moreover, it is much simpler to accelerate electrons and generate photons at the kV level than at the MV level with a simple cathode ray tube, as it is to stop and detect them upon exit. As with the disappearance of MV films, kV films have also almost entirely been replaced with the digital equivalent for the same inconveniences of required processing time and materials. Instead, an onboard imaging system consisting of a kV source and flat panel detector attached on retractable arms situated orthogonal to the gantry head is seen on virtually all modern-day linacs, serving as the major backbone of IGRT, as it stands today [5]. The same amorphous silicon technology is employed in flat panel arrays for kV image generation.

Typically, kV-based IGRT involves acquiring a pair of orthogonal radiographs on the treatment couch and registering the treatment anatomy to DRRs generated from the simulation CT at those same orthogonal angles. Depending on the anatomical site, treatment modality (photons, electrons or protons) and/or treatment technique (3D conformal, intensity-modulated or stereotactic radiation therapy), the 2D kV alignment may be followed by additional 3D imaging or simply by beam-on delivery. The volumetric imaging capabilities of this onboard imager system will be later discussed in section 1.3 on 3D imaging. Although 2D radiographs cannot reveal the same level of soft-tissue detail as volumetric imaging, they are clinically valuable for target alignment relative to bony anatomy and especially to implanted fiducial markers. Furthermore, 2D fluoroscopy, which involves acquiring serial 2D radiographs at a predefined frame rate for a specific duration, enables live monitoring of respiratory-induced motion [6–8]. Fluoroscopy, in particular when combined with fiducial markers implanted into the tumor, enables real-time tumor tracking and even gating and repositioning of the beam (based on live updated modeling) to predict and subsequently account for the intrafractional target motion. It is important to note that the kV x-ray tubes and detectors do not have to be mounted on an arm attached to the linac but can also be positioned within the ceiling and floor, as commercially offered by the ExacTrac system (BrainLab AG, Munich, Germany) and CyberKnife (Accuray Inc., Sunnyvale, CA) [9–12].

1.2.3 Ultrasound

Most prevalent in high-dose-rate (HDR) and low-dose-rate (LDR) brachytherapy treatments [13–20], ultrasound-guided imaging takes advantage of the physics of sound waves and thus does not require any ionizing radiation, nor does it contribute additional dose to the patient. Ultrasound systems come with a variety of probe applications such as transabdominal, transrectal and transvaginal, all of which function via the piezoelectric effect to emit sound waves at a specific frequency. The transmitted sound waves travel into the body and are reflected back by different tissue boundaries with different arrival times back to the crystals within the probe. Using this information to calculate the distance of the different boundaries, based on the time of arrival back to the probe and the speed of sound, a 2D image of grey intensities is generated. Although the image quality and resolution cannot compare to those of kV x-rays, its ability to image internal anatomy without any additional radiation dose has enabled ultrasound imaging to have a strong presence in both LDR and HDR prostate brachytherapy. A transrectal probe guides the insertion of needles either with LDR seeds through the perineum to be implanted throughout the prostate for a permanent implant, or, in the case of HDR, just the needles are implanted for the duration of treatment planning and delivery via a remote afterloader. Furthermore, transabdominal ultrasound imaging has been shown to aid the insertion of intracavitary applicators and prevent perforation of the uterus when treating gynecological malignancies with brachytherapy. Ultrasound imaging has even aided the insertion of brachytherapy applicators and needles to treat breast malignancies as well.

Nevertheless, ultrasound imaging does have its disadvantages. Ultrasound images can be difficult to interpret due to the grainy image quality and less familiarity and experience with this modality, leading to large user variability and thus necessitating larger margin expansions for prostate external beam radiotherapy [21–23]. The main disadvantage of utilizing ultrasound for external beam radiotherapy position verification is the induced anatomical distortions and displacements required to obtain the image by applying pressure with the probe itself [24]. However, significant research efforts remain underway to improve upon the shortcomings of this technology and innovate ways to aid inter- and intrafractional target positioning for radiotherapy applications [25].

1.2.4 MRI

Another imaging modality that does not require the use of ionizing radiation is MRI. This form of electromagnetic radiation exploits the water content of human anatomy with the use of strong magnetic fields. The magnet housed by the scanner forces all of the protons in the body to align with the applied magnetic field, and this is followed by the application of different radiofrequency pulses to purposely disrupt the proton alignment to measure the emitted signal (i.e. this creates the images) of the protons returning back to their rest state of alignment. These relaxation rates are different for different types of tissue, which is what gives rise to the subtle soft-tissue contrast displayed by MRI images. The soft-tissue contrast and lack of additional radiation exposure together would make MRI the most ideal imaging modality if it were not for the high cost of maintaining the MRI machine itself and the required long acquisition scan times. Furthermore, many patients have a difficult time tolerating MRI scanners due to claustrophobia and the loud noises of the gradients switching on and off. Despite this, the benefits of MRI-guided radiotherapy are undeniable in terms of truly achieving adaptive planning and accounting for not just inter- and intrafractional anatomical motion but also for the changes in tumor size and shape that often occur throughout the course of treatment.

In fact, several commercial vendors have capitalized on this benefit and currently offer MRI-guided radiotherapy linacs. The ViewRay MRIdian MRI-guided Linac

(ViewRay Technologies Inc., Oakwood Village, OH) consists of a split 0.35T magnetic resonance (MR) scanner with three Cobalt-60 heads mounted on a ring gantry, each with their own multileaf collimators [26]. The more recent and competing system is the Elekta Unity MR-Linac (Elekta, Stockholm, Sweden), which consists of a 7 MV linac that rotates around a closed bore 1.5T magnet. Although these two systems each have their own strengths and weaknesses that are beyond the scope of this chapter, both share a slew of benefits in regard to achieving real-time MRI-guided radiotherapy. In addition to the vastly superior soft-tissue contrast, MRI is also better able at resolving respiratory-induced motion within the slice of imaged anatomy because the signal resulting from relaxation (i.e. T_1, T_2, etc) can be acquired within a fraction of a second. It is important to note, however, that this only produces a 2D image quickly, and obtaining a volumetric MRI takes a significant amount of acquisition and processing time. The speed of the 2D acquisition allows for dynamic MRI, which is beneficial for extracting respiratory-induced target motion by associating each acquired slice with a respiratory phase. This capability is extremely advantageous in tracking motion of both the tumor and surrounding organs-at-risk in real-time during radiation delivery, particularly for hypofractionated treatments (where extremely large doses are delivered in few fractions), requiring superior accuracy and precision. Onboard MRI imaging also provides the ability to acquire functional information and potentially track and predict treatment response early on in the course of radiation, thereby allowing for treatment modification. Moreover, the ability to capture images without added radiation exposure provides a much larger dataset for further quantitative imaging research [27].

Some of the potential challenges of using MRI online to guide radiotherapy include those inherent to all MRI imaging, such as geometric distortion, chemical shift and other magnetic field inhomogeneities that may cause imaging artifacts, obstructing the anatomy-of-interest. Also, certain patients would not be eligible due to presence of pacemakers or certain other metal implants. Furthermore, the long acquisition times may lead to increased probability of patient motion, thus inducing motion blur in the acquired images. Despite the aforementioned challenges, MRI-guided radiotherapy is a powerful tool, and its measurable clinical impact is yet to be determined with ongoing trials and data collection.

1.3 3D imaging technology

1.3.1 MV-CBCT

As the ability to create highly conformal radiation treatment plans improves, the need for precisely positioning patients also increases. CBCT using a treatment beam (e.g. MV) in combination with the EPID gives physicians access to 3D patient images, improving the accuracy of pretreatment patient positioning. Projection images are acquired using the radiotherapy linear accelerator with photons primarily in the MV energy range. Open-field projection images acquired at different rotational positions around the patient are used to create the CBCT. Unlike conventional CT, which utilizes a fan beam geometry with multiple rows of detectors to form an image by translating the patient, CBCT utilizes a 2D detector array without multiple gantry rotations, table movement or slice artifacts. A convolution back projection algorithm is used to reconstruct the 2D projection data into a 3D volume [28–30]. A 3D rendering of the patient anatomy in the treatment position facilitates more accurate registration to the planning CT prior to treatment. The field of view is governed by the source-to-imager-distance, up to 27 cm with a slice thickness between 0.5 and 10 mm [31]. In half-beam acquisition mode, the reconstruction size of the axial plane can be increased to 40 cm. The typical dose to the patient ranges from 3 to 10 cGy [31].

A major advantage of the MV-CBCT system is the shared use of the linac beam for treatment and imaging, making the system both inexpensive and convenient to use. The shared radiation isocenter also makes quality assurance simpler. MV-CBCT also reduces the image artifacts caused by materials of relatively high atomic number, which are implanted inside the patient's body, allowing improved image visualization. However, as contrast-to-noise ratio (CNR) is determined by the differential attenuation of the beam through different tissues, and since Compton scattering provides the majority of beam attenuation, the resulting contrast is relatively constant in MV imaging. Thus, MV-CBCT has largely been replaced by kV-CBCT.

1.3.2 kV-CBCT

First commercially available for dental imaging, kV-CBCT has rapidly gained a foothold in IGRT [32]. For kV-CBCT–based image guidance, additional hardware is installed on the linac. In almost all modern linacs, an x-ray tube and flat panel imager are mounted to the gantry orthogonally to the treatment beam. This hardware can also be used to acquire 2D x-ray planar images. Prior to treatment, a patient is set up in the same manner as at simulation, and the gantry rotates slowly around the patient to acquire the projection data for CBCT reconstruction. The shape of the x-ray beam is conical, hence the name cone-beam CT. The Varian system uses custom 'bow-tie' filters to improve the homogeneity of the x-ray intensity laterally across the detector for different modes of acquisition. For most systems, the imaging panel can be shifted to create smaller and larger fields of view [33].

The flat panel detector is used in fluoroscopy mode, and multiple projections per second are acquired. The projection data are reconstructed to create a volumetric image. While the contrast is not as good as a diagnostic CT, the image quality is sufficient for bony and most soft-tissue image registration. As a result of the longer imaging acquisition time, blurring due to motion can impact image quality as well. However, the high resolution of the flat panel detector allows for isotropic submillimeter voxels. The dose from kV-CBCT can range from 0.2 to 2 cGy [31].

The acquired images are primarily used for verifying the patient treatment position by registration with the simulation CT. However, the kV images can also be used for adaptive radiotherapy planning if Hounsfield unit values and an electron density calibration curve are obtained for dose calculation. CBCT is used for IGRT for a number of treatment sites: head and neck, lung, prostate, etc. Studies have demonstrated agreement between CBCT imaging and MV portal imaging with fiducial markers in the prostate, which cannot be directly viewed in projection imaging without markers [34]. CBCT can be used to verify tumor motion as a function of respiratory motion in the lungs and abdomen [35, 36]. Four-dimensional CBCT is commercially available from multiple linear accelerator vendors. The different phases of tumor motion can be observed to better ensure the target does not move outside of the planned target volume [37].

Several different types of artifacts can influence kV-CBCT imaging, including ring artifacts, scatter and noise, beam hardening and aliasing. Iterative reconstruction algorithms can be utilized to improve traditional image reconstruction to reduce noise and artifacts [38]. Additional mapping—i.e. flex mapping—is utilized to correct for tube and imaging panel sag to ensure imaging and treatment isocenters match. More quality assurance testing is needed due to the addition of the x-ray source and imaging panel when compared with MV-CBCT.

1.3.3 CT-on-rails

CT-on-rails consists of a diagnostic CT scanner that is integrated into the treatment room. The gantry of the CT scan is mounted on rails so that it can move across the patient instead of the couch moving the patient through the scanner. By rotating the treatment couch, the couch is aligned with the CT gantry motion, and pretreatment imaging can be performed while the patient is immobilized. After imaging is completed, the couch is rotated back for treatment. This IGRT system has been used for frameless stereotactic treatments, fiducial-less prostate localization, studies of organ motion and anatomical changes [39–42]. One major advantage of CT-on-rails is that a diagnostic quality CT image is acquired for registration/confirmation of patient setup against the planning CT. It can also be used directly for adaptive replanning of patient treatment dose if there are interfractional changes to the patient anatomy. However, the quality assurance is more extensive due to the separate treatment unit and imaging isocenters.

1.3.4 MRI

The main benefit to volumetric image acquisition in MRI is improved spatial resolution, making thin-slice imaging possible. As slice thickness decreases, the CNR in MRI also decreases. To acquire really thin slices, rather than acquiring multiple 2D slices with compromised CNR, volumetric imaging is utilized. The idea behind 3D imaging in MR is to simply apply a second phase-encoding axis in the slice direction. In this way, spatial encoding in that dimension is achieved. However, as a result of this second phase encoding, the scan time will also increase with the number of phase-encoding steps [43]. There are a limited number of 3D MR pulse sequences available for IGRT, including gradient readout echo, turbo flash and true fast imaging with steady-state free precession.

References

[1] Droege R T and Bjarngard B E 1979 Influence of metal screens on contrast in megavoltage x-ray imaging *Med. Phys.* **6** 487–93

[2] Boyer A L *et al* 1992 A review of electronic portal imaging devices (EPIDs) *Med. Phys.* **19** 1–16

[3] Antonuk L E 2002 Electronic portal imaging devices: a review and historical perspective of contemporary technologies and research *Phys. Med. Biol.* **47** R31–65

[4] Nguyen N P and Karlsson U L 2015 Editorial: image-guided radiotherapy for effective radiotherapy delivery *Front Oncol.* **5** 253

[5] Jaffray D A *et al* 2002 Flat-panel cone-beam computed tomography for image-guided radiation therapy *Int. J. Radiat. Oncol. Biol. Phys.* **53** 1337–49

[6] Covington E E 1966 Fluoroscopy technique for tumor localization *JAMA* **198** 180–2

[7] Leong J 1986 Use of digital fluoroscopy as an on-line verification device in radiation therapy *Phys. Med. Biol.* **31** 985–92

[8] Shirato H *et al* 2000 Four-dimensional treatment planning and fluoroscopic real-time tumor tracking radiotherapy for moving tumor *Int. J. Radiat. Oncol. Biol. Phys.* **48** 435–42

[9] Murphy M J 2004 Tracking moving organs in real time *Semin. Radiat. Oncol.* **14** 91–100

[10] Nuyttens J J *et al* 2006 Lung tumor tracking during stereotactic radiotherapy treatment with the CyberKnife: marker placement and early results *Acta Oncol.* **45** 961–5

[11] Verellen D *et al* 2010 Gating and tracking, 4D in thoracic tumours *Cancer Radiother.* **14** 446–54

[12] Willoughby T R *et al* 2006 Evaluation of an infrared camera and X-ray system using implanted fiducials in patients with lung tumors for gated radiation therapy *Int. J. Radiat. Oncol. Biol. Phys.* **66** 568–75

[13] Brascho D J, Kim R Y and Wilson E E 1978 Use of ultrasonography in planning intracavitary radiotherapy of endometrial carcinoma *Radiology* **129** 163–7

[14] Holm H H *et al* 1983 Transperineal ^{125}iodine seed implantation in prostatic cancer guided by transrectal ultrasonography *J. Urol.* **130** 283–6

[15] Stock R G *et al* 1997 A new technique for performing Syed-Neblett template interstitial implants for gynecologic malignancies using transrectal-ultrasound guidance *Int. J. Radiat. Oncol. Biol. Phys.* **37** 819–25

[16] Van Dyk S *et al* 2009 Conformal brachytherapy planning for cervical cancer using transabdominal ultrasound *Int. J. Radiat. Oncol. Biol. Phys.* **75** 64–70

[17] Sharma D N *et al* 2010 Use of transrectal ultrasound for high dose rate interstitial brachytherapy for patients of carcinoma of uterine cervix *J. Gynecol. Oncol.* **21** 12–7

[18] Watkins J M *et al* 2011 Ultrasound-guided tandem placement for low-dose-rate brachytherapy in advanced cervical cancer minimizes risk of intraoperative uterine perforation *Ultrasound Obstet. Gynecol.* **37** 241–4

[19] Morton G C 2015 Prostate high-dose-rate brachytherapy: transrectal ultrasound based planning, a technical note *Pract. Radiat. Oncol.* **5** 238–40

[20] Lauche O *et al* 2016 Single-fraction high-dose-rate brachytherapy using real-time transrectal ultrasound based planning in combination with external beam radiotherapy for prostate cancer: dosimetrics and early clinical results *J. Contemp. Brachytherapy* **8** 104–9

[21] Johnston H *et al* 2008 3D ultrasound for prostate localization in radiation therapy: a comparison with implanted fiducial markers *Med. Phys.* **35** 2403–13

[22] Scarbrough T J *et al* 2006 Comparison of ultrasound and implanted seed marker prostate localization methods: Implications for image-guided radiotherapy *Int. J. Radiat. Oncol. Biol. Phys.* **65** 378–87

[23] Van den Heuvel F *et al* 2003 Independent verification of ultrasound based image-guided radiation treatment, using electronic portal imaging and implanted gold markers *Med. Phys.* **30** 2878–87

[24] Dobler B *et al* 2006 Evaluation of possible prostate displacement induced by pressure applied during transabdominal ultrasound image acquisition *Strahlenther. Onkol.* **182** 240–6

[25] Western C, Hristov D and Schlosser J 2015 Ultrasound imaging in radiation therapy: from interfractional to intrafractional guidance *Cureus* **7** e280

[26] Mutic S and Dempsey J F 2014 The viewray system: magnetic resonance–guided and controlled radiotherapy *Semin. Radiat. Oncol.* **24** 196–99

[27] van Herk M *et al* 2018 Magnetic resonance imaging–guided radiation therapy: a short strengths, weaknesses, opportunities, and threats analysis *Int. J. Radiat. Oncol. Biol. Phys.* **101** 1057–60

[28] Chen J *et al* 2006 Dose-guided radiation therapy with megavoltage cone-beam CT *Br. J. Radiol.* **79 Spec No 1** S87–98

[29] Pouliot J *et al* 2005 Low-dose megavoltage cone-beam CT for radiation therapy *Int. J. Radiat. Oncol. Biol. Phys.* **61** 552–60

[30] Morin O *et al* 2006 Megavoltage cone-beam CT: system description and clinical applications *Med. Dosim.* **31** 51–61

[31] Bissonnette J P *et al* 2012 Quality assurance for image-guided radiation therapy utilizing CT-based technologies: a report of the AAPM TG-179 *Med. Phys.* **39** 1946–63

[32] Miracle A C and Mukherji S K 2009 Conebeam CT of the head and neck, part 2: clinical applications *AJNR Am. J. Neuroradiol.* **30** 1285–92

[33] Srinivasan K, Mohammadi M and Shepherd J 2014 Applications of linac-mounted kilo-voltage cone-beam computed tomography in modern radiation therapy: a review *Pol. J. Radiol.* **79** 181–93

[34] Moseley D J *et al* 2007 Comparison of localization performance with implanted fiducial markers and cone-beam computed tomography for on-line image-guided radiotherapy of the prostate *Int. J. Radiat. Oncol. Biol. Phys.* **67** 942–53

[35] Bissonnette J P *et al* 2009 Cone-beam computed tomographic image guidance for lung cancer radiation therapy *Int. J. Radiat. Oncol. Biol. Phys.* **73** 927–34

[36] Hawkins M A *et al* 2006 Assessment of residual error in liver position using kV cone-beam computed tomography for liver cancer high-precision radiation therapy *Int. J. Radiat. Oncol. Biol. Phys.* **66** 610–9

[37] Thengumpallil S *et al* 2016 Difference in performance between 3D and 4D CBCT for lung imaging: a dose and image quality analysis *J. Appl. Clin. Med. Phys.* **17** 97–106

[38] Jarema T and Aland T 2019 Using the iterative kV CBCT reconstruction on the Varian Halcyon linear accelerator for radiation therapy planning for pelvis patients *Phys. Med.* **68** 112–16

[39] Barker J L Jr *et al* 2004 Quantification of volumetric and geometric changes occurring during fractionated radiotherapy for head-and-neck cancer using an integrated CT/linear accelerator system *Int. J. Radiat. Oncol. Biol. Phys.* **59** 960–70

[40] Uematsu M *et al* 1996 A dual computed tomography linear accelerator unit for stereotactic radiation therapy: a new approach without cranially fixated stereotactic frames *Int. J. Radiat. Oncol. Biol. Phys.* **35** 587–92

[41] Wong J R *et al* 2008 Interfractional prostate shifts: review of 1870 computed tomography (CT) scans obtained during image-guided radiotherapy using CT-on-rails for the treatment of prostate cancer *Int. J. Radiat. Oncol. Biol. Phys.* **72** 1396–401

[42] Yenice K M *et al* 2003 CT image-guided intensity-modulated therapy for paraspinal tumors using stereotactic immobilization *Int. J. Radiat. Oncol. Biol. Phys.* **55** 583–93

[43] McRobbie D *et al* 2003 *MRI: From Picture to Proton* 3rd edn (Cambridge: Cambridge University Press)

IOP Publishing

Principles and Practice of Image-Guided Abdominal
Radiation Therapy

Yu Kuang

Chapter 2

History and future of image-guided radiation therapy in abdominal cancer

Gregory P Penoncello, Daniel G Robertson, Ronik S Bhangoo, Yanle Hu, Wei Liu, Yunze Yang, William G Rule and Carlos E Vargas

Image guidance for abdominal tumors has always been challenging for numerous reasons. One challenge is that similar tissue densities result in difficulties creating contrast to distinguish different targets and organs at risk. A second challenge is motion within the abdominal cavity secondary to the diaphragmatic excursion. Motion can introduce target deformation as well as alterations in the proximity of normal tissues and surrounding organs relative to the target. Peristalsis and the normal variability of volume within distensible luminal gastrointestinal structures also complicate image guidance. Throughout the history of radiation therapy in the abdomen, numerous strategies have been developed in an attempt to mitigate these issues. This chapter discusses the evolution of the different strategies as well as advanced and upcoming image guidance techniques such as magnetic resonance imaging linear accelerators, four-dimensional magnetic resonance imaging, proton radiography and computed tomography, surface imaging and the evaluation of interplay effects in proton therapy for patients with abdominal cancer.

2.1 Introduction

Image guidance for abdominal tumors has always been challenging. The difficulties are multiple and related to relatively larger motions than other body sites, deformation and changes in position of normal tissue, and tumor rotations, motion, and deformation, among others. Over time, multiple strategies have been developed to mitigate problems with intra-abdominal target changes. This introductory chapter highlights the evolution of image guidance for abdominal tumors and briefly highlights strategies such as interval tracking, magnetic resonance imaging linear accelerators (MRI-Linacs), four-dimensional MRI (4DMRI) proton computed tomography (CT), surface imaging, evaluation of interplay effect and robustness.

2.2 The history

Radiation therapy is a delicate delivery process. The goal of the treatment has remained the same throughout its evolution: to deliver an effective amount of ionizing radiation to the target treatment site while minimizing the dose to normal, healthy tissues. The advancement of technology has allowed for improvements in delivery with the introduction of complex modalities, such as three-dimensional (3D) conformal radiation therapy, intensity-modulated radiation therapy, stereotactic body radiation therapy (SBRT) and intensity-modulated proton therapy (IMPT), all of which deliver a conformal dose distribution around the target to a varying degree. The improvement of the delivery would be less advantageous without the concurrent improvement of target alignment. Image-guided radiation therapy (IGRT) has improved target alignment and has contributed significantly to reducing the margin size on the target, thus reducing radiotoxicities to the normal tissue [1–4]. Due to its nature, the abdomen in particular is an area in which the evolution of image guidance systems has improved treatments drastically. The increased intrafraction and interfraction motion associated with its anatomical location increases the need to utilize all techniques to ensure the target is stable and to adjust as necessary if changes have occurred.

Image guidance is an idea that has been around since the start of external beam radiation therapy. Fluoroscopy-type image-guidance techniques were explored as described by Nielsen and Jensen using high energy x-rays for treatment [5]. Haus *et al* described using film during treatment using a Co-60 machine [6]. Even the concept of having an extendable diagnostic x-ray tube on a Linac was described by Weissbluth *et al* in the Stanford Linac [7]. Despite all of these ideas and attempts at image guidance early on, the general technique of a simulator, depicted in figure 2.1, and using marks on the skin were prevalent clinically until the late 1990s and early 2000s. The images created by simulators were used to verify the immobilization and were not used for correcting positioning errors the day of treatment [8]. Therefore, although different imaging modalities were used for target delineation and immobilization confirmation, true IGRT is still a relatively new standard of treatment.

The initial images used for image guidance utilized film. Double exposure techniques were used in an attempt to visualize anatomy relative to the treatment field and to confirm beam positioning and shape. These images were typically of poor quality, however, due to compounding factors of the sensitivity of the film, small differences in contrast between tissues to the energy of the megavoltage (MV) beam generating the image, and the source size, particularly for Co-60 machines [6, 9]. There were techniques that could be used to overcome some of these difficulties. Hare *et al* described a method for pelvic patients in which they improved the contrast in the film between specific tissues by injecting air into the bladder and rectum [10, 11]. Another technique to improve image quality in general was to use multiple types of film and lead sheets [11, 12]. Despite attempts to improve image quality the visualization of

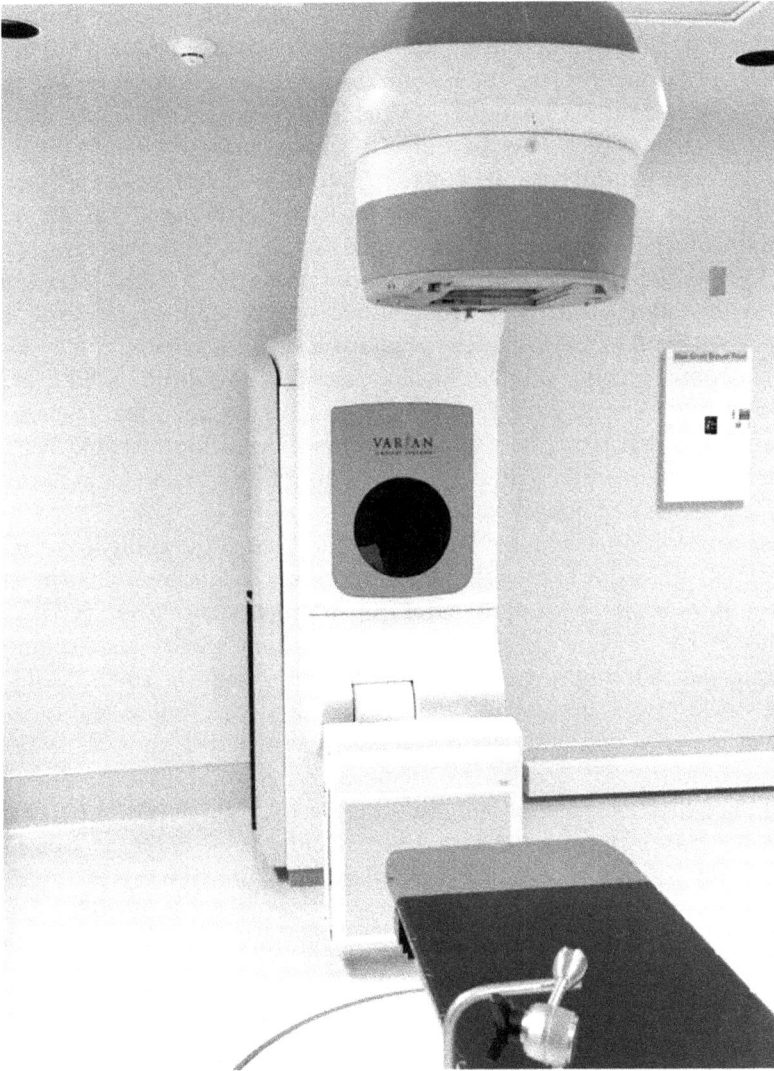

Figure 2.1. An acuity simulator made by Varian at Mayo Clinic in Arizona that was used for initial patient setup and positioning prior to treatment on a Linac.

different tissues was difficult and even more of a challenge for targets in the abdomen than for other sites. Similar electron densities between tissues resulted in poor contrast, making it difficult to isolate the target.

Electronic portal imaging devices (EPID) in conjunction with improved computer speed gave rise to the use of imaging as a means to position patients as part of the setup or online imaging [8, 11, 13]. EPID images still utilized the double-exposure technique of the MV treatment beam to verify the positioning of the patient relative to the surrounding anatomy; however, since the images were electronic and available in an instantaneous manner, the patient's position could be corrected before treatment. The speed of image acquisition and visualization allowed for this ability. An advantage of EPID images using the treatment beam to create the image is that the isocenter of the image is the same as the isocenter for treatment [8, 11]. However, the issue of tissue contrast for targets in the abdomen was not improved with this technological advancement. As treatments became more conformal, the need to visualize the target itself became a higher necessity. Thus, implanted radiopaque fiducials began to be used to visualize the location of the target volume. Gall *et al* described their technique to utilize radiopaque markers, which were used to visualize targets close to radiosensitive normal tissues, targets with high doses and heavy particle treatments [14]. This forced contrast allowed the target location to be represented by the fiducial in patient alignment. Typical fiducial materials include gold, polymer and carbon. Handsfield *et al* described how different fiducial material has an effect on image artifacts and visibility and determined that carbon markers resulted in minimal artifacts but still allowed for a visual contrast [15]. Fiducials for abdominal targets are still commonly used in current treatments and are useful for motion management as described below.

The use of MV beams from the treatment beam lacked the ability to provide good contrast across an image, even with fiducials present. The desire to take images using kilovoltage (kV) beams became apparent for a more accurate setup. One solution was to mount kV imagers in the room. Solutions that utilized this technique were first investigated by numerous groups in the 1990s through the 2000s. Many of these specifically investigated cranial and spinal radiosurgery as well as real-time imaging for motion management [11, 16–20]. Vendors also added an x-ray tube and an imaging panel to the gantry of Linacs, depicted in figure 2.2, to allow for this lower-energy imaging. This addition to the Linac, although it created more quality assurance to ensure the isocentricity and coincidence of both the Linac and the additional arms, allowed for much more versatility for online image guidance. At this point, the imaging was used in real time to correct for positioning differences. The additional arms allowed for better contrast in two-dimensional (2D) images to align to fiducials and bony anatomy, the ability to take combined MV/kV images and use 2/3D imaging, and the ability to arc around the patient for a 3D volumetric comparison of patient positioning. This technique employing 3D volumetric images created from these arms in a rotational image acquisition is known as cone-beam CT (CBCT). The kV CBCT is acquired using an x-ray tube and flat panel detector that is orthogonal to the treatment beam. An x-ray beam in a cone shape is emitted throughout rotation to create projections, which are then reconstructed to create 3D images. The geometrical accuracy of a kV CBCT image is below 1 mm and has a spatial resolution that is submillimeter [21]. This technology allowed for the first image-guidance images with the capability of visualizing some soft tissues and was pivotal for confidence in alignment for dose escalation and SBRT treatments [21, 22]. Although it is the current gold-standard imaging technique for most treatment sites, including the abdomen, kV CBCT imaging still maintains the issue of target contrast relative to the surrounding tissue as depicted by figure 2.3. Fiducials continue to commonly be used as a surrogate in the abdomen for alignment purposes.

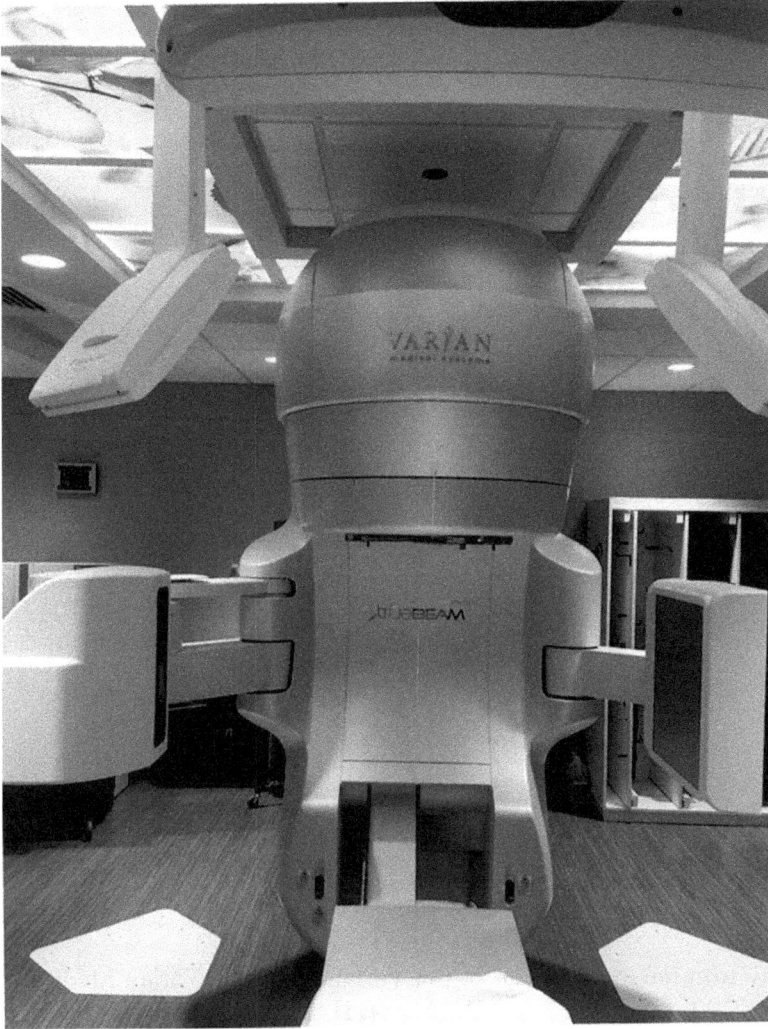

Figure 2.2. Truebeam at the Mayo Clinic in Arizona with all imaging arms extended, EPID for MV imaging, an x-ray and image panel orthogonal to the MV beam and room-mounted kV imaging in the form of ExacTrac imaging.

Other techniques to create 3D imaging have been employed as well. One such method as initially described by Uematsu *et al* was to use a CT scanner in the Linac vault. This design is commonly called a 'CT on rails' in which the CT translates over the patient to create a helical scan of the patient. Although originally used for cranial fixed stereotactic framed treatments, this technique showed to have treatment errors within 1 mm [23]. This provides a CT simulator–type image quality since the 3D image is created using a fan beam instead of a cone beam. Other techniques include MV CT and the MV CBCT. The MV CBCT lost momentum for a period of time once the kV CBCT was created. However, MV CBCT techniques regained

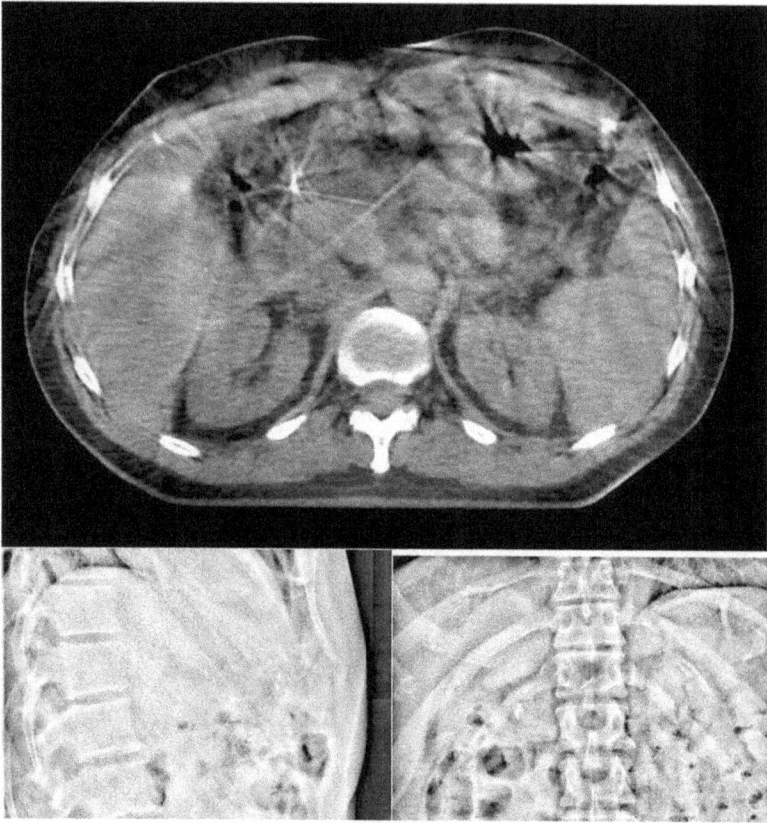

Figure 2.3. Shows a CBCT (top) and kV image orthogonal pair, lateral (lower left) and anterior–posterior (lower right) of the same patient. CBCT clearly gives superior soft-tissue contrast to abdominal structures; however, fiducials are still needed to localize to the target.

popularity with the introduction of the Varian Halcyon (Varian Medical Systems, Palo Alto, CA), which uses a 6 MV beam [11]. MV CT techniques are used by the Accuray TomoTherapy unit, which utilizes a 3.5 MV fan beam to create the image for IGRT. The tomotherapy unit delivers a 6 MV beam in a helical manner for treatments. The imaging for these machines can be of high quality despite the higher energy since the image acquisition is a fan beam rather than a cone beam and also has more accurate Hounsfield unit (HU) values [24]. The increased contrast to noise ratios and more accurate HU values for dose calculation allowed for adaptive planning capabilities; however, targets in the abdomen were still difficult to distinguish without fiducials.

Even with these advancements in image acquisition and image quality, abdominal imaging still maintained the difficulty of targeting an area where limited contrast differences lead to challenges visualizing the target. Fiducials were still utilized in 2D images for alignment, and 3D images still offered little improved visualization of a target directly, although 3D images provided enough contrast to distinguish organs such as the bowel, kidneys and liver. Additionally, abdominal targets create further concerns in that the target is moving both interfractionally and intrafractionally. Instantaneous imaging using fiducials, as well as CBCT images, allowed for correcting the positional interfractional differences in alignment; however, intra-fractional motion is more complicated. Some possible ways to manage the motion described in detail by TG-76 include using an internal target volume (ITV) to encompass the entire motion of the target, using compression on the abdomen to reduce the motion of the target, using respiratory gating techniques to treat when the target is in the correlating position and using breath-hold techniques to minimize the motion [25]. From a planning point of view, to quantify the motion of the ITV and other organs at risk around the target, 4DCT simulation images have been used [8, 26, 27]. For image guidance, CBCT images acquire a slow scan of the patient, capturing the image of numerous breathing cycles. Comparing the average values from the acquired 4DCT to the CBCT for image guidance allows for the most like-for-like comparison. A maximum intensity projection created from the 4DCT can be helpful for target delineation throughout specified breathing phases to take motion into account. For liver targets, a minimum intensity projection can be useful to visualize the respiratory motion of the target.

Advances in volumetric imaging for patient positioning and to further provide a visualization of the change in the target to suggest adaptive planning improved the accuracy of deliveries tremendously; however, for abdominal targets, the issue of contrast between the target and the surrounding tissues remained at large. This continued difficulty led to different imaging modalities being explored. One such modality was ultrasound imaging. Fuss *et al* described a technique to use daily ultrasound imaging for upper abdominal malignancies. This was a technique that was used clinically for prostate and prostate bed targets; however, it could be applied to upper-abdominal targets, albeit with numerous difficulties. These difficulties include targets not usually having a clearly identifiable border, air in the bowel causing challenges when getting a direct view of the target, and difficulties seeing the entire target in one view limiting the success of ultrasound. Fuss *et al* described success in targets in the liver and targets near vascular references [28]. Due to the difficulties noted above, commercial clinical ultrasound systems are still typically restricted to image guidance in the prostate area [11].

The modality that offers the best contrast in soft tissue is MRI as depicted by figure 2.4. MRI has been used in many treatment sites, including abdominal targets, to assist in target delineation through an image-registration process [29]. Recently, the utilization of MRI for image guidance and adaptive treatment has been a high interest of concern. MRI-guided radiation therapy has been investigated and implemented commercially; however, MRI in conjunction with a Linac produces challenges. On the treatment delivery side, one challenge is that the magnetic field used for imaging can affect the acceleration of the electrons as well as affect the dose deposition of electrons created by interactions in matter. Another challenge involves the MRI correlation to the isocenter to that of the treatment beam. The improvement, however, in contrast and the ability to create different soft-tissue contrasts with different MR sequences allows for the ability of direct visualization of the target in the abdomen [29]. The direct visualization of the target, which is often

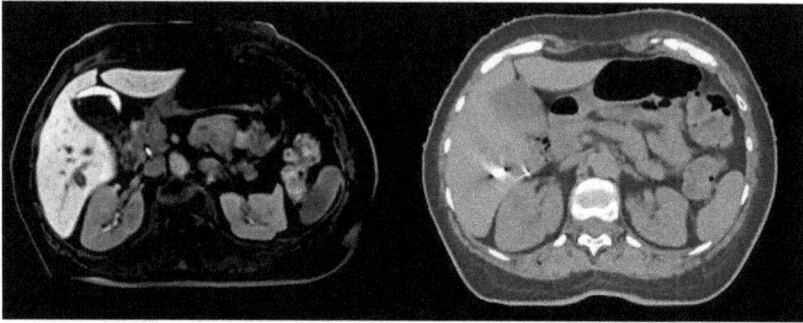

Figure 2.4. Shows an MRI (left) and a CT (right) scan on the same slice to illustrate the difference in visibility of the target in the liver, which is visible in MR due to the soft-tissue contrast but has to be represented by a fiducial in the CT image.

indistinguishable from its surrounding tissues using CT and requires a fiducial to localize the target as described above, allows for a more accurate localization of the target. This can result in decreased margins and, thus, reduced radiotoxicities. MRI also allows for the ability to visualize functional information with different sequences. This direct visualization of the target can also allow for the possibilities of a truly adaptive radiotherapy, allowing one to modify the treatment plan to represent the daily setup and variation of the target interfractionally [30].

IGRT for abdominal tumors has evolved greatly over the years, going from using marks on the skin, to bony anatomy and fiducials, to volumetric imaging of organs and being able to directly visualize the target. This evolution of image quality correlates with the need for accuracy as more precise treatment deliveries are developing with such modalities as volumetric modulated arc therapy, SBRT and IMPT. With this evolution, the treatment goal has remained homogenous: to treat the target as accurately as possible and reduce radiotoxicities. The upcoming section will discuss in more detail the evolution of techniques to manage the motion of abdominal tumors.

2.3 Motion management

2.3.1 4DCT, compression, gating, and breath hold

Optimizing the therapeutic index has always been a goal that has been vigorously pursued by radiation oncologists. The basic challenge of delivering relevant cytotoxic doses to the target volume(s) while at the same time minimizing the dose to the surrounding normal tissues can be approached in a variety of different and often complimentary ways. From the image guidance and motion management perspectives, our field has seen a dramatic evolution over the past several decades. Historically, the default motion management approach was simply to allow the patient to breathe freely, without modification or intervention, with motion compensation integrated into larger radiotherapeutic fields and volumes. Fluoroscopic guidance with or without fiducial marker placement was helpful, but it clearly lacked the nuanced soft-tissue delineation that modern radiotherapy requires.

The need to optimize the therapeutic index is epitomized by the development of SBRT, also known as stereotactic ablative radiotherapy. SBRT was pioneered by the group at Karolinska Hospital in Stockholm, Sweden in the early 1990s [31]. Motion management of their abdominal targets, critical to the safe and efficacious delivery of SBRT, was achieved by an abdominal-compression device integrated into their stereotactic immobilization frame/box. Abdominal compression proved to be quite efficacious at limiting diaphragmatic excursion. This resulted in dampening the downstream respiratory and target motion uncertainties (as evaluated by fluoroscopy) [32].

Several iterations of abdominal-compression devices have been developed, with the general goal, reduction/stabilization of target volume motion with resulting decreases in normal tissue dose exposure, remaining the same [33]. While generally effective, individual patient tolerance to the rigors of abdominal compression, in addition to factors such as target lesion anatomic site, degree of compression, initial magnitude of respiratory motion, body mass index, gender, changes in gastric filling and others, has been associated with a potential decrease in efficacy of abdominal compression [34–37].

With the development and subsequent integration of 4DCT imaging and treatment planning, characterization of target volume and surrounding normal abdominal structure motion became relatively easy for the typical radiation oncology clinic. The acquisition of 4DCT images allowed for the reconstruction and visualization of a representative respiratory cycle, during which target volume motion could be observed and delineated (ITV). In current practice, this is typically achieved by utilizing an external surrogate for internal motion, such as the Varian RPM system (Varian Medical Systems, Palo Alto, CA).

This 4DCT paradigm now allows the radiation oncology team incredible flexibility regarding how to manage motion. For example, with conventional dose fraction approaches, the focus tends to be more on moving target identification and delineation with potentially less concern given to the surrounding normal tissues. For hypofractionated or SBRT approaches, minimizing intermediate and high dose volumes in the normal tissues surrounding the target volume is absolutely critical in allowing for safe and efficacious treatment outcomes [38].

With 4DCT data, free breathing treatments can now be predicated on how much the 3D target volume(s) and/or the surrounding normal tissues are actually moving. Interventions targeted at reducing motion (e.g. abdominal compression, gating, breath hold, etc) can also be applied and evaluated at the time of simulation.

Respiratory gating, based on either respiratory phase or amplitude data gathered at the time of 4DCT simulation, allows for treatment delivery during discrete portions of the respiratory cycle (the gating window) in an otherwise free-breathing patient [39, 40]. Figure 2.5 shows a depiction of a breathing traces and the associated phase gating window. In practice, the gating window that is selected needs to balance target motion minimization and geometric certainty with reasonable treatment times and duty cycles. Image guidance performed at the time of treatment delivery must also be linked, in some way, to the selected gating window.

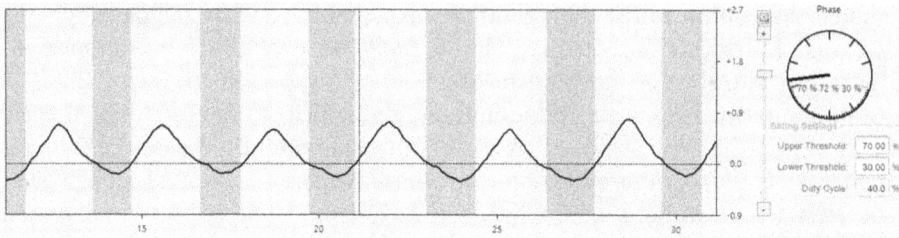

Figure 2.5. Shows a breathing trace in which phase gating is utilizing a duty cycle of 40% between the 30% and 70% phases.

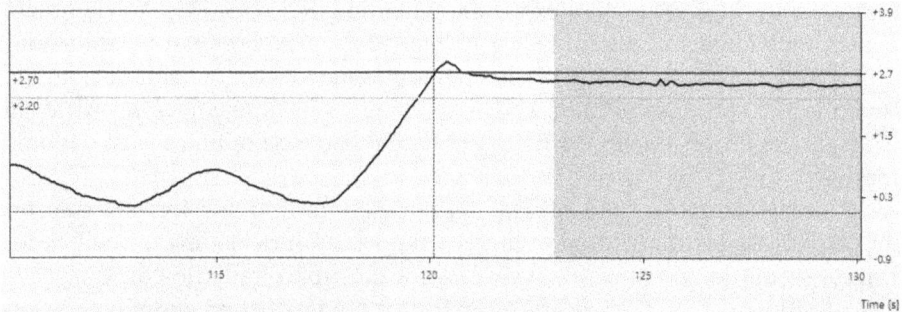

Figure 2.6. Shows a breathing trace in which a breath-hold technique is utilized with an amplitude window restricting between 2.20 and 2.70 cm and a 2 s beam on time delay.

Breath-hold imaging and subsequent treatment delivery can be accomplished via voluntary techniques or via computer-controlled methods such as the ABC System (Elekta AB, Stockholm, Sweden). In general, breath-hold imaging and treatment is accomplished with audio and/or visual feedback to guide the patient into their optimal and reproducible breath-hold window, with assessment utilizing optical tracking information from systems such as RPM (Varian, Palo Alto, CA) and alignRT (VisionRT, London, UK). An example of a breath-hold gating window is depicted in figure 2.6. This methodology clearly requires significant patient compliance during the simulation and treatment delivery phases of care, though advanced technology, such as flattening-filter–free treatment delivery, can minimize treatment times in breath-hold patients [41].

2.3.2 Interval tracking

The challenge of motion management has led to developments in intrafractional tracking of abdominal tumors. Such developments help ensure that motion-management techniques are accurate during each treatment. Generally, interval tracking of abdominal tumors can be accomplished either through real-time x-ray imaging or electromagnetic tracking. These techniques may directly track markers implanted in or near the target or indirectly track surface markers or respiratory monitors correlated to tumor motion. With accurate real-time imaging of tumor motion, real-time treatment adaptations can be implemented, including moving the position of the beam, the multileaf collimator and couch [42–44].

With regard to real-time x-ray imaging, interval tracking techniques may utilize room- or gantry-mounted kV or MV imaging sources and acquire 2D or 3D information. Continuous x-ray imaging or fluoroscopy provided one of the earliest solutions for real-time tracking of tumor motion, but concerns were raised regarding patient dose from continuous x-ray exposure [45]. For example, the real-time tumor-tracking radiation-therapy system uses four room-mounted kV imagers to monitor implanted markers continuously during treatment [46, 47]. More recently, attention has turned to kV–MV stereoscopic imaging, which utilizes onboard imaging already available on conventional Linacs. Another benefit of this approach is a decrease in dose to the patient from intrafractional imaging. Several different techniques for determining 3D tumor position during treatment using onboard kV–MV imaging have been reported in the literature, as this capability is not a standard feature of Linacs [48–50]. Typically, interval kV images orthogonal to the MV treatment beam are taken and this information is used to triangulate the location of implanted markers using 3D modeling. One such system, the kV intrafraction monitoring system, has been successfully implemented in a small series in prostate cancer [51, 52]. However, limited evidence of clinical use exists for similar techniques in abdominal cancers; thus, further investigation is warranted [53, 54].

Another commercial system that provides real-time tumor tracking without using fluoroscopy is the CyberKnife System (Accuray Inc., Sunnyvale, CA). This system employs a coplanar fixed dual kV imager with frequent synchronous stereoscopic imaging using dual kVs taken at the same time. These images can then be correlated to external respiratory signals [4]. Image acquisition, target localization and alignment corrections occur approximately every 30–60 s during treatment with the Linac mounted on a mobile robotic arm [55]. The robotic arm 'moves the beam' so that it can account for changes in tumor position, as opposed to a conventional Linac in which the gantry has a fixed plane of motion. Target accuracy is considered to be less than 1.5 mm for mobile tumors. Clinical data for the use of CyberKnife for abdominal malignancies, particularly pancreatic and hepatic tumors, have demonstrated reasonable treatment outcomes [40].

Initial data published in the early 1990s found that electromagnetic tracking of implanted beacons is a non-ionizing alternative for tumor tracking [56]. The Calypso system (Varian, Palo Alto, CA) is a wireless system in which electromagnetic beacon transponders are implanted within or adjacent to the target. An electromagnetic array is centered over the patient and subsequently excites the transponders. Signals from the transponders are then received by the array panel to identify their locations [57]. Through this process, the Calypso system can provide real-time target tracking, including the ability to gate treatment using these signals. However, practical issues include an additional monitoring system within the treatment room and possible interference from materials like metal prostheses. Similar to above, published data are primarily in the setting of prostate cancer, but use of the Calypso system for pancreatic and hepatic malignancies has also been studied in small series [53, 58, 59].

The next sections will begin to discuss new and future techniques for imaging and treating abdominal targets, including using MRI for image guidance and using 4DMRI to evaluate motion. Additionally, image guidance associated with proton therapy such as proton CT and surface guidance, and handling issues such as interplay effects and robustness, will be covered.

2.4 Newer and upcoming image-guidance techniques

2.4.1 MRI Linac

MRI Linac [60, 61] is one of the most advanced radiotherapy systems. It integrates an MRI with a Linac to provide better on-board imaging for patient setup. Compared to other on-board imaging options like CBCT, kV radiographic imaging or MV portal imaging, MRI is able to provide volumetric imaging with superior soft-tissue contrast, making it very attractive for guiding radiotherapy treatment of abdominal cancer [62–64]. Integrating MRI with radiotherapy, however, is a non-trivial task. On one side, the MRI magnetic field deflects electrons and affects radiation dose deposition, and on the other side, large amounts of metal from the radiotherapy system degrade the MRI performance. To develop MRI-guided radiotherapy systems, a substantial amount of work has been invested. As of today, there are two vendors, ViewRay and Elekta, providing US Food and Drug Administration (FDA)–approved MRI Linacs for patient treatment. ViewRay offers the first commercial MR-guided radiotherapy system which received US FDA 510(k) premarket notification clearance in 2012. The first MRI-guided radiotherapy system from ViewRay combines a 0.35 T MRI with a cobalt radio-therapy system [60]. The use of cobalt instead of a Linac minimizes the interference between MRI and the radiotherapy system, but it does impose significantly more regulatory requirements for radiation safety. ViewRay is continuing its efforts in advancing the field of MRI-guided radiotherapy. In 2017, ViewRay received US FDA 510(k) clearance for its first MRI Linac system (0.35 T/6 MV). Figure 2.7 shows a sketch of the ViewRay MRIdian Linac system. Elekta has collaborated with Philips in development of MRI-guided radiotherapy. The goal is to integrate 1.5 T MRI with a Linac to provide an image quality close to 1.5 T MRI that is commonly used in the diagnostic department [61]. The use of high-field MRI, however, intensifies the interference between MRI and the Linac. To address this, Elekta came up with many solutions including taking the magnetic field into account in the dose calculation, using multiple treatment fields to mitigate the electron return effect, and optimizing active shielding to minimize the effect of magnetic field on Linac operation. Elekta received US FDA 510(k) premarket notification clearance for its MRI Linac (1.5 T/7 MV) in 2018. In addition to ViewRay and Elekta, the Cross Cancer Institute in Alberta, Canada is developing a rotating biplanar MRI Linac system (0.5 T/6 MV) [65, 66] and the Australian MRI-Linac Program is developing a split-bore MRI Linac system (1 T/6 MV) [67].

An MRI Linac has many benefits in radiotherapy treatment of abdominal cancer. For example, the superior image quality makes it feasible to monitor patient anatomy change and enable online adaptive treatment. Since MRI has no radiation,

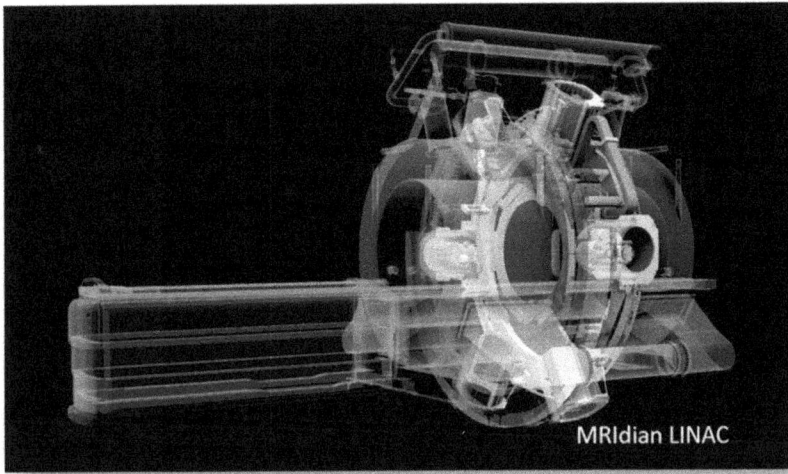

Figure 2.7. Shows a sketch of the ViewRay MRIdian Linac system. Reproduced with the permission of ViewRay Technologies, Inc. All rights reserved.

it is also possible to track tumor motion in real time and implement gated treatment. Currently, an MRI Linac is an active research area. Although commercial systems are available for patient treatment, they still require efforts to develop new functionalities, integrate them into clinical workflow and turn the potential of an MRI Linac into improved patient outcomes.

2.4.2 4DMRI

For radiotherapy treatment of abdominal cancer, it is critical to obtain information regarding how tumors and surrounding organs move with respiration. Currently, the standard-of-care practice is to use a 4DCT to obtain motion information [43], but CT has suboptimal soft-tissue contrast, which makes it challenging to acquire accurate knowledge of tumor motion, while MRI has superior soft-tissue contrast and involves no radiation. It has great potential in visualizing abdominal tumors and providing better tumor motion information. 4DMRI [68–72] shares certain similarities in image acquisition to 4DCT but also has its unique aspects. An obvious distinction between the two is that the table translates through the gantry in 4DCT but remains static in 4DMRI. This is because 4DMRI uses gradient magnetic fields to select the slice being imaged, whereas 4DCT has to move the table so that the slices being imaged are at the center of the CT gantry. Earlier 4DMRI implementations utilize the retrospective binning method in a similar way that mimics 4DCT. It selects one slice at a time and repetitively acquires images from that slice for at least one respiratory cycle to capture motion information. Then it selects the next slice and repeats the process until images from all slices are acquired. During 4DMRI image acquisition, the respiratory signal is being acquired simultaneously, and this information is used subsequently to bin MRI images into appropriate respiratory states. The characteristics of MRI acquisition, however, allow other novel 4DMRI implementations to be explored. For example, it is feasible to implement prospectively triggered 4DMRI in which the respiratory signal is monitored in real time and image acquisition from a given slice and respiratory state starts only when the respiratory signal reaches a predetermined condition. It is also possible to implement retrospective binning based on body area in image space, or even raw data in k-space, instead of reconstructed MRI images. 4DMRI has shown promising potential for treatment of abdominal cancer, but it is relatively immature and still requires significant efforts in developing the technology and integrating it into clinical workflow.

2.4.3 Proton radiography and CT

Proton beam therapy is a highly conformal radiation therapy modality that is rapidly increasing in popularity due to its ability to treat cancerous lesions while sparing healthy tissues. The dosimetric advantage of proton beam therapy derives from the Bragg peak, a physical phenomenon featuring a low entrance dose followed by a high-dose peak at the depth corresponding to the end of the proton beam range, after which the dose drops rapidly to zero (figure 2.8). The shape of the proton beam depth–dose distribution enables the sparing of healthy tissues both proximal and distal to the target volume. Proton therapy has proven advantageous in the treatment of various abdominal cancers, including prostate [73], anal and rectal [74, 75], liver [76], and other gastrointestinal (GI) cancers [77].

While the treatment delivery technology for modern proton beam therapy can provide superior dose distributions to photon-based treatment modalities, the image-guidance technology in proton therapy centers has lagged behind modern Linac-based treatment systems. While some new proton therapy centers incorporate

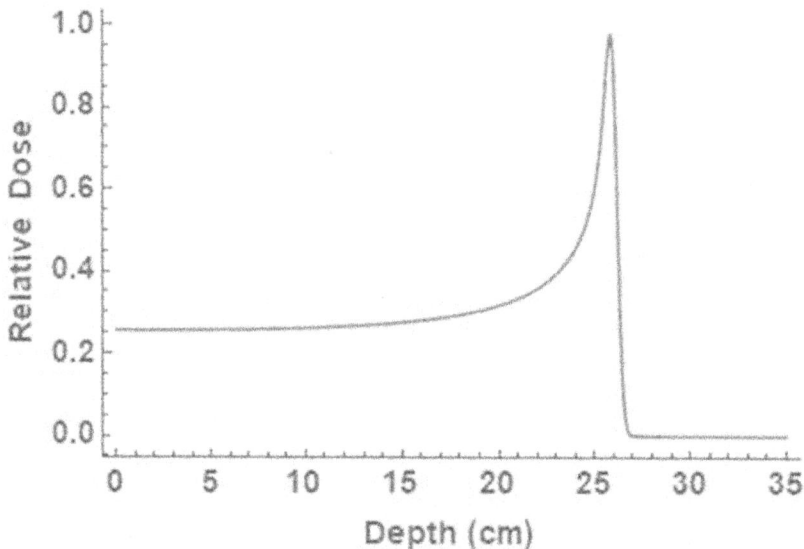

Figure 2.8. The Bragg peak of a 200 MeV proton beam.

on-board CBCT for image guidance, most lack this technology which has long been considered standard for photon therapy.

Many proton centers rely primarily on a pair of orthogonal kV x-ray imagers for patient alignment. Planar x-rays provide inferior image guidance, especially for GI cancers where the target and critical structures are comprised of mobile soft tissues that are relatively distant from bony anatomy. Radio-opaque fiducial markers are sometimes implanted in the tumor to provide a surrogate for the target volume that is visible in planar x-rays [78]. However, implantation of fiducials is not always possible, and it necessitates an invasive procedure. Fiducial markers also provide no information on the locations of nearby critical structures, which may change positions substantially between or even during treatments. Even in centers with CBCT, the image quality may not be sufficient to image GI targets and critical structures due to the considerable x-ray scatter and the radiological similarity of most GI tissues.

Diagnostic CT scanners on rails have been placed in proton treatment rooms to enable high-quality images of GI structures. However, there are problems associated with this approach, including extended imaging time, delays between imaging and treatment (caused by the need to move the patient between the CT bore and the proton treatment isocenter), concerns about patient motion during transit from the CT to the proton isocenter, and the need to carefully validate the alignment between the imaging coordinate system and the treatment coordinate system.

An additional challenge for image guidance for proton therapy is the lack of images from the perspective of the treatment beam. In photon-based radiotherapy, the treatment beam is used in conjunction with an imaging panel to produce port films or beam's-eye-view images of the treatment area. These are commonly used for target alignment and treatment safety. Due to the difficulty of placing an x-ray tube in the proton beamline, most proton therapy centers lack the ability to provide beam's-eye-view images.

Diagnostic CT scanners are the standard tool for virtual simulation of proton-therapy patients. However, because protons and x-rays interact with tissue in different ways, a conversion is required between the x-ray HU provided by the CT scanner and the proton-stopping power ratios used to calculate proton range within the patient. Unfortunately, this conversion process introduces uncertainties in the calculated proton beam range. Consequently, an added proximal and distal margin is usually added to the treatment volume to ensure coverage of the target in the presence of range uncertainties. Range uncertainty margins are typically on the order of 3% of the total proton range [79, 80]. This margin decreases the conformity of the treatment, increases dose to healthy tissues and limits the ability of the proton beam to spare nearby critical structures.

While x-rays are used for most treatment planning and image guidance, it is also possible to use the proton beam itself to image the patient [81]. During treatment, the proton energy is selected to place the proton end-of-range in or near the target inside the patient. However, if a higher proton energy is selected, the protons will travel through the patient and exit the opposite side, having lost a part of their energy proportional to the integral stopping power of the tissues in their path.

Detectors have been developed to measure the protons' path through the patient and their residual energy after exiting the patient [82, 83]. This data can be combined to form a planar image of the integral stopping power or water-equivalent thickness of the patient anatomy. This image is called a proton radiograph. Figure 2.9 shows a schematic of a common design for proton radiography detectors.

The proton beam and detector can also be rotated around the patient, producing many radiographs, which can then be reconstructed into a 3D image of the patient. This is a proton CT. Differential contrast in the proton CT is provided by differences in the relative proton-stopping powers of the patient tissues [84, 85].

Proton radiography and CT can provide several advantages over x-ray imaging modalities in the areas of patient alignment, reduction in margins, and radio-toxicities. Proton radiography provides a beam's-eye-view that can be used for patient alignment and field verification analogous to the port films used in Linac-based photon radiotherapy. Proton CT can provide 3D imaging at the treatment isocenter, even on the many systems lacking a full 360 degree rotating gantry. This is potentially faster than CT-on-rails and decreases concerns about patient motion between imaging and treatment. Additionally, because the treatment beam is used for imaging, proton radiography and CT eliminate uncertainties related to alignment between the imaging and treatment isocenters.

One of the most significant potential impacts of proton radiography and proton CT is their ability to significantly decrease the uncertainty in proton-stopping power ratios of patient tissues. This will substantially decrease proton range uncertainties, allowing a decrease in the proton range uncertainty margins. Smaller margins will translate to less radiotoxicity of healthy structures near the target. It may also enable the use of the sharp distal falloff of the proton Bragg peak for sparing of critical structures, a technique which is not currently practiced due to concerns about range uncertainty. The 3D imaging provided by proton CT will also provide a clearer view

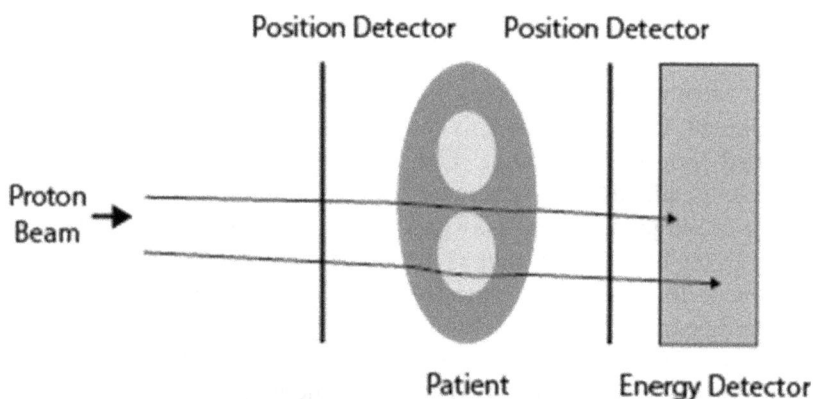

Figure 2.9. Schematic of a typical proton radiography system. Protons pass through an initial position detector, the patient, and a secondary position detector and deposit their remaining energy in an energy detector. The position detectors are used to trace the path of the proton as it traverses the patient, and the energy detector determines the energy lost during transit.

of GI targets and critical structures, allowing for further decreases in treatment margins compared to current image-guidance approaches using 2D x-ray imaging.

Proton radiography and CT are also expected to decrease imaging dose to the patient. Standard photon-based image guidance, especially 3D image guidance, can contribute a non-trivial radiation dose to the patient over a course of radiation treatment. Proton radiographs have been found to decrease imaging dose by a factor of 50–100 relative to x-ray radiographs [86], and proton CT is expected to deliver similar dose savings compared to CBCT. These lower imaging doses can decrease concerns about the frequency of 3D imaging in sensitive patient populations like children, pregnant patients and patients with prior radiation treatments.

Proton radiography and proton CT have yet to be implemented clinically. However, at the time of writing, several research systems and commercial prototypes exist [82, 87–89] and translation to the clinic is expected soon. Several challenges must be overcome before proton radiography and proton CT can be used for image guidance, including the need for high proton beam energy, very low dose rate, and integration with complex proton treatment systems.

The greatest challenge of applying proton imaging to GI cancers lies in the maximum available proton energy of current clinical proton accelerators. Because the proton beam must pass completely through the patient and into the imaging system, higher proton energies are required for imaging than for patient treatment. Most hospital-based proton accelerators provide a maximum energy of ~230 MeV, which translates to a proton range in water of ~33 cm. This beam range is insufficient to pass completely through the abdomen in most adult patients. Higher beam energies are being developed at some centers with the view of proton imaging [90]. It is also possible that limited-angle image reconstruction techniques and hybrid proton–photon imaging techniques could bring the benefits of proton radiography and proton CT to centers with lower maximum proton energies.

2.4.4 Surface imaging in abdominal cancer treatment

Surface-guided radiation therapy (SGRT) utilizes optical imaging system for patients' surface monitoring to help patient alignment and manage motion during radiotherapy treatment. Several optical surface imaging systems are commercially available for clinical use [91, 92]. These systems image the patient's 3D surface in real time and generate difference maps in all six degrees of freedom between the reference surface generated during treatment planning and the captured surface during the treatment. Since surface imaging provides on-bed real-time feedback about patient alignment, it has been employed clinically for patient positioning during the treatment to reduce interfractional setup uncertainties [93, 94]. The accurate tracking of patient surface also allows precise monitoring of patient motion during the treatment, which enables surface imaging as a multi-point respiratory motion surrogate for motion management such as gating or breath-holding during the treatment [95].

Intrafractional target motion has been a big challenge in abdominal cancer treatment. Compared with single-point motion surrogate, such as RPM, SGRT provide 3D topologic information, which could be a better surrogate for motion monitoring [96] and a volumetric measurement of patient respiration [97]. Evaluation of different surface-imaging systems showed similar performances [98–100].

In addition to 3D spatial information, the high temporal resolution and long-term continuous monitoring capability of patients' respiratory motion during the treatment make surface imaging a suitable tool for motion management and evaluation during the treatment. This is particularly important for spot-scanning proton therapy due to the concern of interplay effect in the treatment of mobile tumors. The data acquired by surface imaging during the treatment makes it possible to achieve real-time interplay effect evaluation in spot-scanning proton therapy (for details, please see the next section). Therefore the proton treatment can be intervened in real time if needed, which gives safer and more reliable treatment of spot-scanning proton therapy.

However, the limitation of surface imaging is that only the external surface of the patient is captured. The actual relationship between the external surface and internal tumor motion during the treatment is mostly unclear. This limitation can be partially mitigated by the joint use of other internal surrogates during the treatment, such as fluoroscopic imaging [101]. This cannot be achieved solely by fluoroscopic imaging due to the poor soft-tissue contrast. It is possible to have surface imaging during the 4DCT simulation and then use artificial intelligence to predict the tumor position. The accuracy of the tumor position prediction during the treatment can be refined by fluoroscopic imaging. However, such correlation models could be complicated and need further investigation.

2.4.5 Interplay effect evaluation in patients with abdominal cancer treated by IMPT

IMPT is highly sensitive to uncertainties and respiratory motion [102, 103]. Unfortunately, significant motion is present in patients with abdominal cancer. The interaction between dynamic beamlet delivery and respiratory motion, the so-called interplay effect, may degrade the quality of planned dose distributions [104–116], compromising the safety and efficacy of the proposed treatment.

Different proton machines have different parameters for spot delivery. Here the spot delivery parameters for the Hitachi PROBEAT-V machine at the Mayo Clinic Proton Center in Arizona are presented in table 2.1.

Iso-layer repainting is commonly used in IMPT to mitigate the impact of interplay effects [117–119]. How to implement iso-layer repainting is usually determined by the minimum and maximum MU limits of the proton machine and the maximum MU limit is usually determined by the respiratory motion amplitude. Smaller maximum MU limits thereby enforce a higher number of iso-layer repainting for these patients to mitigate interplay effects. At the Mayo Clinic Proton Center in Arizona, a respiratory motion amplitude threshold of 5 mm is typically used. When the respiratory motion amplitude is over the threshold, 0.003 MU minimum and 0.01 MU maximum MU limits are applied. Otherwise, the limits are set to be 0.003 and 0.04 MU, respectively. For a spot with intensity larger than

Table 2.1. Spot delivery parameters for the Hitachi PROBEAT-V machine at the Mayo Clinic proton center in Arizona.

Spot size (in air at isocenter)	Energy dependent (σ: 2–6 mm)
Spot spacing	Fixed (5 mm)
Minimal MU limit (MU)	0.003
Maximum MU limit (MU)	0.04
Energy layer switching time (s)	1.9
Spill length (s)	7.9
Effective magnet scanning speed in horizontal direction, V_x (m s^{-1})	Medium-energy group: 5.7; low-energy group: 7.0
Effective magnet scanning speed in vertical direction, V_y (m s^{-1})	High-energy group: 17.1; medium-energy group: 18.2; low-energy group: 22.2
Magnet preparation/verification Time (ms)	1.93
Proton spill rate (MU/s)	High-energy group: 9.8; medium-energy group: 8.1; low-energy group: 8.5

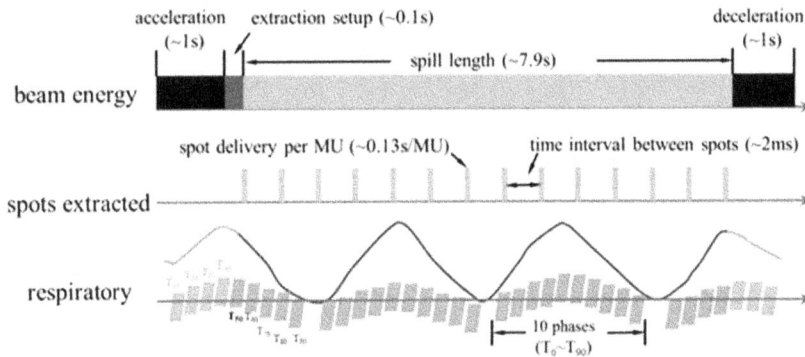

Figure 2.10. Time-dependent beam spot delivery sequence and modelling of interplay effects. A random starting phase (for example, T_{50}) was used for each field per fraction to minimize the impact of the starting phase in the evaluation of interplay effects. [122] © 2019 Wei Liu, Jonathan B. Ashman, Todd DeWees, *et al. Journal of Applied Clinical Medical Physics* published by Wiley Periodicals, Inc. on behalf of American Association of Physicists in Medicine.

the maximum MU limit, it would be split into multiple spots, which would be appended to the end of the spot list in the same energy layer to be repainted ultimately. For a spot with planned intensity less than the minimum MU limit, by checking whether its intensity is larger or smaller than the half of the minimum MU limit, its intensity would be rounded up to be the minimum MU limit or otherwise be dropped.

Dose evaluation software has been developed to assess the plan's ability to retain dose volume objectives when the interplay effect is considered. Within the dose evaluation software, on one hand, the time structure of spot delivery, shown in figure 2.10, is modeled by using the machine parameters listed in table 2.1: spot delivery time per MU (i.e., dose rate), allowable extraction time (i.e., spill length), time for proton acceleration, deceleration and extraction setup (i.e., spill interval length) and time interval between consequent spots within the same energy layer (i.e., spot interval length). On the other hand, the patient-specific respiratory motion is modeled using 4DCTs and the real-time position management (RMP) data, which is obtained in a 4DCT simulation via the Varian RMP system (Varian Medical System, Palo Alto, CA). Then based on these two models, every spot of each field per fraction could be assigned to the corresponding respiratory phases at the relevant delivery time. At last, the spot doses are summed to a reference phase using an in-house–developed deformable image registration software [118, 120]. The influence of starting phases is taken care of and mitigated as well by randomizing the starting phase of each field for each fraction. Unfortunately the current interplay effect evaluation can be only done before the treatment with the assumption that patients will breathe the exact same respiratory pattern as the one during the 4DCT simulation.

However, respiratory motion may vary from respiratory cycle to respiratory cycle within any given treatment session and from day to day during the treatment course. [121] This is defined as irregular respiratory motion. Irregular respiratory motion is very common in patients with abdominal cancer. Therefore the real-time interplay effect evaluation with the irregular respiratory motion considered is very important for IMPT to treat abdominal cancer. This can be achieved by the use of surface imaging and fluoroscopic imaging as discussed in the previous section.

References

[1] Dawson L A and Jaffray D A 2007 Advances in image-guided radiation therapy *J. Clin. Oncol.* **25** 938–46

[2] Perkins C L, Fox T and Elder E *et al* 2006 Image-guided radiation therapy (IGRT) in gastrointestinal tumors *J. Pancreas* **7** 372–81

[3] Hong T S, Tome W A and Chappell R J *et al* 2005 The impact of daily setup variations on head and neck intensity-modulated radiation therapy *Int. J. Radiat. Oncol. Biol. Phys.* **61** 779–88

[4] Balter J M, Brock K K and Litzenberg D W *et al* 2002 Daily targeting of intrahepatic tumors for radiotherapy *Int. J. Radiat. Oncol. Biol. Phys.* **52** 226–71

[5] Nielsen J and Jensen S H 1942 Some experimental and clinical lights on the rotation therapy, its bases and possibilities *Acta Radiol.* **23** 51–66

[6] Haus A G, Pinsky S M and Marks J E 1970 A technique for imaging patient treatment area during a therapeutic radiation exposure *Radiology* **97** 653–6

[7] Weissbluth M, Karzmark C J and Steele R E 1959 The standford medical linear accelerator *Radiology* **72** 242–53

[8] Verellen D, De Ridder M and Storme G A 2008 (Short) history of image-guided radiotherapy *Radiother. Oncol.* **86** 4–13

[9] Marks J E and Haus A G 1976 The effect of immonilisation on localization error in the radiotherapy of head and neck cancer *Clin. Radiol.* **27** 175–7

[10] Hare H F, Lippincott S W Jr and Sawyer D *et al* 1951 Physical and clinical aspects of supervoltage rotational therapy *Radiology* **57** 157–68

[11] Perryman C R, McAllister J D and Burwell J A 1960 Cobalt 60 radiography *Am. J. Roentgenol. Radium Ther. Nucl. Med.* **83** 525–32

[12] Verellen D, De Ridder M, Linthout N, Tournel K, Soete G and Storme G 2007 Innovations in image-guided radiotherapy *Nat. Rev. Cancer* **7** 949–60

[13] Alaei P and Ding G X 2018 *Image Guidance in Radiation Therapy: Techniques, Accuracy, and Limitations* (Madison, WI: The American Association of Physicists in Medicine by Medical Physics Publishing, Inc.)

[14] Gall K P and Verhey L J 1993 Computer-assisted positioning of radiotherapy patients using Implanted radiopaque fiducials *Med. Phys.* **20** 1153–9

[15] Handsfield L L, Yue N J, Zhou J, Chen T and Goyal S 2012 Determination of optimal fiducial marker across image-guided radiation therapy (IGRT) modalities: visibility and artifact analysis of gold, carbon and polymer fiducial markers *J. Appl. Clin. Med. Phys.* **13** 181–9

[16] Adler J R Jr, Murphy M J, Chang S D and Hancock S L 1999 Image-guided robotic radiosurgery *Neurosurgery* **44** 1299–307

[17] Murphy M J and Cox R S 1996 The accuracy of dose localization for an image-guided frameless radiosurgery system *Med. Phys.* **23** 2043–9

[18] Schewe J E, Lam K L, Balter J M and Ten Haken R K 1998 A room-based diagnostic imaing system for measurement of patient setup *Med. Phys.* **25** 2385–7

[19] Shirato H, Shimizu S and Kunieda T *et al* 2000 Physical aspects of a real-time tumor tracking system for gated radiotherapy *Int. J. Radiat. Oncol. Biol. Phys.* **48** 1187–95

[20] Yin F F, Ryu S and Aljouni M *et al* 2002 A technique of intensity-modulated radiosurgery (IMRS) for spinal tumors *Med. Phys.* **29** 2815–22

[21] Bissonnette J-P, Balter P A and Dong L *et al* 2012 Quality assurance for image-guided radiation therapy utilizing CT-based technologies: a report of the AAPM TG-179 *Med. Phys.* **39** 1946–63

[22] Jaffray D A, Siewerdsen J H, Wong J W and Martinez A A 2002 Flat-panel cone-beam computed tomography for image-guided radiation therapy *Int. J. Radiat. Oncol. Biol. Phys.* **53** 1337–49

[23] Uemstsu M, Fukui T and Shioda A 1996 A duel computed tomography linear accelerator unit for stereotactic radiation therapy: a new approach without cranially fixated stereotactic frames *Int. J. Radiat. Oncol. Biol. Phys.* **35** 587–92

[24] Langen K M, Meeks S L and Poole D O *et al* 2005 The use of megavoltage CT (MVCT) images for dose recomputations *Phys. Med. Biol.* **50** 4259–76

[25] Keall P J, Mageras G S and Balter J M *et al* 2006 The management of respiratory motion in radiation oncology report of the AAMP Task Group 76 *Med. Phys.* **33** 3874–900

[26] Ford E C, Margeras G S and Yorke E *et al* 2003 Respiration-correlated spiral CT: a method of measuring respiratory-induced anatomic motion for radiation treatment planning *Med. Phys.* **30** 88–97

[27] Lu W, Parikh P J and Hubenschmidt J P *et al* 2006 A comparison between amplitude sorting and phase-angle sorting using external respiratory measurement for 4D CT *Med. Phys.* **33** 2964–74

[28] Fuss M, Salter B J and Cavanaugh S X *et al* 2004 Daily ultrasound-based image guided targeting for radiotherapy of upper abdominal malignancies *Int. J. Radiat. Oncol. Biol. Phys.* **59** 1245–56

[29] Khoo V S, Dearnaley D P, Finnigan D J, Padhani A, Tanner S F and Leach M O 1997 Magnetic resonance imaging (MRI): considerations and applications in radiotherapy treatment planning *Radiother. Oncol.* **42** 1–15

[30] Metcalfe P, Liney G P and Holloway L *et al* 2013 The potential for an enhanced role for MRI in radiation-therapy treatment planning *Technol. Cancer Res. Treat.* **12** 429–46

[31] Lax I, Blomgren H and Naslund I *et al* 1994 Stereotactic radiotherapy of malignancies in the abdomen: methodological aspects *Acta Oncol.* **33** 677–83

[32] Negoro Y, Nagata Y and Aoki T *et al* 2001 The effectiveness of an immobilization device in conformal radiotherapy for lung tumor: reduction of respiratory tumor movement and evaluation of the daily setup accuracy *Int. J. Radiat. Oncol. Biol. Phys.* **50** 889–98

[33] Lovelock D M, Zatcky J and Goodman K *et al* 2014 The effectiveness of a pneumatic compression belt in reducing respiratory motion of abdominal tumors in patients undergoing stereotactic body radiotherapy *Technol. Cancer Res. Treat.* **13** 259–67

[34] Hu Y, Zhou Y K and Chen Y X *et al* 2017 Magnitude and influencing factors of respiration-induced liver motion during abdominal compression in patients with intra hepatic tumors *Radiat. Oncol.* **12** 9

[35] Mampuya W A, Nakamura M and Matsuo Y *et al* 2013 Interfraction variation in lung tumor position with abdominal compression during stereotactic body radiotherapy *Med. Phys.* **40** 091718

[36] Eccles C L, Dawson L A and Moseley J L *et al* 2011 Interfraction liver shape very ability and impact on GTV position during liver stereotactic radiotherapy using abdominal compression *Int. J. Radiat. Oncol. Biol. Phys.* **80** 938–46

[37] Timmerman R D, Forster K M and Chinsoo Cho L 2005 Extracranial stereotactic radiation delivery *Semin. Radiat. Oncol.* **15** 202–7

[38] Heinzerling J H, Anderson J F and Papiez L *et al* 2008 Four-dimensional computed tomography scan analysis of tumor and organ motion at varying levels of abdominal compression during stereotactic treatment of lung and liver *Int. J. Radiat. Oncol. Biol. Phys.* **70** 1571–8

[39] Ford E C, Mageras G S, Yorke E and Ling C C 2003 Respiration-correlated spiral CT: a method of measuring respiratory-induced anatomic motion for radiation treatment planning *Med. Phys.* **30** 88–97

[40] Dieterich S, Green O and Booth J 2018 SBRT targets that move with respiration *Phys. Med.* **56** 19–24

[41] Boda-Heggeman J, Mai S and Fleckenstein J *et al* 2013 Flattening-filter-free intensity modulated breath-hold image-guided (SABR) (stereotactic ablative radiotherapy) can be applied in a 15-min treatment slot *Radiother. Oncol.* **109** 505–9

[42] Colvill E, Booth J and Nill S *et al* 2016 A dosimetric comparison of real-time adaptive and non-adaptive radiotherapy: a multi-institutional study encompassing robotic, gimbaled, multileaf collimator and couch tracking *Radiother. Oncol.* **119** 159–65

[43] Keall P J, Mageras G S and Balter J M *et al* 2006 The management of respiratory motion in radiation oncology report of AAPM Task Group 76 *Med. Phys.* **33** 3874–900

[44] Caillet V, Booth J T and Keall P 2017 IGRT and motion management during lung SBRT delivery *Phys. Med.* **44** 113–22

[45] Cho B, Poulsen P R and Keall P J 2010 Real-time tumor tracking using sequential kV imaging combined with respiratory monitoring: a general framework applicable to commonly used IGRT systems *Phys. Med. Biol.* **55** 3299–316

[46] Kitamura K, Shirato H and Shimizu S *et al* 2002 Registration accuracy and possible migration of internal fiducial gold marker implanted in prostate and liver treated with real-time tumor-tracking radiation therapy (RTRT) *Radiother. Oncol.* **62** 275–81

[47] Shirato H, Shimizu S and Kitamura K *et al* 2000 Four-dimensional treatment planning and fluoroscopic real-time tumor tracking radiotherapy for moving tumor *Int. J. Radiat. Oncol. Biol. Phys.* **48** 435–42

[48] Wiersma R D, Mao W and Xing L 2008 Combined kV and MV imaging for real-time tracking of implanted fiducial markers *Med. Phys.* **35** 1191–8

[49] Montanaro T, Nguyen D T and Keall P J *et al* 2018 A comparison of gantry-mounted X-ray-based real-time target tracking methods *Med. Phys.* **45** 1222–32

[50] Liu W, Ma X, Yan H, Chen Z, Nath R and Li H 2017 Comparison of 2D and 3D modeled tumor motion estimation/prediction for dynamic tumor tracking during arc radiotherapy *Phys. Med. Biol.* **62** N168–79

[51] Ng J A, Booth J T and Poulsen P R *et al* 2012 Kilovoltage intrafraction monitoring for prostate intensity modulated arc therapy: first clinical results *Int. J. Radiat. Oncol. Biol. Phys.* **84** e655–61

[52] Nguyen D T, O'Brien R and Kim J H *et al* 2017 The first clinical implementation of a real-time six degree of freedom target tracking system during radiation therapy based on kilovoltage intrafraction monitoring (KIM) *Radiother. Oncol.* **123** 37–42

[53] Worm E S, Høyer M, Fledelius W and Poulsen P R 2013 Three-dimensional, time-resolved, intrafraction motion monitoring throughout stereotactic liver radiation therapy on a conventional linear accelerator *Int. J. Radiat. Oncol. Biol. Phys.* **86** 190–7

[54] Vinogradskiy Y, Goodman K A, Schefter T, Miften M and Jones B L 2019 The clinical and dosimetric impact of real-time target tracking in pancreatic SBRT *Int. J. Radiat. Oncol. Biol. Phys.* **103** 268–75

[55] Kilby W, Dooley J R, Kuduvalli G, Sayeh S and Maurer C R Jr 2010 The CyberKnife robotic radiosurgery system in 2010 *Technol. Cancer Res. Treat.* **9** 433–52

[56] Houdek P V, Schwade J G and Serago C F *et al* 1992 Computer controlled stereotaxic radiotherapy system *Int. J. Radiat. Oncol. Biol. Phys.* **22** 175–80

[57] Shah A P, Kupelian P A, Willoughby T R and Meeks S L 2011 Expanding the use of real-time electromagnetic tracking in radiation oncology *J. Appl. Clin. Med. Phys.* **12** 3590

[58] Shinohara E T, Kassaee A and Mitra N *et al* 2012 Feasibility of electromagnetic transponder use to monitor inter- and intrafractional motion in locally advanced pancreatic cancer patients *Int. J. Radiat. Oncol. Biol. Phys.* **83** 566–73

[59] Sandler H M, Liu P Y and Dunn R L *et al* 2010 Reduction in patient-reported acute morbidity in prostate cancer patients treated with 81-Gy intensity-modulated radiotherapy using reduced planning target volume margins and electromagnetic tracking: assessing the impact of margin reduction study *Urology* **75** 1004–8

[60] Mutic S and Dempsey J F 2014 The ViewRay system: magnetic resonance-guided and controlled radiotherapy *Semin. Radiat. Oncol.* **24** 196–9

[61] Raaymakers B W, Jurgenliemk-Schulz I M and Bol G H *et al* 2017 First patients treated with a 1.5 T MRI-linac: clinical proof of concept of a high-precision, high-field MRI guided radiotherapy treatment *Phys. Med. Biol.* **62** L41–50

[62] Bohoudi O, Bruynzeel A M E and Senan S *et al* 2017 Fast and robust online adaptive planning in stereotactic MR-guided adaptive radiation therapy (SMART) for pancreatic cancer *Radiother. Oncol.* **125** 439–44

[63] Feldman A M, Modh A, Glide-Hurst C, Chetty I J and Movsas B 2019 Real-time magnetic resonance-guided liver stereotactic body radiation therapy: an institutional report using a magnetic resonance-linac system *Cureus* **11** e5774

[64] Rudra S, Jiang N and Rosenberg S A *et al* 2019 Using adaptive magnetic resonance image-guided radiation therapy for treatment of inoperable pancreatic cancer *Cancer Med.* **8** 2123–32

[65] Fallone B G 2014 The rotating biplanar linac-magnetic resonance imaging system *Semin. Radiat. Oncol.* **24** 200–2

[66] Keyvanloo A, Burke B St and Aubin J *et al* 2016 Minimal skin dose increase in longitudinal rotating biplanar linac-MR systems: examination of radiation energy and flattening filter design *Phys. Med. Biol.* **61** 3527–39

[67] Keall P J, Barton M and Crozier S 2014 The Australian magnetic resonance imaging-linac program *Semin. Radiat. Oncol.* **24** 203–6

[68] Cai J, Chang Z, Wang Z, Paul Segars W and Yin F F 2011 Four-dimensional magnetic resonance imaging (4D-MRI) using image-based respiratory surrogate: a feasibility study *Med. Phys.* **38** 6384–94

[69] Tryggestad E, Flammang A and Han-Oh S *et al* 2013 Respiration-based sorting of dynamic MRI to derive representative 4D-MRI for radiotherapy planning *Med. Phys.* **40** 051909

[70] Hu Y, Caruthers S D, Low D A, Parikh P J and Mutic S 2013 Respiratory amplitude guided 4-dimensional magnetic resonance imaging *Int. J. Radiat. Oncol. Biol. Phys.* **86** 198–204

[71] Du D, Caruthers S D and Glide-Hurst C *et al* 2015 High-quality T2-weighted 4-dimensional magnetic resonance imaging for radiation therapy applications *Int. J. Radiat. Oncol. Biol. Phys.* **92** 430–37

[72] Paganelli C, Summers P, Bellomi M, Baroni G and Riboldi M 2015 Liver 4DMRI: a retrospective image-based sorting method *Med. Phys.* **42** 4814–21

[73] Mendenhall N P, Hoppe B S and Nichols R C *et al* 2014 Five-year outcomes from 3 prospective trials of image-guided proton therapy for prostate cancer *Int. J. Radiat. Oncol. Biol. Phys.* **88** 596–602

[74] Anand A, Bues M and Rule W G *et al* 2015 Scanning proton beam therapy reduces normal tissue exposure in pelvic radiotherapy for anal cancer *Radiother. Oncol.* **117** 505–8

[75] Colaco R J, Nichols R C and Huh S *et al* 2013 Protons offer reduced bone marrow, small bowel, and urinary bladder exposure for patients receiving neoadjuvant radiotherapy for resectable rectal cancer *J. Gastrointest. Oncol.* **5** 3–8

[76] Dionisi F, Widesott L, Lorentini S and Amichetti M 2014 Is there a role for proton therapy in the treatment of hepatocellular carcinoma? a systematic review *Radiother. Oncol.* **111** 1–10

[77] Verma V, Lin S H, Simone C B and Mehta M P 2016 Clinical outcomes and toxicities of proton radiotherapy for gastrointestinal neoplasms: a systematic review *J. Gastrointest. Oncol.* **7** 644–64

[78] Kulkarni N M, Hong T S, Kambadakone A and Arellano R S 2015 CT-guided implantation of intrahepatic fiducial markers for proton beam therapy of liver lesions: assessment of success rate and complications *Am. J. Roentgenol.* **204** W207–13

[79] Moyers M F, Sardesai M, Sun S and Miller D W 2010 Ion stopping powers and CT numbers *Med. Dosim.* **35** 179–94

[80] Schneider U, Pedroni E and Lomax A 1996 The calibration of CT hounsfield units for radiotherapy treatment planning *Phys. Med. Biol.* **41** 111–24

[81] Poludniowski G, Allinson N M and Evans P M 2015 Proton radiography and tomography with application to proton therapy *Br. J. Radiol.* **88** 20150134

[82] Pemler P, Besserer J and de Boer J *et al* 1999 A detector system for proton radiography on the gantry of the Paul-Scherrer-Institute *Nucl. Instrum. Methods Phys. Res. A* **432** 483–95

[83] Bashkirov V A, Schulte R W and Hurley R F *et al* 2016 Novel scintillation detector design and performance for proton radiography and computed tomography *Med. Phys.* **43** 664–74

[84] Cormack A M and Koehler A M 1976 Quantitative proton tomography: preliminary experiments *Phys. Med. Biol.* **21** 560–9

[85] Zygmanski P, Gall K P, Rabin M S and Rosenthal S J 2000 The measurement of proton stopping power using proton-cone-beam computed tomography *Phys. Med. Biol.* **45** 511–28

[86] Schneider U, Besserer J and Pemler P *et al* 2004 First proton radiography of an animal patient *Med. Phys.* **31** 1046–51

[87] Doolan P J, Testa M, Sharp G, Bentefour E H, Royle G and Lu H M 2015 Patient-specific stopping power calibration for proton therapy planning based on single-detector proton radiography *Phys. Med. Biol.* **60** 1901–17

[88] Bashkirov V A, Johnson R P, Sadrozinski H F and Schulte R W 2016 Development of proton computed tomography detectors for applications in hadron therapy *Nucl. Instrum. Methods Phys. Res. A* **809** 120–9

[89] Tanaka S, Nishio T, Tsuneda M, Matsushita K, Kabuki S and Uesaka M 2018 Improved proton CT imaging using a bismuth germanium oxide scintillator *Phys. Med. Biol.* **63** 035030

[90] Holder D J, Green A F and Owen H L 2014 A compact superconducting 330 MeV proton gantry for radiotherapy and computed tomography *5th Int. Particle Accelerator Conf.* 2202–4

[91] Selzer R H, Hodis H N and Kwongfu H *et al* 1994 Evaluation of computerized edge tracking for quantifying intima-media thickness of the common carotid-artery from B-mode ultrasound images *Atherosclerosis* **111** 1–11

[92] Zhang X D, Li Y P and Pan X N *et al* 2010 Intensity-modulated proton therapy reduces the dose to normal tissue compared with intensity-modulated radiation therapy or passive scattering proton therapy and enables individualized radical radiotherapy for extensive stage IIIB non-small-cell lung cancer: a virtual clinical study *Int. J. Radiat. Oncol. Biol. Phys.* **77** 357–66

[93] Stanley D N, McConnell K A, Kirby N, Gutiérrez A N, Papanikolaou N and Rasmussen K 2017 Comparison of initial patient setup accuracy between surface imaging and three point localization: a retrospective analysis *J. Appl. Clin. Med. Phys.* **18** 58–61

[94] Walter F, Freislederer P, Belka C, Heinz C, Söhn M and Roeder F 2016 Evaluation of daily patient positioning for radiotherapy with a commercial 3D surface-imaging system (Catalyst™) *Radiat. Oncol.* **11** 154

[95] Alderliesten T, Sonke J-J, Betgen A, Honnef J, van Vliet-Vroegindeweij C and Remeijer P 2013 Accuracy evaluation of a 3-dimensional surface imaging system for guidance in deep-inspiration breath-hold radiation therapy *Int. J. Radiat. Oncol. Biol. Phys.* **85** 536–42

[96] Hughes S, McClelland J and Tarte S *et al* 2009 Assessment of two novel ventilatory surrogates for use in the delivery of gated/tracked radiotherapy for non-small cell lung cancer *Radiother. Oncol.* **91** 336–41

[97] Li G, Huang H and Wei J *et al* 2015 Novel spirometry based on optical surface imaging *Med. Phys.* **42** 1690–7

[98] Kauweloa K I, Ruan D and Park J C *et al* 2012 GateCT™ surface tracking system for respiratory signal reconstruction in 4DCT imaging *Med. Phys.* **39** 492–502

[99] Spadea M F, Baroni G, Gierga D P, Turcotte J C, Chen G T and Sharp G C 2011 Evaluation and commissioning of a surface based system for respiratory sensing in 4D CT *J. Appl. Clin. Med. Phys.* **12** 162–9

[100] Schaerer J, Fassi A, Riboldi M, Cerveri P, Baroni G and Sarrut D 2011 Multi-dimensional respiratory motion tracking from markerless optical surface imaging based on deformable mesh registration *Phys. Med. Biol.* **57** 357

[101] Glide-Hurst C K, Ionascu D, Berbeco R and Yan D 2011 Coupling surface cameras with on-board fluoroscopy: a feasibility study *Med. Phys.* **38** 2937–47

[102] Chang J Y, Li H and Zhu X R *et al* 2014 Clinical implementation of intensity modulated proton therapy for thoracic malignancies *Int. J. Radiat. Oncol. Biol. Phys.* **90** 809–18

[103] Berman A T, Teo B-K K and Dolney D *et al* 2013 An in-silico comparison of proton beam and IMRT for postoperative radiotherapy in completely resected stage IIIA non-small cell lung cancer *Radiat. Oncol.* **8** 144

[104] Kraus K M, Heath E and Oelfke U 2011 Dosimetric consequences of tumor motion due to respiration for a scanned proton beam *Phys. Med. Biol.* **56** 6563–81

[105] Phillips M H, Pedroni E, Blattmann H, Boehringer T, Coray A and Scheib S 1992 Effects of respiratory motion on dose uniformity with a charged-particle scanning method *Phys. Med. Biol.* **37** 223–34

[106] Lambert J, Suchowerska N, McKenzie D R and Jackson M 2005 Intrafractional motion during proton beam scanning *Phys. Med. Biol.* **50** 4853–62

[107] Grozinger S O, Bert C, Haberer T, Kraft G and Rietzel E 2008 Motion compensation with a scanned ion beam: a technical feasibility study *Radiat. Oncol.* **3** 34

[108] Seco J, Robertson D, Trofimov A and Paganetti H 2009 Breathing interplay effects during proton beam scanning: simulation and statistical analysis *Phys. Med. Biol.* **54** N283–94

[109] Grozinger S O, Rietzel E, Li Q, Bert C, Haberer T and Kraft G 2006 Simulations to design an online motion compensation system for scanned particle beams *Phys. Med. Biol.* **51** 3517–32

[110] Dowdell S, Grassberger C, Sharp G C and Paganetti H 2013 Interplay effects in proton scanning for lung: a 4D monte carlo study assessing the impact of tumor and beam delivery parameters *Phys. Med. Biol.* **58** 4137–56

[111] Grassberger C, Dowdell S and Lomax A *et al* 2013 Motion interplay as a function of patient parameters and spot size in spot scanning proton therapy for lung cancer *Int. J. Radiat. Oncol. Biol. Phys.* **86** 380–6

[112] Knopf A-C, Hong T S and Lomax A 2011 Scanned proton radiotherapy for mobile targets-the effectiveness of re-scanning in the context of different treatment planning approaches and for different motion characteristics *Phys. Med. Biol.* **56** 7257–71

[113] Li Y, Kardar L and Li X *et al* 2014 On the interplay effects with proton scanning beams in stage III lung cancer *Med. Phys.* **41** 021721

[114] Kardar L, Li Y and Li X *et al* 2014 Evaluation and mitigation of the interplay effects of intensity modulated proton therapy for lung cancer in a clinical setting *Pract. Radiat. Oncol.* **4** e259–68

[115] Bortfeld T, Jokivarsi K, Goitein M, Kung J and Jiang S B 2002 Effects of intra-fraction motion on IMRT dose delivery: statistical analysis and simulation *Phys. Med. Biol.* **47** 2203–20

[116] Liu C, Schild S and Chang J *et al* 2018 Impact of spot size and spacing on the quality of robustly-optimized intensity-modulated proton therapy plans for lung cancer *Int. J. Radiat. Oncol. Biol. Phys.* **101** 479–89

[117] Liu W, Schild S and Chang J *et al* 2015 A novel 4D robust optimization mitigates interplay effect in intensity-modulated proton therapy for lung cancer *Med. Phys.* **42** 3525

[118] Liu W, Liao Z and Schild S E *et al* 2015 Impact of respiratory motion on worst-case scenario optimized intensity modulated proton therapy for lung cancers *Pract. Radiat. Oncol.* **5** e77–86

[119] Liu C, Schild S E and Chang J Y *et al* 2018 Impact of spot size and spacing on the quality of robustly-optimized intensity-modulated proton therapy plans for lung cancer *Int. J. Radiat. Oncol. Biol. Phys.* **101** 479–89

[120] Liu W, Schild S and Chang J *et al* 2015 SU-F-BRD-01: a novel 4D robust optimization mitigates interplay effect in intensity-modulated proton therapy for lung cancer *Med. Phys.* **42** 3525

[121] Chan T C Y, Bortfeld T and Tsitsiklis J N 2006 A robust approach to IMRT optimization *Phys. Med. Biol.* **51** 2567–83

[122] Liu C, Bhangoo R S and Sio T T *et al* 2019 Dosimetric comparison of distal esophageal carcinoma plans for patients treated with small-spot intensity-modulated proton versus volumetric-modulated arc therapies *J. Appl. Clin. Med. Phys.* **20** 15–27

Part II

Principles of image-guided radiation therapy
for abdominal cancer

IOP Publishing

Principles and Practice of Image-Guided Abdominal
Radiation Therapy

Yu Kuang

Chapter 3

Imaging simulation for abdominal cancer image-guided radiation therapy

Emily Hirata and Manju Sharma

This chapter provides an overview of simulation principles for image-guided abdominal radiation therapy. We first discuss how abdominal tumors move and the importance of implementing a motion management approach when treating abdominal tumors. Next, the features of different motion management systems and the corresponding strategies are reviewed. Subsequently, simulation imaging modalities that complement these respiratory motion management systems are described, and practical clinical workflows are detailed as a guideline for implementation. Sample workflows, simulation setup procedures, and immobilization protocols are also given.

3.1 Introduction

The field of radiation oncology relies significantly on imaging innovations for diagnosis, target volume delineation, treatment planning and verification, and quality assurance and assessment of treatment outcomes. The increasing use of computed tomography (CT) imaging for target volume delineation, coupled with the availability of computer-controlled treatment planning and delivery systems, has led to highly sculpted dose distributions with sharp dose fall-off that are targeted to tumor volumes while providing major sparing of normal organs. This increased conformality requires reproducible daily set-up and the ability to accurately image the target volumes for proper alignment. Patient positioning based on bony anatomy is subject to error, especially when the tumor is a moving soft tissue structure, such as liver or pancreatic tumors. Accurate and safe targeted treatment planning and delivery is contingent on appropriate simulation considerations, as well as accurate daily imaging and set up.

Abdominal cancers refer to a variety of cancers that occur in the area between the lower chest and the groin and include liver cancer, pancreatic cancer, renal cell cancer, gastric cancer, and colorectal cancer. The major challenges of highly conformal radiotherapy for abdominal cancers are the mobility of these tumors coupled with their proximity to vital radiosensitive organs. Additionally, these tumors are soft and nonrigid; thus, large tissue deformations are not uncommon. Common organs of concern include the stomach, kidney, duodenum, small intestine, and spinal cord. A geometric miss of the tumor due to unaccounted motion may lead to poor outcomes and introduce significant complications and morbidity if any of these organs are overdosed. The risks and the associated consequences are increasingly severe with dose escalation and stereotactic body radiation therapy (SBRT). Accurate tumor motion assessment along with appropriate immobilization and imaging strategies for simulation are necessary prerequisites in safely delivering conformal, dose escalated radiation treatments for these cancers. In the absence of adequate motion assessment, larger planning margins are required, thus limiting the ability to deliver a curative dose without also exposing neighboring critical organs. This chapter presents a review of the mobility of abdominal cancers and surrounding organs at risk (OARs), as well as an array of current immobilization devices and advances in imaging strategies for simulation.

3.2 Motion of abdominal tumors

To adequately account for the motion of abdominal tumors, it is helpful to understand how they move. The motion of these tumors and surrounding organs are influenced by several factors, including respiration, cardiac motion, skeletal muscular motion, and changes in the filing status. Some of these factors can be mitigated during simulation, but some, such as cardiac motion, are not. The use of immobilization devices and moldable cushions increases patient comfort while keeping the tumor still during treatment. Advising patients to limit food and water intake for several hours before treatment are attempts to reduce tumor motion influenced by filling status. The primary source of motion for abdominal tumors, however, is respiration, and that will be the focus of the motion management strategies in this chapter.

Respiration-induced abdominal tumor motion has been characterized by using a variety of imaging modalities including MRI [1–6], four-dimensional computed tomography (4DCT) [4, 7–11], 4D-CBCT [12], and real-time tracking of implanted fiducials [9, 13].

Liver tumors have been observed to move anywhere from 10 to 21 mm in the superior–inferior direction, 5–8 mm in the anterior–posterior direction, and 3–9 mm in the left–right direction [7, 14–16]. Kitamura *et al* observed tumors that moved an average of 2 ± 1 mm due to cardiac motion [15].

For pancreatic tumors, there seems to be a larger variability in the range of motion: 1–29 mm in the craniocaudal direction, 1–8 mm in the anterior–posterior direction, and 1–13 mm in the left–right direction [2, 3, 11, 17].

Studies of stomach tumors also show wide variability. Uchinami *et al* found that stomach lymphomas moved on the order of 4–17 mm superior–inferior, 3–5 mm anterior–posterior, and 3–5 mm left to right [18]. A study by Wysocka *et al* looked at interfraction and breathing organ motion for gastric cancers at different breathing states and found that the stomach moved a median of 16.4 mm (0.5–41.5 mm) craniocaudal, 8.8 mm (0–29.2 mm) anterior–posterior, and 1.7 mm (0.3–11.5 mm) left to right [19].

Some studies have shown an even larger extent of motion, such as in deep inspiration kidney motion, for which the maximum extent observed was close to 4 cm [20]. A brief summary of some recent motion studies for abdominal tumors is presented in table 3.1.

The published data reinforce that the pattern and degree of motion of abdominal tumors is patient-specific, that it is often unpredictably variable, and that direct assessment of respiratory motion in every case should be performed.

3.2.1 Use of fiducial or internal markers

A commonly used surrogate for identifying abdominal tumors is implanted fiducials or markers. Abdominal tumors are often indistinguishable from surrounding normal tissue on CT, and the use of implanted fiducial markers near or within the tumor can significantly aid the detection and targeting of the diseased area. The fiducial implant requires an invasive procedure, however, and the clinical team must decide on the best approach to both achieve clinical goals and consider what is feasible and reproducible for the patient's status.

Additionally, lipiodol [25], a tumor-seeking embolic agent used in transarterial chemoembolization (TACE), is also used for the visualization of tumors in CT and CBCT. Yue *et al* [25] performed a feasibility study on the use of lipiodol as an internal image-guidance surrogate for patients with unresectable liver tumors after two to three TACE procedures that were performed by infusion of 5–10 ml iodized oil contrast medium. It was found that using lipiodol as a direct surrogate for image guidance with CBCT improved the localization accuracy for liver tumors.

3.3 Respiratory monitoring systems

Respiratory motion management describes a set of strategies that monitor, evaluate, and compensate for tumor motion in treatment planning and daily setup. The first step in respiratory management is to monitor and correlate the patient's breathing with tumor position. Common methods involve monitoring and tracking the motion of an external surrogate, such as chest or abdominal position, as the patient inhales and exhales. The resulting respiratory signal often looks like a sinusoidal waveform with peaks and valleys that correspond to inhalation and exhalation, respectively. Simultaneous image acquisition can correlate the internal tumor location at each point of the respiratory signal. There are different respiratory monitoring technologies available, and a few of the common ones are briefly described below.

Table 3.1. Summary of abdominal tumor motion observed during simulation from literature. AP, anterior–posterior; COM, center of mass; LR, left to right; RL, right to left; SI, superior–inferior.

Tumor site	Observed motion	Method used	Reference
Liver	LR: 0.6 ± 3.0 mm; AP: 2.3 ± 2.4 mm; SI: 5.7 ± 3.4 mm	4DCT, free breathing; segmented liver into eight functional segments (Couinaud classification); Segment 7 showed largest average amplitude in SI: 8.6 ± 3.4 mm	Tsai et al [14]
	LR: 4 ± 4 mm (range 1–12 mm); AP: 5 ± 3 mm (range 2–12 mm); SI: 9 ± 5 mm (range 2–19 mm)	2 mm gold marker implanted near tumor; tracked using fluoro; motion of lesions in left lobe < right lobe for AP and RL directions; patients with cirrhosis also had larger amplitude in LR and AP	Kitamura et al [15]
	LR: 4.19 ± 2.46 mm; AP: 7.23 ± 2.96 mm; SI: 15.98 ± 6.02 mm	4DCT; 6 patients with implanted fiducials, all in the right lobe; patients instructed to do regular shallow breathing	Nishioka et al [16]
	AP: 3.8 ± 1.3 mm; SI: 7.8 ± 2.6 mm	4DCT; 16 patients, 4DCT scan; free breathing with Varian RPM; Motion parameters evaluated included COM motion, edge motion, and volume	Hallman et al [21]
Pancreas	LR: 3.2 mm (range 0.1–13.7 mm); AP: 3.8 mm (range 0.2–7.6 mm); SI: 9.2 mm (range 0.9–28.8 mm)	4DCT; 20 patients, free breathing with Varian RPM; 2–5 gold markers implanted around tumor	Minn et al [17]
	LR <2 mm; AP: 3.8 ± 1.5 mm; SI: 7.4 ± 2.8 mm	Fluoroscopy; 7 patients with 3–10 clips; 30 s fluoro video for AP and Lat views with 30 frames per second	Gierga et al [22]
	LR: 0.7 ± 0.6 mm; AP: 2 ± 0.6 mm; SI: 5.9 ± 2.8 mm	4DCT, free breathing without coaching; 15 patients; Varian RPM	Tai et al [11]
	LR: 2 mm (1–7 mm); AP: 6 mm (0–18 mm); SI: 6 mm (1–15 mm)	3 CT scans: free breathing, inhale breath hold, exhale breath hold; 11 patients (7 pancreatic, 4 liver); 2–5 radio-opaque markers implanted	Yang et al [23]
	Ant border: 8 ± 3 (3–13 mm); Post border: 6 ± 2 (3–9 mm); Sup border: 20 ± 10 (13–42 mm); Inf border: 20 ± 8 (13–38 mm)	Cine 3T MRI; 17 patients; shallow free breathing; evaluated motion of tumor borders	Feng et al [2]
Stomach	LR: 2.9 ± 1.3 mm; AP: 4.1 ± 1.4 mm; SI: 10.1 ± 4.5 mm	4DCT, free breathing; 10 patients; Varian RPM	Uchinami et al [18]
	LR: 2.4 ± 7 mm; AP: 4.6 ± 8 mm; SI: 12.1 ± 12.8 mm	Multiple CT scans at mid-inhale, mid-exhale, shallow free breathing; 6 patients	Watanabe et al [24]

3.3.1 Respiratory belts

Several respiratory monitoring systems use a belted device to detect abdominal movement as a surrogate for respiratory motion. These belts are positioned around the patient's diaphragm and are connected to a sensor that detects abdominal movement as the patient breathes. One system is the bellows device (Philips Medical Systems, Netherlands), which uses a rubber air bellows attached to a pressure transducer. When the abdomen stretches and contracts during respiration, the transducer detects air pressure changes within the bellows, and the resulting signal is digitized and saved as a respiratory waveform to correlate with acquired images [26].

The Anzai Respiratory Gating System (Anzai Medical Co. Ltd, Tokyo, Japan; figure 3.1) is another belt-based system that consists of a pressure sensor with a belt that attaches the sensor to the patient. Using a strain gauge bridge circuit, the sensor detects pressure changes at the abdomen, and the motion of the body surface is displayed as the respiratory waveform. There are two options for the sensors: one for normal breathing and the other for deep breathing [27]. The exhalation phase is reflected with a lower signal amplitude, and the inhalation phase corresponds to the higher amplitude signal. Both belt options interface with certain CT scanners and can be an integrated accessory for 4DCT acquisitions.

3.3.2 Optical tracking systems

Another popular method for tracking surface motion during respiration uses camera-based technologies. The Varian Real-time Position Management (RPM) system (Varian Medical Systems, Inc., Palo Alto, CA; figure 3.2) consists of an infrared (IR) tracking camera, a lightweight plastic box with reflective markers, and a computer console that visualizes the patient's breath signal. The IR tracking camera is a video camera equipped with an array of light emitting diodes (LEDs) that emit IR light in the direction where the camera is pointing. The plastic box is

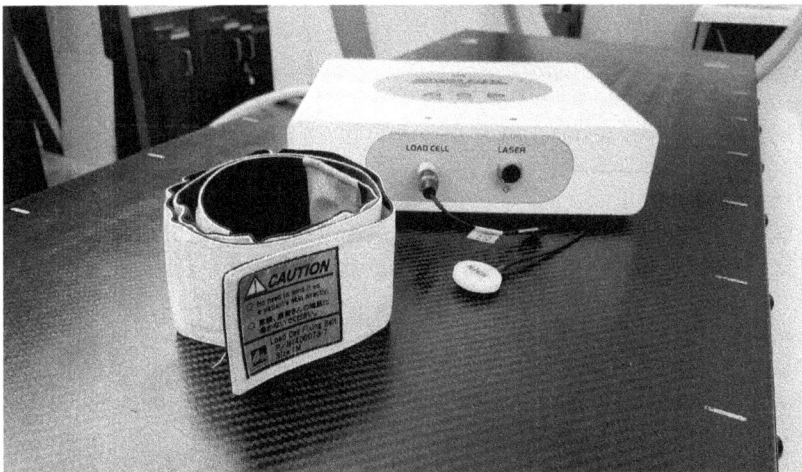

Figure 3.1. Anzai respiratory gating system. Photo from University of California, San Francisco (UCSF).

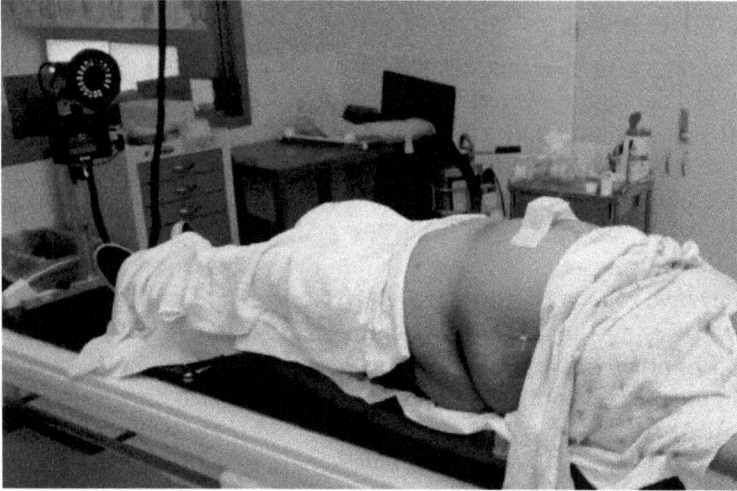

Figure 3.2. Varian real time position management system. Optical block on patient abdomen with optical camera at foot of the table. Photo from UCSF.

placed on the patient's abdomen within view of the tracking camera, typically between the umbilicus and the xiphoid, and the markers on the box reflect the IR light back to the camera, which captures the signal. As the patient breathes, the box moves up and down, and the reflected signal is used to track and analyze the motion. This system is widely used in several respiratory motion management applications, including liver [14, 28, 29], pancreas [11, 17, 30], and stomach [18].

3.3.3 Surface guided imaging systems

Other camera-based tracking systems use light that is reflected directly off the patient's surface. The C-RAD Sentinel (Uppsala, Sweden) system is a laser based optical surface scanning system [31] and consists of two components: a projective unit composed of LED lights and a charge-coupled device camera mounted to the ceiling in the CT room. This system projects a fast sequence of near-visible light patterns onto the patient's surface using wavelengths of 450 nm (blue), 528 nm (green), and 624 nm (red). The camera then detects the reflected light, and the software generates a three-dimensional model of the patient's surface. The user will specify a region of interest on the patient surface to track, and through optical triangulation of the reflected light, a surrogate signal of the patient's respiration can be acquired. No markers or other devices are placed on or around the patient, and both thoracic and abdominal respiratory motion can tracked in simultaneously. Threshold values for the respiratory signal can also be specified. An audio-visual coaching feature is available, and video goggles can be used to provide visual feedback to the patient about their detected respiration. This system can interface with various CT scanners to correlate inhalation and exhalation phases or amplitudes to the CT acquisition.

Another surface guided imaging system is the AlignRT 3D system (Vision RT Ltd, London, United Kingdom; figure 3.3). This system uses three 3D camera pods

Figure 3.3. AlignRT optical camera system. Photo from UCSF.

mounted to the ceiling. Each pod consists of several components: a texture camera, two stereo cameras, a speckle flash and clear light flash, and a speckle projector [32]. A speckled optical pattern is projected onto the patient and detected by the three cameras to generate a 3D surface model. Surface registration is calculated by comparing the distance of a user-defined ROI, and tolerances can be set such that warnings or beam off triggers can be set when the patient moves significantly from the intended treatment position. Compared with CBCT, the system has a published accuracy of ± 0.25 mm and $\pm 0.2°$ [33].

3.3.4 BrainLab ExacTrac system

The BrainLab ExacTrac system integrates (1) 2D x-ray radiographs and (2) an IR optical positioning system with a (3) 6D robotic couch. The CT simulation is an essential step in the patient preparation for image-guided radiation therapy using the ExacTrac system. The ExacTrac IR system consists of two IR cameras, reflective markers of various sizes placed on the patient's surface, and the reference device (the reference star). Typically, two IR systems are installed both in the CT Sim and treatment room. BrainLab provides two different types of markers—the 8 mm aluminum core markers for CT scan visualization are replaced by the plastic core markers to eliminate the dosimetric effects. The patient preparation during simulation involves shaving off (if applicable) and cleaning of a small area on patient skin and for marker placement. The markers are attached with the single use marker sockets and placed in an unambiguous arrangement on stable parts of the body (e.g., hip or sternum) that are not affected by respiratory motion or skin shift. Each marker socket is outlined by a permanent marker or tattoo to outline at the center of the socket onto the patient's skin. The ExacTrac system uses the IR

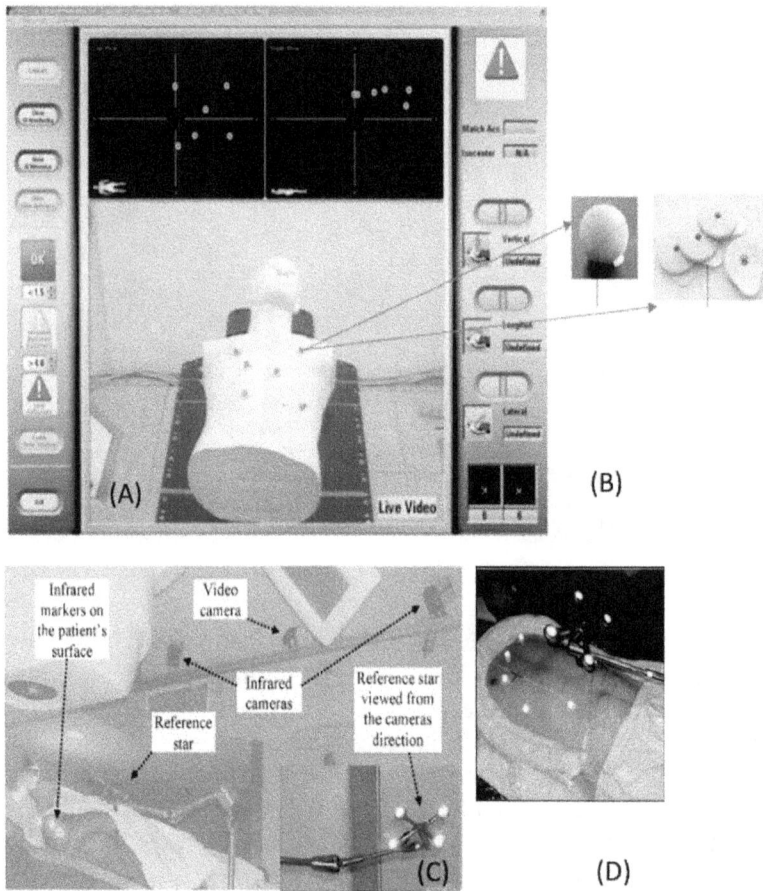

Figure 3.4. The ExacTrac system consisting of (a) the IR video camera system (b) IR markers, (c) Exactrac IR and imaging system with reference star, and (d) the reference star with the body IR markers. Image credit for (a), (b), and (d): Brainlab AG. Panel (c): Reprinted from [36], Copyright (2008), with permission from Elsevier.

markers to control the motion of the robotic couch with an accuracy of 0.3 mm [34]. The IR system can be used for initial patient setup verification and precise adjustment of the couch positioning based on the results of x-ray radiographic imaging. Some studies have used the IR subsystem for respiratory motion management [35]. The ExacTrac system also has an option of using reference star and markers for positioning at the treatment delivery. The reference star consists of four reflective circles as shown in figure 3.4 and is attached to the 6D couch for patient positioning. The marker position at the treatment is compared with the stored reference information from the CT scan.

3.4 Respiratory motion management

Any of the respiratory monitoring devices described above can be used to acquire the patient's respiratory signal and correlate the information to internal tumor location. Given that abdominal tumors frequently move 1–2 cm or more with uncontrolled respiration (also referred to as 'free breathing'), a common strategy to reduce tumor motion is to control respiration by having patients hold their breath during imaging and treatment ('breath hold') or encourage shallow frequent breaths by compressing the abdomen ('abdominal compression.') This section describes breath hold (BH) and abdominal compression strategies and some commonly used technologies for implementation.

3.4.1 Breath hold

Breath hold (BH) strategies help immobilize tumors by reducing the respiratory motion; however, since imaging and treatment is delivered only during the BH, the duration of the simulation and treatment appointments will be extended. It is typical that daily CBCT acquisition will require two BHs, and each arc in a VMAT plan will require two to three BHs to complete. For this reason, it is important that the BH is consistent, reproducible, and sufficiently long to be feasible for daily imaging and treatment. Patients are encouraged to practice holding their breath for approximately 30 s, at either inhale or exhale depending on the anticipated strategy. For abdominal tumors, exhalation BHs seem to result in improved organ stability compared with inhalation BHs. Lens *et al* [37] used MRI to compare BHs at exhalation and inhalation for pancreatic tumors and found that the mean magnitude of motion for the diaphragm and pancreatic head were larger for inhalation BHs than exhalation BHs. Additionally, the first 10 s of the BH illustrated more organ motion compared with the remainder, suggesting that delaying the beam on for treatment until after the first 10 s of BH may yield a more stable state [37]. Taniguchi *et al* [38] reported a significant overlap of pancreatic head tumors and the duodenum in the inspiration phase compared with the expiration phase for 17 consecutive patients, leading to inferior dosimetry.

Particularly in the context of SBRT, it is recommended to use a system that monitors the patient's respiration in real time, has tools to evaluate consistency of the breath hold, and allows user-defined thresholds that trigger alerts or possibly shut off the beam should the patient cease to sustain an adequate breath hold. The respiratory monitoring systems described above can be used to monitor the patient's breath hold, however additional devices are frequently employed depending on whether the BH will be voluntary (VBH) or involuntary (IBH) [39]. During VBH, the patient holds his/her breath at a comfortable level during CT acquisition and treatment delivery, often either at inhalation or exhalation. For IBH, external equipment will force a patient to hold his/her breath at a specific phase of respiration. A BH of 30 s or longer is preferred, although a BH as short as 15–20 s may be acceptable. These techniques are discussed in further detail below.

3.4.1.1 Voluntary breath hold

The VBH technique is relatively simple and economical with minimal equipment needs. The key point for the success of VBH is the patient's comfort. Candidate selection depends on the patient's ability to reproducibly hold their breath at a

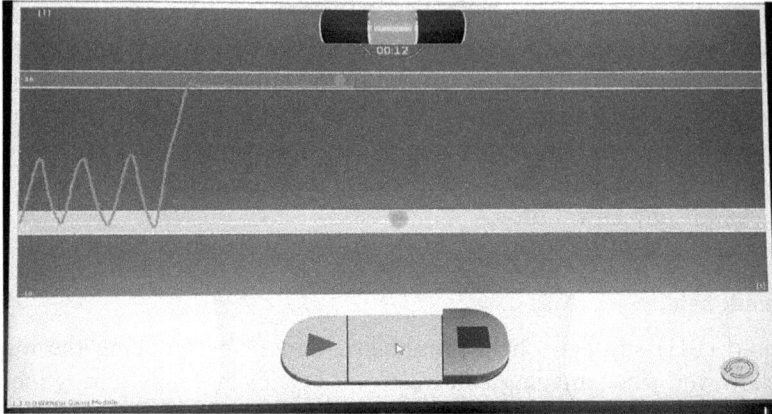

Figure 3.5. Example of visual feedback used in SDX treatments to aid patients in achieving and maintaining consistent BHs. The horizontal green bar indicates the inhale BH window that the patient tries to achieve and maintain. Photo from UCSF.

comfortable level for approximately 30 s and requires a practice and coaching session to assess whether this is an appropriate strategy for the patient [40]. Although patients can hold their breath at any phase of respiration, several studies suggest that end-expiration is more reproducible than full-inspiration [37]. Additionally, systems that provide patients with real-time visual feedback about their BHs (figure 3.5) increase patient BH compliance and reproducibility [41]. Visual feedback systems commonly use goggles that display a simple graphic to show patients their respiratory signal and the BH goal in real time. Nakamura *et al* [30] established reproducible positioning of pancreatic tumors using voluntary end-exhalation BH in conjunction with visual feedback. Patients were given simple audio instructions such as 'breathe in, breathe out, and hold your breath' and used video goggles to observe their abdominal motion signal detected from the Varian RPM system. Pairing the BH strategy with daily cone beam CT for daily localization resulted in average interfraction shifts within 5 mm.

3.4.1.1.1 Spirometry motion management

A spirometry-based system designed for VBH is the spirometry motion management (SDX) system (figure 3.6). The SDX system was developed by DYN'R (Toulouse, France) and is used to control respiratory motion by measuring the pulmonary volume of each BH. The SDX system has a nose clip to ensure respiration through the mouth, a mouthpiece attached to a bacterial filter, and a sensor that detects the volume of air the patient inhales and exhales during respiration. The system also includes video goggles to provide visual feedback for reproducible BHs. To start, the patient is instructed to breathe normally for a few sets of inhale-exhale cycles, then take a deep inhale breath for the system to measure the inspiratory capacity, and then breathe normally. This inspiratory capacity measurement is repeated at least three times to assess the reproducibility of the inhale and stability of the capacity measurement. Exhale BHs are typically set to a threshold of 20% of the inspiratory capacity, and deep inspiration BHs are set to a threshold of 75% of the inspiratory capacity. SDX can be

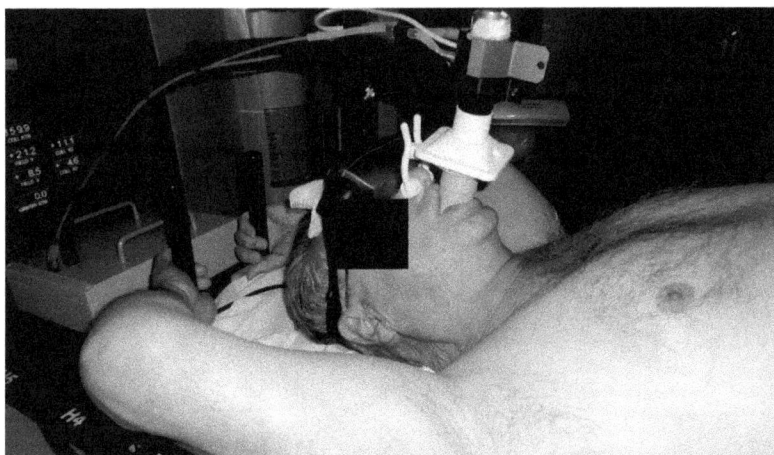

Figure 3.6. SDX respiratory management system. Photo from UCSF.

used for both VBH and in free-breathing motion management applications. The contrast-enhanced scans using BH with the SDX device are exceedingly difficult to obtain because of challenges with timing the contrast with patient BHs.

3.4.1.2 Involuntary breath hold

With IBH, the patient is forcibly made to hold their breath at a specific respiratory phase. At the time of simulation, the patient is first coached on how to hold their breath to assess the feasibility of the technique and to estimate the respiratory level for the BH based on patient comfort. The patient is then introduced to the IBH system, and several practice BHs are performed to continue coaching and finalize the appropriate BH settings. Commonly used IBH technologies include the active breathing coordinator (ABC, Elekta, Stockholm, Sweden) and the SDX system (DYN'R, Toulouse, France). The dosimetric benefits of using breath hold techniques with ABC and SDX systems have been widely reported in literature [42]. Below we describe further details on these systems.

3.4.1.2.1 Active breathing coordinator

The ABC (figure 3.7) is a system developed at William Beaumont Hospital and commercialized by Elekta Inc (Norcross, GA), that is designed to hold a patient's respiration at a specific phase to control internal respiratory motion of the tumor [43]. A clamp is placed on the patient's nose to prevent accidental respiration through the nose, and a mouthpiece is attached to a breathing tube. The device consists of two pairs of flow monitors and scissor valves; one controls the inspiration flow, and the other controls the expiration flow to and from the patient. The respiratory signal is processed continuously, and the system software displays the changing lung volume in real-time as the patient breathes through the mouthpiece. For inspiration BHs, the patient will inhale up to a predetermined volume and then hold their breath, at which point the valves will close so no additional air can enter the lungs during the breath hold. For expiration BHs, the patient will exhale a

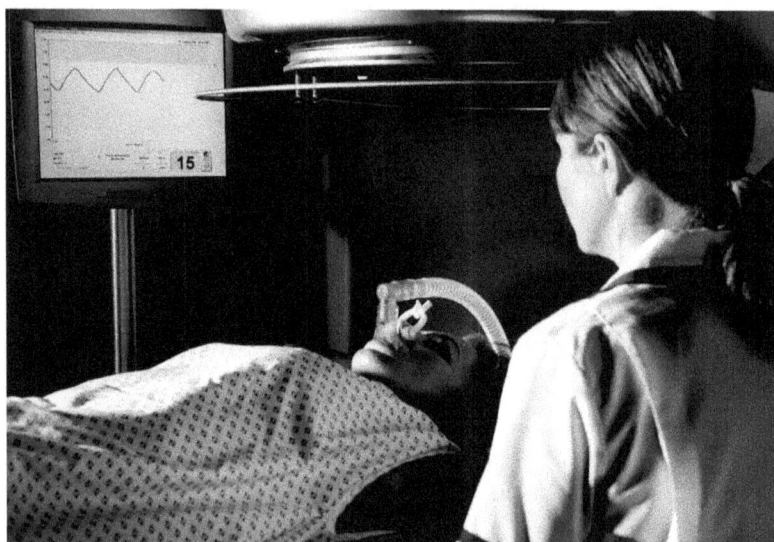

Figure 3.7. Active breathing coordinator (ABC) with three parts: (1) nose clamp, (2) mouthpiece, and (3) tubing, and video monitor. The panic button is not shown in the figure. Image is adapted from Swedish Medical Center. Reprinted from [44], Copyright (2001), with permission from Elsevier.

predetermined volume of air and then hold their breath with the valves closed. Several practice BHs are typically performed to assess the appropriate BH level. If the patient is unable to hold their breath, there is a pressure switch that the patient can release to automatically open the valves to allow free breathing. The ABC system can also be interfaced with some modern linear accelerators such that releasing the BH valves will trigger the treatment beam to turn off. One advantage of ABC over SDX is that with ABC, the time of BH is controlled by the therapist rather than the patient. This enables the use of IV contrast during simulation since the timing of the injection can be ensured by the clinical team at simulation.

A general clinic practice is to acquire three consecutive BH CT scans to evaluate the maximum extent of motion. The best of three or an average of three CT scans can be used for treatment planning.

Several studies have validated the ABC to provide reproducible BHs for abdominal tumor treatments. Dawson *et al* [45] evaluated the use of ABC on patients with unresectable intrahepatic tumors who could comfortably maintain a BH of at least 20 s. The BH positions of the diaphragm and radiopaque microcoils implanted in the liver were visualized relative to bony anatomy using fluoroscopy. They determined the intrafraction reproducibility was 1.6 mm, whereas interfraction reproducibility was 3.8 mm, indicating that ABC does not preclude the need for daily online imaging and repositioning at the start of each fraction.

3.4.2 Abdominal compression and immobilization devices

Abdominal compression is a widely used motion compensation method in radiotherapy of abdominal cancers [46]. The process involves application of a constant force to the abdomen until the maximum tolerable level of the patient. There are various commercially available devices [46–53] such as stereotactic body frames, pressure plates, and screws. The use of an abdominal compression system can improve reproducibility and provide the immobilization required for precise high dose treatments.

In 2014, Lovelock *et al* [47] used an in-house abdominal compression belt with an inflatable air bladder. In this study, 42 patients had two to three radiopaque markers implanted or had surgical clips near the tumor, ranging from liver, adrenal glands, pancreas, and lymph nodes. The pneumatic pressure level was determined by patient feedback, and the cranio-caudal (CC) motion was determined by using fluoroscopy. When using abdominal compression, it is important to minimize the variability in abdominal deformation caused by the compression device. The device should be placed at the same position each day, including indexing and tightness. The location of the belt or compression paddle on the abdomen can be verified by using x-ray port films, verification CT scans, and/or onboard CBCT if available.

Different immobilization devices are used for motion management, including Qfix arm shuttle, BodyFIX with and without wrap [48], alpha cradle of varying lengths, and aquaplast casts [49]. Figure 3.8 summarizes a few commonly used devices. In a study on 3D treatment, the different immobilization methods showed comparable reduction in patient movement [49], but significant differences were observed in SBRT treatments, with the positioning uncertainty ranging from 2 to 6 mm with BodyFIX [19], 3.7–5.7 mm with Elekta body frame [50], and 1.8–4.4 mm with Leigbinger body frame [51]. The thermoplastic masks offered better reproducibility with significantly less interfraction set-up displacement in comparison to vacuum cushions. The stereotactic body frame with vacuum pillow extending from head to thighs had acceptable repositioning accuracy of 5–8 mm for 90% of patients [50]. The modified extracranial frame setup used by Herfarth *et al* [51] included a metal arch rigidly mounted to a carbon-fiber compound, and three V-shaped indicators (lateral and anterior) were attached to the arch, with the posterior indicator attached to the base plate. The metal wires were visible in the CT scan and used to stereotactic target point calculations. Dreher *et al* [52] studied two different immobilization methods for liver SBRT patients immobilized by a vacuum couch with low pressure foil (Medical Intelligence Medizintechnik GmbH, Schwabmünchen, Germany) and abdominal compression belt (Fa. ITV, Innsbruck, Österreich). Both the online adjustments decreased tumor motion significantly and led to higher accuracy in patient positioning, and the abdominal compression led to better reduction in tumor motion.

3.5 Imaging technologies

The goal of simulation is to capture the extent of tumor motion to estimate the likely position of the target during treatment. This requires the ability to visualize and accurately distinguish the target separate from surrounding healthy tissue throughout the patient's respiratory cycle. By partnering an advanced imaging system with a respiratory monitoring device, respiratory management strategy, and appropriate immobilization devices, safely and effectively localizing abdominal tumors with

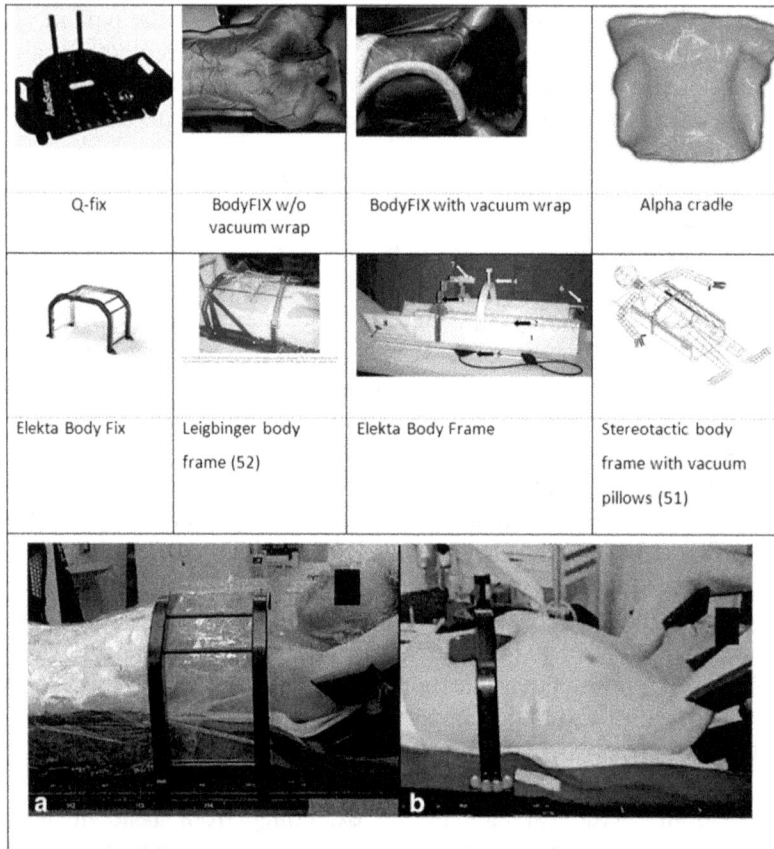

Q-fix	BodyFIX w/o vacuum wrap	BodyFIX with vacuum wrap	Alpha cradle
Elekta Body Fix	Leigbinger body frame (52)	Elekta Body Frame	Stereotactic body frame with vacuum pillows (51)

Figure 3.8. Immobilization setups. (a) Immobilization with a vacuum couch with low pressure foil. (b) Immobilization with abdominal compression [52]. Image credits: Qfix, BodyFIX without vacuum wrap, and BodyFIX with vacuum wrap [48], © The Authors. *Journal of Medical Radiation Sciences* published by John Wiley & Sons Australia, Ltd on behalf of Australian Society of Medical Imaging and Radiation Therapy and New Zealand Institute of Medical Radiation Technology. Elekta Body Fix image reprinted with permission of Elekta. Leibinger body frame, Copyright (2000), reprinted with permission from Elsevier. Elekta Body Frame [53] John Wiley & Sons. Steretactic body frame with vacuum pillows [50], reprinted by permission of the publisher (Taylor & Francis Ltd, http://www.tandfonline.com). (a) and (b) with permission of Springer.

appropriate margins is achievable. CT imaging remains the primary modality for radiation therapy; however, in recent years, significant development has been made in 4D imaging modalities, which will be explored in this section.

Two imaging strategies that have been used for respiratory motion management for lung tumors are fluoroscopy and slow CT; however, significant limitations exist with these approaches for abdominal tumors. Fluoroscopy is readily integrated with CT scanners and can be used to assess internal motion in real-time. The disadvantages of fluoroscopy include being limited to 2D motion and being applicable only to tumors that are visible. With fluoroscopy, abdominal tumors are challenging to visualize and differentiate from the surrounding healthy tissue and would require implanted fiducials near or within the target. Without fiducial markers, these limitations generally preclude the use of fluoroscopy for abdominal tumor motion management and simulation.

Slow CT involves scanning patients using a slow gantry rotation speed to capture tumor motion during each slice acquisition. The acquisition should be slow enough such that a full respiratory cycle is acquired at each couch position [55–63]. Patients are imaged while breathing freely which is one advantage over other methods. A significant disadvantage, however, is the loss of resolution due to motion blurring. For this reason, this approach is not recommended for abdominal tumors, for which the target will not be identifiable.

3.5.1 Breath-hold CT

BH CT involves CT image acquisition while the patient holds their breath at a particular phase, typically inhalation or exhalation, and can be readily implemented in most clinics. Since the patient is holding their breath during the acquisition, any respiration-related blurring is significantly reduced. This approach relies on the patient's ability to maintain a constant BH and to perform the BH reproducibly and is generally more successful with coaching and practice. A method for respiratory monitoring to ensure BH constancy is important, and those systems that incorporate patient feedback, such as a visual graph, have been shown to increase consistency [54]. BH CT scans can be used to plan the patient in the BH state or to generate an estimated image target volume (ITV) such that the patient will breathe freely during treatment. If the patient can hold their breath consistently and reproducibly for a period of 20 s or more, treatments may be given using only the inhale or exhale BH scans. This significantly reduces the volume of the target for treatment and the dose to surrounding normal tissue. If the patient will be treated with free breathing, the inhale and exhale scans can be co-registered, and a contour that encompasses the target in both scans can be defined as the ITV.

3.5.2 4DCT imaging

4DCT is now a commonly used approach for obtaining high -quality CT images in the presence of respiratory motion. A 4DCT image can be generated using a multislice CT scanner and can be performed in two modes: cine or helical. With cine mode, the couch stays in the same position for each scan and then indexes to the next position, repeating for each section of the CT scan. In helical mode, the couch is continuously moving during the acquisition. Additionally, knowledge of the respiratory phase for each projection is required and must be correlated with the scan. For this purpose, a respiratory trace is acquired simultaneously with the CT data acquisition, and any devices described above under 'Respiratory monitoring systems' can be employed. Correlating the respiratory cycle with the internal tumor motion assumes that a change in the amplitude of the respiratory signal will have a corresponding impact on the internal tumor motion. Image acquisition without an external surrogate can also be performed using an internal fiducial or marker, with either deformable image registration or image segmentation.

Sorting the images into their corresponding respiratory phases can be done retrospectively or prospectively. Retrospective sorting is when each projection is correlated with a point in the respiratory cycle. Prospective sorting is when projections are acquired at a defined point in the respiratory cycle. When acquiring a retrospective 4DCT data set, the respiratory cycle is divided into 8–10 bins per cycle, and the corresponding images are sorted by these bins, then reconstructed into a 3D data set. The 4D motion can then be observed by looping through the 3D images for each bin, providing a visualization of the captured range of motion. Inhalation, exhalation, and time-averaged scans are examples of the types of 3D data sets that can be reconstructed. Significant artifacts can result in 4DCT studies if the patient breathes irregularly during acquisition, and patients may require verbal reminders to breathe regularly during the scan.

There are two ways to determine the respiratory bins: phase-based or amplitude-based binning. In phase-based binning, the temporal relationship of the cycle is what determines the appropriate bin. In this mode, the peak inhalation points of the respiratory signal are designated as the 0% phase, and the other phases are determined by the elapsed time between detected respiratory maxima on the waveform. This approach works well for respiratory patterns that are periodic, or when the breathing frequency is variable but the shape of the respiratory cycle from inhale to exhale is constant. The internal motion for inhaling, however, is not the same with exhaling, and the tumor is found to follow an elliptical path, called hysteresis [55].

In amplitude-based binning, the fraction of maximum amplitude determines the bins. This approach can account for hysteresis because identical amplitudes of inhale and exhale will be binned separately. A common artifact of amplitude binning, however, is when the inhale amplitude is inconsistent throughout the scan. Additionally, amplitude binning will not reconstruct data outside the normal amplitude of the respiratory signal, and there may be situations in which the clinical team will want to evaluate those images to assess the full range of motion. In general, phase-based binning is more commonly used.

3.5.3 4D magnetic resonance imaging

Magnetic resonance imaging (MRI) in radiation therapy has increased over the past decade because of its ability to generate both anatomical and functional contrast, often providing clear distinction between tumors and normal tissue. Tumor motion can also be characterized through MRI. Recently, there has been significant interest in the development of a clinically useful 4D-MRI sequence for radiotherapy [6, 60–70]. Although 4DCT only acquires images in the transverse orientation, 4D-MRI can visualize soft tissue anatomy in any orientation, which would be particularly useful for abdominal tumors when both tumors and organs at risk are visible, unlike 4DCT with limited soft-tissue contrast. Similar to 4DCT, external surrogates can be used to acquire the respiratory signal, such as pneumatic belts [6], although the logistics of setup and MRI bore interference can present challenging considerations. Unique to 4D-MRI is the use of internal surrogate signals with pencil-beam navigators [56, 57] or self-gating methods [5].

Current and potential applications of 4D-MRI include establishing 4D deformation vector fields that could be used to align MR images of different respiratory phases [58]. Moreover, there is significant work focused on generating synthetic CT images for treatment planning [59].

3.5.4 4D positron emission tomography–CT

Improved accuracy in tumor identification using fluorodeoxyglucose (FDG)–positron emission tomography (PET) has resulted in its widespread use for target volume delineation. The accuracy of PET-CT is negatively affected by respiratory motion by which the smearing effect can lead to the reduction of measured standard uptake value (SUV) and the overestimation of the apparent tumor size [60]. These respiratory motion artifacts may be mitigated by the application of 4D PET-CT imaging strategies [29]. Respiratory-gated PET-CT scans are obtained in several steps. Aristophanous *et al* described their process which included performing a whole-body 3D PET scan with FDG (between 18.7 and 22 mCi given 100 min prior to scan) [61]. Using the Anzai AZ-733V respiratory gating system, a respiratory gated 4D PET scan of the whole lung was acquired, with binning into five phases (0%, 20%, 40%, 60%, and 80%). Attenuation correction for both the 3D and 4D PET scans was performed using the 3D CT scan. Scarsbrook *et al* investigated the feasibility of 4D-PET-CT in 15 patients with distal esophageal carcinoma [62]. Patients were injected with 400 MBq of FDG prior to the simulation appointment. Using a 64-slice GE Discovery 690 PET-CT scanner coupled with the Varian RPM system, a planning CT was obtained, followed by a static PET acquisition. The 4D PET scan was acquired approximately 60 min post-FDG injection, with real-time monitoring of respiration, including breaths per minute (BPM) and the average breathing period, followed by 4DCT. Both the 4D PET and 4DCT scans were binned into 10 phases, and an average 3D PET was created using post-processing software. The authors found tumor volume delineation differences when using 4D PET-CT compared with 4DCT along or 4DCT with 3D PET; however, because of the small cohort, the results remain inconclusive, and continuing investigation is needed.

3.6 Simulation strategies and clinical workflows

With respiratory monitoring systems, a motion management approach, and advanced imaging, management of tumor and organ motion commonly employs one of three main strategies: motion encompassing, motion control, and motion tracking or gating [63, 64].

Motion encompassing approaches allow the patients to breathe as they normally do while the imaging technology attempts to capture all possible positions of the moving tumor to define an ITV. The physician or treatment planner will often expand the ITV to generate the planning target volume (PTV) for treatment planning. Although this approach is easy on the patient, if the tumor moves significantly during respiration, the resulting plan will irradiate a larger volume of healthy tissue compared with other strategies. For this reason, many clinics will first acquire a free-breathing 4DCT and evaluate the tumor motion. If the extent of tumor motion is large, typically greater than 1 cm, motion-control or motion tracking strategies may be used.

Motion control or motion reduction strategies involve application of compression devices to restrict the respiratory motion or actively manage how the patient breathes. The goal of these approaches is to limit the magnitude of the respiratory motion such that smaller margins can be used. These strategies require the patient's cooperation, and depending on the devices used, some patients may not be able to tolerate such breathing controls. Simpler devices include straps or paddles that compress the abdomen to force the patient to take shallow breaths. More complex devices involve machines that maintain a fixed volume of air in the lungs.

Gating and motion tracking strategies aim to synchronize the treatment delivery with the tumor motion. The beam on and off times or the beam shaping and positioning are triggered and automatically controlled only when the tumor (or surrogate) is in the correct location. These strategies are also better suited for patients who exhibit regular respiratory patterns. Typically, these systems involve frequent imaging during treatment and use tighter treatment margins but also require that the tumor or surrogate is visible and unobstructed.

Prior to the simulation appointment, the strategy for immobilization, respiratory motion management, imaging, and treatment delivery should be determined based on the available technology and resources. Table 3.2 summarizes the overall respiratory motion management strategies and the options for imaging, immobilization, and treatment approaches. Table 3.3 summarizes the available respiratory motion compensation techniques and associated imaging strategies.

3.7 Clinical workflows

The simulation strategy for radiation therapy of abdominal tumors requires assessment of multiple domains, including tumor location and nearby normal organs, tumor visibility and localization, respiratory motion considerations and whether the patient can tolerate BH or abdominal compression, and available immobilization tools. This assessment will further inform the dosimetric strategy such as whether to use SBRT and how much dose escalation is safe and appropriate.

The following examples highlight the considerations and clinical decision tree when preparing for simulation of abdominal tumors.

3.7.1 Example 1. Liver SBRT with SDX BH

Prior to the simulation, the physician and clinical team consult with the patient to discuss the possible use of BH with SDX for simulation and treatment. To assess the feasibility of this technique, the patient performs a timed exhale BH. If the patient can hold their breath for 30 s or more, they are a candidate for a BH treatment approach. Instructions are given to the patient to practice the exhale BH for a

Table 3.2. Summary of respiratory motion management strategies and treatment approaches.

Strategy	Imaging technique	Breathing at simulation	Devices at simulation	Breathing at treatment	Devices at treatment	Beam on
Motion encompassing	Fluoroscopy	Free or compression	Implanted fiducial	Free or compression	Implanted fiducial	Continuously, uncorrelated with respiratory cycle
	3D CT	Free		Free		Continuously, uncorrelated with respiratory cycle
	4D (CT, MRI, PET)	Free	Respiratory monitoring device	Free		Continuously, uncorrelated with respiratory cycle
Motion control	4D (CT, MRI, PET)	Compression	Compression device; respiratory monitoring device	Compression	Compression device	Continuously, uncorrelated with respiratory cycle with compression
	3D CT	VBH	Respiratory monitoring device; patient feedback regarding BH level; SDX	VBH	Respiratory monitoring device; patient feedback regarding BH level; SDX	During BH only
	3D CT	IBH	ABC; display/feedback regarding BH level	IBH	ABC; display/feedback regarding BH level	During BH only
Gating	4D (CT, PET, MRI); fluoroscopy	Free	Respiratory monitoring device	Free	Respiratory monitoring device	Only during specific phases/amplitude of breath cycle ('Gating')
Tracking	4D (CT, PET, MRI)	Free	Respiratory monitoring device; fiducial	Free	Respiratory monitoring device; fiducial	Continuously on only while fiducial is tracked ('tracking')

Table 3.3. Characteristics of motion compensation and imaging strategies.

Motion compensation method	DIBH	Spontaneous breathing gating	Real-time tracking	ITV/margins
Available techniques	Free DIBH; computer controlled DIBH (spirometry, surface tracking with markers or markerless)	Spirometry, surface tracking with markers or markerless	Couch tracking	TP with 4DCT, potentially with abdominal compression
Imaging	All imaging under DIBH: planning CT, CBCT, ultrasound surveillance in BH, simultaneous VMAT-CT during treatment	Dynamic planar or ultrasound; 4D CBCT immediately before treatment	Dynamic planar or ultrasound imaging	4DCT, 4D MRI for treatment planning, 4DCBCT immediately before treatment.

minimum of 30 s at home, along with a link to a video explaining the SDX system and the BH procedure.

On the day of the simulation, the therapists build and mold a customized body cushion such as a vac-lok cushion (Civco Radiotherapy, Coralville, Iowa) and then begin coaching the patient through several BHs. At this time, the patient is oriented to the nose clip and mouthpiece to see if they are able to tolerate use of these devices. They are introduced to the video goggles and the graphic illustrating the BH threshold they should aim for and maintain. Several practice BHs are performed to assess the reproducibility of the BH and ensure that the patient is comfortable with the procedure. When ready, the patient performs the exhale BH, and CT scanning commences. If the patient is unable to sustain a BH for a minimum of 20 s during the practice and coaching sessions, the physician is notified, and typically a 4DCT scan is acquired under free breathing. Details of the decision tree and set up instructions are summarized in figure 3.9.

3.7.2 Example 2. Pancreas SBRT with 4DCT

Patients who are unable to tolerate an SDX BH will be scheduled for a 4DCT scan, with non-SDX VBH or free breathing. During simulation, a 4DCT is obtained. After reconstruction, the tumor motion is evaluated (figure 3.10). If the tumor motion is acceptably small, an average IP scan is reconstructed and used for treatment planning. If the tumor moves more than 1 cm, abdominal compression is attempted. A compression belt or paddle is placed on the patient's abdomen and systematically tightened with guidance from the patient. The patient should exhibit quick and shallow breaths. A second 4DCT scan is obtained, and tumor motion is evaluated.

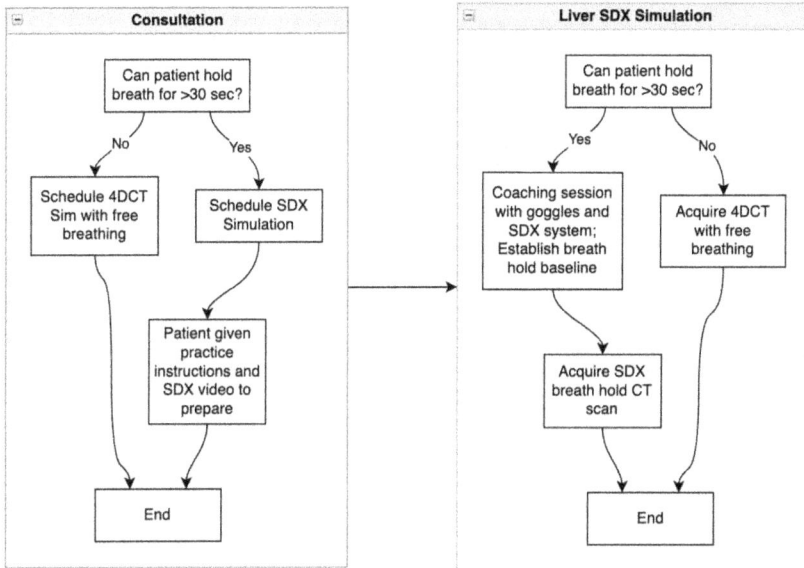

Figure 3.9. Decision tree diagrams for BH considerations.

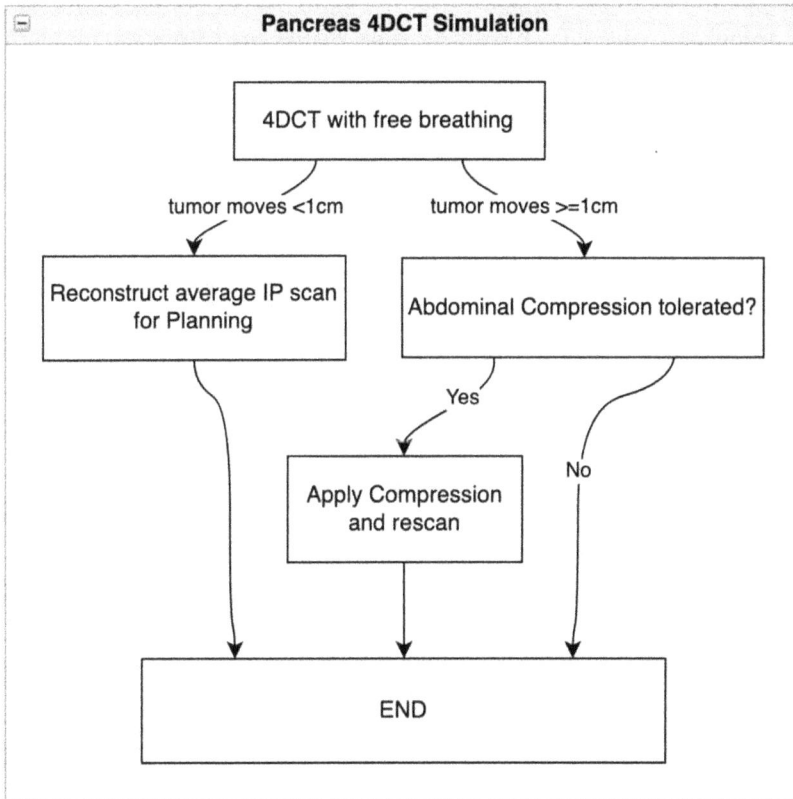

Figure 3.10. Decision tree diagram for abdominal compression considerations.

3.7.3 Example 3. Liver SBRT with ABC BH

Figure 3.11 shows the CT simulation workflow in a sample radiation oncology department. The prominent regions that distinguish an SBRT CT simulation workflow from the routine CT simulation workflow are shaded gray. The SBRT candidates get a 4DCT scan to access the extent of motion. If the tumor motion is greater than 15 mm, an ABC procedure is recommended. This is followed by patient coaching and assessment as described in section 3.4.2.1 (breath hold ⩾25 s). An ideal ABC candidate is brought back to the CT simulation room, and three consecutive CT scans are performed at BH. The physician reviews the CT scans and selects one to be used for treatment planning.

3.8 Simulation setup and immobilization protocols

Tables 3.4–3.11 show simulation setup and immobilization protocols.

3.9 Conclusion

Safe and effective radiotherapy for abdominal cancers requires the application of respiratory monitoring systems, breath control strategies, and advanced imaging techniques. Particularly in the case of SBRT, in which radiation is more focal, accurate definition and reproducible localization of the tumor is critical. Whenever feasible, using MR or PET imaging to help define the tumor extent, implanting fiducial markers, and patient coaching for BH strategies all help minimize margins and errors. For patients who are unable to tolerate a BH, a 4DCT scan can be obtained for which the motion of the tumor during respiration is visualized and used to define an ITV. An abdominal compression device should also be considered to minimize breathing motion if necessary.

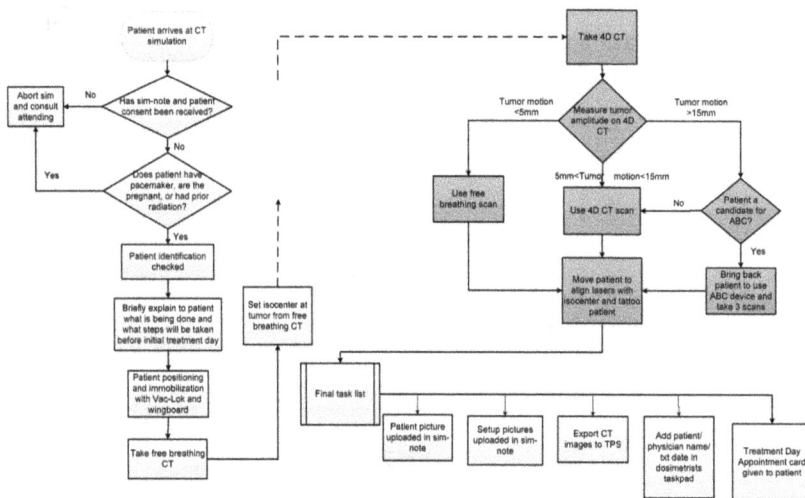

Figure 3.11. Simulation workflow for ABC BH considerations.

Table 3.4. Liver SBRT.

Pre-CT preparation	NPO (Latin *nil per os*; nothing by mouth) 4 h before simulation appointment; CT with SDX exhale breath hold (90 min appt); oral contrast may be requested *(Ex. 3 oz oral contrast 10 min before appointment)*
Patient setup	Head first, supine (HFS); neutral head position; both arms up; legs straight with wedge
Immobilization devices	Custom vac-lok bag; head rest, sponge, knee fix
Scanning protocol	Reference point on xiphoid process—marked, tattooed, and BB placed; scout at exhale; scan 16 slices in treatment region to review for body rotation; upper border at 7 cm above diaphragm; lower border at 2 cm below pelvic brim
Motion management	Exhale BH with SDX (threshold set to 20% full inspiration volume) (if SDX not tolerated, perform 4DCT with free breathing)
MR simulation option	NPO 4 h prior to appointment; HFS, arms up, straight legs, vac-lok bag, head rest, knee fix; *T1-vibe with BH; arterial portal venous contrast (Eovist per weight); bolus tracking; scan whole liver + 5 cm sup and inf beyond borders; cine images for tumor motion assessment

* *Evaluated for MR compatibility and safety*

Table 3.5. Liver fractionated IMRT.

Pre-CT preparation	NPO 2 h before simulation appointment; 4DCT with free breathing; oral contrast may be requested *(Ex. 3 oz oral contrast 10 min before appointment)*
Patient setup	HFS; neutral head position; both arms up; legs straight with wedge
Immobilization devices	Custom vac-lok bag; head rest, sponge, knee fix
Scanning protocol	Reference point on xiphoid process—marked, tattooed, and BB placed; scout at exhale; scan 16 slices in treatment region to review for body rotation; upper border at 7 cm above diaphragm; lower border at 2 cm below pelvic brim
Motion management	Voluntary exhale BH (non-SDX) followed by 4DCT with free breathing

Table 3.6. Pancreas SBRT.

Pre-CT preparation	NPO 4 h before simulation appointment; CT with SDX BH (90 min appt); IV contrast—extra 30 min for pre-sim nursing appointment for IV placement
Patient setup	HFS; neutral head position; both arms up; legs straight with wedge
Immobilization devices	Custom vac-lok bag; head rest, sponge, knee fix
Scanning protocol	Reference point midline at level of xiphoid tip—marked, tattooed, and BB placed; upper border at 2 cm above diaphragm; lower border at pelvic brim; IV contrast for scan (80 s delay)
Motion management	Exhale breath hold with SDX (threshold set to 20% full inspiration volume) (if SDX not tolerated, switch to non-SDX voluntary exhale BH, followed by 4DCT with free breathing)

Table 3.7. Pancreas fractionated EBRT.

Pre-CT preparation	NPO 4 h before simulation appointment; IV contrast—extra 30 min for pre-sim nursing for IV placement
Patient setup	HFS; neutral head position; both arms up; legs straight with wedge
Immobilization devices	Wing board; foot loc
Scanning protocol	Reference point midline at level of xiphoid tip—marked, tattooed, and BB placed; upper border at top of diaphragm; lower border at bottom of pelvis; IV contrast for voluntary exhale—timing 160–180 s, then immediate 4DCT
Motion management	Free breathing CT scan; voluntary exhale (no SDX) with IV contrast and 4DCT

Table 3.8. Esophagus fractionated EBRT.

Pre-CT preparation	Oral contrast: 1oz Scan C Oran contrast immediately before starting CT scan
Patient setup	HFS; neutral head position; both arms up; legs straight with wedge
Immobilization devices	Custom vac-lok bag; head rest, knee fix
Scanning protocol	Physician to place Iso at simulation; CT reference point: midline at level of xiphoid tip; upper border at C2; lower border through kidneys
Motion management	Free breathing CT scan (4DCT if stomach involvement)

Table 3.9. Gastric fractionated EBRT.

Pre-CT preparation	NPO 2 h before simulation appointment; Oral contrast: 3oz Scan C Oran contrast immediately before starting CT scan
Patient setup	HFS; neutral head position; both arms up; legs straight with wedge
Immobilization devices	Custom vac-lok bag; head rest, knee fix
Scanning protocol	CT Reference point: midline at level of xiphoid tip; upper border at C2; lower border through kidneys
Motion management	4DCT with free breathing

Table 3.10. Rectal cancer fractionated EBRT.

Pre-CT preparation	Empty rectum
Patient setup	Head first, prone (HFP); neutral head position; both arms up; legs straight with wedge; bowel positioned anteriorly, tape buttocks apart
Immobilization devices	Prone belly board; reverse wedge under ankles; if female, patient to place vaginal dilator
Scanning protocol	CT reference point: close to pubic symphysis; upper border at top of L3; lower border at distal ⅓ of femurs; place BB on anal verge
Motion management	Free breathing CT scan

Table 3.11. Anal cancer fractionated EBRT.

Pre-CT preparation	Empty rectum
Patient setup	HFS; neutral head position; arms on chest; frogged leg position, vac-lok should push upper thighs apart and provide stable position for vaginal dilator if used; for male patients, position penis superiorly
Immobilization devices	Large vac-lok bag; head rest; if female, patient to place vaginal dilator
Scanning protocol	CT reference point: close to pubic symphysis; upper border at 2 cm above diaphragm; lower border at halfway through femurs; place BB on anal verge
Motion management	Free breathing CT scan

References

[1] Cai J, Chang Z, Wang Z, Paul Segars W and Yin F-F 2011 Four-dimensional magnetic resonance imaging (4D-MRI) using image-based respiratory surrogate: a feasibility study: 4D-MRI using body area as respiratory surrogate *Med. Phys.* **38** 6384–94

[2] Feng M, Balter J M, Normolle D, Adusumilli S, Cao Y and Chenevert T L *et al* 2009 Characterization of pancreatic tumor motion using cine MRI: surrogates for tumor position should be used with caution *Int. J. Radiat. Oncol.* **74** 884–91

[3] Heerkens H D, van Vulpen M, van den Berg C A T, Tijssen R H N, Crijns S P M and Molenaar I Q *et al* 2014 MRI-based tumor motion characterization and gating schemes for radiation therapy of pancreatic cancer *Radiother. Oncol.* **111** 252–7

[4] Li G, Citrin D, Camphausen K, Mueller B, Burman C and Mychalczak B *et al* 2008 Advances in 4D medical imaging and 4D radiation therapy *Technol. Cancer Res. Treat.* **7** 67–81

[5] Oar A, Liney G, Rai R, Deshpande S, Pan L and Johnston M *et al* 2018 Comparison of four dimensional computed tomography and magnetic resonance imaging in abdominal radiotherapy planning *Phys. Imaging Radiat. Oncol.* **7** 70–5

[6] Paganelli C, Summers P, Bellomi M, Baroni G and Riboldi M 2015 Liver 4DMRI: a retrospective image-based sorting method: image-based 4DMRI *Med. Phys.* **42** 4814–21

[7] Ge J, Santanam L, Noel C and Parikh P J 2013 Planning 4-dimensional computed tomography (4DCT) cannot adequately represent daily intrafractional motion of abdominal tumors *Int. J. Radiat. Oncol.* **85** 999–1005

[8] Kruis M F, van de Kamer J B, Sonke J-J, Jansen E P M and van Herk M 2013 Registration accuracy and image quality of time averaged mid-position CT scans for liver SBRT *Radiother. Oncol.* **109** 404–8

[9] Lens E, van der Horst A, Kroon P S, van Hooft J E, Dávila Fajardo R and Fockens P *et al* 2014 Differences in respiratory-induced pancreatic tumor motion between 4D treatment planning CT and daily cone beam CT, measured using intratumoral fiducials *Acta Oncol. Stockh. Swed.* **53** 1257–64

[10] Moorrees J and Bezak E 2012 Four dimensional CT imaging: a review of current technologies and modalities *Australas. Phys. Eng. Sci. Med.* **35** 9–23

[11] Tai A, Liang Z, Erickson B and Li X A 2013 Management of respiration-induced motion with 4-dimensional computed tomography (4DCT) for pancreas irradiation *Int. J. Radiat. Oncol.* **86** 908–13

[12] Wang M, Sharp G C, Rit S, Delmon V and Wang G 2014 2D/4D marker-free tumor tracking using 4D CBCT as the reference image *Phys. Med. Biol.* **59** 2219–33

[13] Ge J, Santanam L, Yang D and Parikh P J 2013 Accuracy and consistency of respiratory gating in abdominal cancer patients *Int. J. Radiat. Oncol. Biol. Phys.* **85** 854–61

[14] Tsai Y-L, Wu C-J, Shaw S, Yu P-C, Nien H-H and Lui L T 2018 Quantitative analysis of respiration-induced motion of each liver segment with helical computed tomography and 4-dimensional computed tomography *Radiat. Oncol.* **13** 59

[15] Kitamura K, Shirato H, Seppenwoolde Y, Shimizu T, Kodama Y and Endo H *et al* 2003 Tumor location, cirrhosis, and surgical history contribute to tumor movement in the liver, as measured during stereotactic irradiation using a real-time tumor-tracking radiotherapy system *Int. J. Radiat. Oncol.* **56** 221–8

[16] Nishioka T, Nishioka S, Kawahara M, Tanaka S, Shirato H and Nishi K *et al* 2009 Synchronous monitoring of external/internal respiratory motion: validity of respiration-gated radiotherapy for liver tumors *Jpn J. Radiol.* **27** 285–9

[17] Minn A Y, Schellenberg D, Maxim P, Suh Y, McKenna S and Cox B *et al* 2009 Pancreatic tumor motion on a single planning 4D-CT does not correlate with intrafraction tumor motion during treatment *Am. J. Clin. Oncol.* **32** 364–8

[18] Uchinami Y, Suzuki R, Katoh N, Taguchi H, Yasuda K and Miyamoto N *et al* 2019 Impact of organ motion on volumetric and dosimetric parameters in stomach lymphomas treated with intensity-modulated radiotherapy *J. Appl. Clin. Med. Phys.* **20** 78–86

[19] Wysocka B, Kassam Z, Lockwood G, Brierley J, Dawson L A and Buckley C A *et al* 2010 Interfraction and respiratory organ motion during conformal radiotherapy in gastric cancer *Int. J. Radiat. Oncol.* **77** 53–9

[20] Langen K M and Jones D T L 2001 Organ motion and its management *Int. J. Radiat. Oncol.* **50** 265–78

[21] Hallman J L, Mori S, Sharp G C, Lu H-M, Hong T S and Chen G T Y 2012 A four-dimensional computed tomography analysis of multiorgan abdominal motion *Int. J. Radiat. Oncol.* **83** 435–41

[22] Gierga D P, Chen G T Y, Kung J H, Betke M, Lombardi J and Willett C G 2004 Quantification of respiration-induced abdominal tumor motion and its impact on IMRT dose distributions *Int. J. Radiat. Oncol. Biol. Phys.* **58** 1584–95

[23] Yang W, Fraass B A, Reznik R, Nissen N, Lo S and Jamil L H *et al* 2014 Adequacy of inhale/exhale breathhold CT based ITV margins and image-guided registration for free-breathing pancreas and liver SBRT *Radiat. Oncol.* **9** 11

[24] Watanabe M, Isobe K, Uno T, Harada R, Kobayashi H and Ueno N *et al* 2011 Intrafractional gastric motion and interfractional stomach deformity using CT images *J. Radiat. Res. (Tokyo)* **52** 660–5

[25] Yue J, Sun X, Cai J, Yin F-F, Yin Y and Zhu J *et al* 2012 Lipiodol: a potential direct surrogate for cone-beam computed tomography image guidance in radiotherapy of liver tumor *Int. J. Radiat. Oncol.* **82** 834–41

[26] Philip Medical Systems 2013 *Respiratory Motion Management for CT* Philip Medical Systems https://healthmanagement.org/uploads/Respiratory%20Motion%20Management%20for%20CT_white%20paper.pdf

[27] Heinz C, Reiner M, Belka C, Walter F and Söhn M 2015 Technical evaluation of different respiratory monitoring systems used for 4D CT acquisition under free breathing *J. Appl. Clin. Med. Phys.* **16** 334–49

[28] Choi G W, Suh Y, Das P, Herman J, Holliday E and Koay E *et al* 2019 Assessment of setup uncertainty in hypofractionated liver radiation therapy with a breath-hold technique using automatic image registration–based image guidance *Radiat. Oncol.* **14** 154

[29] Crivellaro C, De Ponti E, Elisei F, Morzenti S, Picchio M and Bettinardi V *et al* 2018 Added diagnostic value of respiratory-gated 4D 18F–FDG PET/CT in the detection of liver lesions: a multicenter study *Eur. J. Nucl. Med. Mol. Imaging.* **45** 102–9

[30] Nakamura M, Akimoto M, Ono T, Nakamura A, Kishi T and Yano S *et al* 2015 Interfraction positional variation in pancreatic tumors using daily breath-hold cone-beam computed tomography with visual feedback *J. Appl. Clin. Med. Phys.* **16** 108–16

[31] Stanley D N, McConnell K A, Kirby N, Gutiérrez A N, Papanikolaou N and Rasmussen K 2017 Comparison of initial patient setup accuracy between surface imaging and three point localization: a retrospective analysis *J. Appl. Clin. Med. Phys.* **18** 58–61

[32] Bert C, Metheany K G, Doppke K and Chen G T Y 2005 A phantom evaluation of a stereo-vision surface imaging system for radiotherapy patient setup: evaluation of a stereo-imaging patient setup system *Med. Phys.* **32** 2753–62

[33] Oliver J A, Kelly P, Meeks S L, Willoughby T R and Shah A P 2017 Orthogonal image pairs coupled with OSMS for noncoplanar beam angle, intracranial, single-isocenter, SRS treatments with multiple targets on the varian edge radiosurgery system *Adv. Radiat. Oncol.* **2** 494–502

[34] Wang L T, Solberg T D, Medin P M and Boone R 2001 Infrared patient positioning for stereotactic radiosurgery of extracranial tumors *Comput. Biol. Med.* **31** 101–11

[35] Chang Z, Liu T, Cai J, Chen Q, Wang Z and Yin F-F 2011 Evaluation of integrated respiratory gating systems on a novalis Tx system *J. Appl. Clin. Med. Phys.* **12** 71–9

[36] Jin J-Y, Yin F-F, Tenn S E, Medin P M and Solberg T D 2008 Use of the BrainLAB ExacTrac x-ray 6D system in image-guided radiotherapy *Med. Dosim.* **33** 124–34

[37] Lens E, Gurney-Champion O J, Tekelenburg D R, van Kesteren Z, Parkes M J and van Tienhoven G *et al* 2016 Abdominal organ motion during inhalation and exhalation breath-holds: pancreatic motion at different lung volumes compared *Radiother. Oncol.* **121** 268–75

[38] Taniguchi C M, Murphy J D, Eclov N, Atwood T F, Kielar K N and Christman-Skieller C *et al* 2013 Dosimetric analysis of organs at risk during expiratory gating in stereotactic body radiation therapy for pancreatic cancer *Int. J. Radiat. Oncol. Biol. Phys.* **85** 1090–5

[39] Lu L, Diaconu C, Djemil T, Videtic G M, Abdel-Wahab M and Yu N *et al* 2018 Intra- and inter-fractional liver and lung tumor motions treated with SBRT under active breathing control *J. Appl. Clin. Med. Phys.* **19** 39–45

[40] Poitevin-Chacón M A, Ramos-Prudencio R, Rumoroso-García J A, Rodríguez-Laguna A and Martínez-Robledo J C 2020 Voluntary breath-hold reduces dose to organs at risk in radiotherapy of left-sided breast cancer *Rep. Pract. Oncol. Radiother.* **25** 104–8

[41] Nakamura M, Shibuya K, Shiinoki T, Matsuo Y, Nakamura A and Nakata M *et al* 2011 Positional reproducibility of pancreatic tumors under end-exhalation breath-hold conditions using a visual feedback technique *Int. J. Radiat. Oncol. Biol. Phys.* **79** 1565–71

[42] Giraud P, Morvan E, Claude L, Mornex F, Le Pechoux C and Bachaud J-M *et al* 2011 Respiratory gating techniques for optimization of lung cancer radiotherapy *J. Thorac. Oncol.* **6** 2058–68

[43] Wong J W, Sharpe M B, Jaffray D A, Kini V R, Robertson J M and Stromberg J S *et al* 1999 The use of active breathing control (ABC) to reduce margin for breathing motion *Int. J. Radiat. Oncol. Biol. Phys.* **44** 911–9

[44] Dawson LA, Brock K K, Kazanjian S, Fitch D, McGinn C J, Lawrence T S, Ten Haken R K and Balter J 2001 The reproducibility of organ position using active breathing control (ABC) during liver radiotherapy *Int. J. Radiat. Oncol. Biol. Phys.* **51** 1410–21

[45] Dawson L A, Eccles C, Bissonnette J-P and Brock K K 2005 Accuracy of daily image guidance for hypofractionated liver radiotherapy with active breathing control *Int. J. Radiat. Oncol.* **62** 1247–52

[46] Eccles C L, Dawson L A, Moseley J L and Brock K K 2011 Interfraction liver shape variability and impact on GTV position during liver stereotactic radiotherapy using abdominal compression *Int. J. Radiat. Oncol.* **80** 938–46

[47] Lovelock D M, Zatcky J, Goodman K and Yamada Y 2014 The effectiveness of a pneumatic compression belt in reducing respiratory motion of abdominal tumors in patients undergoing stereotactic body radiotherapy *Technol. Cancer Res. Treat.* **13** 259–67

[48] Hubie C, Shaw M, Bydder S, Lane J, Waters G and McNabb M *et al* 2017 A randomised comparison of three different immobilisation devices for thoracic and abdominal cancers *J. Med. Radiat. Sci.* **64** 90–6

[49] Song P Y, Washington M, Vaida F, Hamilton R, Spelbring D and Wyman B *et al* 1996 A comparison of four patient immobilization devices in the treatment of prostate cancer patients with three dimensional conformal radiotherapy *Int. J. Radiat. Oncol. Biol. Phys.* **34** 213–9

[50] Lax I, Blomgren H, Näslund I and Svanström R 1994 Stereotactic radiotherapy of malignancies in the abdomen: methodological aspects *Acta. Oncol.* **33** 677–83

[51] Herfarth K K, Debus J, Lohr F, Bahner M L, Fritz P and Höss A *et al* 2000 Extracranial stereotactic radiation therapy: set-up accuracy of patients treated for liver metastases *Int. J. Radiat. Oncol. Biol. Phys.* **46** 329–35

[52] Dreher C, Oechsner M, Mayinger M, Beierl S, Duma M-N and Combs S E *et al* 2018 Evaluation of the tumor movement and the reproducibility of two different immobilization setups for image-guided stereotactic body radiotherapy of liver tumors *Radiat. Oncol.* **13** 15

[53] Pinelo R C, Alfonso R, Pérez Y G, García A A and Rubio A 2016 Moving stereotactic fiducial system to obtain a respiratory signal: proof of principle *J. Appl. Clin. Med. Phys.* **17** 80–91

[54] Cervino L I, Gupta S, Rose M A, Yashar C and Jiang S B 2009 Using surface imaging and visual coaching to improve the reproducibility and stability of deep-inspiration breath hold for left-breast-cancer radiotherapy *Phys. Med. Biol.* **54** 6853–65

[55] Seppenwoolde Y, Shirato H, Kitamura K, Shimizu S, van Herk M and Lebesque J V *et al* 2002 Precise and real-time measurement of 3D tumor motion in lung due to breathing and heartbeat, measured during radiotherapy *Int. J. Radiat. Oncol. Biol. Phys.* **53** 822–34

[56] Li G, Wei J, Olek D, Kadbi M, Tyagi N and Zakian K *et al* 2017 Direct comparison of respiration-correlated four-dimensional magnetic resonance imaging reconstructed using concurrent internal navigator and external bellows *Int. J. Radiat. Oncol.* **97** 596–605

[57] Stemkens B, Tijssen R H N, de Senneville B D, Heerkens H D, van Vulpen M and Lagendijk J J W *et al* 2015 Optimizing 4-dimensional magnetic resonance imaging data sampling for respiratory motion analysis of pancreatic tumors *Int. J. Radiat. Oncol.* **91** 571–8

[58] Freedman J N, Collins D J, Bainbridge H, Rank C M, Nill S and Kachelrieß M *et al* 2017 T2-weighted 4D magnetic resonance imaging for application in magnetic resonance–guided radiotherapy treatment planning. *Invest. Radiol.* **52** 563–73

[59] Freedman J N, Bainbridge H E, Nill S, Collins D J, Kachelrieß M and Leach M O *et al* 2019 Synthetic 4D-CT of the thorax for treatment plan adaptation on MR-guided radiotherapy systems *Phys. Med. Biol.* **64** 115005

[60] Nehmeh S A and Erdi Y E 2008 Respiratory motion in positron emission tomography/computed tomography: a review *Semin. Nucl. Med.* **38** 167–76

[61] Aristophanous M, Berbeco R I, Killoran J H, Yap J T, Sher D J and Allen A M *et al* 2012 Clinical utility of 4D FDG-PET/CT scans in radiation treatment planning *Int. J. Radiat. Oncol.* **82** e99–105

[62] Scarsbrook A, Ward G, Murray P, Goody R, Marshall K and McDermott G *et al* 2017 Respiratory-gated (4D) contrast-enhanced FDG PET-CT for radiotherapy planning of lower oesophageal carcinoma: feasibility and impact on planning target volume *BMC Cancer* **17** 671

[63] Keall P J, Mageras G S, Balter J M, Emery R S, Forster K M and Jiang S B *et al* 2006 The management of respiratory motion in radiation oncology report of AAPM Task Group 76a: respiratory motion in radiation oncology *Med. Phys.* **33** 3874–900

[64] Abbas H, Chang B and Chen Z J 2014 Motion management in gastrointestinal cancers *J. Gastrointest. Oncol.* **5** 223–35

[65] Heinzerling J H, Anderson J F, Papiez L, Boike T, Chien S and Zhang G *et al* 2008 Four-dimensional computed tomography scan analysis of tumor and organ motion at varying levels of abdominal compression during stereotactic treatment of lung and liver *Int. J. Radiat. Oncol. Biol. Phys.* **70** 1571–8

[66] Eccles C L, Patel R, Simeonov A K, Lockwood G, Haider M and Dawson L A 2011 Comparison of liver tumor motion with and without abdominal compression using cine-magnetic resonance imaging *Int. J. Radiat. Oncol. Biol. Phys.* **79** 602–8

[67] Lagerwaard F J, Van Sornsen de Koste J R, Nijssen-Visser M R J, Schuchhard-Schipper R H, Oei S S and Munne A *et al* 2001 Multiple 'slow' CT scans for incorporating lung tumor mobility in radiotheraphy planning *Int. J. Radiat. Oncol.* **51** 932–7

[68] van Sörnsen de Koste J R, Lagerwaard F J, Schuchhard-Schipper R H, Nijssen-Visser M R J, Voet P W J and Oei S S *et al* 2001 Dosimetric consequences of tumor mobility in radiotherapy of stage I non-small cell lung cancer – an analysis of data generated using 'slow' CT scans *Radiother. Oncol.* **61** 93–9

[69] Nakamura M, Narita Y, Matsuo Y, Narabayashi M, Nakata M and Yano S *et al* 2008 Geometrical differences in target volumes between slow CT and 4D CT imaging in stereotactic body radiotherapy for lung tumors in the upper and middle lobe: Comparison of slow CT and 4D CT imaging *Med. Phys.* **35** 4142–8

[70] Li G, Liu Y and Nie X 2019 Respiratory-correlated (RC) vs. time-resolved (TR) four-dimensional magnetic resonance imaging (4DMRI) for radiotherapy of thoracic and abdominal cancer *Front Oncol.* **9** 1024

[71] Zhang J, Srivastava S, Wang C, Beckham T, Johnson C and Dutta P *et al* 2019 Clinical evaluation of 4D MRI in the delineation of gross and internal tumor volumes in comparison with 4DCT *J. Appl. Clin. Med. Phys.* **20** 51–60

[72] Du D, Mutic S, Li H H and Hu Y 2018 An efficient model to guide prospective T2-weighted 4D magnetic resonance imaging acquisition *Med. Phys.* **45** 2453–62

[73] Glide-Hurst C K, Kim J P, To D, Hu Y, Kadbi M and Nielsen T *et al* 2015 Four dimensional magnetic resonance imaging optimization and implementation for magnetic resonance imaging simulation *Pract. Radiat. Oncol.* **5** 433–42

[74] Stemkens B, Paulson E S and Tijssen R H N 2018 Nuts and bolts of 4D-MRI for radiotherapy *Phys. Med. Biol.* **63** 21TR01

IOP Publishing

Principles and Practice of Image-Guided Abdominal Radiation Therapy

Yu Kuang

Chapter 4

Treatment planning for abdominal cancer

Bingqi Guo and Peng Qi

This chapter introduces the principles and practice of image-guided abdominal radiation therapy treatment planning. Treatment planning for abdominal cancer therapy aims to deliver a sufficient dose to the planning target volume and minimize the toxicity to organs at risk in the abdominal cavity. Image registration is essential for accurate target delineation. Treatment modalities and techniques include intensity-modulated radiation therapy, stereotactic body radiation therapy, and proton therapy. Auto-contouring and auto-planning techniques have been used to improve the efficiency and quality of treatment planning.

4.1 Introduction

Abdominal tumors are tumors in the abdominal cavity, which include lower esophageal, gastric, hepatic, pancreatic, renal, and colorectal cancer. Depending on the origin, the site, and the stage of the disease, the treatment intent of radiation therapy can be definitive, palliative, salvage, pre-operative (neoadjuvant), or post-operative (adjuvant). Treatment planning for image-guided radiation therapy for abdominal cancer aims to deliver sufficient doses to the tumor and microscopic cancer cells while minimizing doses to organs at risk (OARs). The most common treatment techniques for abdominal radiotherapy are three-dimensional (3D) conformal radiotherapy and intensity-modulated radiation therapy (IMRT), either by static gantry IMRT or volumetric modulated arc therapy (VMAT). The treatment fractionation regime includes conventional fractionation for definitive (25–30 fractions) and palliative (~10 fractions) treatments and stereotactic body radiation therapy (SBRT; 1–5 fractions).

4.2 Image registration for treatment planning

Simulation computed tomography (CT) is the standard imaging modality for treatment planning for abdominal tumors. Additional diagnostic images such as magnetic resonance imaging (MRI) scans, positron emission tomography (PET) images, and contrast-enhanced CT images improve the accuracy of target delineation because of their superior soft tissue contrast and/or functional image capability. Aligning the diagnostic images with simulation CT is done through the process of image registration. Besides target delineation, dose accumulation and contour propagation for re-treatment or adaptive planning require good image registration.

There are two types of image registrations used in radiation therapy: rigid and deformable [1]. Rigid registration has six degrees of freedom: three transitional and three rotational. Rigid registration assumes that the patient's anatomy does not change and tissue does not deform between two images, which may not be sufficient for abdominal radiation therapy due to the deformable nature of the soft tissue in the abdomen. Deformable registration allows spatial variation transformation. Its degree of freedom can be as large as thrice the number of voxels. While deformable registration can account for the anatomical change between two images and tissue deformation, it is known to potentially produce non-realistic and inaccurate deformations.

Many commercial software platforms include image registration tools that are designed based on different algorithms [2]. Each image registration software platform used in the clinic should be evaluated for its accuracy regarding the type of deformation (rigid or deformable), the treatment sites, and the application of the image registration (for tumor segmentation, dose accumulation, and/or adaptive planning) using visual examination, point evaluation, and contour propagation [1, 3].

For abdominal tumors, a common challenge of image registration is change of breathing status. While four-dimensional CT (4DCT) is widely used for motion management during CT simulation, diagnostic images are most commonly fast free-breathing or voluntary breath-hold (BH) scans. Different patient positions may contribute to errors in image registration too. Therefore, it is recommended that one use the same immobilization device for the diagnostic scans to keep the same patient position. Another common challenge of abdominal image registration is anatomical change caused by the filling status of different organs. Changes in stomach and bowel filling affect the registration accuracy in the upper abdomen while changes in bladder and rectum filling affect the registration accuracy in the lower abdomen.

4.2.1 CT and MRI fusion

Figure 4.1 shows an example of target delineation based on a simulation CT and MRI image fusion for a liver tumor. The simulation CT was acquired with an involuntary BH using an active breathing coordinator (ABC) device (Elekta). No contrast was given in the simulation. A T1 post-contrast MRI scan was acquired using a 3D vibe acquisition technique and a voluntary BH. Rigid registration was performed with the region of interest (ROI) focused on the tumor and the liver boundary nearby. The gross tumor volume (GTV) was delineated based mainly on MRI and was propagated to the CT for planning.

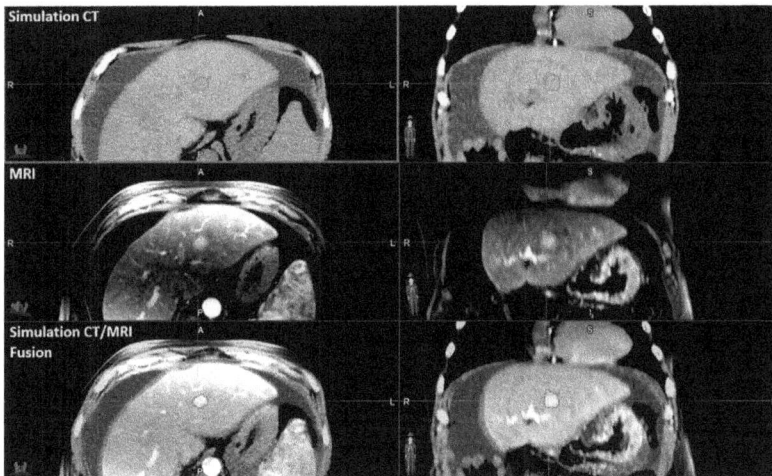

Figure 4.1. Simulation CT/MRI fusion for liver tumor delineation. GTV was outlined.

4.2.2 CT and PET/CT fusion

PET is a functional imaging modality that uses radioactive tracers to measure the metabolism of the body, and it is often used to identify positive lymph nodes in addition to primary tumors in radiation therapy. Figures 4.2(a) and (b) show the use of a fluorodeoxyglucose PET/CT and simulation CT fusion in delineation of a primary liver tumor and a positive lymph node, respectively. In each scenario, the simulation CT was the 4DCT-derived average intensity projection CT with intravascular (IV) contrast. PET/CT is inherently an average intensity acquisition due to its slow acquisition speed. Rigid registration was performed to align the PET/CT with the simulation CT. GTV was delineated based on both simulation CT and PET/CT.

4.2.3 CT and CT fusion

Simulation CT and diagnostic CT fusion is often used to assist in tumor or tumor bed delineation. Simulation CT may also be fused with a previous planning CT in the consideration of re-treatment or adaptive planning. Depending on the purpose of image registration and the extent of anatomical change between two CT images, either rigid or deformable registration may be used. Figure 4.3(a) shows an example of a simulation CT rigidly registered with diagnostic CT with IV contrast to delineate a pancreas tumor. Both the planning CT and diagnostic CT were acquired while the patient was under free-breathing conditions. Often, an abdominal compression device is used during simulation to reduce the tumor motion. Figure 4.3(b) shows a simulation CT fused with a previous planning CT to evaluate the impact of the dose delivered to critical structures from previous treatment. Although the anatomical change at the previous treatment site was significant, the site to be treated has similar anatomy; therefore, the two images were rigidly

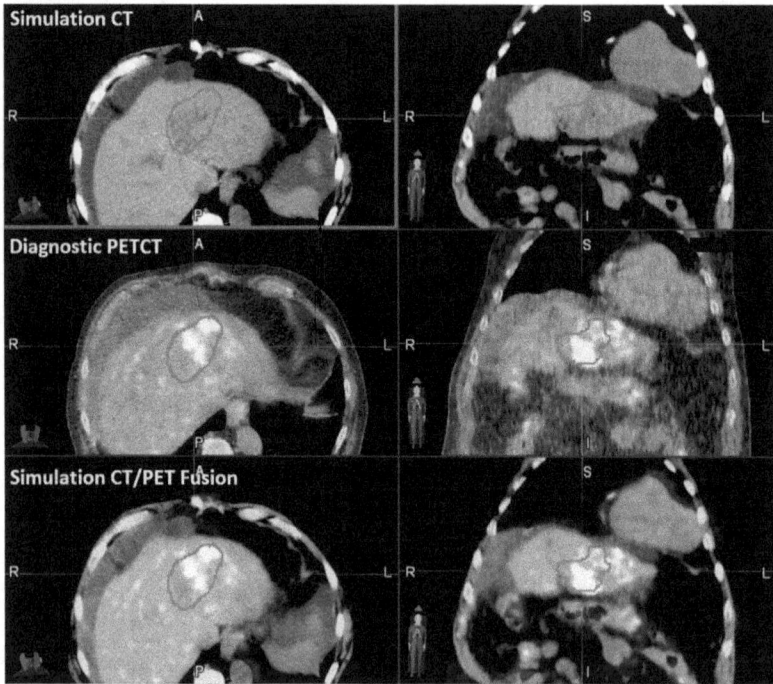

Figure 4.2(a). Simulation CT and PET/CT fusion for liver tumor delineation. GTV was outlined.

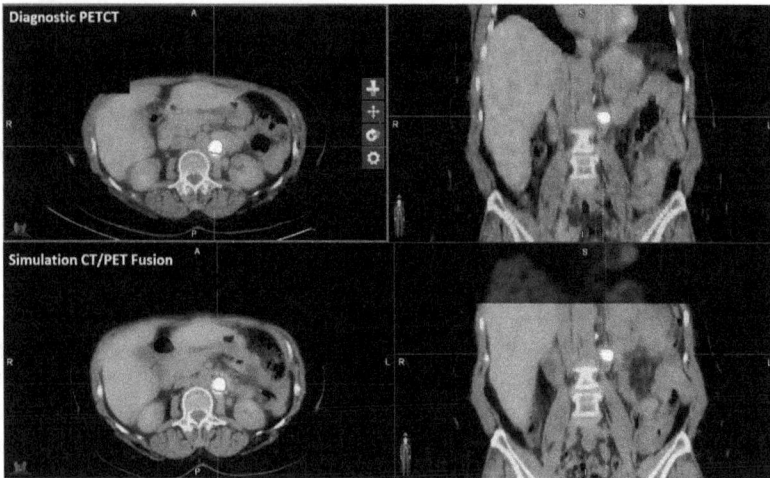

Figure 4.2(b). Simulation CT and PET/CT fusion to delineate a positive lymph node. The lymph node GTV was outlined.

registered with the ROI focused on the area to be treated. In this case, the 50% isodose line (IDL) from previous treatment was converted to a contour and was transferred to the new simulation CT as an avoidance structure for treatment planning. Figure 4.3(c) shows an example of deformable image registration for

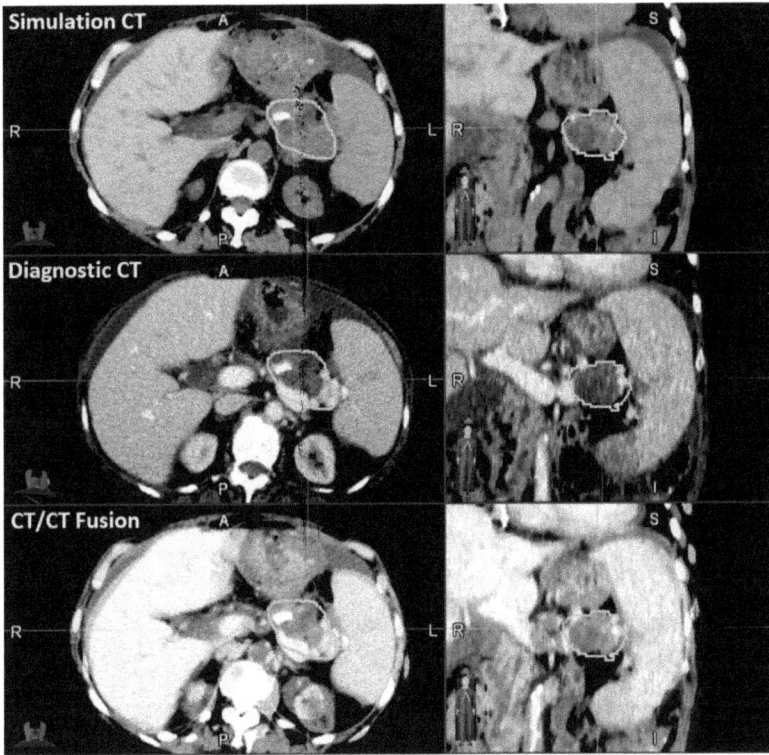

Figure 4.3(a). CT/CT fusion for pancreas tumor delineation. The GTV was outlined.

adaptive planning. The previous planning CT was registered to the new simulation CT for contour propagation and dose accumulation. The deformed images and contours were visually examined for quality assurance of deformable registration. Deformed contours were reviewed and edited by a physician before being used for adaptive planning.

4.3 Target volume and OAR definition

According to the *International Commission on Radiation Units and Measurements* reports 50 and 62 [4, 5], the GTV is defined as the gross demonstrable tumor extent and location of the malignant growth, which consists of the primary tumor, possible metastatic lymphadenopathy, and other metastases. The clinical target volume (CTV) is a tissue volume that contains a demonstrable GTV and/or subclinical malignant disease that must be eliminated. Internal target volume (ITV) is defined as CTV expanded by a margin to account for the internal motion caused by physiological activities such as breathing, bladder and rectum filling, bowel movement, heartbeat, etc. The planning target volume (PTV), also called the target volume, is a geometrical concept used for treatment planning; it is usually defined as CTV (or ITV) with a setup margin.

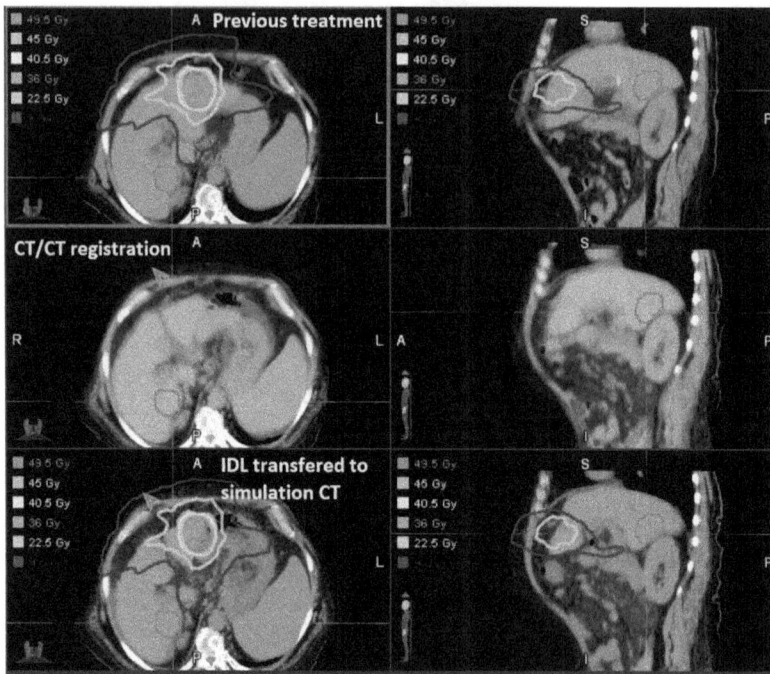

Figure 4.3(b). CT/CT fusion to evaluate dose effect from previous treatment. GTV from the new simulation CT was outlined.

4.3.1 Esophageal cancer

Lower esophageal cancer, including cancer of the gastroesophageal junction, is usually treated with definitive chemoradiotherapy, neoadjuvant radiotherapy, or adjuvant radiotherapy. GTVs in the esophagus and involved lymph nodes are delineated based on CT with IV contrast, esophagogastro-duodenoscopy, endoscopic ultrasound, and PET/CT. For adjuvant radiotherapy, a pre-operative image is usually used to delineate the tumor bed. Depending on the purpose of radiotherapy and the stage of the disease, the CTV may include the esophageal GTV with 4 cm of superior and inferior expansion and 1 cm of radial expansion or the tumor bed and regional lymph node basin at risk. PTV expands 1 cm uniformly from CTV. The prescription for esophageal cancer includes 5040 cGy with 180 cGy/fraction or 5000 cGy with 200 cGy/fraction for definitive radiotherapy, 4140–5040 cGy with 180–200 cGy/fraction for neoadjuvant radiotherapy, and 4500–5040 cGy with 180–200 cGy/fraction for adjuvant radiotherapy.

4.3.2 Gastric cancer

Gastric cancer locates in the stomach and is most commonly treated with adjuvant chemoradiotherapy. The GTV of gastric cancer includes any remaining tumor after surgery and positive lymph nodes. The CTV, depending on the stage of the disease, may include the tumor bed, remaining stomach, and involved lymph nodes. MRI,

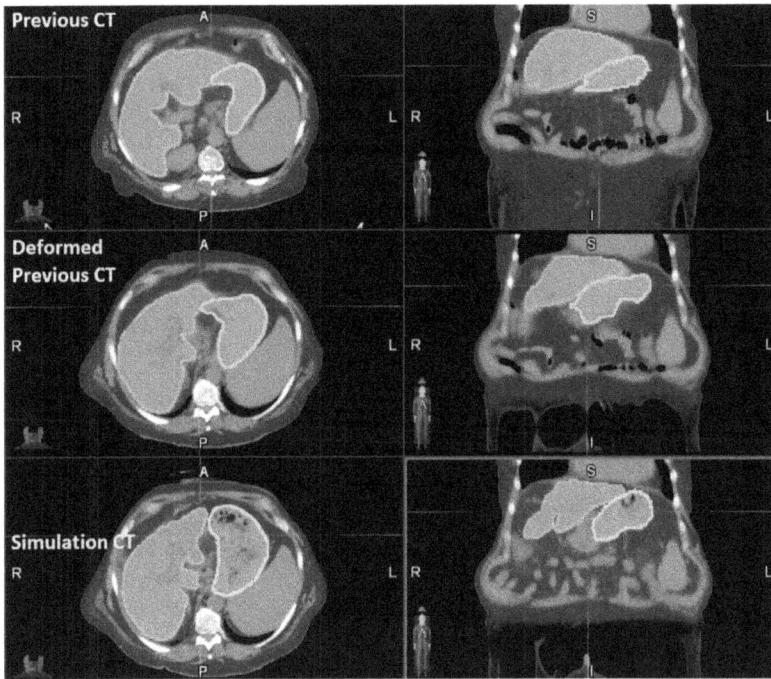

Figure 4.3(c). CT/CT deformable registration for adaptive planning contour propagation. The liver and stomach were outlined.

esophagogastro-duodenoscopy, endoscopic ultrasound, and contrast CT may be used to delineate the tumor. For adjuvant therapy, the CTV should receive 4500 cGy with 180 cGy/fraction and a boost of 540 cGy to 900 cGy with 180 cGy/fraction for positive margins or gross residual disease without violating normal tissue constraints.

4.3.3 Pancreatic cancer

Pancreatic cancer may locate in the head, neck, body, or tail of the pancreas, with the majority of the cases locating in the pancreas head. Radiotherapy can be used as a definitive treatment for unresectable tumors or as neoadjuvant or adjuvant treatments. The GTV of pancreatic cancer includes the gross tumor for neoadjuvant radiotherapy or unresectable disease and clinically positive lymph nodes. The CTV, depending on the purpose of radiotherapy, may include 0.5–1 cm expansion from the GTV, tumor bed, and primary lymph node drainage. GTV and CTV were delineated using contrast-enhanced MRI and/or CT with IV contrast. Specific use of CT pancreatic contrast timing protocols will improve resolution. If the planning CT is 4DCT-derived CT or BH CT, the PTV expands 0.5 cm from the ITV. Otherwise, the PTV expands 1 cm from the CTV.

A typical prescription for pancreatic cancer adjuvant radiotherapy is to deliver 4500 cGy with 180 cGy/fraction to CTV volume and boost tumor bed and pathologically involved lymph node region to 5040 cGy in 180 cGy/fraction. For definitive radiotherapy of unresectable disease or pre-operative radiotherapy, a prescription of 5040 to 5400 cGy in 180 cGy/fraction is often used. Neoadjuvant radiotherapy for pancreatic cancer typically delivers 4500 cGy to the CTV and boosts the GTV prescription to 5040–5400 cGy with 180 cGy/fraction.

4.3.4 Hepatobiliary cancer

Liver cancer can be divided into primary liver cancer and metastatic liver cancer. For primary liver cancer, radiotherapy is an option for unresectable, medically inoperable tumors. Radiotherapy can be definitive to the tumor or palliative to the whole liver. The GTV of primary liver cancer is usually delineated using contrast CT. The CTV expands 1 cm from the GTV and the PTV expands 1 cm from the CTV or 0.5 cm from the ITV if 4DCT was used. The dose prescription is usually determined individually.

4.3.5 Metastatic liver cancer

The liver is the most common site for metastasis from tumors in the gastrointestinal (GI) tract. Unresectable liver metastases are often treated with SBRT. The GTV and CTV are delineated from contrast CT and MRI. Four-dimensional CT is commonly used in liver SBRT to create the ITV. The PTV expands from the ITV with a 5 mm margin. The dose prescription varies among 3000 cGy in one fraction, 4500 cGy in three fractions, and 5000 Gy in five fractions.

4.3.6 Rectal and anal cancer

Rectal cancer may be treated with neoadjuvant chemoradiotherapy. The GTV of rectum cancer is usually delineated as the primary tumor and clinically positive lymph nodes using PET/CT and/or MRI fused with simulation CT. Regional lymph nodes may be included in CTV. The PTV expands 1 cm from the CTV. A typical dose prescription for rectal cancer is to deliver 4500 Gy in 180 cGy/fraction to the CTV and boost the dose for the GTV and involved lymph nodes by an additional 540 Gy in 180 cGy/fraction.

Primary management of anal cancer is definitive chemoradiotherapy with the GTV being defined as the primary tumor with clinically positive lymph nodes using PET/CT and/or MRI fused with planning CT. The CTV includes the GTV with 1 cm expansion and regional lymph nodes at risk. The PTV expands the CTV by 1 cm. A typical prescription for anal cancer is to deliver 4500 cGy in 180 Gy/fraction to the CTV, with a 900–1440 cGy boost to the GTV in 180 cGy/fraction for T3, T4, or T2 lesions with residual disease.

4.3.7 OARs in abdomen

OARs from abdominal radiotherapy are listed in table 4.1. They are delineated from the simulation CT. Contrast may be used to delineate some normal structures (e.g., using oral contrast to delineate the small bowel [SM bowel]). For serial organs such as the stomach, duodenum, spinal cord, SM bowel, and large bowel (LG bowel), using a planning organ at risk volume (PRV) margin of 5 mm (PRV5) to account for

Table 4.1. OARs from abdomen radiotherapy.

Liver	Stomach	SM bowel
LG bowel	Esophagus	Duodenum
Spinal cord	Kidneys	Gallbladder
Ureters	Bile duct	—

the inter- and intra-fraction motion is a common practice in treatment planning. The contouring guideline for normal structures was defined by various RTOG protocols. Contour atlases are available from the RTOG website: https://www.rtog.org/.

Table 4.2 lists the dose constraints for various OARs in the abdomen used at our institution.

For rectal and anal cancer, OARs also include structures in the pelvis, such as the bladder, femoral head, external genitalia, etc.

4.3.8 Auto-segmentation of OARs in abdomen

Auto-segmentation of OARs for radiation therapy has the potential to reduce contouring time and improve contouring consistency [6]. There are two types of auto-segmentation techniques commonly used in radiotherapy: atlas-based [7, 8] and artificial intelligence-based [9–12].

Atlas-based segmentation utilizes an atlas, which is a library of previously delineated contours with images. Deformable image registration was used to transfer contours from an atlas subject to a new image. While atlas-based segmentation has been proven to reduce contouring time [7], its accuracy is limited by deformable registration. Figure 4.4 compares the atlas-based auto-segmentation with clinically approved contours for an example patient.

Artificial intelligence-based auto-segmentation has evolved from statistical models [11, 12] to deep learning models using a convolutional neural network [9, 10]. Deep learning uses deep neural network architectures with multiple hidden layers to learn features (contours) from an image by modeling complex nonlinear relationships. Compared with atlas-based segmentation, deep learning models improve the accuracy of auto-segmentation for abdominal structures [13].

4.4 Treatment planning techniques

4.4.1 Three-dimensional conformal radiation therapy

Three-dimensional conformal radiation therapy (3D-CRT) has been widely used for treating abdominal cancer. Figure 4.5 shows the beam setup, digitally reconstructed radiographs, and isodose distribution for whole-liver irradiation, which is a palliative treatment technique for patients with multiple liver metastases and liver-related symptoms who are not candidates for other therapy. Many palliative dose prescriptions may be used, but two common ones are fractionated treatments to deliver 3000 cGy in 20 fractions twice daily or a single-fraction dose of 800 cGy. In this case, a pair of 15 MV, equally weighted anterior/posterior–posterior/anterior (AP–PA) photon beams were used to deliver 800 cGy in one fraction. The multiple

Table 4.2. Dose constraints for abdominal OARs.[a]

Table 4.2(a). Dose constraints for esophagus IMRT.

Structure/ROI	Type	Primary goal dose (cGy)	Primary goal volume	Secondary goal dose (cGy)	Secondary goal volume
GTV-5040	Min DVH	5040	≥99%	—	—
CTV-5040	Min DVH	5040	≥98%	—	—
PTV-5040	Min DVH	5040	≥95%	—	—
PTV-5040	Max DVH	6000	<10%	—	—
GTV-4500	Min DVH	4500	≥99%	—	—
CTV-4500	Min DVH	4500	≥98%	—	—
PTV-4500	Min DVH	4500	≥95%	—	—
Heart	Max DVH	5200	<0.03 cc	—	—
Heart	Max DVH	4000	<50%	—	—
Heart	Mean dose	<3200	—	<3400	—
Whole lung	Max DVH	5544	0.03 cc	<5695	<0.03 cc
Whole lung	Max DVH	3000	<20%	<3000	<25%
Whole lung	Max DVH	2000	<25%	2000	<30%
Whole lung	Max DVH	1000	<40%	1000	<50%
Whole lung	Max DVH	500	<50%	2000	<55%
Whole lung	Mean dose	<2000	—	2100	—
Whole kidney	Max DVH	4500	<0.03 cc	5000	<0.03 cc
Whole kidney	Max DVH	2000	<30%	2000	<40%
Spinal cord	Max DVH	4500	<0.03 cc	5000	<0.03 cc
Liver	Max DVH	3000	<30%	3000	<40%
Liver	Mean dose	2100	—	2500	—

Table 4.2(b). Dose constraints for pancreas IMRT.

Structure/ROI	Type	Primary goal dose (cGy)	Primary goal volume	Secondary goal dose (cGy)	Secondary goal volume
PTV-5600	Min DVH	5600	≥95%	—	—
CTV-5040	Min DVH	5040	≥99%	—	—
PTV-5040	Min DVH	5040	≥95%	—	—
Stomach	Max DVH	5300	<0.03 cc	—	—
Stomach	Max DVH	4500	<25%	—	—
SM bowel	Max DVH	5300	<0.03 cc	—	—
SM bowel	Max DVH	4500	<25%	—	—
LG bowel	Max DVH	5300	<0.03 cc	—	—
LG bowel	Max DHV	4500	<50%	—	—
Duodenum	Max DVH	5300	<0.03 cc	—	—
Duodenum	Max DVH	4500	<33%	—	—
Spinal cord	Max DVH	4500	<0.03 cc	—	—
Liver	Max DVH	3000	<30%	—	—

Liver	Mean dose	<2800	—	—	—
Kidney, left	Max DVH	1500	<15%	2000	<20%
Kidney, right	Max DVH	1500	<15%	2000	<20%

Table 4.2(c). Dose constraints for SBRT liver (five fractions).

Structure/ROI	Type	Primary goal dose (cGy)	Primary goal volume	Secondary goal dose (cGy)	Secondary goal volume
PTV-4000	Min DVH	4000	≥95%	—	—
Liver-GTV	Mean dose	<1500	—	—	—
Esophagus	Max DVH	3200	<0.5 cc	—	—
Stomach	Max DVH	3000	<0.5 cc	—	—
Stomach	Max DVH	2500	<5 cc	—	—
SM bowel	Max DVH	3000	<0.5 cc	—	—
SM bowel	Max DVH	2500	<5 cc	—	—
LG bowel	Max DVH	3200	<0.5 cc	—	—
Duodenum	Max DVH	3200	<0.5 cc	—	—
Duodenum	Max DVH	4500	<33%	—	—
Spinal cord	Max DVH	2500	<0.5 cc	—	—
Whole kidney	Mean dose	<1000	—	—	—
Liver	Mean dose	<2800	—	—	—
Heart	Max DVH	3000	<30 cc	—	—
Gallbladder	Max DVH	5500	<0.5 cc	—	—

Table 4.2(d). Dose constraints for SBRT pancreas (five fractions).

Structure/ROI	Type	Primary goal dose (cGy)	Primary goal volume	Secondary goal dose (cGy)	Secondary goal volume
PTV	Min DVH	3000	≥95%	—	—
Duodenum	Max DVH	3000	<0.5 cc	—	—
Duodenum	Max DVH	2500	<5 cc	2650	<5 cc
Duodenum	Max DVH	1250	<10 cc	—	—
SM bowel	Max DVH	3000	<0.5 cc	—	—
SM bowel	Max DVH	1950	<5 cc	1950	<10 cc
LG bowel	Max DVH	3200	<0.5 cc	—	—
Stomach	Max DVH	3000	<0.5 cc	—	—
Stomach	Max DVH	2500	<5 cc	—	—
Spinal cord	Max DVH	2000	<0.5 cc	—	—
Liver	Mean dose	<2800	—	—	—
Kidney, left	Mean dose	<1000	—	—	—
Kidney, right	Mean dose	<1000	—	—	—

ᵃAbbreviation: DVH, dose–volume histogram.

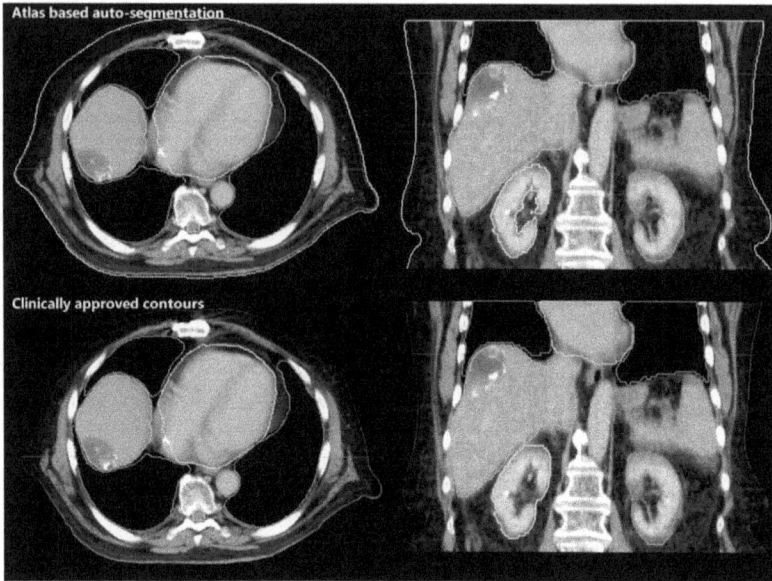

Figure 4.4. Comparing atlas-based auto-segmentation (top figures) with clinically approved contours (bottom figures).

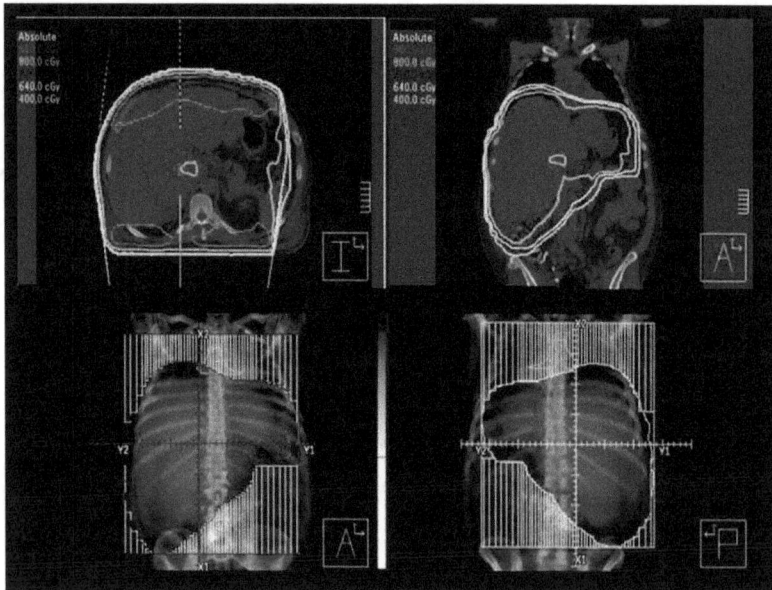

Figure 4.5. Whole-liver irradiation.

leaf collimator of each beam was set to conform to the liver with a 1 cm block margin in the beam's eye view. The dose was prescribed to the midplane in the liver.

Figure 4.6 shows an example of a 3D conformal esophagus treatment plan to deliver 5040 cGy in 28 fractions. Three beams were used: a pair of equally weighted

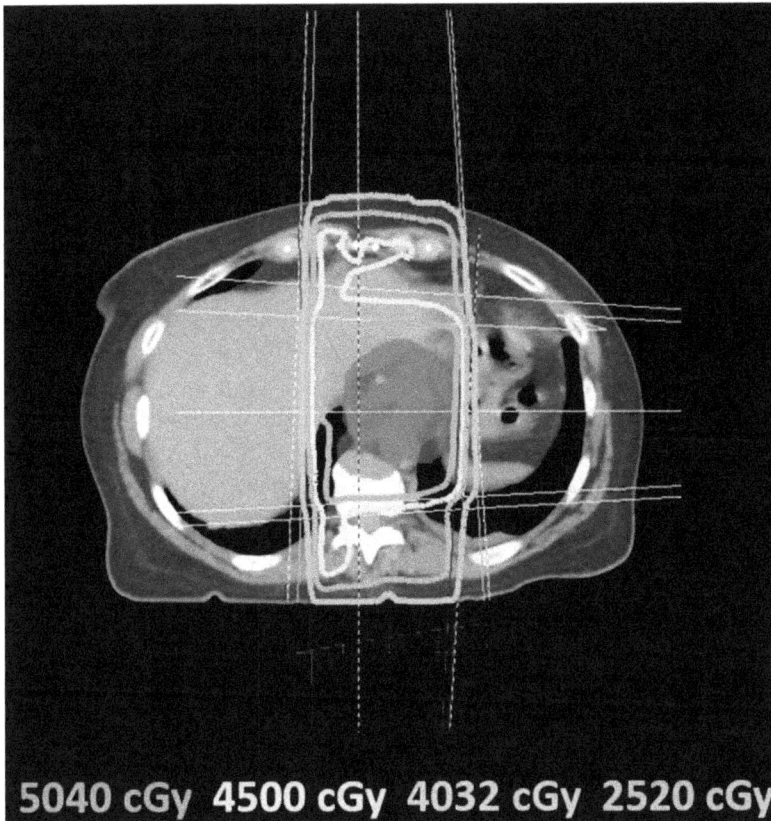

5040 cGy 4500 cGy 4032 cGy 2520 cGy

Figure 4.6. An example 3D-CRT for esophagus cancer.

AP–PA beams and a lightly weighted left lateral beam (~20% weight) to steer the 45 Gy dose away from the cord. An enhanced dynamic wedge was used to create a uniform dose distribution and eliminate hotspots (>110% of the prescription dose). All beams used a multiple leaf collimator conformed to the PTV with a 1 cm block margin. Both high- and low-energy photons can be used for esophagus treatment planning depending on the depth of the tumor.

Any tumors in the abdomen can have 3D-CRT planning applied to them. However, due to the need to spare normal tissues, abdominal tumors are increasingly treated with IMRT nowadays.

4.4.2 Intensity-modulated radiation therapy

Compared to forward-planned 3D-CRT, the inversely optimized IMRT normally achieves more conformal target coverage and reduced doses to OARs. IMRT plans can be delivered by static beams (step-and-shoot or sliding window) or volumetric modulated arcs. For static IMRT, an adequate number of beams (six or more) are typically required to avoid large low-dose spread-out. The selection of collimator angles [14] and segment sizes [15, 16] are also important to achieve desired plan qualities.

Figure 4.7 shows the isodose distributions and beam arrangement of a step-and-shoot IMRT plan for a patient with stomach cancer. The IMRT plan was designed to deliver 4500 cGy in 25 fractions to the PTV (pink color). The treatment plan was composed of seven coplanar step-and-shoot beams with equally spaced gantry angles and collimator angles optimized to cover target regions and spare nearby OARs. The treatment planning goals include an adequate and uniform dose to the PTV (D95 (minimum dose delivered to at least 95% of the volume) > 4500 cGy or V100 (percentage of volume receiving dose equal or higher than prescription) > 95%, and maximum dose [D_{max}] < 110%) and a reduced dose to surrounding OARs, including the LG bowel, liver, kidneys, and spinal cord. An in-house score card (table 4.3) was used to evaluate the plan, including the CTV and PTV coverage, normal tissue max, mean doses, and max DVH.

Whether to use IMRT or VMAT depends on the equipment, planning system, and experience of the staff.

Figure 4.8 shows the isodose distributions and beam arrangement of a VMAT plan for the treatment of pancreatic cancer. The treatment goals are to deliver two different dose levels, 5600 cGy and 5040 cGy, in 28 fractions. The low-dose PTV (PTV-5040) was defined as the CTV with 5 mm expansion, and the high-dose PTV (PTV-5600) was defined as PTV-5040 excluding the expanded section (1 cm) of the nearby GI tract OARs (duodenum, SM bowel, LG bowel, stomach, and esophagus) and the expanded section (5 mm) of the kidneys. Two 10 MV full-arc VMAT beams

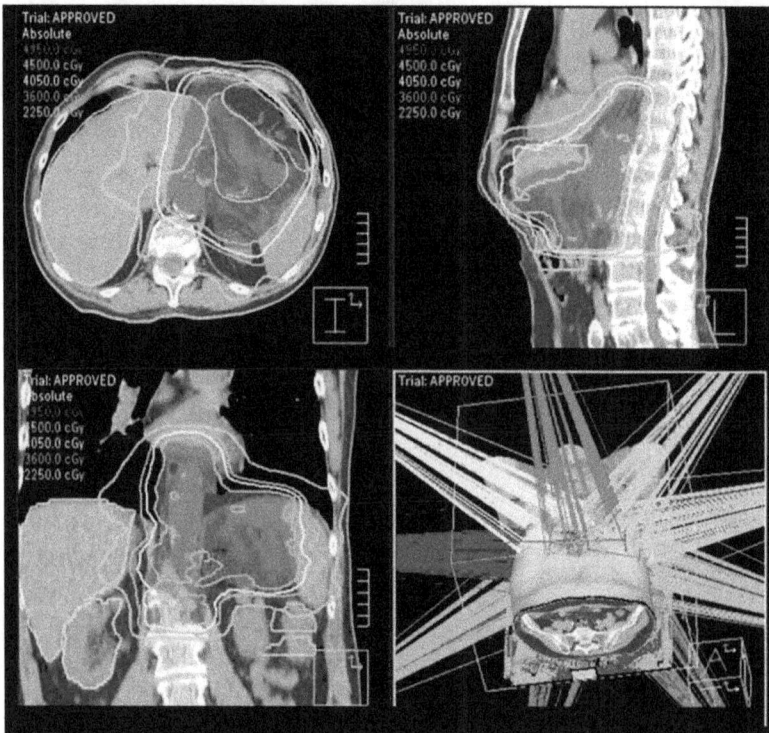

Figure 4.7. Dose distributions and beam configuration of a stomach IMRT plan.

Table 4.3. Planning objectives for the stomach IMRT plan.

Structure/ROI	Type	Primary goal dose (cGy)	Primary goal volume	Secondary goal dose (cGy)	Secondary goal volume
PTV-4500	Min DVH	4500	≥95%	—	—
CTV-4500	Min DVH	4500	≥98%	—	—
Liver	Max DVH	3000	<30%	—	—
Liver	Mean dose	<2800	—	—	—
LG bowel	Max DVH	5300	<0.03 cc	—	—
LG bowel	Max DVH	3800	<30%	—	—
Spinal cord	Max DVH	3500	<0.03 cc	—	—
Kidney, left	Mean dose	<2000	—	—	—
Kidney, left	Max DVH	1500	<15%	—	—

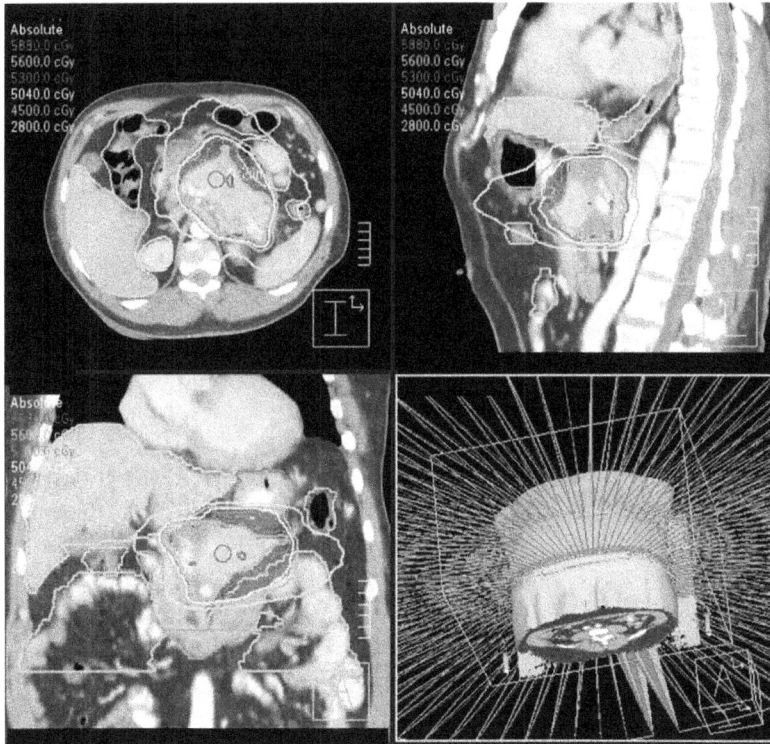

Figure 4.8. Pancreas VMAT plan.

were used with the collimator off zero to reduce the tongue and groove effect. The plan was optimized to ensure adequate and uniform dose coverage of the two PTVs and a reduced dose to the surrounding OARs. An in-house score card (table 4.4) was used to evaluate the plan quality.

Table 4.4. Planning objectives for the pancreas VMAT plan.

Structure/ROI	Type	Primary goal dose (cGy)	Primary goal volume	Secondary goal dose (cGy)	Secondary goal volume	Volume/ dose at primary goal dose
PTV-5600	Min DVH	5600	≥95%	—	—	95.5%
CTV-5040	Min DVH	5040	≥99%	—	—	100.0%
PTV-5040	Min DVH	5040	≥95%	—	—	97.6%
Stomach	Max DVH	5300	<0.03 cc	—	—	0.01 cc
Stomach	Max DVH	4500	<25%	—	—	12.2%
SM bowel	Max DVH	5300	<0.03 cc	—	—	0.002 cc
SM bowel	Max DVH	4500	<25%	—	—	4.2%
LG bowel	Max DVH	5300	<0.03 cc	—	—	0.0 cc
LG bowel	Max DHV	4500	<50%	—	—	0.0%
Duodenum	Max DVH	5300	<0.03 cc	—	—	0.0 cc
Duodenum	Max DVH	4500	<33%	—	—	6.8%
Spinal cord	Max DVH	4500	<0.03 cc	—	—	0.0 cc
Liver	Max DVH	3000	<30%	—	—	1.3%
Liver	Mean dose	<2800	—	—	—	683 cGy
Kidney, right	Max DVH	1500	<15%	2000	<20%	0.3%
Kidney, left	Max DVH	1500	<15%	2000	<20%	22.5%[a]

[a] Meets the secondary goal.

4.4.3 Stereotactic body radiation therapy

There is an increasing trend to use the technique of SBRT to treat primary and metastasized tumors in the abdominal cavity. Compared to IMRT, SBRT requires higher precision for patient immobilization, respiratory motion management, and daily imaging guidance. For abdominal tumors, a typical setup may use a vacuum bag for the patient immobilization, 4DCT, or BH technique to encompass or reduce breathing motion of tumor, and daily cone-beam CT (CBCT) for patient setup. If 4DCT is used, the derived average intensity projection CT should be used as the planning CT. If a BH technique is used, the reproducibility of BH needs to be assessed. Typically, the assessment can be done by comparing the tumor positions on two fused BH CTs.

Planning of SBRT is similar to planning for IMRT except that the dose is prescribed to a lower IDL (70%–80%). Therefore, SBRT plans allow hotspots within the PTV and achieve a sharper dose fall-off from the edge of the PTV. Nowadays, although not required, flattening filter-free (FFF) beams are recommended for SBRT due to their high dose rates (e.g., 1400 MU/min for 6 MV FFF and 2,400 MU/min for 10 MV FFF on Varian TrueBeam linear accelerator systems [Varian Medical Systems, Inc.]). Furthermore, 10 MV FFF beams are preferred to 6 MV FFF beams for abdominal SBRT planning because of better beam penetration and higher dose rate.

When deciding whether a center is equipped for SBRT treatment or not, imaging to assess the breathing motion (e.g., 4DCT) and image-guided radiation therapy (e.g., on-board CBCT or on-rail CT) is a must. For a treatment planning system to be able to plan SBRT, beam models must be commissioned to include the smallest field sizes that can possibly be used in SBRT. End-to-end tests using phantoms provided by calibration laboratories are recommended to evaluate the overall accuracy of the SBRT from simulation to treatment.

To implement an SBRT program, another consideration is the dose grid size and dose calculation algorithm. American Association of Physicists in Medicine TG-101 recommends using a dose grid size of no more than 2 mm for SBRT and a dose algorithm which can accurately account for the heterogeneity of tissue, such as a convolution superposition algorithm [17].

4.4.3.1 Liver SBRT

Figure 4.9 shows an example of an SBRT plan for the treatment of a liver metastasis. This patient was simulated using the BH technique with an ABC device. Three repeated BH scans were acquired at the time of simulation and were fused to evaluate the reproducibility of the BH. One of three BH scans was used as the planning CT. The GTV was delineated based on the fusion of the planning CT with a T1 post-contrast MRI. Based on the fusions between BH CTs, the ITV was drawn to encompass the GTVs from other BH CTs. The PTV was expanded 5 mm from the ITV. Two partial-arc VMAT beams were used instead of full-arc beams to avoid beams entering through the bowels and to avoid possible table/patient and gantry collision. Considering the patient's short BHs (20 s), 10 MV FFF photon beams were used for improved delivery efficiency. The plan was optimized to provide sufficient PTV coverage (cover PTV with the prescription of 5000 cGy in five fractions), control hotspots (keeping $D_{max} \leqslant 130\%$ of the prescription dose), and achieve a high dose gradient near the edge of the PTV for reduced doses to the surrounding OARs, including the normal liver (liver-GTV), heart, bowels, esophagus, stomach, kidneys, and spinal cord. An auto-planning technique was used. The planner put in objective goals such as the target coverage, hotspot control, dose fall-off, and OAR avoidance (dose constraints and qualitative weighting factors [low, medium, high]). The optimizer will progressively add optimization criteria and assign constraints during optimization. An in-house score card (table 4.5) was used to evaluate the plan quality.

4.4.3.2 Pancreas SBRT

SBRT for the pancreas follows the same workflow as liver SBRT. Figure 4.10 shows a pancreas SBRT with a prescription of 4000 cGy in five fractions. The patient was simulated using an ABC device for the BH. The GTV was delineated on the planning CT based on the fusion with two diagnostic CT scans at arterial and venous phases. The ITV was created from three repeated BH scans at simulation. Two PTVs were created. The low-dose PTV (PTV-2500) expands the ITV by 5 mm uniformly and the high-dose PTV (PTV-4000) excludes the normal GI structures with a 7 mm margin from the low-dose PTV.

Two full arcs were used because of the central location of the tumor, and a 10 MV FFF photon beam was used for fast delivery. Auto-planning was used to simplify the planning process. Table 4.6 lists the score card used to evaluate the plan quality.

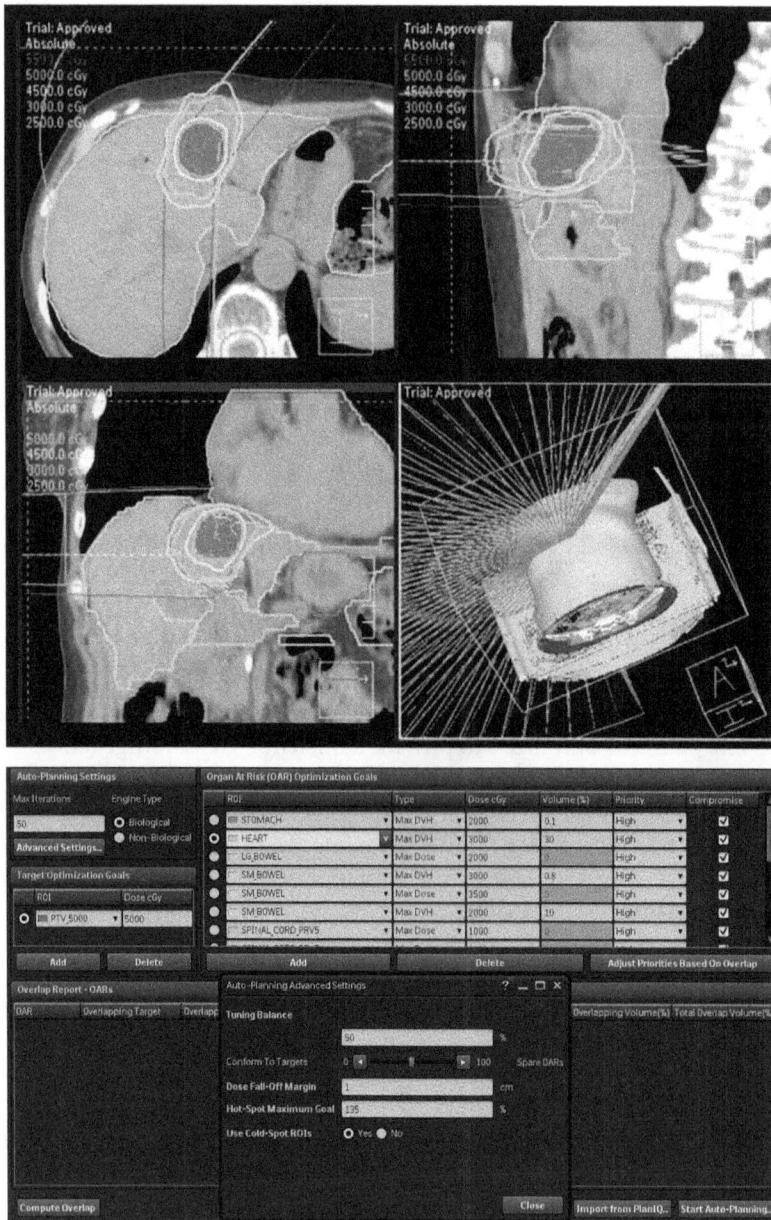

Figure 4.9. Liver SBRT treatment planning with an auto-planning technique.

4.4.3.3 Adrenal SBRT

Figure 4.11 shows the isodose distributions and beam arrangement for an adrenal SBRT. During simulation, a free-breathing CT and a 4DCT were acquired. The GTV was delineated on the free-breathing CT. The 4DCT (ten phases) was used to draw the ITV and the average intensity projection CT was used as the planning CT.

Table 4.5. Planning objectives for the liver SBRT plan.

Structure/ROI	Type	Primary goal dose (cGy)	Primary goal volume	Secondary goal dose (cGy)	Secondary goal volume	Volume/ dose at primary goal dose
PTV-5000	Min DVH	5000	⩾95%	—	—	95.8%
Stomach	Max DVH	2000	<0.03 cc	—	—	0.0 cc
SM bowel	Max DVH	3000	<0.5 cc	—	—	0.006 cc
SM bowel	Max DVH	2000	<5 cc	—	—	0.2 cc
LG bowel	Max DVH	2000	<0.03 cc	—	—	0.0 cc
Heart	Max DVH	5400	<0.03 cc	5500	<0.03 cc	0.05 cc[a]
Heart	Max DVH	3000	<30 cc	—	—	0.0%
Spinal cord	Max DVH	1000	<0.03 cc	—	—	0.0 cc
Spinal cord, PTV5	Max DVH	1000	<0.03 cc	—	—	0.0 cc
Liver-GTV	Mean dose	<1300	—	—	—	867.9 cGy
Whole kidney	Mean dose	1000	—	—	—	10.5 cGy
Esophagus	Max DVH	2000	<0.03 cc	5500	<0.03 cc	0.0%

[a] Meets the secondary goal.

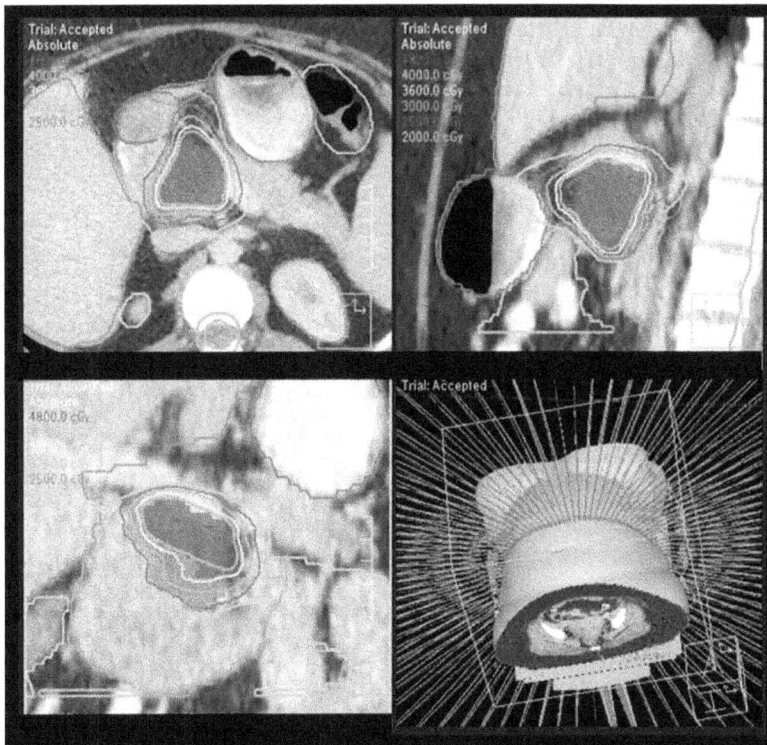

Figure 4.10. Pancreas SBRT treatment planning.

Table 4.6. Planning objectives for the pancreas SBRT plan.

Structure/ROI	Type	Primary goal dose (cGy)	Primary goal volume	Secondary goal dose (cGy)	Secondary goal volume	Volume/ dose at primary goal dose
ITV-2500	Min DVH	2500	≥99%	—	—	99.9%
PTV-2500	Min DVH	4000	≥95%	—	—	99.8%
PTV-4000	Min DVH	4000	≥95%	—	—	95.1%
Duodenum	Max DVH	3200	<0.03 cc	—	—	0.0 cc
Duodenum	Max DVH	3000	<0.5 cc	—	—	0.004 cc
Duodenum	Max DVH	2500	<5 cc	—	—	4.3 cc
Stomach	Max DVH	3200	<0.03 cc	—	—	0.0 cc
Stomach	Max DVH	3000	<0.5 cc	—	—	0.0 cc
Stomach	Max DVH	2500	<5 cc	—	—	2.7 cc
SM bowel	Max DVH	3200	<0.03 cc	—	—	0.0 cc
SM bowel	Max DVH	3000	<0.5 cc	—	—	0.006 cc
SM bowel	Max DVH	2000	<5 cc	—	—	0.2 cc
LG bowel	Max DVH	3500	<0.03 cc	—	—	0.0 cc
LG bowel	Max DVH	3200	<0.5 cc	—	—	0.0 cc
LG bowel	Max DVH	2500	<5 cc	—	—	0.0 cc
Kidney, right	Mean dose	<1000	—	—	—	318.6 cGy
Kidney, left	Mean dose	<1000	—	—	—	412.3 cGy
Spinal cord, PTV5	Max DVH	2250	<0.03 cc	—	—	0.0 cc
Liver	Mean dose	<1000	—	—	—	9.0 cGy

The PTV was defined as the ITV, expanded 5 mm. Two 10 MV FFF partial arcs were used to deliver 4500 cGy in five fractions. The goal of the plan was to cover the PTV with the highest dose achievable without violating normal tissue constraints such as those for the kidneys, ureters, bowel, and aorta. Auto-planning was used to simplify the planning process. The 3000 cGy IDL was converted to a structure and used during daily CBCT alignment. Table 4.7 lists the score card used to evaluate the plan quality.

4.4.3.4 Single lymph node SBRT

Single lymph nodes can also be treated with SBRT. Depending on the location of the lymph node, OARs may vary, including the GI structures, spinal cord, ureters, kidneys, and aorta. Figure 4.12 shows a metastatic retroperitoneal lymph node (the primary was prostate cancer) treated to 4000 cGy in five fractions. The GTV was contoured on the simulation CT. A 4DCT (ten phases) was used to contour the ITV. The PTV was defined as the ITV, expanded 5 mm. A low-dose PTV was created by subtracting the high-dose PTV with the expanded bowel (5 mm). The density in the SM bowel and both kidneys were overridden to 1 g cm^{-3} to override

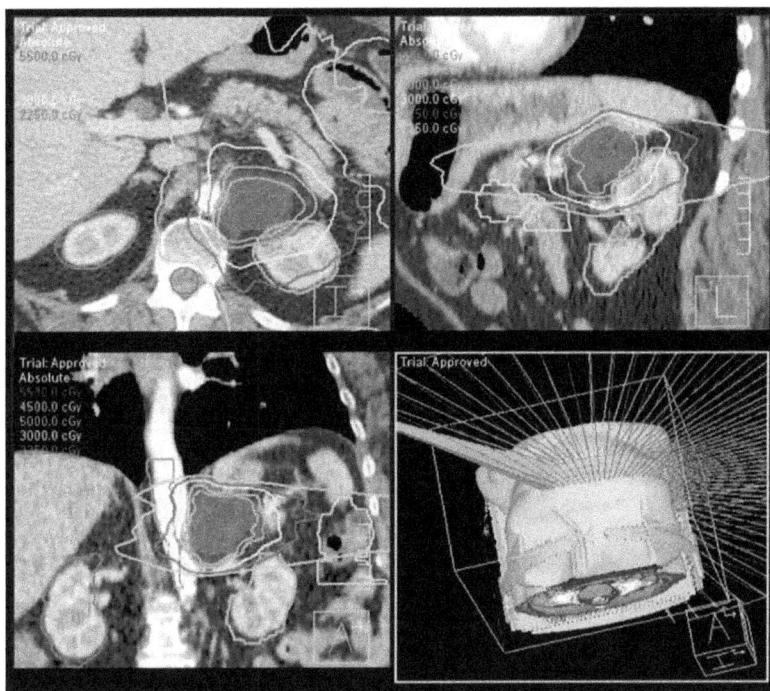

Figure 4.11. Adrenal SBRT treatment planning.

Table 4.7. Planning objectives for the adrenal SBRT plan.

Structure/ROI	Type	Primary goal dose (cGy)	Primary goal volume	Secondary goal dose (cGy)	Secondary goal volume	Volume/ dose at primary goal dose
PTV-4500	Min DVH	4500	≥95%	4500	90%	90.8%[a]
SM bowel	Max DVH	3000	<0.5 cc	—	—	0.0 cc
SM bowel	Max DVH	1950	<5 cc	1950	<10 cc	5.7 cc[a]
Stomach	Max DVH	4200	<0.03 cc	—	—	0.0 cc
Stomach	Max DVH	3000	<0.5 cc	—	—	0.48 cc
Stomach	Max DVH	2500	<5 cc	—	—	1.7 cc
Spinal cord	Max DVH	1180	<0.03 cc	—	—	0.0 cc
Aorta	Max DVH	5250	<0.03 cc	—	—	0.0 cc
Kidney, right	Mean dose	<1000	—	—	—	10.5 cGy
Kidney, left	Mean dose	<1000	—	—	—	0.0%

[a] Meets the secondary goal.

the contrast. The treatment plan was optimized to cover the two PTVs with the prescription dose, reduce the dose to surrounding bowel structures, and avoid a hotspot in the aorta nearby. Table 4.8 lists the score card used to evaluate the plan quality.

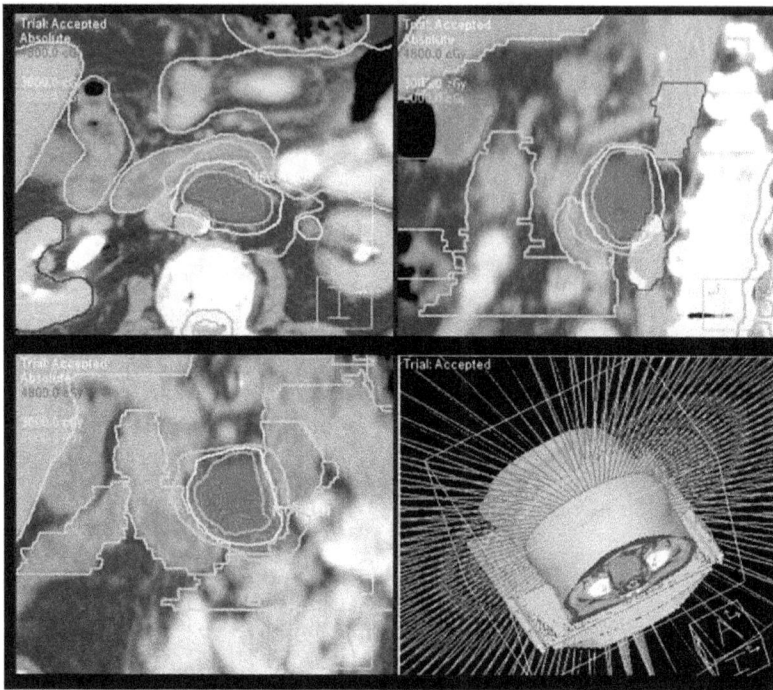

Figure 4.12. SBRT to a single abdominal lymph node.

4.4.4 Assisting structures for CBCT to CT alignment

A common difficulty in abdominal treatment is aligning the CBCT with the CT for patient setup. Target visibility in CBCT is low due to the lack of soft tissue contrast and the artifact caused by breathing motion and air in the abdominal cavity. Normal structure contours such as the liver, bowel, and kidneys may be used as surrogates to align the tumor. For pancreas radiation therapy, the presence of a stent may help the CBCT to CT alignment although caution has to be taken as a stent may move relative to anatomy [18]. Blood vessels in the abdominal cavity, including the aorta, inferior vena cava, and superior mesenteric artery, may help with alignment as well. Figure 4.13 shows an example. The blood vessels were delineated in the CT and superimposed on the CBCT for alignment.

4.4.5 Use of spacers in abdominal radiotherapy

Using a spacer in the abdomen to separate the intestine and treatment area was introduced in the 1980s, but its application has been limited due to the risk of surgical procedures to insert and remove the spacer [19, 20]. Recently, with the improvement in placement techniques [21] and spacer material development [22], this technique has been applied in particle therapy and showed promising results in reducing the dose to critical structures [23].

Table 4.8. Planning objectives for the single lymph node SBRT plan.

Structure/ROI	Type	Primary goal dose (cGy)	Primary goal volume	Secondary goal dose (cGy)	Secondary goal volume	Volume/ dose at primary goal dose
PTV-2500	Min DVH	4000	≥95%	—	—	99.5%
PTV-4000	Min DVH	4000	≥95%	4000	≥90%	90.1%[a]
Duodenum	Max DVH	3200	<0.03 cc	—	—	0.002 cc
Duodenum	Max DVH	3000	<0.5 cc	—	—	0.3 cc
Duodenum	Max DVH	2500	<5 cc	2500	<7 cc	6.7 cc
Stomach	Max DVH	3200	<0.03 cc	—	—	0.0 cc
Stomach	Max DVH	3000	<0.5 cc	—	—	0.0 cc
Stomach	Max DVH	2500	<5 cc	—	—	0.0 cc
SM bowel	Max DVH	3200	<0.03 cc	—	—	0.0 cc
SM bowel	Max DVH	3000	<0.5 cc	—	—	0.004 cc
SM bowel	Max DVH	2000	<5 cc	—	—	0.4 cc
LG bowel	Max DVH	3500	<0.03 cc	—	—	0.0 cc
LG bowel	Max DVH	3200	<0.5 cc	—	—	0.0 cc
LG bowel	Max DVH	2500	<5 cc	—	—	0.0 cc
Kidney, right	Mean dose	<1000	—	—	—	393.4 cGy
Kidney, left	Mean dose	<1000	—	—	—	536.0 cGy
Spinal cord, PTV5	Max DVH	2250	<0.03 cc	—	—	0.0 cc
Liver	Mean dose	<1000	—	—	—	129.6 cGy

[a] Meets the secondary goal.

4.4.6 Auto-planning for abdominal tumors

Auto-planning techniques have been introduced in radiation therapy treatment planning to improve the efficiency and quality of treatment planning [24]. Two types of auto-planning algorithms exist: kernel based [25] and knowledge based [26, 27]. Kernel-based algorithms use dose kernels to estimate the spread of a dose outside of the targets and predict the achievable doses to normal structures while ensuring target coverage. Knowledge-based auto-planning uses a library of previous treatment plans to estimate dose distributions and/or dosimetric endpoints by statistical modeling or deep learning. For both types of algorithms, the predicted doses serve as input for automatic inverse treatment planning [28] to create deliverable treatment plans. For abdominal radiation therapy, auto-planning has been shown to improve the plan quality compared with manual planning [25, 27].

Figure 4.14 gives an example of using PlanIQ (Elekta Inc.), a kernel-based auto-planning tool for a liver SBRT plan and a pancreas VMAT plan. The feasibility dose and DVH to an example OAR are shown. Liver SBRT has a prescription of 4000 cGy in five fractions and pancreas VMAT has a prescription of 5600 cGy in 28 fractions. The green area in the DVH graph means 'easy to achieve' and the red area means 'unachievable'.

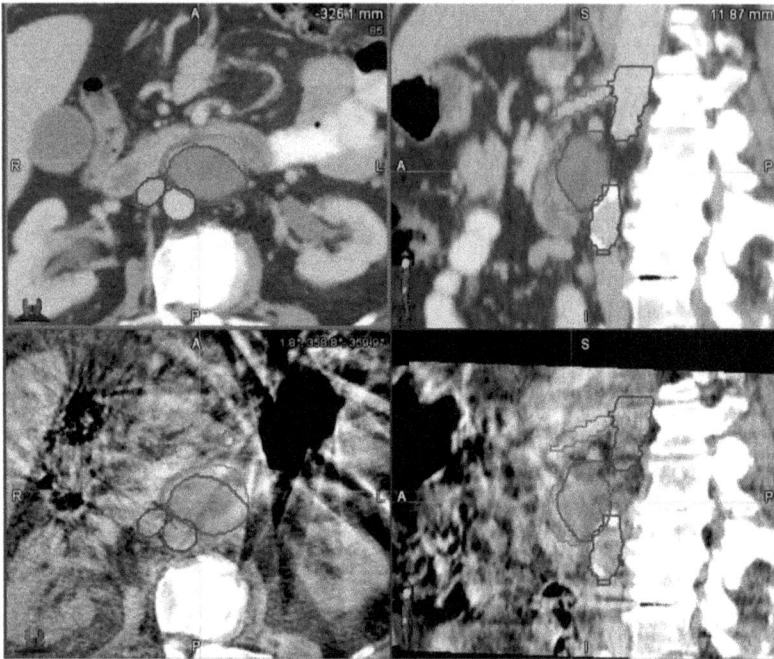

Figure 4.13. Blood vessel contours used to assist CBCT (lower figures) to CT (upper figures) alignment.

4.4.7 Re-treatment considerations

In-field reoccurrence after radiation therapy or second primary tumors occurring in previously irradiated areas present challenges for radiotherapy [29]. While it has shown that re-irradiation in the abdomen is effective and the benefits would outweigh the risks [30, 31], special considerations need to be taken for re-treatment planning [32, 33]. Highly conformal radiotherapy such as IMRT and SBRT should be used for re-treatment. The dose (or dose biologically equivalent to 2 Gy fractions [EQD2]) to normal structures from previous treatment(s) should be estimated and dose constraints for OARs should be updated accordingly. The accumulated dose and/or EQD2 from previous and current treatments should be calculated for plan evaluation [34].

In the abdomen, changes in anatomy and soft tissue deformation present challenges for estimating the previously delivered dose to normal structures and dose accumulation. Deformable image registration is often used for contour propagation and dose transfer [31], but the accuracy of deformable image registration needs to be carefully examined.

Figure 4.15 shows a liver re-treatment example. The patient received two previous treatments to the liver: a TheraSphere (Novant Health) treatment to three different areas of the liver 10 months ago and a liver SBRT treatment that delivered 4000 cGy in five fractions to a tumor in the caudate lobe of the liver 6 months ago. A new liver SBRT treatment to deliver 4000 cGy in five fractions to a newly found tumor in segment VII was planned. A 'good liver' contour was created to exclude the areas

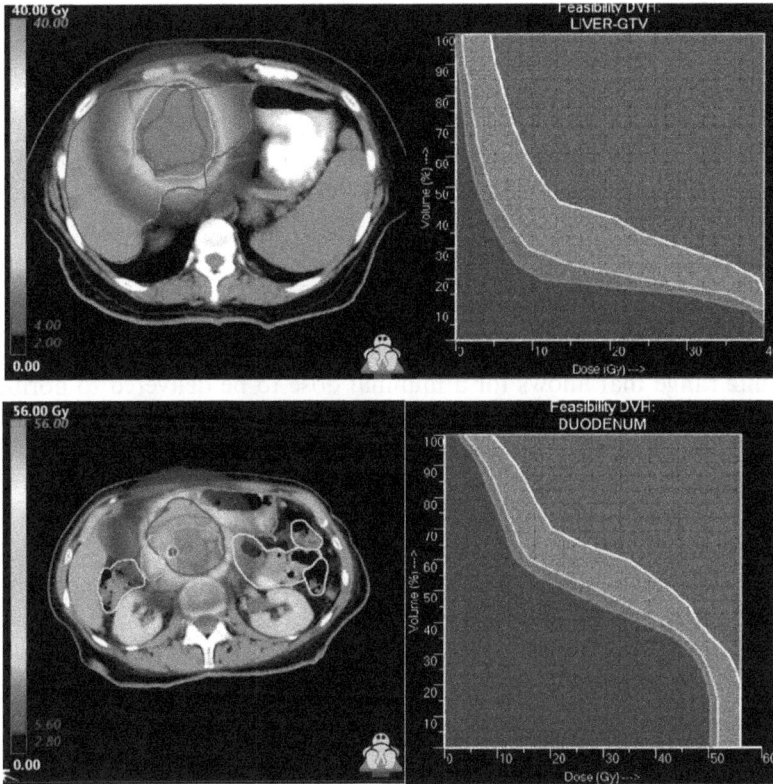

Figure 4.14. Feasibility dose and DVH for a liver SBRT (top figures) and a pancreas VMAT (bottom figures) treatment plan.

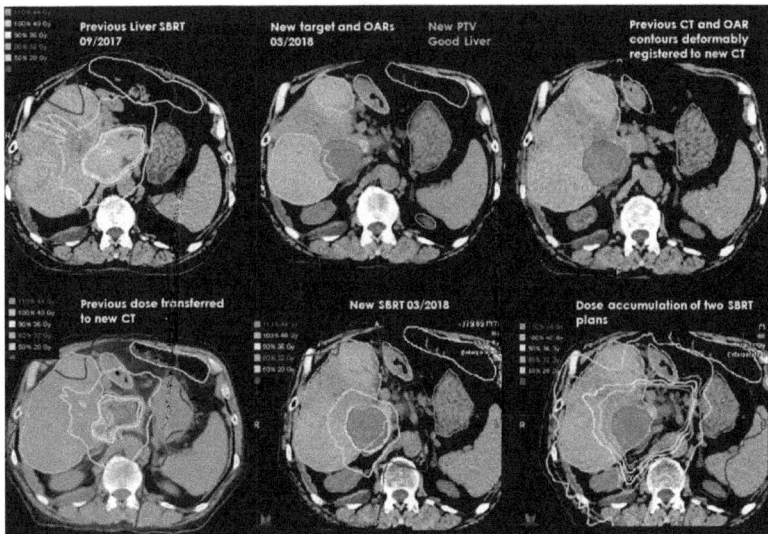

Figure 4.15. Re-treatment dose accumulation for an example liver SBRT in a patient.

treated with TheraSphere. The mean dose to the liver and 'good liver' structures from the previous SBRT treatment was estimated and used to adjust the mean dose constraints of the liver and the 'good liver' for the new SBRT plan. For composite dose accumulation of the two SBRT plans, a hybrid deformable registration algorithm based on the images' intensity and structure contours was used to transfer the previous SBRT plan dose to the new simulation CT. The accumulated dose was included in the treatment report.

4.4.8 Proton therapy for abdominal tumors

Proton radiotherapy has dosimetric advantages over photon radiotherapy. Protons have a finite range that allows for a minimal dose to be delivered to normal tissue distal to the tumor. This therapy is gaining popularity in treating abdominal tumors, including liver, pancreas, esophagus, and GI tumors [35–38]. Motion-robust planning methods [39–41] and breathing motion management techniques [41, 42] have been introduced for modern intensity-modulated proton therapy (IMPT) to minimize the effect of inter- and intra-fraction motion.

Figure 4.16 compares the dose distribution of a liver SBRT planned with IMPT and VMAT. The VMAT plan used two partial arcs with 10× FFF beams. The IMPT plan used three intensity-modulated proton beams. While the proton plan

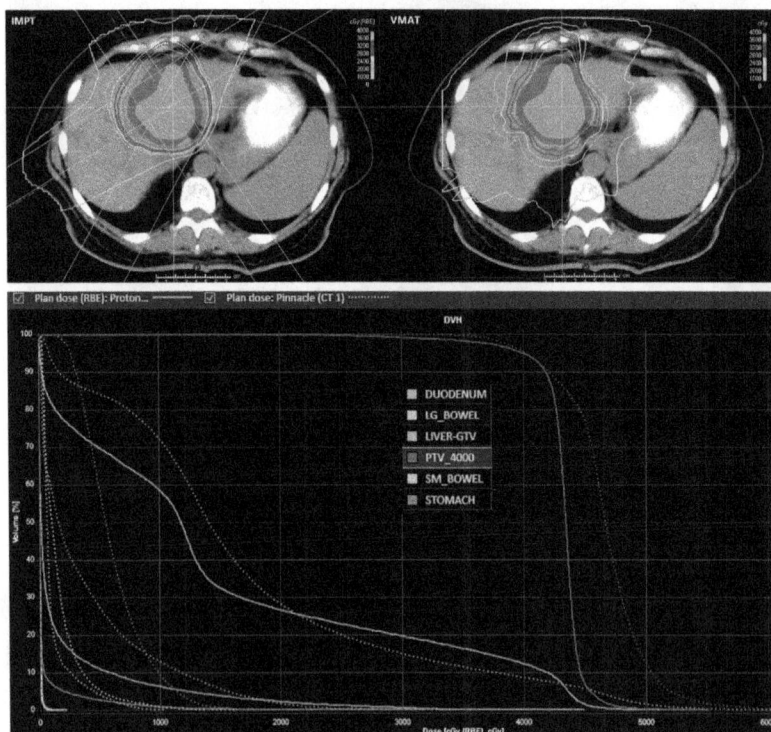

Figure 4.16. Comparing IMPT with VMAT for an example liver SBRT in a patient.

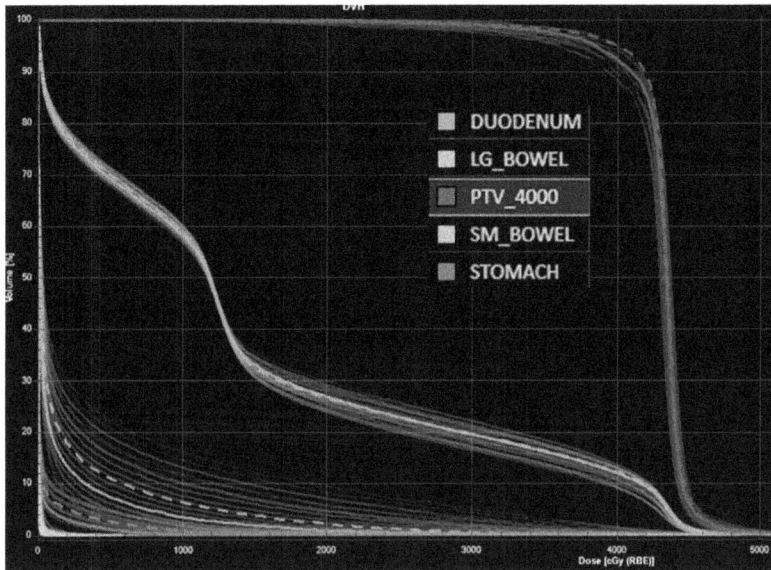

Figure 4.17. Motion robustness evaluation for a liver IMPT SBRT plan.

showed slightly less dose conformity to the PTV, it reduced the mean dose to normal liver and the D_{max} to the stomach, duodenum, and bowel.

Proton plans have to be evaluated for motion robustness due to the sharp dose gradient near the range of the protons. Figure 4.17 shows the variation of DVH to PTV and normal structures for the IMPT plan with a 5 mm position uncertainty and a 3.5% range uncertainty. Although the target coverage was acceptable under all scenarios, the dose to adjacent normal structures may vary largely. Under the worst-case scenario, a D_{max} of 0.5 cc to the duodenum increased the value from the planned value of 2950 cGy to 4410 cGy.

4.5 Conclusion

Treatment planning for abdominal cancer therapy aims to deliver a sufficient dose to the planning target volume and minimize the toxicity to organs at risk in the abdominal cavity. Image registration is essential for accurate target delineation. A variety of treatment modalities and techniques, including intensity-modulated radiation therapy, stereotactic body radiation therapy, and proton therapy, are used.

References

[1] Brock K K, Mutic S, McNutt T R, Li H and Kessler M L 2017 Use of image registration and fusion algorithms and techniques in radiotherapy: report of the AAPM Radiation Therapy Committee Task Group No. 132 *Med. Phys.* **44** e43–76

[2] Fukumitsu N *et al* 2017 Registration error of the liver CT using deformable image registration of MIM Maestro and Velocity AI *BMC Med. Imaging* **17** 30

[3] Zhe L, Deng D and Guang-Zhi W 2012 Accuracy validation for medical image registration algorithms: a review *Chin. Med. Sci. J.* **27** 176–81

[4] Landberg T, Chavaudra J, Dobbs J, Hanks G, Johansson K-A, Möller T and Purdy J 1993 Report 50 *J. Int. Comm. Radiat. Units Meas.* **os26** iii–72

[5] Landberg T *et al* 1999 Report 62 *J Int Comm Radiat Units Meas* **os32** iii–52

[6] Cardenas C E, Yang J, Anderson B M, Court L E and Brock K B 2019 Advances in auto-segmentation *Semin. Radiat. Oncol.* **29** 185–97

[7] Hu Y *et al* 2019 Implementing user-defined atlas-based auto-segmentation for a large multi-centre organisation: the Australian Experience *J. Med. Radiat. Sci.* **66** 238–49

[8] Schreibmann E, Marcus D M and Fox T 2014 Multiatlas segmentation of thoracic and abdominal anatomy with level set-based local search *J. Appl. Clin. Med. Phys.* **15** 4468

[9] Kim H, Jung J, Kim J, Cho B, Kwak J, Jang J Y, Lee S-W, Lee J-G and Yoon S M 2020 Abdominal multi-organ auto-segmentation using 3D-patch-based deep convolutional neural network *Sci. Rep.* **10** 6204

[10] Mohagheghi S and Foruzan A H 2020 Incorporating prior shape knowledge via data-driven loss model to improve 3D liver segmentation in deep CNNs *Int. J. Comput. Assist. Radiol. Surg.* **15** 249–57

[11] Zhang X, Tian J, Deng K, Wu Y and Li X 2010 Automatic liver segmentation from CT scans based on a statistical shape model *Annu. Int. Conf. IEEE Eng. Med. Biol.* 5351–4

[12] Zhang X, Tian J, Deng K, Wu Y and Li X 2010 Automatic liver segmentation using a statistical shape model with optimal surface detection *IEEE Trans. Biomed. Eng.* **57** 2622–6

[13] Ahn S H *et al* 2019 Comparative clinical evaluation of atlas and deep-learning-based auto-segmentation of organ structures in liver cancer *Radiat. Oncol.* **14** 213

[14] Tobler M, Leavitt D D and Watson G 2004 Optimization of the primary collimator settings for fractionated IMRT stereotactic radiotherapy *Med. Dosim.* **29** 72–9

[15] Wu Q J, Wang Z, Kirkpatrick J P, Chang Z, Meyer J J, Lu M, Huntzinger C and Yin F-F 2009 Impact of collimator leaf width and treatment technique on stereotactic radiosurgery and radiotherapy plans for intra- and extracranial lesions *Radiat Oncol* **4** 3

[16] Qi P and Xia P 2013 Relationship of segment area and monitor unit efficiency in aperture-based IMRT optimization *J. Appl. Clin. Med. Phys.* **14** 4056

[17] Benedict S H *et al* 2010 Stereotactic body radiation therapy: the report of AAPM Task Group 101 *Med. Phys.* **37** 4078–101

[18] Lee S L, Velec M, Munoz-Schuffenegger P, Stanescu T and Dawson L 2019 Extensive unpredictable pancreas cancer inter-fraction motion: a case report *Cureus* **11** e5047

[19] Fu Y, Liu S, Li H H, Li H and Yang D 2018 An adaptive motion regularization technique to support sliding motion in deformable image registration *Med. Phys.* **45** 735–47

[20] Tang Q, Zhao F, Yu X, Wu L, Lu Z and Yan S 2018 The role of radioprotective spacers in clinical practice: a review *Quant. Imaging Med. Surg.* **8** 514–24

[21] Nagai S, Nagayoshi K, Mizuuchi Y, Fujita H, Ohuchida K, Ohtsuka T, Imai R and Nakamura M 2020 Laparoscopic spacer placement for recurrent sacral chordoma before carbon ion radiotherapy: a case report *Asian J. Endosc. Surg.* **13** 582–5

[22] Kubo N *et al* 2020 An abdominal spacer that does not require surgical removal and allows drainage of abdominal fluids in patients undergoing carbon ion radiotherapy *PLoS One* **15** e0234471

[23] Yamada M *et al* 2019 In silico comparison of the dosimetric impacts of a greater omentum spacer for abdominal and pelvic tumors in carbon-ion, proton and photon radiotherapy *Radiat. Oncol.* **14** 207

[24] Amaloo C, Hayes L, Manning M, Liu H and Wiant D 2019 Can automated treatment plans gain traction in the clinic? *J. Appl. Clin. Med. Phys.* **20** 29–35

[25] Perumal B, Sundaresan H E, Ranganathan V, Ramar N, Anto G J and Meher S R 2019 Evaluation of plan quality improvements in PlanIQ-guided autoplanning *Rep. Pract. Oncol. Radiother.* **24** 533–43

[26] Zhang Y, Li T, Xiao H, Guo M, Zeng Z and Zhang J 2018 A knowledge-based approach to automated planning for hepatocellular carcinoma *J. Appl. Clin. Med. Phys.* **19** 50–9

[27] Kaderka R, Mundt R, Li N, Ziemer B, Bry V N, Cornell M and Moore K L 2019 Automated closed- and open-loop validation of knowledge-based planning routines across multiple disease sites *Pract. Radiat. Oncol.* **9** 257–65

[28] Mihaylov I B, Mellon E A, Yechieli R and Portelance L 2018 Automated inverse optimization facilitates lower doses to normal tissue in pancreatic stereotactic body radio-therapy *PLoS One* **13** e0191036

[29] Abusaris H, Hoogeman M and Nuyttens J J 2012 Re-irradiation: outcome, cumulative dose and toxicity in patients retreated with stereotactic radiotherapy in the abdominal or pelvic region *Technol. Cancer Res. Treat.* **11** 591–7

[30] Gkika E *et al* 2019 Repeated SBRT for in- and out-of-field recurrences in the liver *Strahlenther Onkol.* **195** 246–53

[31] Lee S *et al* 2018 Evaluation of hepatic toxicity after repeated stereotactic body radiation therapy for recurrent hepatocellular carcinoma using deformable image registration *Sci. Rep.* **8** 16224

[32] Bandyopadhyay A, Patro K C, Basu P and Roy K 2018 Approach towards re-irradiation of common cancers *J. Curr. Oncol.* **1** 29–34

[33] Das S, Patro K C and Mukherji A 2018 Recovery and tolerance of the organs at risk during re-irradiation *J. Curr. Oncol.* **1** 23–8

[34] Paradis K C *et al* 2019 The special medical physics consult process for reirradiation patients *Adv. Radiat. Oncol.* **4** 559–65

[35] Rutenberg M S and Nichols R C 2020 Proton beam radiotherapy for pancreas cancer *J. Gastrointest. Oncol.* **11** 166–75

[36] Barsky A R, Reddy V K, Plastaras J P, Ben-Josef E, Metz J M and Wojcieszynksi A P 2020 Proton beam re-irradiation for gastrointestinal malignancies: a systematic review *J. Gastrointest. Oncol.* **11** 187–202

[37] Chuong M, Kaiser A, Molitoris J, Romero A M and Apisarnthanarax S 2020 Proton beam therapy for liver cancers *J. Gastrointest. Oncol.* **11** 157–65

[38] Raldow A, Lamb J and Hong T 2020 Proton beam therapy for tumors of the upper abdomen *Br. J. Radiol.* **93** 20190226

[39] Engwall E, Fredriksson A and Glimelius L 2018 4D robust optimization including uncertainties in time structures can reduce the interplay effect in proton pencil beam scanning radiation therapy *Med. Phys.* **45** 4020–9

[40] Poulsen P R, Eley J, Langner U, Simone C B, II and Langen K 2018 Efficient interplay effect mitigation for proton pencil beam scanning by spot-adapted layered repainting evenly spread out over the full breathing cycle *Int. J. Radiat. Oncol. Biol. Phys* **100** 226–34

[41] Gelover E, Deisher A J, Herman M G, Johnson J E, Kruse J J and Tryggestad E J 2019 Clinical implementation of respiratory-gated spot-scanning proton therapy: an efficiency analysis of active motion management *J. Appl. Clin. Med. Phys.* **20** 99–108

[42] Dolde K *et al* 2019 Comparing the effectiveness and efficiency of various gating approaches for PBS proton therapy of pancreatic cancer using 4D-MRI datasets *Phys. Med. Biol.* **64** 085011

IOP Publishing

Principles and Practice of Image-Guided Abdominal Radiation Therapy

Yu Kuang

Chapter 5

Treatment delivery and verification

Seng Boh Lim and Grace Tang

Treatment delivery and verification are two key inter-dependent components crucial for the accuracy and safety of modern day radiation therapy. In this chapter, we explored the personnel required to build a successful and safe stereotactic body radiation therapy (SBRT) program and provided the background of different treatment modalities with several current state of the art SBRT delivery systems, such as C-arm linear accelerator (Linac) and magnetic resonance Linac. Different intrafractional monitoring strategies, which are critical in ensuring the fidelity of SBRT delivery, were discussed to enhance the understanding of the strengths and weaknesses of these techniques. An overview of the imperative dosimetric commissioning of a SBRT program was provided as a guide. A summary of the current patient-specific quality assurance (PSQA) techniques, ranging from pre-treatment PSQA to *in vivo* transit dosimetry, was also presented.

A successful SBRT program involves multiple elements to ensure efficient, safe, and effective treatment delivery. This chapter will discuss the necessary elements, from personnel and equipment requirements to patient verification strategies, to achieve safe SBRT delivery.

5.1 Personnel requirements for a stereotactic body radiation therapy program

It is very important to clearly define the minimum requirements and responsibilities of each member of the stereotactic body radiation therapy (SBRT) team. An overview of the minimum set as recommended by the American College of Radiology (ACR) and the American Society for Radiation Oncology (ASTRO) Practice Parameter for Performance of Stereotactic Body Radiation Therapy [1] will be presented below.

5.1.1 Radiation oncologist

A radiation oncologist (RO) is responsible for managing the overall treatment care. It is recommended that the RO has board certification in radiation oncology or therapeutic radiology given by the American Board of Radiology (ABR) as specified by the ACR–ASTRO Practice Parameter for the Performance of Stereotactic Body Radiation Therapy [1] and ACR–ASTRO Practice Parameter for Radiation Oncology [2]. The RO is responsible for working collaboratively with qualified medical physicists (QMPs) to determine the proper treatment regimen, such as choosing the method of immobilization and motion management, mitigating toxicity to organs at risk (OARs), and ensuring the robustness of treatment delivery. The RO is also responsible for supervising patient simulations, delineating and contouring the target volumes, and generating case-specific prescriptions. During the patient treatment session, it is recommended that the RO be present to direct the actual treatment process.

5.1.2 Medical physicist

As a member of the SBRT program, the medical physicist should be trained in stereotactic radiosurgery (SRS)/SBRT treatment and satisfy the qualification of being a QMP as defined in the American Association of Physicists in Medicine (AAPM) procedure and policy [3]. The QMP should be certified in the therapeutic medical physics subfield by the ABR, American Board of Medical Physics, or Canadian College of Physicists in Medicine. The QMP should also possess a valid local practice license when applicable. The main responsibilities of a QMP are to provide oversight of the technical aspect of the program and consultation throughout the treatment process [4]. The technical oversight includes the following:
  i) Equipment and devices purchase and planning.
  ii) Acceptance and commissioning of treatment planning and delivery systems.
  iii) Establishment and maintenance of a comprehensive quality assurance (QA) program with the following components:
    a) Treatment delivery system
    b) Ancillary system
    c) Treatment planning system (TPS)
    d) Immobilization devices
    e) Patient-specific QA (PSQA)
    f) End-to-end (E2E) test
    g) Quality control checklist for treatment workflow.
  iv) Supervision and monitoring of the treatment planning process.
  v) Supervision and monitoring of the treatment delivery process.
    a) In-person supervision of the first treatment session.
  vi) Establishment of a standard operating procedure.

5.1.3 Medical dosimetrist

A certified medical dosimetrist (CMD) should be the certified by the Medical Dosimetry Certification Board and should have training in SRS/SBRT planning [3, 4].

The planning competency of the CMD should be credentialed by a QMP. It is recommended that the CMD participate in the simulation session to gain clinical insights, such as insight on patient immobilization and motion effects, related to the treatment delivery. Under the supervision of the RO and the QMP, the CMD is responsible for contouring normal tissues. The CMD is also responsible for the documentation of the approved treatment plan and the wellbeing of data transfer prior to the beginning of the treatment session.

5.1.4 Radiation therapist

A radiation therapist (RTT) should carry a valid radiation therapy certification given by the American Registry of Radiologic Technologists and a practice license given by the government agency [4]. The RTT should receive appropriate SBRT training before carrying out SBRT treatment. During the patient simulation session, the RTT is responsible for preparing immobilization devices and performing image acquisition, following the SBRT procedure of the clinic. During the treatment session, the RTT is responsible for preparing the treatment setup based on the patient-specific instructions prescribed by the RO and for performing treatment delivery upon approvals from the RO and the QMP.

5.1.5 Others

The RO, as the primary person overseeing and managing a patient's treatment, may choose to enlist other specialists for consultation [1].

5.2 Treatment machine

The typical dose per fraction of a SBRT treatment is 6–30 Gy, which is three to ten times higher than the dose of conventional fractionated treatment [5]. Thus, the accuracy and precision of treatment delivery machines are critical [1, 4, 5], and image guidance is necessary for accurate tumor localization. SBRT treatments can be achieved by using megavoltage (MV) x-rays or proton [1] delivery systems. For MV units in particular, because maintaining a high conformality of small targets while minimizing the dose to the nearby OAR is often required, the leaf size of the multi-leaf collimators (MLCs) used in the delivery system should be small enough to provide a sharp dose drop-off. A MLC with a leaf width of 1.0 cm is considered inadequate [1]. This section will discuss a few MV x-ray machines that are currently available for SBRT delivery.

5.2.1 Isocentric delivery

Isocentric treatment involves beams delivered from multiple gantry angles using a constant source-to-axis distance [6]. This is usually achieved by rotating the linear accelerator (Linac) around the patient while the patient is stationary on the treatment couch. The rotation is typically achieved by mounting the Linac on a C-arm or a ring-mount gantry.

5.2.1.1 C-arm

Figure 5.1 shows an example of a C-arm radiation therapy machine for image-guided SRS/SBRT. The white arrow shows the MV treatment beam direction. The Varian TrueBeam® STx (Varian Medical Systems, Palo Alto, CA) is equipped with a HD120 MLC, which is comprised of 60 leaf pairs, including 32 pairs with leaves that are 2.5 mm wide at the central region and 28 pairs with leaves that are 10 mm wide at the outer region. To facilitate image-guided treatment, several imaging devices are also installed. The red arrows show the onboard kilovoltage (kV) imaging unit that allows the acquisition of radiographs and cone-beam computed tomography (CBCT) images. The yellow arrows indicate the ExacTrac system, with the two kV imagers mounted on the ceiling and two kV sources housed underground (one of the kV sources cannot be seen in figure 5.1). The green arrow points to the electronic portal imaging device (EPID) or MV imager. The blue arrow points to the couch with six degrees of freedom (6DOF) that can be used to correct patient setup errors in the superior–inferior, left–right, anterior–posterior, row, pitch, and yaw directions. In addition to the imaging devices demonstrated in figure 5.1, the treatment room can be equipped with non-ionizing imaging devices, such as those using infrared cameras, to monitor intrafractional patient motion. Figure 5.2 shows the infrared cameras of multiple different systems used to monitor intrafractional motion.

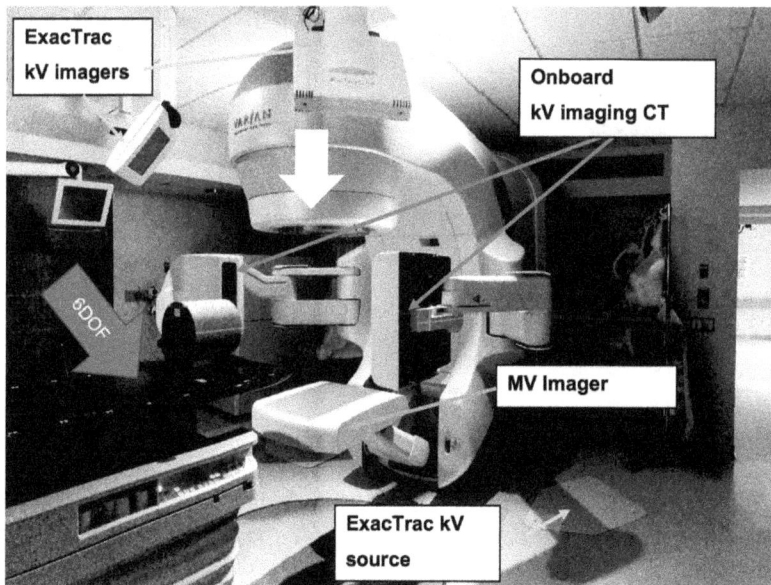

Figure 5.1. A C-arm image-guided radiation therapy (IGRT) SRS/SBRT machine, Varian TrueBeam® STx, equipped with multiple imaging devices: onboard kV imaging system enabled for both radiography and CBCT (red arrows), ExacTrac system (yellow arrows), and MV imager (green arrow).

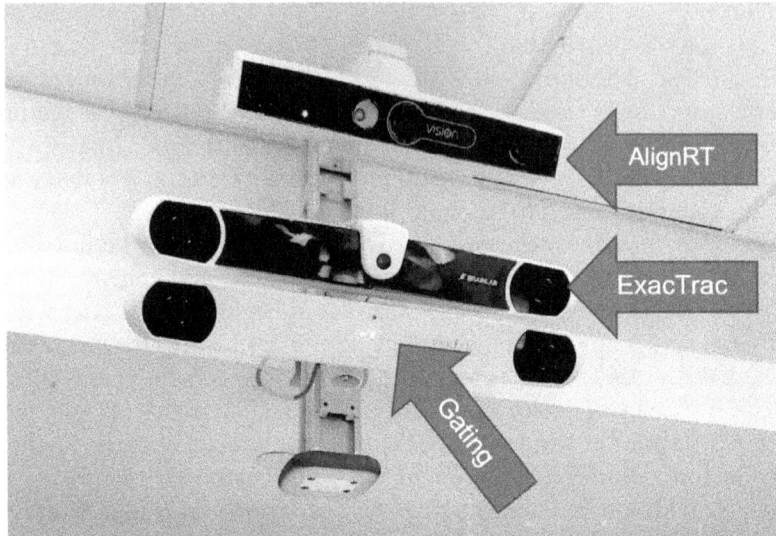

Figure 5.2. Infrared cameras mounted in the treatment room that are used to monitor intrafractional patient motion during treatment.

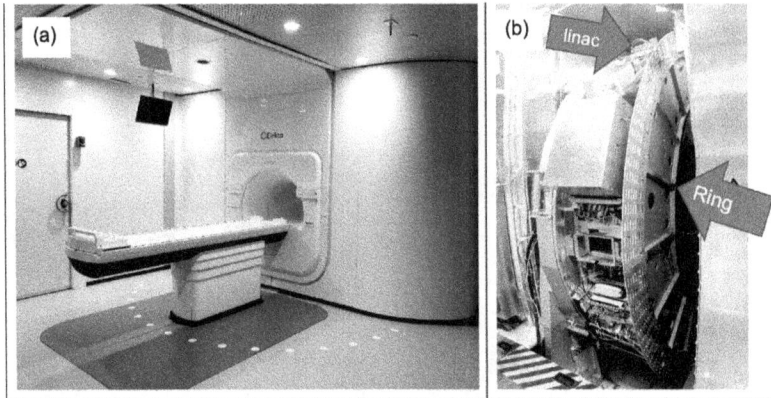

Figure 5.3. (a) View of the treatment room of Elekta Unity, which is a ring-type IGRT SBRT machine with a MV Linac mounted on a ring. (b) Linac mounted on a ring in a separate machine room that is directly behind the treatment room.

5.2.1.2 Ring-type machine

A ring-type machine, as the name implies, has the MV treatment machine mounted on a ring. The imaging device can be mounted on the same ring or on a different concentric ring. Figure 5.3 shows a 1.5T Unity (Elekta, Stockholm, Sweden) magnetic resonance (MR) Linac. Figure 5.3(a) shows the treatment room view of the machine. The gantry and MR system are situated in the machine room located directly behind the treatment room and are hidden from view. Figure 5.3(b) shows the ring on which the Linac is mounted in the machine room. Instead of kV imaging,

Unity uses MR imaging (MRI) to localize and monitor the tumor during treatment. The MRI device is mounted concentrically with the MV ring. Because of the superior soft tissue contrast of MRI, soft tissue targets can be better delineated with MRI than with the conventional kV imaging system, which is especially useful in abdominal cases.

5.2.2 Non-isocentric delivery

Unlike isocentric machines, robotic radiosurgery machines deliver treatment by using multiple non-isocentric and non-coplanar beams [7]. CyberKnife® (Accuray, Sunnyvale, CA) is a dedicated radiosurgery treatment unit with a Linac mounted on an industrial robotic arm and continuous imaging feedback from the in-room x-ray imaging system (see figure 5.4).

5.3 Treatment planning system

Treatment planning is an important component of the radiation therapy workflow. It is a process in which QMPs, CMDs, and ROs plan and optimize the treatment course prior to the actual delivery. Patient simulation can be performed with multiple modalities, including computed tomography (CT), positron emission tomography/CT, and MRI. These imaging systems must be carefully commissioned, especially for accurate electron density and mass density determination for heterogeneous dose calculation. A phantom with multiple inserts with known electron and mass densities should be used for commissioning. Details of recommendations can be found in AAPM TG-66 [8]. If a MR-only simulation workflow is used, care must be performed to verify the geometric and dosimetric accuracy of the synthetic CT images [9].

The modeling of the treatment beams in a treatment planning system (TPS) is critical in the whole treatment workflow. About 31% of the Imaging and Radiation Oncology Core (IROC) E2E tests failed to meet the dosimetric requirement of the credentialing standard (i.e., dose error < 5%), in which insufficient TPS commissioning can be a main attributing factor. This weakness has been found in a significant number of clinics [10]. In this section, a guideline for commissioning and verification will be presented.

5.3.1 TPS commissioning

5.3.1.1 Algorithm choice
Prior to TPS commissioning, the QMP should first determine the appropriate calculation algorithm used for SBRT. For accurate heterogeneous dose calculation, algorithms that account for three-dimensional (3D) scatter integration, such as convolution-superposition and Monte Carlo algorithms, are recommended. As it is possible that the beams can pass through the lung during abdominal SBRT, the use of a pencil beam algorithm is not recommended [11].

5.3.1.2 Commissioning
Recently, the AAPM has published Medical Physics Practice Guideline (MPPG) 5a, which outlines the recommended steps of TPS commissioning. Figure 5.5 shows a

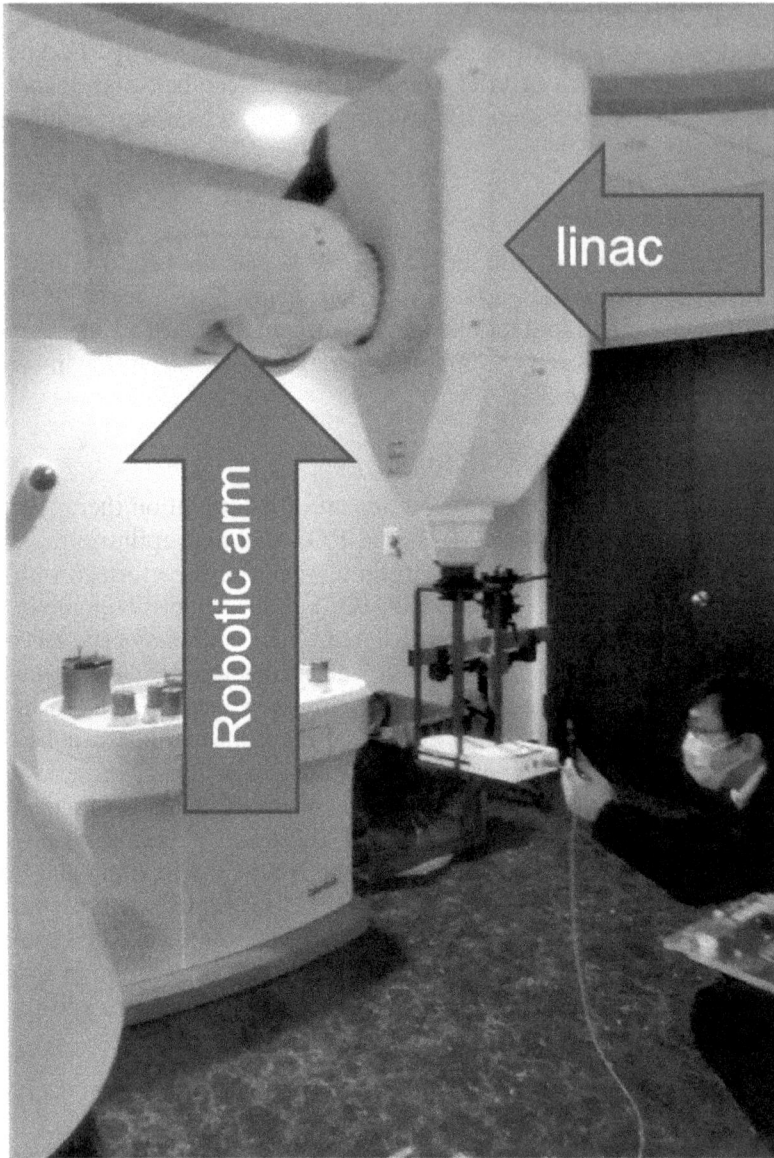

Figure 5.4. Performing QA on a robotic radiosurgery machine, CyberKnife (Accuray, Sunnyvale, CA).

schematic of the commissioning process, which can be broken down into the following main steps:

1. Open beam modeling and verification
2. MLC configuration
3. Intensity-modulated radiotherapy (IMRT)/volumetric modulated arc therapy (VMAT) verification
4. E2E test

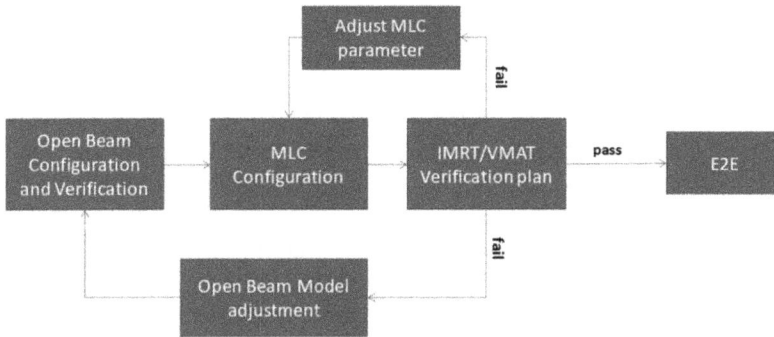

Figure 5.5. A schematic outlining the commissioning of a TPS with MLC parameter optimization.

Table 5.1. Selected minimum requirements for TPS commissioning as recommended by MPPG 5a, comparing output of TPS to measured data.

Items	Tolerance
TPS reference dosimetry	0.5%
Percentage depth dose and off-axis ratio factors	2.0%
Penumbra	3.0 mm
VMAT/IMRT area average within the region of interest: low-gradient, high-dose region (e.g., PTV[a]) and lower-dose region (e.g., OAR)	2.0%–3.0%
E2E test	5.0% of prescribed dose

[a]PTV: planning target volume.

Note that if verifications fail to satisfy the tolerances [12] of minimum requirements, the open beam parameters and/or MLC parameters may need to be modified and re-verified (figure 5.5). Table 5.1 shows a few selected items and the corresponding minimum tolerances as recommended by MPPG 5a [12].

5.3.2 Verification

In the case that a TPS already exists (and has been commissioned) prior to the implementation of a new SBRT program, it is critical for the QMP to verify that the existing TPS meets the required dosimetric accuracy for SBRT plans (table 5.1). If the TPS does not meet the tolerance, the QMP must investigate and remedy the issues. Typical sources of errors include miscalibration of the MLC of the treatment machine, the MLC parameters set in the TPS, or the basic beam modeling itself. An E2E test must also be performed to ensure workflow integrity after all pre-clinical dosimetry tests are satisfied. It is important that the QMP understands the limitations and accuracy of the TPS.

5.4 Imaging verification

5.4.1 Pre-treatment verification

For the past couple of decades, various radiation treatment techniques have been developed and clinically implemented. Traditional patient setup using skin markers no longer suffices for these complex treatments. Image guidance in radiation therapy has become part of the standard of care, and most Linacs are now equipped with onboard systems for both kV and MV imaging. Specifically, for liver and pancreas cases, which are typically treated with SBRT or an ablative technique, CBCT is used for accurate patient setup prior to treatment. For MR Linac, MRI would be used.

5.4.2 Intrafractional monitoring

Similar to thoracic cancers, the treatment delivery quality for abdominal diseases can be affected by intrafractional motion in addition to interfractional motion. Respiratory motion, peristaltic motion, and gastrointestinal filling attribute to intrafractional motion. For liver cases, a nominal motion magnitude of 14.7 $\pm$ 5.3 mm (superior–inferior), 5.2 $\pm$ 3.1 mm (anterior–posterior), and 2.9 $\pm$ 1.8 mm (left–right) has been observed, while the magnitude is 5.5 $\pm$ 2.0 mm (superior–inferior), 3.0 $\pm$ 1.7 mm (anterior–posterior), and 3.0 $\pm$ 1.8 mm (left–right) for pancreas cases [13]. To mitigate any dosimetric error caused by tumor motion, a variety of motion management methods have been developed. This includes respiratory gating, deep-inspiration breath hold, compression, and MLC/couch tracking. All of these treatment techniques require accurate real-time tumor local-ization. Various tumor motion-monitoring techniques have been developed and used clinically, but owing to the similarity of tissue densities in the abdominal region, these techniques are often facilitated by using fiducial markers implanted in the patients.

5.4.2.1 kV imaging

5.4.2.1.1 Intrafractional motion review

Varian TrueBeam Linacs are equipped with intrafractional motion review (IMR), in which a user can define an automatic kV imaging trigger based on the time, monitor units (MU), gantry angle, or Real-time Position Management™ (RPM) gating level during treatment delivery. Table 5.2 shows an example of the imaging trigger settings for SBRT. Fiducial markers on the two-dimensional (2D) kV images are segmented and instantaneously compared to digitally reconstructed radiographs. The user can define tolerance of the fiducial position difference based on the diameter of the expected position or contour. The contour is typically an expansion from the fiducial marker defined in planning CT. The expansion margin is essentially the tolerance of marker positional errors (e.g., a 2 mm expansion margin is referred to a 2 mm tolerance). At present, IMR does not provide quantitative analysis online. Therefore, it is somewhat helpful to use contour-based tolerance as positional errors can be visually identified, assuming therapists constantly monitor the system. Note

Table 5.2. Example of imaging trigger settings for Varian IMR.

IMR trigger settings for SBRT treatment	
VMAT	Every 20°
IMRT	Non-FFF: every 200 MU
	6FFF: every 300–500 MU

FFF = flattening filter free.

that IMR is a monoscopic imaging technique and one of the three dimensions of the 3D marker position is unresolved based on the individual 2D planar images. Instead, IMR only provides 2D information of the fiducial marker locations at each imaging trigger instance (i.e., the superior–inferior direction is always known while the anterior–posterior and left–right directions are only resolved at certain gantry angles).

5.4.2.1.2 kV intrafractional monitoring

To resolve the third dimension of the fiducial marker positions using 2D planar images from monoscopic kV imaging and enable triangulation, 2D to 3D transformation is needed. Prior to treatment delivery, kV images are acquired at 5–11 Hz over an angular range of >120°. A 3D Gaussian probability density function (PDF) is generated based on maximum likelihood estimation using the kV images collected. Subsequent kV images acquired during treatment delivery are used to recompute the PDF, which determines the 3D target position based on the fiducial markers segmented from the images. This approach is known as kV intrafractional monitoring (KIM) [14–17] and is a non-commercial product developed at University of Sydney. KIM was first clinically implemented for the prostate and has since begun expanding its use for liver cases.

5.4.2.2 MV-kV imaging

Memorial Sloan Kettering Cancer Center developed the simultaneous MV/kV imaging technique for VMAT delivery [18]. During treatment, a pair of MV and kV images are acquired simultaneously every 20°. MV imaging control points are added in the optimized VMAT plan. These imaging control points are copies of the adjacent plan control points, but MLCs are minimally retracted to expose the fiducial markers. These imaging control points add a minimum dose to the treatment plan while renormalization of up to 5% is allowed if necessary. Matching templates are created and are used as reference for image registration.

During treatment, MV and kV images are acquired and extracted automatically with iTools Capture (Varian Medical Systems, Palo Alto, CA), which is a frame grabber connected to the Linac. The images are then transferred to an in-house software package for analysis. Fiducial markers are segmented and registered automatically. The software provides quantitative analysis, and an audible alert is triggered when a shift of >1.5 mm is recorded for two consecutive control points.

Currently, MV/kV imaging is only used for prostate cases but the application to abdominal diseases is under investigation.

5.4.2.3 Hybrid methods

Infrared imaging systems, such as the Varian RPM system, and surface imaging systems, such as AlignRT® (Vision RT, London, United Kingdom), monitor the external motion during treatment. These systems may not be sufficient for accurate internal target motion monitoring. The RPM system only provides one-dimensional motion, and AlignRT is not appropriate for surfaces that are relatively symmetrical, such as the abdominal area. More critically, the external motion is merely a surrogate for tumor motion monitoring, and it is well known that the external motion often does not correlate with internal motion [19]. Hybrid methods combine the use of external motion monitoring with kV imaging for better internal target localization.

5.4.2.3.1 CyberKnife

The CyberKnife Synchrony® system obtains the external motion from a special vest that is worn by the patient [20]. The patient's external motion is monitored by three ceiling-mounted cameras based on the light-emitting diode markers on the vest. During patient setup, a series of kV image pairs are acquired at different stages of the breathing cycle, triangulating the implanted fiducial markers. Meanwhile, the external motion is continuously recorded and an external correlation model is built. During delivery, the treatment is adapted (i.e., beam re-alignment) based on the external motion, while kV images are acquired regularly to crosscheck the validity of the external correlation model. The model can be adjusted online if necessary.

5.4.2.3.2 ExacTrac

Similar to the CyberKnife system, ExacTrac® (Brainlab, Munich, Germany) utilizes an optical system in conjunction with stereoscopic kV imaging. Multiple infrared-reflective markers are placed on the patient surface or immobilization device, and the external motion is detected by the ceiling-mounted cameras. A kV image pair is acquired when the external motion is at the set gating level and the triangulated position of the fiducial markers is identified in the kV images and compared to the reference position.

5.4.2.3.3 Combined optical and sparse monoscopic imaging with kilovoltage x-rays

Combined optical and sparse monoscopic imaging with kilovoltage x-rays (COSMIK) is a non-commercial motion-monitoring method developed by Bertholet et al [21] and has been clinically implemented for liver cases. With an RPM signal, an external correlation model is built using the fiducial marker positions extracted from the pre-treatment CBCT that is acquired during patient setup. During treatment delivery, the 3D trajectory of the fiducial markers is estimated by the correlation model based on the continuous external motion monitoring from RPM. The kV images are sparsely acquired every 3 s, and the 3D position of the fiducial markers is estimated. This is then verified against the external correlation model, and an update can be made online if necessary.

5.4.2.4 Electromagnetic transponders

The imaging dose is an ongoing concern for radiotherapy, even though a high level of radiation is delivered as the treatment intent. Depending on the technique used (i.e., settings such as the kV peak and pulse duration) and imaging frequency, the accumulated skin dose to the patient over the course of treatment can be significant and can cause unintentional skin toxicity. Calypso® (Varian Medical Systems, Palo Alto, CA) is a motion-monitoring system that does not use ionizing radiation, and unlike the aforementioned intrafractional motion monitoring techniques, the 3D tumor motion is continuously detected. The patient is implanted with two to three transponders or beacons, each with different resonance frequencies. An electromagnetic panel hangs above the patient to excite and detect these beacons. The centroid of the beacons is calculated and compared to the Linac isocenter, which is determined by three ceiling-mounted cameras tracking the infrared markers on the panel. While it has been demonstrated to be a promising intrafractional motion-monitoring method for the liver and pancreas [22–26], one critical drawback of Calypso is the severe MRI artifact induced by the beacons.

5.4.2.5 Ultrasound

While Calypso is a good solution for intrafractional motion monitoring, the need to implant beacons into the patient may not be desirable. Ultrasound systems monitor tissue motion without the need of fiducial markers. Currently, the Clarity® Autoscan Ultrasound System (Elekta, Stockholm, Sweden) is the only commercial system available. The Clarity system was designed for the prostate, but application to the pancreas and liver has been investigated [27, 28] . The ultrasound probe is held by a fixation device that is locked onto the treatment couch. To correlate to room coordinates, a ceiling-mounted camera tracks the probe position using the infrared markers on it. Under the fixated housing, the probe continuously sweeps back and forth to provide four-dimensional images. Instead of using the pancreas body as the monitor structure, blood vessels such as the portal vein, inferior vena cava, and superior mesenteric artery are tracked instead, depending on the location of the tumor. For example, the portal vein would be used as the landmark if the tumor is located at the pancreatic head. While preliminary reports demonstrated promising results, further investigation is ongoing, including investigation of probe placement, the anatomical deformation induced by the probe, and the interaction of treatment beam and probe. [27–32]

5.4.2.6 MRI

Compared to kV imaging, MRI provides excellent soft tissue contrast and is especially useful in abdominal sites [33, 34], eliminating the need to implant fiducial markers in the patient. Intrafractional motion monitoring using MRI is currently feasible only on dedicated MR Linac units. At present, 2D cine MR images are used to monitor tumor motion and as input signals for gated treatments, while 3D cine MR is not yet clinically feasible and warrants further investigation.

5.5 Delivery verification

Delivery verification is an important aspect of the treatment delivery process that ensures patient safety and treatment effectiveness. This is even more critical for SBRT. The accuracy of delivery should satisfy the minimum criteria recommended by MPPG 9a [4] or AAPM TG-142 [35]. The QMP may want to maintain tighter tolerances on certain aspects based on clinical experience and the risk of the treatment sites. For details on machine QA, please refer to the chapter on QA. In this section, the verification process is examined in two aspects: commissioning and PSQA.

5.5.1 Commissioning

As mentioned in section 5.3.2, it is important to validate against delivery accuracy for a new TPS. The same logic applies to a new treatment unit or treatment modality. In addition, a SBRT program requires the imaging system to work in concert with the MV delivery system to deliver the dose at the intended location with high precision.

5.5.1.1 Pre-clinical verification

Pre-clinical verification should be completed prior to any E2E test. A basket of relevant test plans with a range of plan complexity, such as the SBRT plans here, should be delivered in phantom and measured with a 2D detector. These plans should include all the delivery modalities that are intended for clinical use. For example, if both IMRT and VMAT are to be clinically used, the basket of plans should contain both treatment modalities. In terms of the detector, ideally, films should be used because of the high spatial resolution and angular independence. However, if a 2D detector array, such as Sun Nuclear MapCheck®, is used, tighter analytical tolerance, such as γ (3%, 2 mm) with a 10% threshold and absolute dose criteria [36, 37], must be used to understand the limitation of the system. The passing rate should be above 90%–95% [37]. It is the responsibility of the QMP to understand the limitations of the TPS and delivery system and to provide consultation for the RO. As mentioned in section 5.3.2, the QMP must also investigate and take necessary steps to mitigate any issues of the TPS and/or the delivery system if the pre-clinical verification test results fail to meet tolerance. Figure 5.6 shows an example of pre-clinical verification with a SBRT field using Gafchromic EBT2® film in a solid water phantom. Here, a more stringent and sensitive analysis of the average dose difference was used in conjunction with γ-analysis.

5.5.1.2 E2E test verification

An E2E test should be performed when all the verifications of the TPS and machines are completed. An anthropomorphic phantom with a hidden target should be used in this process [5]. If such a phantom is not available at the facility, the QMP should consider ordering one from IROC [38] or other sources to perform the test. Figure 5.7 shows an example of anthropomorphic spine phantom with a hidden target.

As the name of E2E implies, this phantom should follow a standard patient treatment workflow. It should be scanned at the CT simulator with an

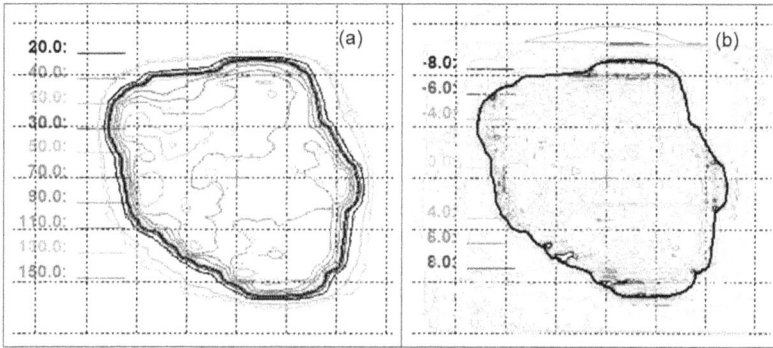

Figure 5.6. A pre-clinical SBRT test field measured with Gafchromic EBT2® film. (a) The overlay of film and the TPS calculation with a 1.25 mm dose grid. (b) The dose difference (film – TPS) distribution with an average dose difference of −0.4% within a region of interest (indicated with thick black line) that approximately corresponds to a 50% isodose line. The γ (3%, 3 mm) passing rate is 100%.

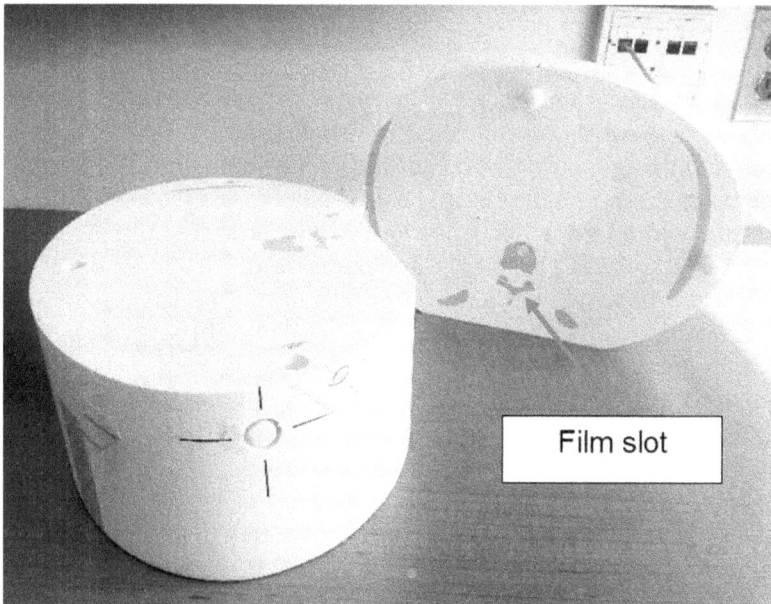

Figure 5.7. An example of anthropomorphic spine phantom (CIRS, Norfolk, VA) with a film slot.

immobilization device by a RTT, and a RO should contour the target on the planning CT. This should be then passed onto a CMD to prepare a treatment plan with the commissioned TPS. The RO and QMP should review and approve the final treatment plan. The plan should be delivered on a SBRT machine with the appropriate clinical imaging protocol by a RTT. The QMP should analyze and review the localization and dosimetry results based on a tolerance of 1.0 mm and 5.0%, respectively, as recommended by TG-101 and MPPG 9a.

5.5.2 Patient-specific delivery verification

Patient-specific verification is an important element of the treatment verification process for ensuring safe delivery of the treatment plan. This can be achieved by pre-treatment QA measurement, machine log file analysis, and *in vivo* measurements. For SBRT, more than one technique should be used to verify the integrity of treatment plan delivery.

5.5.2.1 Pre-treatment PSQA measurement

Pre-treatment PSQA measurement is often the last line of defense against potential errors prior to treatment delivery. Two-dimensional array detectors are commonly used for these measurements. Figure 5.8(a) shows a 2D diode array with a spatial resolution of 0.7 cm and an inherent build-up of 2.0 cm. Additional build-up can be added to simulate deeper delivery. Figure 5.8(b) shows an a-Si–based 2D detector with 0.3 mm spatial resolution. These measurement devices require proper calibration prior to clinical use. For EPID-based PSQA, the QMP must verify the accuracy of the calibration and applicable algorithms. Furthermore, some detectors may have significant angular dependency and spectral effects. The QMP must thoroughly assess the limitations and behavior of these detectors prior to clinical use.

The protocols used in analyzing PSQA vary quite significantly from institution to institution. In practice, γ-analysis [39] is probably the most widely used metric in the community. The AAPM recommends [37] using the absolute dose with global normalization for the γ-analysis. The normalization should not be less than 90% of the maximum dose point in the field. The dose difference and distance to agreement parameters should be 3.0% and 2.0 mm with a 10.0% low-dose threshold. In general, a 95.0% passing score is considered acceptable with 90.0% as the action level. Figures 5.9 and 5.10 show two examples of commercial packages for PSQA with a 2D diode array and an EPID, respectively. The (a) and (b) panes in figures 5.9 and 5.10 show the measurements and TPS calculations. Panes (c) show the analysis parameters for the measurements. As each software package can be quite different from others, the QMP should get familiar with the software package to ensure proper analysis parameters are used for the PSQA.

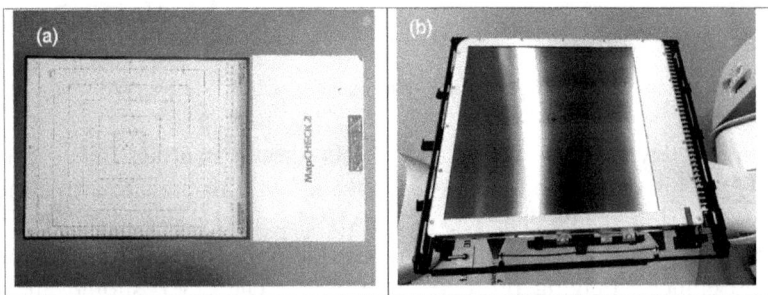

Figure 5.8. (a) Typical 2D detectors: MapCheck 2 (Sun Nuclear, Melbourne, FL). (b) EPID (Varian Medical Systems, Palo Alto, CA).

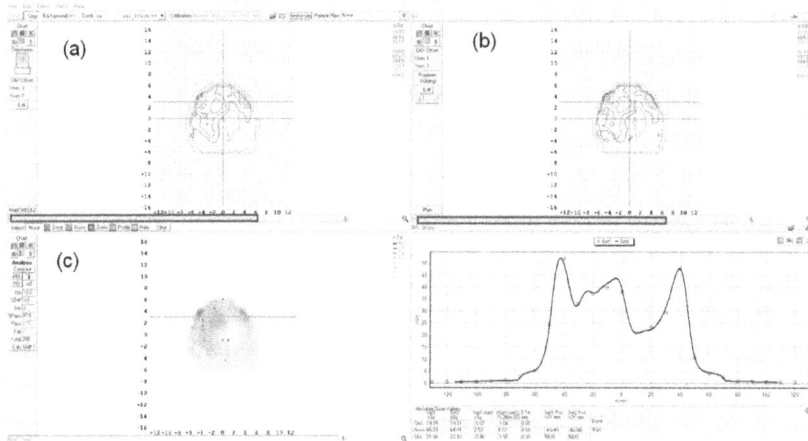

Figure 5.9. Analysis of an IMRT PSQA measurement using a 2D diode array and γ (3%, 2 mm) with a 10% threshold is shown. (a) Measurement. (b) TPS calculation. (c) Analysis results.

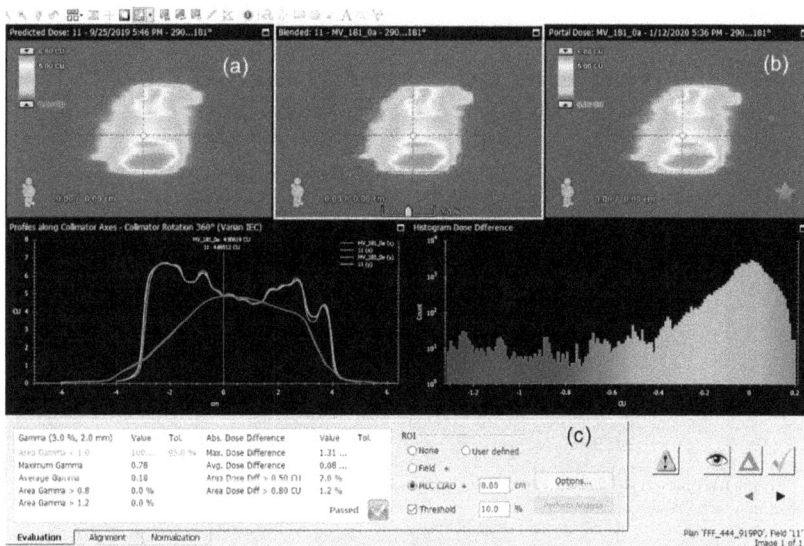

Figure 5.10. A PSQA analysis using EPID with γ (3%, 2 mm) and a 10% threshold. (a) Measurement. (b) TPS calculation. (c) Analysis results.

If the plan fails to pass these criteria, the QMP must investigate to determine whether the failure points are clinically significant. If a resolution cannot be reached, the QMP should consider using another detector to confirm the validity of the results. For further information, the recent AAPM Task Group report 218 [37] is a valuable resource for practicing QMPs.

5.5.2.2 *Machine log file analysis*

Modern treatment machines typically generate log files, which contain the time series parameters, such as the MU, gantry angles, and MLC positions, during treatment delivery. It has been shown to be an efficient way of performing QA [40]. Figure 5.11 shows an example of the log-file-generated fluence comparison with the plan after a delivery using LinacView (Standard Imaging, Middleton, WI). As this is not limited to pre-treatment PSQA, it provides additional delivery verification during actual patient treatment. As with any other QA tools, the QMP must understand the definition and limitations of log file analysis. For example, the definition of the MLC leaf position may represent physical or radiological positions depending on the vendors' convention. With proper commissioning and routine QA, log file analysis can be a reliable PSQA tool. Furthermore, the whole analysis process can be automated and expedite heavy clinical workflows.

5.5.2.3 In vivo *measurement and dose reconstruction*

A transit dosimetry system that uses exit radiation from patients during treatments has been investigated and implemented for both real-time and offline *in vivo* delivery monitoring [41–44]. As these systems use exit radiation from patients, no additional radiation or time burden are incurred. Figure 5.12 shows the real-time analysis interface of an *in vivo* transit dosimetry measurement for SBRT delivery using

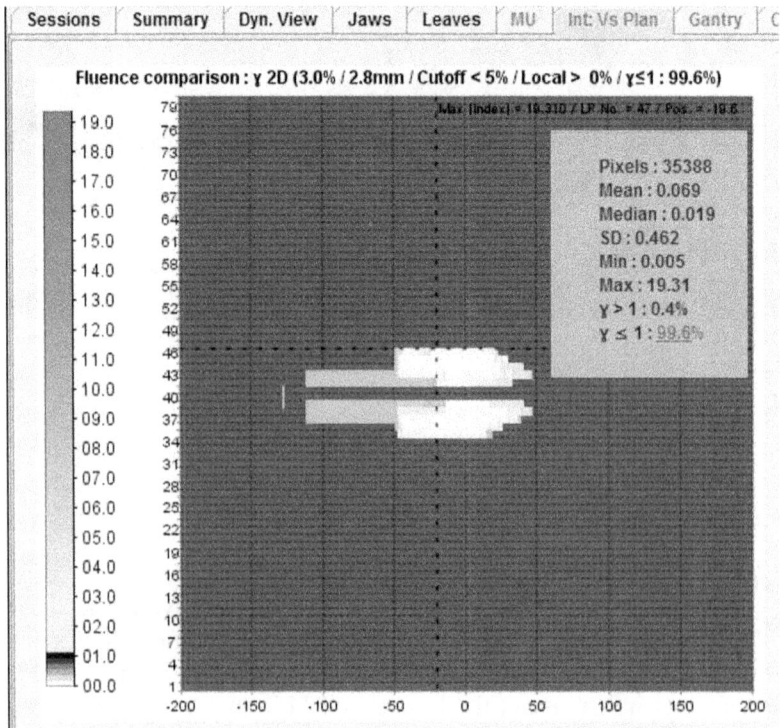

Figure 5.11. Comparison of log-file-generated fluence and plan fluence.

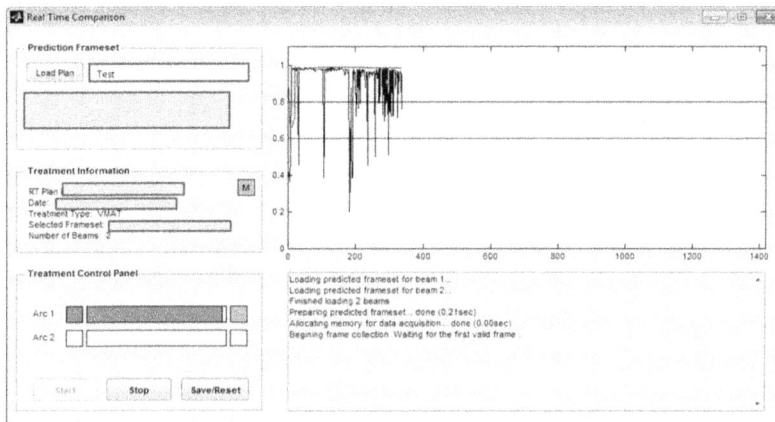

Figure 5.12. The real-time *in vivo* transit dosimetry PSQA of an SBRT treatment using WatchDog. The blue and black traces represent the cumulative and instantaneous χ-analysis, respectively.

WatchDog [44], which is an EPID-based *in vivo* transit dosimetry tool. The blue trace represents the cumulative χ-analysis [15]. The black trace represents the instantaneous analysis that shows a significantly noisier signal than the cumulative trace using a dose difference and distance to agreement of 4% and 2 mm, respectively.

The systematic relationship between the measured EPID signals, χ- or γ-metrics [45, 46], and the clinical decision metrics, such as the dose–volume histogram, of the OAR and PTV [43] are not well established. Rozendaal *et al* [42] reconstructed the 3D dose distributions using the back-projected fluence derived from the measured patient exit dose and the planning CT. The results showed that the γ results from this approach have a statistically significant correlation with the PTV changes based on a cohort of 20 patients with head and neck cancer. Similar PSQA techniques can be applied to gastrointestinal SBRT, which warrants further investigation.

References

[1] Chao S T, Dad L K and Dawson L A *et al* 2020 ACR–ASTRO practice parameter for the performance of stereotactic body radiation therapy *Am. J. Clin. Oncol.* **43** 545–52

[2] American College of RadiologyAmerican Society for Radiation Oncology 2018 *ACR–ASTRO Practice Parameter for Radiation Oncology*

[3] American Association of Physicists in Medicine 2018 Definition of a qualified medical physicist https://aapm.org/org/policies/details.asp?id=449&type=PP

[4] Halvorsen P H, Cirino E and Das I J *et al* 2017 AAPM-RSS medical physics practice guideline 9.a. for SRS-SBRT *J. Appl. Clin. Med. Phys.* **18** 10–21

[5] Benedict S H, Yenice K M and Followill D *et al* 2010 Stereotactic body radiation therapy: the report of AAPM Task Group 101 *Med. Phys.* **37** 4078–101

[6] Parker W and Patrocinio H 2005 Clinical Treatment Planning in External Photon Beam Radiotherapy *Review of Radiation Oncology Physics: A Handbook for Teachers and Students* (Vienna: IAEA)

[7] Dieterich S and Pawlicki T 2008 Cyberknife image-guided delivery and quality assurance *Int. J. Radiat. Oncol. Biol. Phys.* **71** S126–30

[8] Mutic S, Palta J R and Butker E K *et al* 2003 Quality assurance for computed-tomography simulators and the computed-tomography-simulation process: report of the AAPM Radiation Therapy Committee Task Group No. 66 *Med. Phys.* **30** 2762–92

[9] Tyagi N, Fontenla S and Zelefsky M *et al* 2017 Clinical workflow for MR-only simulation and planning in prostate *Radiat. Oncol.* **12** 119

[10] Molineu A, Hernandez N, Nguyen T, Ibbott G and Followill D 2013 Credentialing results from IMRT irradiations of an anthropomorphic head and neck phantom *Med. Phys.* **40** 022101

[11] Papanikolaou N, Battista J and Boyer A *et al* 2004 *Tissue Inhomogeneity Corrections for Megavoltage Photon Beams* AAPM Report No. 85 American Association of Physicists in Medicine

[12] Smilowitz J B, Das I J and Feygelman V *et al* 2015 AAPM medical physics practice guideline 5.a.: commissioning and QA of treatment planning dose calculations—megavoltage photon and electron beams *J. Appl. Clin. Med. Phys.* **16** 14–34

[13] Abbas H, Chang B and Chen Z J 2014 Motion management in gastrointestinal cancers *J. Gastrointest. Oncol.* **5** 223–35

[14] Keall P J, Aun N J and O'Brien R *et al* 2015 The first clinical treatment with kilovoltage intrafraction monitoring (KIM): a real-time image guidance method *Med. Phys.* **42** 354–8

[15] Ng J A, Booth J T and Poulsen P R *et al* 2012 Kilovoltage intrafraction monitoring for prostate intensity modulated arc therapy: first clinical results *Int. J. Radiat. Oncol. Biol. Phys.* **84** e655–61

[16] Poulsen P R, Cho B and Keall P J 2008 A method to estimate mean position, motion magnitude, motion correlation, and trajectory of a tumor from cone-beam CT projections for image-guided radiotherapy *Int. J. Radiat. Oncol. Biol. Phys.* **72** 1587–96

[17] Poulsen P R, Cho B, Langen K, Kupelian P and Keall P J 2008 Three-dimensional prostate position estimation with a single X-ray imager utilizing the spatial probability density *Phys. Med. Biol.* **53** 4331–53

[18] Hunt M A, Sonnick M and Pham H *et al* 2016 Simultaneous MV-kV imaging for intrafractional motion management during volumetric-modulated arc therapy delivery *J. Appl. Clin. Med. Phys.* **17** 473–86

[19] Zeng C, Xiong W and Li X *et al* 2019 Intrafraction tumor motion during deep inspiration breath hold pancreatic cancer treatment *J. Appl. Clin. Med. Phys.* **20** 37–43

[20] Ozhasoglu C, Saw C B and Chen H *et al* 2008 Synchrony–CyberKnife respiratory compensation technology *Med. Dosim.* **33** 117–23

[21] Bertholet J, Toftegaard J and Hansen R *et al* 2018 Automatic online and real-time tumour motion monitoring during stereotactic liver treatments on a conventional Linac by combined optical and sparse monoscopic imaging with kilovoltage X-rays (COSMIK) *Phys. Med. Biol.* **63** 055012

[22] James J, Cetnar A and Dunlap N E *et al* 2016 Technical note: validation and implementation of a wireless transponder tracking system for gated stereotactic ablative radiotherapy of the liver *Med. Phys.* **43** 2794–801

[23] Metz J M, Kassaee A and Ingram M *et al* 2009 First report of real-time tumor tracking in the treatment of pancreatic cancer using the Calypso System *Int. J. Radiat. Oncol. Biol. Phys.* **75** S54–5

[24] Poulsen P R, Worm E S, Hansen R, Larsen L P, Grau C and Høyer M 2015 Respiratory gating based on internal electromagnetic motion monitoring during stereotactic liver radiation therapy: first results *Acta Oncol.* **54** 1445–52

[25] Shinohara E T, Kassaee A and Mitra N *et al* 2012 Feasibility of electromagnetic transponder use to monitor inter- and intrafractional motion in locally advanced pancreatic cancer patients *Int. J. Radiat. Oncol. Biol. Phys.* **83** 566–73

[26] Worm E S, Høyer M and Hansen R *et al* 2018 A prospective cohort study of gated stereotactic liver radiation therapy using continuous internal electromagnetic motion monitoring *Int. J. Radiat. Oncol. Biol. Phys.* **101** 366–75

[27] Omari E A, Erickson B and Ehlers C *et al* 2016 Preliminary results on the feasibility of using ultrasound to monitor intrafractional motion during radiation therapy for pancreatic cancer *Med. Phys.* **43** 5252

[28] Schlosser J, Gong R H and Bruder R *et al* 2016 Robotic intrafractional US guidance for liver SABR: system design, beam avoidance, and clinical imaging *Med. Phys.* **43** 5951

[29] Bazalova-Carter M, Schlosser J, Chen J and Hristov D 2015 Monte Carlo modeling of ultrasound probes for image guided radiotherapy *Med. Phys.* **42** 5745–56

[30] Schlosser J and Hristov D 2016 Radiolucent 4D ultrasound imaging: system design and application to radiotherapy guidance *IEEE Trans. Med. Imaging* **35** 2292–300

[31] Şen H T, Lediju Bell M A and Zhang Y *et al* 2015 System integration and preliminary in-vivo experiments of a robot for ultrasound guidance and monitoring during radiotherapy *Proc. Int. Conf. Adv. Robot.* **2015** 53–9

[32] Zhong Y, Stephans K, Qi P, Yu N, Wong J and Xia P 2013 Assessing feasibility of real-time ultrasound monitoring in stereotactic body radiotherapy of liver tumors *Technol. Cancer Res. Treat.* **12** 243–50

[33] Boldrini L, Cusumano D, Cellini F, Azario L, Mattiucci G C and Valentini V 2019 Online adaptive magnetic resonance guided radiotherapy for pancreatic cancer: state of the art, pearls and pitfalls *Radiat. Oncol.* **14** 71

[34] Feldman A M, Modh A, Glide-Hurst C, Chetty I J and Movsas B 2019 Real-time magnetic resonance-guided liver stereotactic body radiation therapy: an institutional report using a magnetic resonance-Linac system *Cureus* **11** e5774

[35] Klein E E, Hanley J and Bayouth J *et al* 2009 Task Group 142 report: quality assurance of medical accelerators *Med. Phys.* **36** 4197–212

[36] Ezzell G A, Burmeister J W and Dogan N *et al* 2009 IMRT commissioning: multiple institution planning and dosimetry comparisons, a report from AAPM Task Group 119 *Med. Phys.* **36** 5359–73

[37] Miften M, Olch A and Mihailidis D *et al* 2018 Tolerance limits and methodologies for IMRT measurement-based verification QA: recommendations of AAPM Task Group No. 218 *Med. Phys.* **45** e53–e83

[38] MD Anderson Imaging and Radiology Oncology Core (IROC) Houston Quality Assurance Center http://rpc.mdanderson.org/RPC/home.htm

[39] Low D A and Dempsey J F 2003 Evaluation of the gamma dose distribution comparison method *Med. Phys.* **30** 2455–64

[40] Rangaraj D, Zhu M and Yang D *et al* 2013 Catching errors with patient-specific pretreat-ment machine log file analysis *Pract. Radiat. Oncol.* **3** 80–90

[41] Fuangrod T, Greer P B and Woodruff H C *et al* 2016 Investigation of a real-time EPID-based patient dose monitoring safety system using site-specific control limits *Radiat. Oncol.* **11** 106

[42] Rozendaal R A, Mijnheer B, Hamming-Vrieze O, Mans A and van Herk M 2015 Impact of daily anatomical changes on EPID-based in vivo dosimetry of VMAT treatments of head-and-neck cancer *Radiother. Oncol.* **116** 70–4

[43] Rozendaal R A, Mijnheer B J, van Herk M and Mans A 2014 In vivo portal dosimetry for head-and-neck VMAT and lung IMRT: linking gamma-analysis with differences in dose-volume histograms of the PTV *Radiother. Oncol.* **112** 396–401

[44] Woodruff H C, Fuangrod T and Van Uytven E *et al* 2015 First experience with real-time EPID-based delivery verification during IMRT and VMAT sessions *Int. J. Radiat. Oncol. Biol. Phys.* **93** 516–22

[45] Bakai A, Alber M and Nüsslin F 2003 A revision of the gamma-evaluation concept for the comparison of dose distributions *Phys. Med. Biol.* **48** 3543–53

[46] Low D A, Harms W B, Mutic S and Purdy J A 1998 A technique for the quantitative evaluation of dose distributions *Med. Phys.* **25** 656–61

Chapter 6

Quality assurance of image-guided radiation therapy

Bo Liu, Ke Nie and Taoran Cui

In this chapter, we discuss the quality assurances (QAs) used in the different aspects of abdominal image-guided radiation therapy (IGRT). We first introduce the QAs of the motion mitigation and motion assessment techniques commonly used during simulation. We then detail the necessary QAs of target delineation, image registration, and treatment planning system in order to generate accurate and precise treatment plans for abdominal IGRT. We also review the appropriate QAs of various IGRT systems to ensure accurate treatment delivery of IGRT, such as radiographic, non-radiographic, and hybrid systems. Finally, the QAs of IGRT techniques used in the emerging MRI-based radiation therapy and proton therapy are mentioned.

6.1 Introduction

Abdominal malignancies represent nearly the most challenging cases in radiation treatment. The major issues are organ movement due to respiratory motion, gastrointestinal filling or peristalsis, and the proximity of target to normal organs that are highly sensitive to radiation. The requirements of high precision in dose delivery during motion, and high conformity to spare surrounding organs, have driven the integration of the advanced imaging technologies and tracking systems into abdominal radiation.

While image-guided radiation treatment (IGRT) substantially reduces positioning errors and enables more conformal treatment, its application also heavily relies on the consistent performance of the system. This in turn imposes rigorous protocols of quality assurance (QA) to meet the high level of accuracy that surpasses the

doi:10.1088/978-0-7503-2468-7ch6

requirements of conventional radiation treatment. Hence, in this chapter, we give an overview of common IGRT technologies used in abdominal radiation treatment and their associated QA programs.

6.2 Imaging technology in simulation

As one of the major challenges for abdominal radiation therapy, organ motion should be mitigated in the simulation process to reduce its effects on image acquisition and dose distribution. The extent of motion should be evaluated with simulation images as well and must be accounted for in treatment planning. Hence, comprehensive QAs of the motion mitigation and motion assessment techniques should be performed to ensure their implementations with high accuracy.

6.2.1 Motion mitigation

6.2.1.1 Abdominal compression

Abdominal compression (AC) is a method to mitigate respiratory motion by applying pressure to the abdomen to reduce diaphragmatic excursion. It is widely applied in radiation therapy for thoracic and abdominal lesions, especially in stereotactic body radiation therapy (SBRT) [1–5]. Yet, the dosimetric impact of AC should be carefully evaluated because of beam perturbation. This includes additional beam attenuation, increased skin dose for photons beams, and reduced distal range for charged particle beams. Although a common approach is to prevent beams directly passing through AC devices, this may not always be possible because of the relative location of targets to the AC device. As recommended by AAPM report No. 176 [6], immobilization devices should be included within the scanning field of view (FOV) during CT acquisition and for dose calculation. The real CT number instead of a bulk density should be used for dose calculation.

In addition, actual physical measurements are highly recommended to evaluate the dosimetric impact of an AC device prior to the clinical use. For photon beams, a point measurement using an ionization chamber in a cylindrical phantom can be used to determine the attenuation of an AC device. The measurement geometries should include most probable gantry angles and have beams passing through the maximum attenuation parts of an AC device [6]. The measured data should be compared with the calculated one using the same geometry to validate the treatment planning system (TPS) modeling of the AC device, as shown in figure 6.1. Regarding proton beams, an AC device may reduce the distal range. Hence, it is critical for the TPS to accurately model the water equivalent thickness (WET) of an AC device and to determine the distal range of proton beams in the presence of the AC device. The measurement can be performed by scanning the depth variation of a proton beam with a parallel plate chamber. The range difference between the measurements with and without the AC device is equivalent as the measured WET of the AC device, which can be used to validate the calculated WET in the TPS. The WET of an AC device can also be determined using radiochromic films or multilayer ionization chambers (e.g., Zebra, IBA Dosimetry GmbH., Schwarzenbruck, Germany).

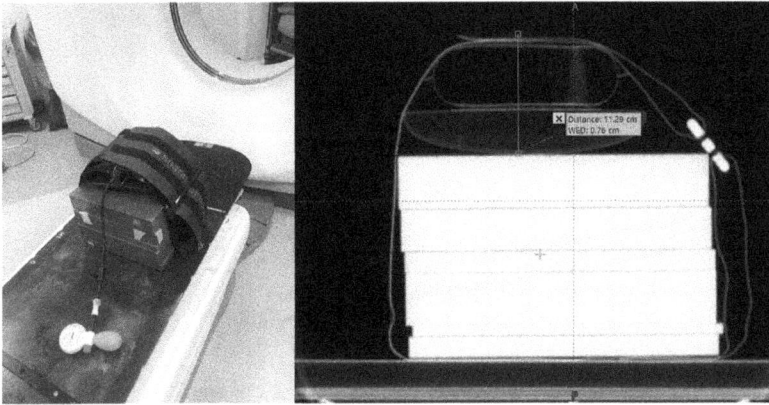

Figure 6.1. ZiFix Abdominal Compression belt (Qfix, Avondale, PA) and its WET modeled in treatment planning system.

6.2.1.2 ABC

An active breathing control (ABC) device [7] is an effective method to manage a patient's respiratory motion. The ABC device enables clinicians to better help a patient to hold his/her breath at a desired phase, usually at the end of exhalation for liver [8–11] or pancreas radiotherapy [12, 13], in order to achieve a reproducible management of both intrafraction and interfraction respiratory motions.

One important component of the QA program for the ABC device is to ensure the reproducibility of breath-hold through patient-related QA. It is recommended to provide patients an education session prior to the simulation in order to familiarize patients with the equipment and procedure and to establish the baselines for the amount of exhaled air and the breath-hold duration [11, 14]. The baselines should be chosen so that the breath-hold duration is long enough to reduce the number of interruptions during the treatment, but not so long that it distresses the patients. The integrity of mouthpiece and nose clip should be verified prior to each ABC session to ensure that patients can only breathe through the mouthpiece. Multiple image studies with breath-hold should be acquired at the simulation to examine the breath-hold reproducibility. KV fluoroscopy, CT, or fast cine-MRI are commonly used. The positions of tumor or implanted fiducials should be compared among different acquisitions. If the tumor is not visible due to its location or image contrast, other nearby surrogates, such as diaphragm and liver, could also be used for reproducibility evaluation.

Similar to other devices used in radiation therapy, a rigorous equipment-related QA program consisting of commissioning and routine QAs should also be established to ensure the accuracy of ABC devices. AAPM Task Group 76 [15] provides detailed recommendations on what should be included in equipment-related QA. The main goal is to calibrate the key functionalities of an ABC device: the accurate measurement of airflow using a spirometer and the ability to stop and resume airflow with a balloon valve. The spirometer and balloon valve can be calibrated either by manually withdrawing a known amount of air with a syringe or

Figure 6.2. One commercially available QA device (Aktina Medical, Congers, NY) for ABC that controls the amount of air injected to the spirometer of the ABC device. © Aktina Medical Group.

by using a vendor-provided calibration device with a piston driven by a linear actuator. One commercially available QA device made by Aktina Medical is shown in figure 6.2. As recommended in AAPM Task Group 76, the baseline calibrations of spirometer with different airflow rates should be performed, to reflect physiological airflow rates seen in patients [15].

6.2.2 Motion assessment

6.2.2.1 4D-CT

4D-CT has become the standard of practice for the treatment planning of tumors in the thorax and abdomen in most clinics. 4D-CT can be reconstructed by acquiring multiple 3D CT scans at different respiratory phases and resorting images according to the breathing cycles. 4D-CT can be used for target delineation, tumor motion assessment, and respiratory gating window determination, as well as dose calculation. To ensure the accuracy of 4D-CT and quantify the potential errors of clinically implementing this technology, routine QA program of 4D-CT should be established.

The routine QA of 4D-CT usually involves the use of a motion phantom with programmable moving amplitude and period. The QA program should include an analysis of target volume at various respiratory phases, CT number consistency between moving and static phantom, target position at certain respiratory phases, and measured motion amplitude against the known value. In addition, 4D-CT image quality should also be routinely checked [16]. A commercially available phantom designed for 4D-CT QA is shown in figure 6.3.

While there is a lack of guidelines on 4D-CT quality assurance from AAPM, the Canadian Organization of Medical Physicists (COMP) and the Canadian Partnership of Quality Radiotherapy (CPQR) published technical quality control guidelines for CT simulators with specific recommendations on routine quality assurance of 4D-CT [16]. In this COMP report, the functionality of respiratory monitoring system and audio/video coaching systems (if applicable) need to be checked daily. For systems that use external respiratory signals, the ability of the

Figure 6.3. Quasar respiratory motion phantom designed for 4D-CT QA (Modus QA, London, Ontario, Canada). The insert of the phantom can be driven with a motor to simulate respiratory motion. © Modus Medical Devices.

respiratory monitoring system to accurately monitor the external surrogate should be checked with a moving phantom quarterly. The quarterly QA should include tests on the amplitude of a moving target measured with 4D-CT, the spatial integrity and positioning of moving targets, the mean CT number and standard deviation of moving targets at each phase, and post-processed image creation for treatment planning (e.g., Average Intensity projection, Maximum Intensity Projection). 4D-CT image quality, which includes 4D low-contrast resolution, 4D high-contrast spatial resolution, and 4D slice thickness at each respiratory phase, should be checked annually [16].

6.3 Imaging technology in treatment planning

While IGRT provides high accuracy for target localization and reduced target margin for inter-/intrafraction motions, it is also essential to that IGRT cooperate with conformal treatment plans to precisely focus radiation to the target. Compared with conventional 3D conformal techniques used in pre-IGRT eras, advanced intensity-modulated radiotherapy (IMRT) or volumetric modulated arc therapy

(VMAT) plans are created with high conformality and steep dose gradient to reduce excessive radiation to organs at risk (OARs). However, owing to their complexity, intensity-modulated plans are more error-prone and thus require comprehensive QA programs, especially when integrating with IGRT.

6.3.1 Target delineation and definition

Given the presence of both tumor and adjacent OAR motions in abdominal radio-therapy caused by either respiration or changes in the filling status or the internal structure of OARs [17], ICRU 62 introduced the concept of the internal margin to account for the anatomical variation of a target and recommended to create an internal target volume (ITV) by expanding a clinical target volume (CTV) with the internal margin for treatment planning. The definition of ITV and internal margin are usually performed on 4D-CTs, which reflect any anatomical variation due to respiratory motion. Therefore, the QA of 4D-CTs should be rigorously performed in order to achieve optimal image quality for the evaluation of patient-specific respiratory motion and therefore to determine the appropriate ITV and internal margin for treatment planning. The details have been discussed in the previous chapter.

An ITV can be created either by combining CTVs contoured on each individual phase of 4DCT or by directly contouring an ITV on the maximum intensity projection (MIP) scan of 4D-CT [18, 19]. Either approach requires the TPS to accurately generate the combination of CTVs or the MIP of 4D-CT. Therefore, it is recommended to contour a target of known geometric undergoing a well-defined motion pattern and compare the delineation with the ground truth. While geometric systematic errors of the TPS can be investigated and managed through rigorous QA programs, it has been shown that inter-observer variations in target delineation may introduce more errors [20], and studies have reported these inter-observer variations which may impair the outcomes of radiotherapy [21].

CT has been widely used as the image modality for target delineation and definition in radiotherapy treatment planning due to its high geometry fidelity to provide accurate localization of targets and OARs, as well as its ability to create electron density map for radiation dose calculation [22]. Given the extensive involvements of soft issue, such as liver, stomach, duodenum, and pancreas during abdominal radiotherapy, CT without contrast, known for suboptimal soft tissue contrast due to the similar electron densities to water, is usually acquired in conjunction with other image modalities. Besides CT with barium-based contrast, which is often used to enhance the image quality of stomach, esophagus, and small/large intestines, MRI has been routinely used to assist target and OAR delineation because of its superior soft contrast, especially when treating hepatic and pancreatic cancers [23]. However, it is known that MRI suffers from its inherent geometric inaccuracies caused by magnetic field inhomogeneity, gradient nonlinearity, etc. Therefore, both ACR and AAPM published several guidelines for the QA program of MRI. Specifically, the upcoming AAPM TG-117 report focuses on the application of MRI in RT TPS and proposes the proper QA program required to achieve high spatial accuracy.

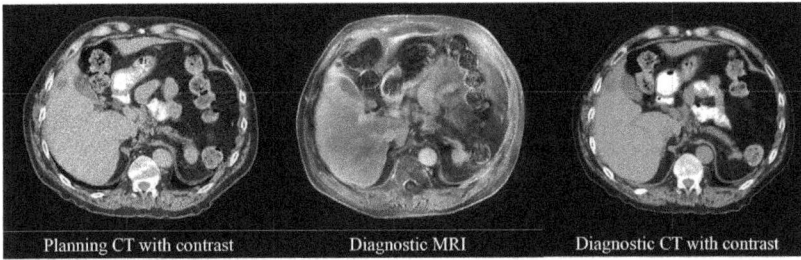

| Planning CT with contrast | Diagnostic MRI | Diagnostic CT with contrast |

Figure 6.4. Image registrations between the planning CT, the prior diagnostic MRI, and the diagnostic CT for a patient with liver cancer.

6.3.2 Image registration

Despite involving efforts towards MRI-only simulation, the standard clinical practice still relies on CT for planning purposes, when electron densities of different tissues are well known for dose computation. Soft tissue structures delineated on other diagnostic modalities such as MRI or PET then need to be transferred to planning CT, as demonstrated in figure 6.4. This transferring requires image registrations to establish voxel-to-voxel correlations between two images.

Rigid image registration is commonly used and clinically acceptable with the assumptions that a patient is a rigid body and there are no internal anatomical deformations [24]. On the contrary, it is common to observe large anatomical deformation when registering diagnostic images with CT. Diagnostic images are often acquired prior to planning CT with a curved couch top and different patient positions. These inconsistencies between the diagnostic images and planning CT may result in difficulties when performing rigid image registration. Alternatively, deformable image registration can be used for image registration in the presence of large deformations However, since there is yet to be a consensus on whether it should be clinically accepted, deformable image registration should be used with extra caution [25, 26].

The result of registration can be visually examined by verifying the alignment of the same anatomical structures between two images. The image registration is usually performed with TPSs or commercial image management application. The performance of the software should be investigated to ensure consistent accuracy. AAPM TG-132 provides a detailed recommendation on the QA program for image registration and suggests end-to-end tests of image registration system using a physical phantom [27]. Also, it is recommended to evaluate any potential registration errors during a comprehensive commissioning of image registration system and to account for the errors in the setup margin during treatment planning.

6.3.3 Treatment planning system

Because of the complex involvements of multiple OARs in abdominal IGRT, highly modulated IMRT and VMAT plans are often required to achieve OARs sparing while maintaining target coverage. Unlike the traditional 3D-CRT plans, these

IMRT/VMAT plans contain more degrees of freedom and can be only generated through an inversed planning process with the optimization of an objective function to determine the optimal balanced dose coverage and OARs sparing. In order to assure the quality of IMRT/VMAT plans, a comprehensive QA program of TPS should be developed for each step of treatment planning in the TPS, including (1) accurate machine commissioning and beam modeling and (2) verification of TPS dose calculation and optimization algorithms. AAPM has published a series of guidelines on the QA program of TPS, from the general recommendations on commissioning, TG-106 [28] and TG-157 [29], to specific IMRT commissioning, TG-119 [30] and dose calculation [31].

6.4 Imaging technology in treatment delivery

Accurate dose delivery during abdominal radiotherapy is challenging due to interfraction motions such as day-to-day setup and anatomical variations due to weight loss or tumor progression/shrinkage. In addition, intrafraction changes caused by bladder filling/peristalsis, or fast motions due to respiratory or cardiac activities, are also frequently present during the treatments. The demand for accurate localization of radiation to account for abdominal motion has led to the development of numerous image-guided techniques for treatment delivery.

6.4.1 Radiographic system

Image-based methods using kilovoltage (kV) and/or megavoltage (MV) x-ray are a natural development of IGRT extending the use of in-room imaging to during-treatment situations. Commonly used radiographic systems include (1) electronic portal imaging detectors (EPIDs); (2) cone-beam CT with KV or MV imaging; (3) fan-beam KVCT; and (4) fan-beam MVCT.

Conventional linear accelerator (Linac) systems consist of retractable amorphous silicon x-ray detectors mounted either orthogonal to the treatment beam axis (e.g., Eleka Versa HD, Varian Truebeam for KV) or along the treatment beam axis (e.g., Varian Truebeam, Siemens Artise for MV). All these commercial configurations are capable of radiographic, fluoroscopic, and volumetric CBCT. Commissioning and routine QA recommendations are given in the report of AAPM TG-179 [32] and AAPM TG-58 [33]. Another in-room imaging system is CT-on-rail, which moves a CT scanner across a patient instead of moving patients into the scanner. The commissioning procedure for an in-room CT scanner is similar to those described in the AAPM Report No.39 [34] and AAPM TG-66 [35]. Routine QA recommendations for in-room CT scanners are described in the report of AAPM TG-179 as well [32]. The TomoTherapy® system has a unique design in that a single linear accelerator is used to generate both imaging beam and treatment beams. It is capable of acquiring megavoltage CT (MVCT) images of patients in the treatment position. The AAPM TG-148 [36] and TG-179 [32] contain the recommendations on MVCT-specific QA tests and their frequencies.

The primary checks of an imaging system include a geometric accuracy check, which includes the alignment between imaging and radiation isocenters, and an image quality check, which includes QAs of scale and distance accuracy, spatial resolution, uniformity, signal-to-noise ratio, etc. If the images acquired by a system are also used for dose calculation, the accuracy of CT numbers should be investigated. Since ionizing radiation images are frequently used to guide abdominal radiation, they may result in excessive radiation dose to patients and potentially increase the risks of radiation injury and secondary cancer [37]. Therefore, the imaging dose should be managed to reach a cost–benefit balance for its usage in radiotherapy. As recommended by AAPM TG-180 [38], the imaging dose should be included in the total dose received by patients if the imaging dose is anticipated to exceed 5% of the prescribed dose. Planar imaging dose measurement can be performed in air following the AAPM protocol for 40–300 kV x-ray beam dosimetry [39]. Imaging dose of MV imaging systems, including portal MV imaging, MV CT, and MV CBCT, can be performed following the AAPM MV dosimetry protocol [40, 41]. Consistency checks on imaging dose and imaging parameters should be performed annually and after system upgrades per AAPM TG-142 [42].

6.4.2 Non-radiographic system

The non-radiographic motion tracking systems currently used in clinics include (1) infrared system, (2) optical system, and (3) radiofrequency system. Additionally, some of these systems provide integration with radiation delivery machines, for example, to gate treatment beams or track a tumor using a moving MLC or couch. Hence, tracking system functionality and its integration with treatment and verification systems should be included when designing QA protocols for these systems.

An infrared system, such as the Varian RPM system (Varian, CA), is used to track tumor motion in the thorax and abdomen. The basic idea is that a camera captures the breathing signals from one or more reflector indicators placed on patient surface. These positioning systems are connected to a 6 degree-of-freedom couch and are capable of beam interruption and patient repositioning during treatment with submillimeter accuracy [43, 44]. In addition to a points-based system, a surface-based monitoring system has also been developed. One of the examples is the optical surface monitoring system. It consists of three ceiling-mounted pods (each containing two camera sensors and a projector) to project a red light speckle onto the patient, as shown in figure 6.5. With signals detected from all three camera pods, a 3D surface image can be reconstructed. By progressively comparing the acquired surface with reference surface, patients' shifts can be reported and updated. Some surface-based monitoring systems are able to automatically trigger beam-hold when the acquired surface and the reference surface is beyond a preset tolerance. In addition, these systems can be used to guide patient repositioning without the need for x-ray imaging. Yet, one major concern is that this system only tracks the external motion, and an additional system may be necessary to determine correlation between the alignment of the patient surface and the

Figure 6.5. A cube phantom exposed by the red speckle patterns to align the surface imaging coordinate to the radiographic coordinates.

alignment of the internal target. Additionally, the accuracy of surface mapping may be affected by in-room lighting and/or the reflectivity and color of patients' clothing or skin tone [45].

Another system that can provide continuous monitoring is a radiofrequency (RF) system such as the Calypso® 4D Localization System™ (Calypso Medical Technologies, Seattle, WA). Its use for internal target localization is based on the system's detection of electromagnetic signals generated by markers called Beacon transponders. Those markers, with the size of a small grain of rice, are implanted into a patient's body before radiation. When Calypso detects transponder motion, it sends out an alert and holds the treatment beam until the target is back in range.

It is important to note that all mentioned systems are of high precision and require stringent QA checks to ensure proper operations. AAPM Task group 147 gives guidance/recommendations in the QA for non-radiographic radiotherapy localization and positioning systems [46]. Briefly, the QA program should include:

(1) identification of localizing field of view by identifying the regions where the system may work/fail to detect motion;
(2) consistency check to avoid measurement drifting during repeated measurements;
(3) detection accuracy in both static and dynamic situation; the localization accuracy should be within 2 mm of isocenter for standard dose

fractionations and 1 mm for SBRT and SRS treatments in following recommendations from TG-142 [42]; and

(4) integration with peripheral equipment such as transferring data from/into treatment record and verification system; and, very importantly, if the system is used in real-time gating, a full-scale end-to-end test is needed to ensure the synchronization.

6.4.3 Hybrid system

CyberKnife is a hybrid system that uses both a radiograph imaging system and an infrared camera system to monitor and track motion. The system can monitor the target position by co-registering two simultaneously acquired intra-treatment radiographs every 10–20 s to CT-generated digital reconstructed radiographs. The Synchrony system also provides a vest fitted with light-emitting diode (LED) markers that are monitored by three ceiling-mounted cameras at 20–40 Hz [47]. A motion correlation model is built that relates the internal fiducial marker motion to the LED markers' motion. The LED markers' motion is continuously recorded during radiation delivery to infer the internal marker positions [48].

The recommended IGRT QA tests and frequencies for CyberKnife are documented in AAPM TG-135 [49]. In brief, the QA program to evaluate the image system should consider geometry accuracy, exposure reproducibility, positioning reproducibility, and image quality checks. Regarding x-ray generator and sources, because the x-ray machines used for targeting in the CyberKnife® system are essentially unmodified conventional x-ray generators and x-ray tube configurations, the QA principles and procedures described in AAPM Report of Task Group No.14 [50] and Report of Task Group No. 61 [39] can be applied. The imaging dose can be checked using the method described in AAPM Report of Task Group No. 75 [37] and Task Group 135 [49] for the CyberKnife® system.

6.5 New evolving technologies

6.5.1 MRI-based radiation system

Recently, MRI has been incorporated into radiotherapy treatment units with MR-guided radiotherapy (MRgRT) systems for abdominal radiotherapy. This increases the targeting precision particularly for soft tissues in abdomen to which CT provides insufficient contrast. Several hybrid MRgRT systems that integrate either Linac or Co-60 for radiation delivery are commercially available for clinical use. Both systems rely critically on high image qualities to provide adequate image guidance; therefore, a rigorous QA of the MRI system is essential. Yet to date, few data have been published on the imaging performance of MRgRT systems, nor do there exist any formal guidelines. Recently, formal guidelines on the use of MRI in radiotherapy are being developed by working groups within the AAPM, IPEM, and NCS [51]. Until these guidelines are published, the recommended QC and QA tests for diagnostic MRI described in the guidance documents of AAPM Report No.100 [52] and the ACR Magnetic resonance imaging quality control manual [53] should be considered with necessary adaptions to meet radiotherapy-specific requirements. For MRgRT systems, it is vital that the static magnetic field strength is accurately known, spatially homogeneous, and stable over time, to accurately determine the spatial localization. A thorough geometrical distortion of the scanner should also be characterized. The scanning orientation needs to match the TPS, and the amount of ghosting needs to be checked. Also, very importantly, the interferences of the RF signals with radiation pulsing need to be validated [54].

6.5.2 IGRT in proton treatment

In proton therapy, inline motion and anatomical changes along the beam path may result in large dosimetric errors that cannot be completely compensated with the use of margins [48]. As compared with photon therapy, proton is more vulnerable to sources of uncertainties due to its sharp distal fall-off [55]. Therefore, the image guidance system plays a critical role in the accurate delivery of proton treatments, and thus, it needs to be carefully checked to ensure geometrical accuracy and anatomical reproducibility. Planar kV imaging has been the most widely used image guidance technology for proton therapy. Most recently, volumetric imaging, which includes kV-CBCT and CT-on-rails, became available for proton therapy, especially for the more advanced pencil beam scanning technology. Although there are no guidelines specifically on the QA of image guidance systems for proton therapy, recommendations for kV planar imaging, kV-CBCT and CT-on-rails from AAPM TG-142 [42], AAPM TG-179 [32], and MPPG 2a [56] are comprehensive and can be directly applied to a proton therapy setting.

References

[1] Eccles C L, Dawson L A, Moseley J L and Brock K K 2011 Interfraction liver shape variability and impact on GTV position during liver stereotactic radiotherapy using abdominal compression *Int. J. Radiat. Oncol. Biol. Phys.* **80** 938–46

[2] Mampuya W A, Nakamura M, Matsuo Y, Ueki N, Iizuka Y and Fujimoto T *et al* 2013 Interfraction variation in lung tumor position with abdominal compression during stereotactic body radiotherapy *Med. Phys.* **40** 091718

[3] Hu Y, Zhou Y K, Chen Y X, Shi S M and Zeng Z C 2016 4D-CT scans reveal reduced magnitude of respiratory liver motion achieved by different abdominal compression plate positions in patients with intrahepatic tumors undergoing helical tomotherapy *Med. Phys.* **43** 4335

[4] Hu Y, Zhou Y K, Chen Y X and Zeng Z C 2017 Magnitude and influencing factors of respiration-induced liver motion during abdominal compression in patients with intrahepatic tumors *Radiat. Oncol.* **12** 9

[5] Lin L, Souris K, Kang M, Glick A, Lin H and Huang S *et al* 2017 Evaluation of motion mitigation using abdominal compression in the clinical implementation of pencil beam scanning proton therapy of liver tumors *Med. Phys.* **44** 703–12

[6] Olch A J, Gerig L, Li H, Mihaylov I and Morgan A 2014 Dosimetric effects caused by couch tops and immobilization devices: report of AAPM Task Group 176 *Med. Phys.* **41** 061501

[7] Wong J W, Sharpe M B, Jaffray D A, Kini V R, Robertson J M and Stromberg J S *et al* 1999 The use of active breathing control (ABC) to reduce margin for breathing motion *Int. J. Radiat. Oncol. Biol. Phys.* **44** 911–9

[8] Dawson L A, Brock K K, Kazanjian S, Fitch D, McGinn C J and Lawrence T S *et al* 2001 The reproducibility of organ position using active breathing control (ABC) during liver radiotherapy *Int. J. Radiat. Oncol. Biol. Phys.* **51** 1410–21

[9] Dawson L A, Eccles C, Bissonnette J P and Brock K K 2005 Accuracy of daily image guidance for hypofractionated liver radiotherapy with active breathing control *Int. J. Radiat. Oncol. Biol. Phys.* **62** 1247–52

[10] Eccles C, Brock K K, Bissonnette J P, Hawkins M and Dawson L A 2006 Reproducibility of liver position using active breathing coordinator for liver cancer radiotherapy *Int. J. Radiat. Oncol. Biol. Phys.* **64** 751–9

[11] Lu L, Diaconu C, Djemil T, Videtic G M M, Abdel-Wahab M and Yu N *et al* 2018 Intra- and inter-fractional liver and lung tumor motions treated with SBRT under active breathing control *J Appl. Clin. Med. Phys.* **19** 39–45

[12] Nakamura M, Shibuya K, Shiinoki T, Matsuo Y, Nakamura A and Nakata M *et al* 2011 Positional reproducibility of pancreatic tumors under end-exhalation breath-hold conditions using a visual feedback technique *Int. J. Radiat. Oncol. Biol. Phys.* **79** 1565–71

[13] Su L, Iordachita I, Zhang Y, Lee J, Ng S K and Jackson J *et al* 2017 Feasibility study of ultrasound imaging for stereotactic body radiation therapy with active breathing coordinator in pancreatic cancer *J Appl. Clin. Med. Phys.* **18** 84–96

[14] Jiang S B, Wolfgang J and Mageras G S 2008 Quality assurance challenges for motion-adaptive radiation therapy: gating, breath holding, and four-dimensional computed tomography *Int. J. Radiat. Oncol. Biol. Phys.* **71** S103–7

[15] Keall P J, Mageras G S, Balter J M, Emery R S, Forster K M and Jiang S B *et al* 2006 The management of respiratory motion in radiation oncology report of AAPM Task Group 76 *Med. Phys.* **33** 3874–900

[16] Després P and Gaede S 2018 COMP report: CPQR technical quality control guidelines for CT simulators *J Appl. Clin. Med. Phys.* **19** 12–17

[17] Abbas H, Chang B and Chen Z J 2014 Motion management in gastrointestinal cancers *J. Gastrointest. Oncol.* **5** 223–35

[18] Underberg R W M, Lagerwaard F J, Slotman B J, Cuijpers J P and Senan S 2005 Use of maximum intensity projections (MIP) for target volume generation in 4DCT scans for lung cancer *Int. J. Radiat. Oncol. Biol. Phys.* **63** 253–60

[19] van Dam I E, van Sörnsen de Koste J R, Hanna G G, Muirhead R, Slotman B J and Senan S 2010 Improving target delineation on 4-dimensional CT scans in stage I NSCLC using a deformable registration tool *Radiother. Oncol.* **96** 67–72

[20] Gwynne S, Spezi E, Sebag-Montefiore D, Mukherjee S, Miles E and Conibear J *et al* 2013 Improving radiotherapy quality assurance in clinical trials: assessment of target volume delineation of the pre-accrual benchmark case *Br. J. Radiol.* **86** 20120398

[21] Cox S, Cleves A, Clementel E, Miles E, Staffurth J and Gwynne S 2019 Impact of deviations in target volume delineation – time for a new RTQA approach? *Radiother. Oncol.* **137** 1–8

[22] Davis A T, Palmer A L and Nisbet A 2017 Can CT scan protocols used for radiotherapy treatment planning be adjusted to optimize image quality and patient dose? A systematic review *Br. J. Radiol.* **90** 20160406

[23] Lukovic J, Henke L, Gani C, Kim T K, Stanescu T and Hosni A *et al* 2020 MRI-based upper abdominal organs-at-risk atlas for radiation oncology *Int. J. Radiat. Oncol. Biol. Phys.* **106** 743–53

[24] Rosenman J G, Miller E P, Tracton G and Cullip T J 1998 Image registration: an essential part of radiation therapy treatment planning *Int. J. Radiat. Oncol. Biol. Phys.* **40** 197–205

[25] Velec M, Moseley J L, Hårdemark B, Jaffray D A and Brock K K 2017 Validation of biomechanical deformable image registration in the abdomen, thorax, and pelvis in a commercial radiotherapy treatment planning system *Med. Phys.* **44** 3407–17

[26] Paganelli C, Meschini G, Molinelli S, Riboldi M and Baroni G 2018 Patient-specific validation of deformable image registration in radiation therapy: overview and caveats *Med. Phys.* **45** e908–22

[27] Brock K K, Mutic S, McNutt T R, Li H and Kessler M L 2017 Use of image registration and fusion algorithms and techniques in radiotherapy: report of the AAPM Radiation Therapy Committee Task Group No. 132 *Med. Phys.* **44** e43–e76

[28] Das I J, Cheng C W, Watts R J, Ahnesjö A, Gibbons J and Li X A *et al* 2008 Accelerator beam data commissioning equipment and procedures: report of the TG-106 of the Therapy Physics Committee of the AAPM *Med. Phys.* **35** 4186–215

[29] Ma C M C, Chetty I J, Deng J, Faddegon B, Jiang S B and Li J 2020 Beam modeling and beam model commissioning for Monte Carlo dose calculation-based radiation therapy treatment planning: report of AAPM Task Group 157 *Med. Phys.* **47** e1–e18

[30] Ezzell G A, Burmeister J W, Dogan N, LoSasso T J, Mechalakos J G and Mihailidis D *et al* 2009 IMRT commissioning: multiple institution planning and dosimetry comparisons, a report from AAPM Task Group 119 *Med. Phys.* **36** 5359–73

[31] Ezzell G A, Galvin J M, Low D, Palta J R, Rosen I and Sharpe M B *et al* 2003 Guidance document on delivery, treatment planning, and clinical implementation of IMRT: report of the IMRT Subcommittee of the AAPM Radiation Therapy Committee *Med. Phys.* **30** 2089–115

[32] Bissonnette J P, Balter P A, Dong L, Langen K M, Lovelock D M and Miften M *et al* 2012 Quality assurance for image-guided radiation therapy utilizing CT-based technologies: a report of the AAPM TG-179 *Med. Phys.* **39** 1946–63

[33] Herman M G, Balter J M, Jaffray D A, McGee K P, Munro P and Shalev S *et al* 2001 Clinical use of electronic portal imaging: report of AAPM Radiation Therapy Committee Task Group 58 *Med. Phys.* **28** 712–37

[34] Lin P J P, Beck T J, Borras C, Cohen G, Jucius R A and Kriz R J *et al* 1993 *Specification and Acceptance Testing of Computed Tomography Scanners* 39 American Association of Physicists in Medicine https://www.aapm.org/pubs/reports/rpt_39.pdf

[35] Mutic S, Palta J R, Butker E K, Das I J, Huq M S and Loo L N D *et al* 2003 Quality assurance for computed-tomography simulators and the computed-tomography-simulation process: report of the AAPM Radiation Therapy Committee Task Group No. 66 *Med. Phys.* **30** 2762–92

[36] Langen K M, Papanikolaou N, Balog J, Crilly R, Followill D and Goddu S M *et al* 2010 QA for helical tomotherapy: report of the AAPM Task Group 148 *Med. Phys.* **37** 4817–53

[37] Murphy M J, Balter J, Balter S, BenComo J A Jr, Das I J and Jiang S B *et al* 2007 The management of imaging dose during image-guided radiotherapy: report of the AAPM Task Group 75 *Med. Phys.* **34** 4041–63

[38] Ding G X, Alaei P, Curran B, Flynn R, Gossman M and Mackie T R *et al* 2018 Image guidance doses delivered during radiotherapy: quantification, management, and reduction: report of the AAPM Therapy Physics Committee Task Group 180 *Med. Phys.* **45** e84–e99

[39] Ma C M, Coffey C W, DeWerd L A, Liu C, Nath R and Seltzer S M *et al* 2001 AAPM protocol for 40–300 kV x-ray beam dosimetry in radiotherapy and radiobiology *Med. Phys.* **28** 868–93

[40] Almond P R, Biggs P J, Coursey B M, Hanson W F, Huq M S and Nath R *et al* 1999 AAPM's TG-51 protocol for clinical reference dosimetry of high-energy photon and electron beams *Med. Phys.* **26** 1847–70

[41] McEwen M, DeWerd L, Ibbott G, Followill D, Rogers D W O and Seltzer S 2014 Addendum to the AAPM's TG-51 protocol for clinical reference dosimetry of high-energy photon beams *Med. Phys.* **41** 041501

[42] Klein E E, Hanley J, Bayouth J, Yin F F, Simon W and Dresser S *et al* 2009 Task Group 142 report: quality assurance of medical accelerators *Med. Phys.* **36** 4197–212

[43] Bova F J, Buatti J M, Friedman W A, Mendenhall W M, Yang C C and Liu C 1997 The University of Florida frameless high-precision stereotactic radiotherapy system *Int. J. Radiat. Oncol. Biol. Phys.* **38** 875–82

[44] Meeks S L, Bova F J, Friedman W A, Buatti J M, Moore R D and Mendenhall W M 1998 IRLED-based patient localization for linac radiosurgery *Int. J. Radiat. Oncol. Biol. Phys.* **41** 433–9

[45] Moore C, Lilley F, Sauret V, Lalor M and Burton D 2003 Opto-electronic sensing of body surface topology changes during radiotherapy for rectal cancer *Int. J. Radiat. Oncol. Biol. Phys.* **56** 248–58

[46] Willoughby T, Lehmann J, Bencomo J A, Jani S K, Santanam L and Sethi A *et al* 2012 Quality assurance for nonradiographic radiotherapy localization and positioning systems: report of Task Group 147 *Med. Phys.* **39** 1728–47

[47] Ozhasoglu C, Saw C B, Chen H, Burton S, Komanduri K and Yue N J *et al* 2008 Synchrony–CyberKnife respiratory compensation technology *Med. Dosim.* **33** 117–23

[48] Bertholet J, Knopf A, Eiben B, McClelland J, Grimwood A and Harris E *et al* 2019 Real-time intrafraction motion monitoring in external beam radiotherapy *Phys. Med. Biol.* **64** 15TR01

[49] Dieterich S, Cavedon C, Chuang C F, Cohen A B, Garrett J A and Lee C L *et al* 2011 Report of AAPM TG 135: quality assurance for robotic radiosurgery *Med. Phys.* **38** 2914–36

[50] Rossi R P, Lin P J P, Rauch P L and Strauss K J 1985 *Performance specifications and acceptance testing for x-ray generators and automatic exposure control devices* 14 American Association of Physics in Medicine https://www.aapm.org/pubs/reports/rpt_14.pdf

[51] Tijssen R H N, Philippens M E P, Paulson E S, Glitzner M, Chugh B and Wetscherek A *et al* 2019 MRI commissioning of 1.5 T MR-linac systems – a multi-institutional study *Radiother. Oncol.* **132** 114–20

[52] Jackson E F, Bronskill M J, Drost D J, Och J, Pooley R A and Sobol W T *et al* 2010 *Acceptance Testing and Quality Assurance Procedures for Magnetic Resonance Imaging Facilities* 100 American Association of Physicists in Medicine 23–5 https://www.aapm.org/pubs/reports/rpt_100.pdf

[53] Price R, Allison J and Clarke G 2015 *Magnetic Resonance Imaging Quality Control Manual* American College of Radiology

[54] Kurz C, Buizza G, Landry G, Kamp F, Rabe M and Paganelli C *et al* 2020 Medical physics challenges in clinical MR-guided radiotherapy *Radiat Oncol* **15** 93

[55] Mohan R, Das I J and Ling C C 2017 Empowering intensity modulated proton therapy through physics and technology: an overview *Int. J. Radiat. Oncol. Biol. Phys.* **99** 304–16

[56] Fontenot J D, Alkhatib H, Garrett J A, Jensen A R, McCullough S P and Olch A J *et al* 2014 AAPM medical physics practice guideline 2.a: commissioning and quality assurance of X-ray-based image-guided radiotherapy systems *J Appl. Clin. Med. Phys.* **15** 4528

Part III

Practice of image-guided radiation therapy for abdominal cancer

IOP Publishing

Principles and Practice of Image-Guided Abdominal
Radiation Therapy

Yu Kuang

Chapter 7

Linear accelerator image-guided radiation therapy for abdominal cancers

Yulin Song, Boris Mueller and Maria F Chan

Linear accelerators (Linacs) have been used to treat abdominal cancers for decades. They are the most important medical equipment in radiation oncology. In this chapter, we first review the history of abdominal cancer treatments with Linacs. We also review the current market landscape of modern Linacs in terms of photon energy level, total yearly production, and main manufacturers. We then discuss the main technical specifications of Varian TrueBeam Linacs, including energy choices, on-board imaging devices, and built-in quality assurance (QA) tools. We then present the common forms of abdominal cancer treated with Linacs, their treatment planning techniques, and dose tolerances for the target and organs at risk (OARs).

7.1 Introduction

7.1.1 History of abdominal cancer treatment with Linacs

Radiation therapy has a history of more than a century. It can be divided into four major phases chronologically in terms of milestone technological breakthroughs: (1) the discovery phase; (2) the orthovoltage phase; (3) the megavoltage (MV) phase; and (4) the beam intensity modulation phase. The discovery phase began with the discovery of x-rays by Roentgen, a German physicist, in November 1895 and lasted into the late 1920s. During this phase, several important discoveries occurred. Becquerel discovered the phenomenon of radioactivity and the Curies discovered Radium in 1898. It was during this phase that biologists began to explore the interplay between time and dose on cell survival. Because of his contribution, Roentgen received the first Nobel Prize in Physics in 1901 [1, 2].

doi:10.1088/978-0-7503-2468-7ch7

The orthovoltage phase started in the late 1920s and ended around 1950. The demand for the treatment of deeply seeded tumors prompted the development of deeply penetrating external sources in the supervoltage range of ~500 kV to 2 MV. The first such a supervoltage tube was designed and built by Coolidge in the 1920s. At the same time, a significant breakthrough in radiobiology took place in France, where it was discovered that daily doses of radiation over several weeks greatly improved the patient's chance for survival. After this seminal discovery, radiation treatment methods and the machines that delivered radiation beam have steadily improved.

The MV phase covered the period from 1950 to 1995. The continuing aspiration of treating deeply seeded tumors stimulated the development of Cobalt teletherapy machines and MV Linacs [3, 4]. Higher MV energies made it possible to treat deep lesions, provide greater skin-sparing, and improve patients' survival. Simultaneously, the concept of multi-field treatment planning was developed and implemented. The beam intensity modulation phase started in the mid-1990s with the successful development of multi-leaf collimator (MLC) as a beam shaping and beam intensity modulation device [5, 6].

The rapid advances in beam intensity modulation algorithms, leaf sequencing algorithms, and computer graphics in the late 1990s propelled radiation oncology from three-dimensional (3D)–conformal radiation therapy (CRT) into the era of intensity-modulated radiation therapy (IMRT) [7–13]. Based on the modes of intensity modulation, IMRT can be classified as either step-and-shoot (segmental MLC) or sliding window (dynamic MLC), with the sliding window technique being more time efficient [14]. Currently, most commercial treatment planning systems (TPSs) and mainstream Linacs support both delivery modalities. Because of its greater OAR sparing capability, IMRT was rapidly adopted for the treatment of abdominal cancers shortly after its inception.

However, owing to the beamlet-based intensity modulation optimization algorithm, IMRT tends to produce excessive intensity modulation at the junctions between the planning target volume (PTV) and OARs. This phenomenon becomes particularly striking when the prescribed PTV dose and OAR dose tolerances are hugely different. This technical deficiency ultimately leads to isolated intensity peaks, hot spots, greater dose inhomogeneity, long treatment time, and high radiation leakage. These issues can become very pronounced in the treatment of abdominal cancers in which the PTV often abuts the OARs or even overlaps with them. To overcome these technical deficiencies, aperture-based rotational modalities, including RapidArc and volumetric-modulated arc therapy (VMAT), were introduced clinically in 2008 [15–22]. Today, VMAT has become the main treatment modality for the treatment of various abdominal cancers. With rapid advances in on-board imaging systems and motion monitoring systems, four-dimensional (4D) radiation therapy has also emerged as an important treatment modality in modern radiation oncology [23].

7.1.2 Current market landscape of modern Linacs

Based on the photon energy level, modern Linacs can be divided into low-energy Linacs and high-energy Linacs. According to the latest market study report, low-energy Linacs lead the market consumption, with a market share of 76.48%. The report projects that the total low-energy Linacs production worldwide will be 1737 units in 2022, accounting for 71.45% of the total market. The total high-energy Linacs will be 694 units with a market share of 28.55% [24]. Based on competitive landscape analysis, currently, the major manufacturers are Varian Medical Systems, Elekta, Accuray, and Siemens Healthineers. As of 2016, these four companies controlled 94% of the Linac market. There are several smaller, but rising competitors with the potential to make a big impact on the market in some regions. The most noticeable one is LinaTech (Sunnyvale, CA, USA). Its latest Linac, VenusX, is doing clinical trials at Tianjin Tumor Hospital and Beijing 301 Hospital as part of its pre-market Chinese regulatory certification process. Its triple-layer MLC design, binocular built-in 3D optical surface imaging system, and dual-ring kV imaging system are some of the novel technical features that could reshape the market landscape in China and other emerging markets in Asia, Latin America, and Africa.

7.2 Mainstream modern Linacs

7.2.1 Varian TrueBeam Linacs

Table 7.1 lists the major technical specifications of TrueBeam Linacs that are relevant to clinical treatments. It is these technical parameters that determine TrueBeam Linacs' market competitiveness and its leading position in the world. Figure 7.1 is a photo of the newest TrueBeam Linac being installed at our institution, Memorial Sloan Kettering Cancer Center (MSKCC), during the COVID-19 pandemic in April 2020. It took approximately 2 weeks for the Linac to be fully installed and another week for fine-tuning and adjustment. Figure 7.2 is a photo of the TrueBeam Linac after the installation was completed in May 2020.

7.3 Common forms of abdominal cancers treated with Linacs

7.3.1 Gastroesophageal junction and stomach cancers

According to the latest United States government statistics, the estimated new cases and deaths from esophageal cancer in 2020 are 18 440 and 16 170, respectively. Depending on the stage and histological type, the 5 year relative survival rate was 19.9% from 2010 to 2016 [25]. The rate of new cases and the death rate per 100 000 people have remained stable in the United States for the past 30 years. The management of non-metastatic esophageal and esophagogastric junction (GE junction) cancer has undergone significant changes over the past two decades. Most patients currently receive some form of combined modality therapy rather than local therapy alone. For early stage disease, chemoradiation therapy followed by surgery provides the best treatment outcome. For more locally advanced and or unresectable disease, chemoradiation therapy is usually the predominant treatment modality.

Most esophageal cancers occur in the distal esophagus and often involve the gastroesophageal junction. Some, however, also grow in the upper and middle

Table 7.1. Major technical specifications of TrueBeam Linacs.

Parameters	Technical Specifications	Note
Power source	Klystron	
Photon energy configuration	6X, 6X-FFF, 10X, 10X-FFF, 15X, and 18X	Users have the options to select different energy configurations
Electron energies	4 MeV to 25 MeV	TrueBeam Linacs have a wide range of electron energies
MLC	M120 (60 pairs)	Max field size: 40 × 40 cm, central 40 pairs: 5 mm leaf width, outer 20 pairs: 1 cm leaf width. Max IMRT field size: 15 cm in x-direction and 40 cm in y-direction without jaw tracking option.
Portal imager (electronic portal imaging device, EPID)	aS1000	Max FOV at isocenter plane: 40 × 40 cm, image matrix size: 2048 × 2048. The lowest available energy for MV imaging is 2.5 MV. Calibrated EPID can be used for portal dosimetry measurement.
Dose rate	Conventional photons: 100–600 monitoring unit (MU)/min, electrons: 100–1000 MU/min, 6X-FFF photons: 400–1400 MU/min; 10X-FFF photons: 400–2400 MU/min	For IMRT delivery, the dose rate is fixed. For VMAT delivery, the dose rate can be continuously modulated. For dynamic conformal arcs (DCA) delivery, the dose rate is also fixed.
kV imaging	On-board imaging system	Max FOV at isocenter plane: 40 × 40 cm, image matrix size: 2048 × 2048. kVp is continuously adjustable. It offers IMR capability and provides a variety of trigger mechanisms, including MU, time gantry angle, and respiratory motion.
CBCT	kV CBCT, including phase-averaged 3D-CBCT, gated 3D-CBCT, and 4D-CBCT	There are many built-in imaging protocols to choose from. 4D-CBCT provides maximum intensity projection and ITV only. Individual phases are only available in the Service Mode.
	Machine performance check	The portion of mechanical checks may be used to replace the mechanical check part of the monthly machine QA or used as a reference for SRS machine QA, including 6 degrees of freedom (DOF) couch.

Built-in SRS cylinder cones	No	Externally mounted cones are available to customers as an add-on option.
Types of treatment delivery	3D-CRT, IMRT, DCA, RapidArc, VMAT, HyperArc, SBRT, SRS, and gated treatments for both IMRT and VMAT	Plans computed by most mainstream commercial TPSs can be delivered.
3D optical surface imaging	Oprical surface monitoring system	Ceiling mounted 3-pod imaging system
Treatment couch	6-DOF exact IGRT	Precision in translations: 0.1 mm, precision in rotations: 0.1°.

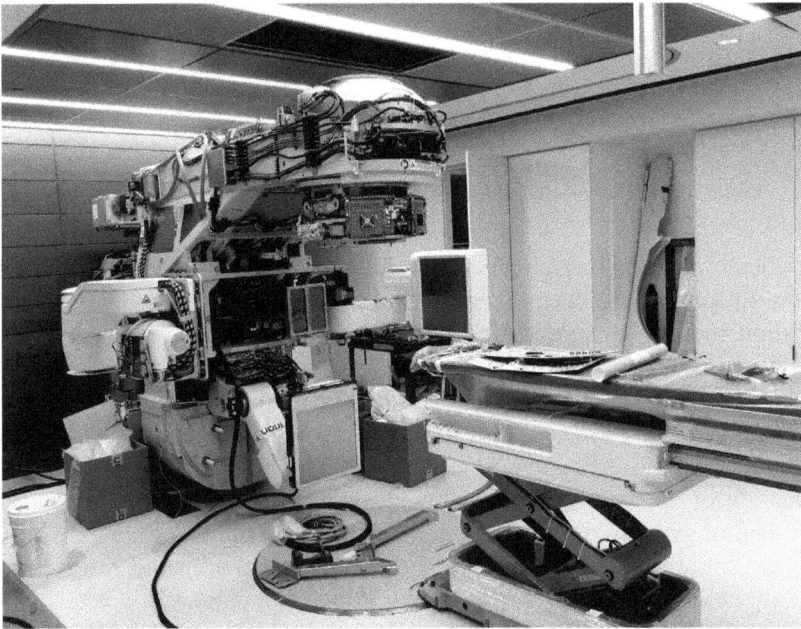

Figure 7.1. A TrueBeam Linac was being installed at MSKCC in April 2020.

portions of the esophagus, forming a long and asymmetric lesion. For this type of gross tumor volume (GTV), the PTV delineated on CT scan for radiation treatment planning can be as long as 35 cm in the superior–inferior direction and as wide as 18 cm in the left–right direction. This imposes a significant technical challenge for conventional radiation therapy.

Unlike GE junction cancer, the rate of new stomach cancer cases has been steadily falling for the last 30 years. The estimated new cases and deaths from stomach cancer in the United States in 2020 are 27 600 and 11 010, respectively. The 5 year relative survival rate was 32.0% from 2010 to 2016 [26]. For early and locoregional stomach

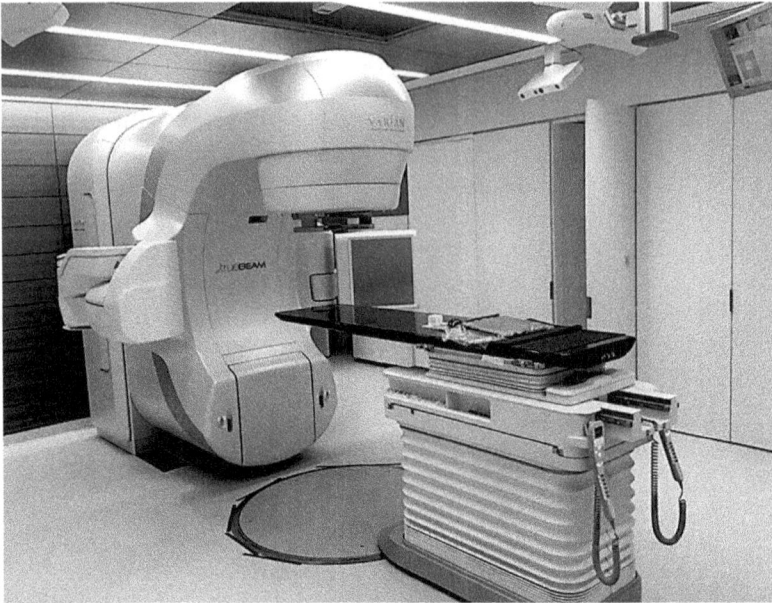

Figure 7.2. The installation of the TrueBeam Linac was completed in May 2020.

cancers, chemoradiation is often given first, followed by gastrectomy with lymph node dissection. In some cases, chemotherapy is also given after surgery.

Radiation therapy techniques for GE junction and stomach cancers vary from institution to institution, but the basic principles and general clinical workflow are similar. The treatment recommendations should be made by a multi-disciplinary team including pathologists, radiologists, surgical oncologists, medical oncologists, and radiation oncologists. All diagnostic scans should be carefully reviewed for the determination of the extent of the treatment volume and the estimation of the field borders. Before simulation and treatment, the patient should follow an nil per os (nothing by mouth) directive for 3 h. For most cases, oral and intravenous (IV) contrast are needed for accurate target delineation. The patient is typically simulated in the supine position with the arms above the head. However, many elderly and frail patients are unable to tolerate the arms-up position for extended periods. This creates a significant challenge for certain types of radiation therapy, such as stereotactic body radiation therapy (SBRT), deep inspiration breath hold (DIBH), and respiratory gating. From the perspective of treatment delivery accuracy, patient comfort level and stability are the most crucial factors. For these patients, simulation should be performed with the arms by the body sides and palms up under buttocks. Both arms should be positioned tightly against the body for stability and reproducibility. A robust immobilization device should be used for daily patient setup reproductivity and infra-fractional motion reduction [27]. A good example is the custom Aquaplast body mask. If organ motion with respiration is significant, a 4D-CT scan should be acquired to aid the delineation of the internal target volume (ITV). The PTV should have enough margin to account for daily variation in stomach content.

Abdominal organs and tissues are radiosensitive. Therefore, every effort should be made to reduce doses to the OARs. The applicable OARs include the cord, esophagus, heart, larynx, liver, kidneys, duodenum, large bowel, small bowel, lungs, and stomach. They should be carefully delineated by the radiation oncologist. Additional anatomical structures, such as the aorta, celiac artery, portal vein, and superior mesenteric artery, are also typically contoured. If the PTV abuts any of the OARs, then an intermediate planning structure should be created by subtracting the PTV from that OAR. Given proper dose constraints, this structure can serve as a dose tuning tool in the inverse optimization. Table 7.2 lists our institutional dose

Table 7.2. Dose constraints for normal tissues and plan acceptance criteria for targets used at MSKCC for GE junction and stomach cancers.

Anatomical Structure Name	Dosimetric Parameter	DVH Constraint	Dose Limit
PTV	D95%	⩾100	
	V100%	⩾90%	
	Dmax	⩽110%	
GTV	V100%	⩾90%	⩾80%
	Dmax	⩽110%	⩽115%
Brachial plexus	Dmax	⩽6500 cGy	
Cord	Dmax	⩽4500 cGy	
Common bile duct	Dmax	⩽8000 cGy	
Duodenum	Dmax		⩽6000 cGy
	V5000 cGy		⩽40 cm3
Esophagus	Dmax	⩽6500 cGy	
Heart	V3000 cGy	⩽20%	
	V3000 cGy	⩽30%	
Kidney (each)	V2000 cGy	⩽33%	
Kidneys (both)	V2000 cGy		⩽50%
Large bowel	Dmax		⩽6500 cGy
Liver-GTV	V (<2800 cGy)		⩾700 cm^3
	Dmcan		⩽2800 cGy
Lung-GTV	V1000 cGy	⩽35%	
	V1000 cGy	⩽40%	
	V1000 cGy	⩽65%	
	V2000 cGy	⩽20%	
	V3000 cGy	⩽15%	
	V4000 cGy	⩽10%	
Skin	Dmax	⩽110%	
Small bowel	Dmax		⩽5500 cGy
	V4500 cGy	⩽40 cm^3	
Stomach	Dmax		⩽6000 cGy
	V5000 cGy		⩽40 cm^3

constraints for normal tissues and plan acceptance criteria for targets for GE junction and stomach cancers. In our clinical practice, dose constraints are used as planning guidelines, and doses exceeding limit require peer review.

There are several dose-fractionation schemes for the treatment of GE junction and stomach cancers. The National Comprehensive Cancer Network (NCCN) recommends a dose range from 4500 cGy to 5040 cGy in 180 cGy/fraction [28]. At MSKCC, we also implement a dose painting fractionation protocol. Currently, the common treatment planning techniques for GE junction and stomach cancers are IMRT and VMAT. Depending on the extent of the PTV, an IMRT plan with five to seven fields should be able to achieve the desired dosimetry. A plan with more fields can improve the PTV dose conformality, but it also elevates the integral dose to the uninvolved normal tissue. The improvement in conformality becomes diminishing once the number of fields is greater than nine. If VMAT is used, a plan with two to four full arcs should be adequate. To enhance intensity modulation, the control point sequence should be interleaved in gantry angles [29]. For example, if the first arc is from 179° to 181° in the counterclockwise (CCW) direction, then the second arc should be from 180° to 178° in the clockwise (CW) direction. The collimator angle should be alternated between 0° and 90° between arcs to further augment intensity modulation [30]. Other collimator angles should be avoided for VMAT because they will increase the field sizes, resulting in increased MLC radiation leakage. Daily kV, cone-beam computed tomography (CBCT), or other combinations of respiration gated imaging options can be used for patient setup verification and target tracking. These include gated kV imaging, gated 3D-CBCT, and 4D-CBCT. 3D-CRT with enhanced dynamic wedges should only be used when insurance does not cover IMRT and VMAT. Because of the lack of intensity modulation, the overall quality of 3D-CRT plans is much inferior to those computed by IMRT and VMAT [31–33]. The dose inhomogeneity of 3D-CRT plans can be improved when combined with the field-in-field technique. The manual field-in-field optimization is an effective way to mitigate the out-of-PTV hot spots and improve target dose conformality. Six to ten segments per field may be enough to suppress the hot spots and steer the higher isodose lines towards to the PTV. The shortcoming of this approach is its low efficiency. The planner may need to try many combinations of segments of different shapes and locations to achieve the desired dose distribution. Table 7.3 lists common prescription doses, fractionation schemes, treatment modalities, and imaging techniques used at our institution.

A middle-aged female patient was diagnosed with locally advanced invasive adenocarcinoma of the GE junction (T3N1). A PET/CT scan revealed that the disease had extended to the celiac nodes and cardia. The patient selected tri-modality therapy: preoperative concurrent chemoradiation therapy, followed by surgery. The patient was simulated in a supine position, immobilized with a custom Alpha Cradle. A 4D-CT scan was acquired with oral and IV contrasts for respiratory management. The radiation oncologist prescribed 180 cGy × 25 to the PTV, with daily kV imaging for patient setup and weekly CBCT for a re-evaluation of the PTV. Because it was a preoperative radiation treatment, IMRT was utilized to

Table 7.3. Common treatment planning techniques used at MSKCC for GE junction and stomach cancers.

Prescription	Planning Techniques	Imaging
4500–5040 cGy, 180 cGy/ fraction in 25–28 fractions	IMRT with 5–7 fields, VMAT with 2–3 full arcs. If respiratory motion management is required: (1) respiratory gating (30%– 70% or 40%–60% phases); (2) free breathing with ITV generated from 4DCT; (3) DIBH; or (4) abdominal compression belt. Use 6X if a significant portion of lungs is in the fields. Otherwise, use 15X	Daily kV/CBCT or daily kV/ weekly CBCT. For respiratory gating, use gated kV imaging and gated 3D-CBCT
Dose painting: 5000–5600 cGy, 200 cGy/fraction to primary tumor, 4500–5040 cGy, 180 cGy/fraction to nodal regions in 25–28 fractions	Four dose painting levels: 4500, 5000, 5040, 5600 cGy. Field arrangements, see above	See above

minimize the normal tissue volume receiving a low dose. The plan consisted of five fields, with a gantry angle of 155°, 70°, 0°, 290°, and 205°, respectively. Our clinical experience indicates that this beam arrangement is effective in reducing the dose to the liver, lungs, and kidneys for GE junction cases. The collimator angle was not rotated because of the elongated shape of the PTV in the superior-to-inferior direction. Because the inferior 1/3 lungs were in the treatment fields, 6X photons were used for treatment planning. The plan was delivered using the sliding window technique, with each field having 166 shoot-only segments. Thus, this approach not only increased the intensity modulation levels but also improved treatment efficiency.

Figure 7.3 shows the dose distribution of the treatment plan with dose color wash rendition. The red contour represents the PTV. The yellow straight lines in figure 7.3(a) are field borders. Figure 7.3(d) is the digitally reconstructed radiograph (DRR) at the gantry angle of 0°. The white area inside the beam aperture is the oral contrast. The plan met all institutional dose constraints and plan acceptance criteria. Table 7.4 shows the plan dosimetry statistics. The PTV conformality index was 1.01, indicating a highly conformal plan. Figure 7.4 shows two verification kV images at the kV x-ray source angle of 0° and 90°, respectively. The images were acquired with kVp = 110 and rendered with the content filter so that the bony structures and organ silhouettes were more clearly visible. The patient tolerated radiation therapy well and did not require a treatment break. He did experience some expected side effects,

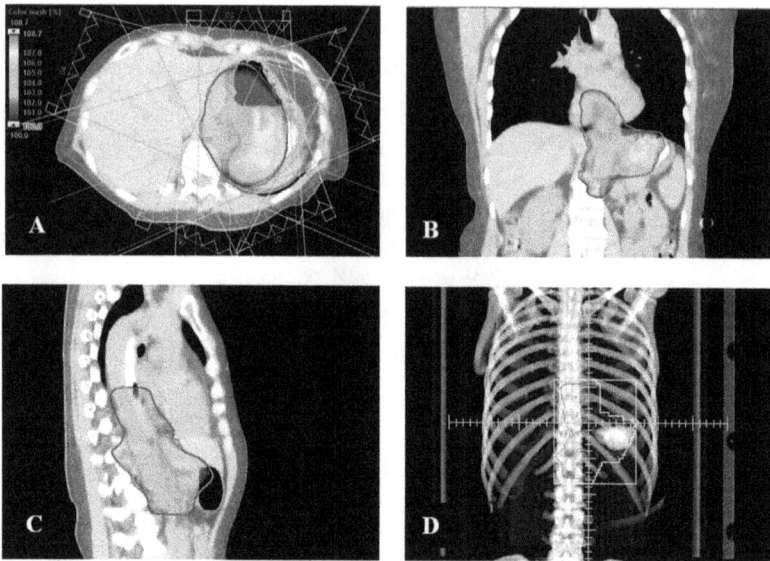

Figure 7.3. The dose distribution of the IMRT plan in representative planes in dose color wash rendition: (a) axial plane, (b) coronal plane, (c) sagittal plane, and (d) the DRR of the field at the gantry angle of 0°. The red contour represents the PTV.

including fatigue and decreased appetite. The patient was counseled to increase oral fluids and daily caloric intake.

7.3.2 Liver cancer

The liver is the largest internal organ by mass in the human body. It is in the right upper portion of the abdomen, right beneath the diaphragm. The most common types of primary liver cancer are hepatocellular carcinoma (HCC) that originates in hepatocytes and biliary tract cancers. However, liver metastases are far more common than primary liver cancers and cases continue to rise worldwide. It is estimated that 30%–70% of cancer patients die of liver metastases [34]. In 2020, 42 810 new cases of liver metastases are projected, which will account for 2.4% of all new cancer cases in the United States. The estimated deaths are 30 160, accounting for 5% of all cancer-related deaths in 2020.

Treatment options for liver cancer include liver transplantation, surgery, chemo-therapy, targeted therapy, and radiation therapy. Liver transplantation is only feasible when the patient meets the United Network for Organ Sharing criteria: (1) small tumor size (2–5 cm in diameter), (2) no macrovascular involvement, and (3) no distant metastases [35]. Surgery is preferred if the patient and tumor have the following characteristics: (1) Child-Pugh Class A or B, (2) absence of portal hypertension, (3) suitable tumor location, (4) adequate liver reserve, and (5) suitable liver remnant [36]. Chemotherapy plays an important role in the treatment of liver cancer. It is the preferred systemic therapy when the tumor is unresectable and has not responded to various locoregional therapies, including

Table 7.4. The IMRT plan dosimetry statistics.

Anatomical Structure Name	Dosimetric Parameter	Plan Value
PTV	D95%	100.1%
	V100%	95.3%
	Dmax	108.7%
GTV	V100%	100%
	Dmax	108.7%
Cord	Dmax	3538 cGy
Duodenum	Dmax	4668 cGy
	V5000 cGy	
Esophagus	Dmax	4843 cGy
Heart	V3000 cGy	19.7%
	V3000 cGy	19.7%
Kidney (each)	V2000 cGy	0.1% (R), 26.8% (L)
Kidneys (both)	V2000 cGy	12.9%
Large bowel	Dmax	4510 cGy
Liver-GTV	V (<2800 cGy)	783.3 cm^3
	Dmean	1810.6 cGy
Lung-GTV	V1000 cGy	24.6%
	V1000 cGy	24.6%
	V1000 cGy	24.6%
	V2000 cGy	10.8%
	V3000 cGy	4.1%
	V4000 cGy	1.7%
Skin	Dmax	108.7%
Small bowel	Dmax	4690 cGy^3
	V4500 cGy	⩽0.7 cm^3
Stomach	Dmax	4893 cGy
	V5000 cGy	0.0 cm^3

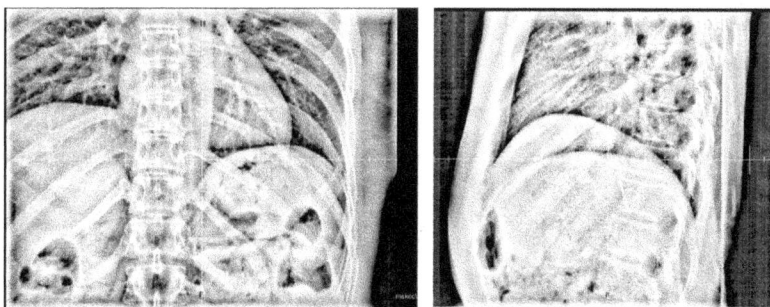

Figure 7.4. Two verification kV images acquired at the kV x-ray source angle of 0° and 90°, respectively, before each treatment.

radiofrequency ablation and transarterial embolization therapies. Radiation therapy is a treatment option for patients who are not candidates for surgical curative treatments. It is also used as a part of a comprehensive treatment regimen to bridge the patient to other curative therapies. All liver cancers regardless of location can be treated with various radiation modalities, including 3D-CRT, IMRT, VMAT, and SBRT. Clinical evidence has shown that SBRT can be used as an alternative to the ablation and embolization therapies when they have failed or are contraindicated [37–40].

There are several dose fractionation schemes for the treatment of liver cancer. The NCCN recommends an SBRT dose of 3000–5000 cGy in 1000 cGy/fraction in 3–5 fractions [36]. At MSKCC, liver cancer is often treated with dose painting fractionation protocols. There are two commonly used fractionation schemes used for liver cancer at MSKCC: a 15-fraction regimen and a 25-fraction regimen with different dose painting levels. Table 7.5 shows the treatment planning techniques. The use of a higher dose is at the radiation oncologist's discretion.

VMAT is preferred over IMRT for better target dose conformality. Because the liver is right beneath the diaphragm, its position is constantly changing with respiration. In many cases, the extent of liver motion is underestimated by visualization of the skin surface when the patient lies on the CT simulation couch. This could

Table 7.5. Common treatment planning techniques for liver cancer used at MSKCC.

Prescription	Planning Techniques	Imaging
15 fractions protocol: Microscopic dose: 3750 cGy, 250 cGy/fraction Ablative dose: 6000–6750 cGy, 400–450 cGy/fraction Kicker dose: 7500–9000 cGy, 500–600 cGy/fraction	VMAT with 2–4 ipsilateral partial arcs covering about 210°. If respiratory motion management is required: (a) respiratory gating with a gating window width of 30%–70% or 40%–60% phases; (b) free breathing with ITV generated from 4DCT; (c) DIBH; or (d) abdominal compression belt. Use 6X photons for dose calculation.	Daily kV/CBCT or daily kV/weekly CBCT. For respiratory gating, use gated kV imaging and gated 3D-CBCT
25 fractions protocol: Microscopic dose: 4500 cGy, 180 cGy/fraction Ablative dose: 7500 cGy, 500 cGy/fraction	See above	See above
SBRT Protocols: 1200 cGy × 3900 cGy × 5	Use 6X FFF photons for dose calculation. Use 1.25 mm grid size for dose calculation. See above for the number of VMAT arcs and respiratory motion management.	See above

give the simulation therapists and the radiation oncologist a false impression that motion management is not needed. Based on our clinical observations, a patient's internal organs move with a much greater amplitude during respiration than the movement of the external abdominal skin surface. Thus, it is strongly recommended that one of the respiratory motion management techniques be implemented for liver plan delivery regardless of the motion amplitude of the abdominal skin surface.

At MSKCC, the most commonly used technique is DIBH, delivered with either Varian RPM or AlignRT [VisionRT, United Kingdom]. Both have their advantages and disadvantages. For example, Varian RPM is fully integrated with TrueBeam Linacs. The system is robust and requires much less preparatory work. One of its shortcomings is the position instability of the reflector box. Ideally, the reflector box should be placed vertically on a flat skin surface for stable tracking. This is the normal posture when the RPM is calibrated. Frequently, a sudden deep inspiration or cough could tilt the reflector box. This could result in the detected breathing amplitude being greatly underestimated. Contrary to Varian RPM, AlignRT requires a lot of preparatory work, such as exporting the plan to the AlignRT system and creating a suitable region of interest for motion tracking. This requires extra time and special training for treatment planners. Because it is a third-party add-on device, the gating signal to the Linac could get interrupted in the middle of treatment by the gating board inside the AlignRT workstation. If AlignRT fails due to any technical reason, the TrueBeam Linac will no longer function. To resolve this issue, the Linac must be disconnected internally from AlignRT and resynchronized. The affected system sub-nodes must be rebooted and MLC must be reinitialized. The entire process takes a long time. Another common issue is the poor 3D surface image quality. Due to low point cloud density and low DLP light intensity, the 3D surface image appears to be less continuous and even has many holes in it. With AlignRT v6.2 being released in the United States in March 2020, some of the technical deficiencies are either greatly improved or eliminated.

The results of the initial tests of AlignRT v6.2 at our institution are very promising [41]. Figure 7.5 shows two 3D surface images acquired with the current AlignRT v5.1 (right) and the latest AlignRT v6.2 (left), respectively. The volunteer

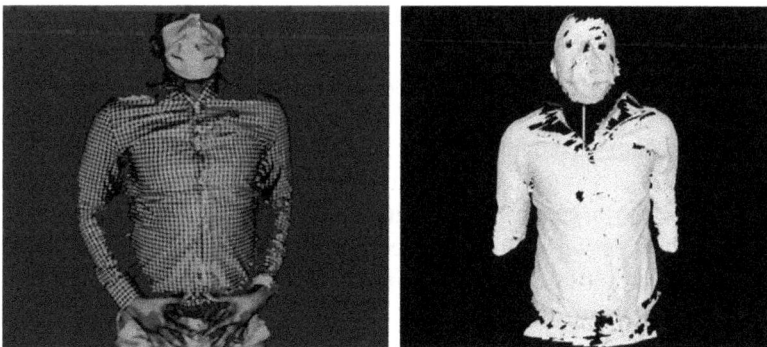

Figure 7.5. Two 3D surface images acquired with the latest AlignRT v6.2 (left) and the current AlignRT v5.1 (right). The quality of the left image has been greatly improved.

was wearing a shirt with a black and white square pattern. The size of each small square is 0.5 × 0.5 cm. The volunteer was also wearing a face mask as required by the institutional directive. The useful field of view (FOV) of the image was increased. The quality of the left image has been greatly enhanced in three aspects: resolution, continuity, and texture details. The image acquired with AlignRT v6.2 has a higher point of cloud density and fewer holes. The current AlignRT v5.1 is unable to capture any surface texture, while AlignRT v6.2 provides exquisite detail of the surface texture. As SGRT is gaining great momentum, AlignRT v6.2 will become invaluable for both patient setup and real-time motion monitoring.

Depending on the PTV size and location, the most common beam arrangement for liver targets consists of two to four ipsilateral partial arcs optimized with 6X photons. The VMAT arcs should cover a sector span of approximately 210°, alternating between gantry rotation directions (CW and CCW) and collimator angles of 0° and 90°. Control point sequences should be gantry-angle interleaved to increase intensity modulation. For a much larger PTV, more arcs may be used, but the number should not be more than 6. VMAT plans with more than 6 arcs have a marginal impact on plan quality but could greatly increase the plan delivery time. To reduce radiation leakage, the jaw tracking technique should be used during plan optimization and final dose calculation if it is available on the Linac.

On TrueBeam Linacs, photon beam intensity modulation is achieved by two devices, MLC and jaws. MLC serves as the primary intensity modulator and moves in one dimension only. Along this dimension, MLC moves unidirectionally for IMRT plans and bidirectionally for VMAT plans. It primarily modulates in-field beam intensity pattern. Jaws serve as the secondary intensity modulators and move in two dimensions bidirectionally. They only modulate out-of-field beam intensity pattern by immediately following the trajectories of the most lagging leaves. The speeds of jaws are a function of MLC leaf speed. Jaws define the maximum beam aperture size for a given control point. Through secondary intensity modulation, the beam intensity outside the useful beam aperture is greatly reduced. Traditional VMAT plans suffer from heavy MLC radiation leakage. This includes radiation leakage through MLC leaves, between neighboring MLC leaves, and between opposing MLC leaf heads. This adverse effect is particularly pronounced when the collimator is not at 0° or 90° and the isocenter is not placed at the geometric center of the PTV. The maximum radiation leakage occurs at the collimator angles of 45° and 315°, where the MLC-defined treatment field size becomes maximal [42]. Jaw tracking optimization can effectively mitigate these negative impacts. Table 7.6 lists our institutional dose constraints for normal tissues and plan acceptance criteria for targets for liver cancer SBRT 900 cGy × 5 protocol. Table 7.7 lists our institutional dose constraints for normal tissues and plan acceptance criteria for targets for dose painting protocol in 15 fractions.

A middle-aged male with multiple medical problems was incidentally found to have a multifocal liver tumor. An abdominal MRI revealed a 2.4 cm left hepatic lobe lesion abutting segment 2a/4a. An ultrasound-guided biopsy revealed HCC. The patient was recommended to undergo fiducial marker placement for ablative radiotherapy delivered with a DIBH protocol. Two dose levels of 250/450 cGy × 15 were prescribed to the PTV and GTV, respectively, and delivered with the dose

Table 7.6. Dose constraints for normal tissues and plan acceptance criteria for targets for liver cancer SBRT 900 cGy × 5 protocol.

Anatomical Structure Name	Dosimetric Parameter	DVH Constraint	Dose Limit
PTV	D95%	⩾100%	
	V100%	⩾90%	
	Dmax	⩽115%	
GTV	V100%	⩾90%	⩾80%
Cord	Dmax	⩽1800 cGy	
	V1500 cGy	⩽10 cm3	
Common bile Duct	Dmax	⩽5500 cGy	
Duodenum	Dmax	⩽2800 cGy	⩽3000 cGy
	D5 cm^3		⩽2500 cGy
Esophagus	Dmax	⩽3000 cGy	
Heart	V4000 cGy	⩽10%	
Kidney (each)	V1500 cGy	⩽67%	⩽50%
Kidneys (both)	V1000 cGy		<33%
Single functioning kidney	V1000 cGy		
Large bowel	Dmax	⩽3000 cGy	⩽3300 cGy ⩽3000 cGy
	D5 cm^3		
Liver-GTV	V (<1500 cGy)		⩾700 cm^3
	Dmean		⩽1600 cGy
	Dmean (HCC)		⩽1200 cGy
Lung-GTV	V2000 cGy		⩽12%
Small bowel	Dmax	⩽2800 cGy	⩽3000 cGy
	V2000 cGy	⩽100 cm^3	⩽2500 cGy
	D5 cm^3		
Stomach	Dmax	⩽2800 cGy	⩽3000 cGy
	D5 cm^3		⩽2500 cGy

painting technique. Imaging guidance included daily kV and CBCT with matching on the fiducial. The patient was simulated with IV contrast in a custom Alpha Cradle and CT-scanned with the DIBH protocol. The GTV and CTV were delineated by the attending radiation oncologist. The PTV was created by adding

Table 7.7. Dose constraints for normal tissues and plan acceptance criteria for targets for liver cancer for dose painting protocol in 15 fractions.

Anatomical Structure Name	Dosimetric Parameter	DVH Constraint	Dose Limit
PTV	D95%V100%Dmax	≥100%≥90% ≤110%	≤110%
GTV	V100%	≥90%	≥80%
Cord	Dmax	≤3500 cGy	
Common bile duct	Dmax	≤7000 cGy	
Duodenum	DmaxV3750 cGy	40 cm^3	≤4500 cGy
Esophagus	Dmax	≤5000 cGy	
Heart	V3000 cGy	≤20%	
Kidney (each) Kidneys (both) Single functioning kidney	V2000 cGy V2000 cGy V2000 cGy	≤33%	≤50% <33%
Large bowel	Dmax		≤5000 cGy
Liver-GTV	V (<2400 cGy) DmeanDmean (HCC)	<33%	≥700 cm^3 ≤2400 cGy ≤1600 cGy
Lung-GTV	V2000 cGy		≤2000%
Small bowel	DmaxV3750 cGy	≤40 cm^3	≤4000 cGy
Stomach	DmaxV3750 cGy	≤40 cm^3	≤4500 cGy

a 0.5 cm 3D uniform margin to the CTV. Relevant OARs were contoured including the bile duct, spinal cord, duodenum, esophagus, heart, kidneys, lungs, small bowel, large bowel, and stomach. A protective 0.3 cm 3D margin was added to the small bowel, large bowel, duodenum, and stomach to better spare these critical organs. Also, four concentric dose turning structures with an incremental thickness of 1 cm were created. By assigning proper dose constraints, these turning structures were able to drive the higher isodose lines towards the PTV in a very effective way.

In our clinical practice, we normally adopt the strategy of combining the NTO and rinds in plan optimization simultaneously. This planning methodology has been proven to be highly effective in achieving steep dose gradients for dose painting and SBRT plans. This technique was particularly valuable for this case. The spatial separation between the PTV3750 and PTV6750 was only 1.0 cm whereas the prescription required a steep dose gradient of 80% between the two. Thus, the combination of the NTO and rinds was indispensable. Otherwise, extensive hot spots and even hot bands would appear outside the PTV3750 and the target dose conformality would be extremely poor. This phenomenon often occurs in the dose painting plans with a short spatial separation between different dose level PTVs. The plan was optimized with the jaw tracking technique. Figure 7.6 shows the delineation of the PTV3750 (pink) and PTV6750 (red), with a volume of 250.8 cm^3 and 177.5 cm^3, respectively.

The VMAT plan was computed with four ipsilateral partial arcs and 6X photons. Each arc covered an arc sector of 179° with 114 control points. Eclipse automatically

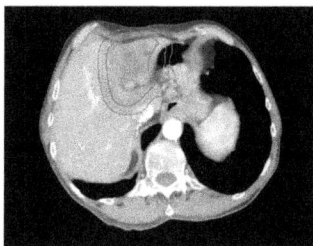

Figure 7.6. The delineation of the PTV3750 (pink) and PTV6750 (red).

Table 7.8. The VMAT plan dosimetry statistics.

Anatomical Structure Name	Dosimetric Parameter	Plan Value
PTV	D95%V100%Dmax	100.0%95.0%111.1%
GTV	V100%Dmax	100%111.1%
Cord	Dmax	1172 cGy
Duodenum	DmaxV3750 cGy	4494 cGy25 cm^3
Esophagus	Dmax	4031 cGy
Heart	V4000 cGy	0.0%
Kidney (right)	V2000 cGy	0.0%
Kidneys (both)	V2000 cGy	0.0%
Large bowel	Dmax	1629 cGy
Liver-GTV	V (<2400 cGy)Dmean	829.2 cm^{3}1586.1 cGy
Lung-GTV	V2000 cGy	1.0%
Small bowel	DmaxV3750 cGy	1439 cGy0.0 cm3
Stomach	DmaxV3750 cGy	4494 cGy25.0 cm^3

increases the control point density in relationship to the gantry angle (number of control points/gantry angle) when a short arc is used. This aims at compensating for the loss in the number of intensity modulation levels caused by a reduced arc length. A shorter arc always provides a higher control point density than a longer one. For example, a full arc of 360° has a control point density of 0.5 control points per degree, whereas a short arc of less than 180° could have a control point density of >1.5 control points per degree. This inherent correlation between the control point density and arc length plays an important role in balancing between plan quality and plan delivery efficiency. The full arc approach in this case was not desirable because the PTV was shallow and also not centrally located. This could result in the beam path length varying greatly at different control points. It would deliver a long tail of low dose to a large volume of the uninvolved normal tissue in both posterior and contralateral sides, making lower isodose lines extremely non-conformal. All the dosimetric parameters in the plan met our institutional plan acceptance criteria (table 7.8). In particular, the uninvolved part of the liver was well spared with V

($<$2400 cGy) = 829.2 cm^3. The MU for each arc was 381, 277, 315, and 338, respectively, yielding an IMRT factor of 2.91. Thus, this was a relatively efficient plan. Using a dose rate of 600 MU/min, each arc was delivered in 2 DIBH sessions. There were no significant breaks during delivery or between fields.

Figure 7.7 shows the dose coverage of the VMAT plan for the PTV3750 in three orthogonal planes. The pink contour represents PTV3750. The inner red area is the PTV6750. The green cross represents the isocenter of the plan. There were no near large hot bands or visible hot spots around the PTV3750. There were also no distant hot spots away from the PTV3750. Particularly, there were no hot bands on the ipsilateral side as commonly observed in dose painting plans when the spatial separation between PTVs of different dose levels was short. As the colors revealed, there were steep dose gradients between the PTV6750 and PTV3750. Each dose level conformed to its corresponding target quite well. Figure 7.7(d) shows a representative control point aperture with its dose contribution projected onto the DRR. Several interesting phenomena were observed. (1) The isocenter of the plan was placed very close to the geometric center of the PTVs. This is an important practice for VMAT planning. If the isocenter had been placed away from the geometric center of the PTVs, the field size defined by jaws would have become unnecessarily large and the MLC leaf travel distance would have become longer. These would have not only increased radiation leakage but also compromised intensity modulation efficiency. (2) The control point aperture size was much larger than those of an IMRT plan. This was due to the fact the VMAT adopts the aperture-based inverse

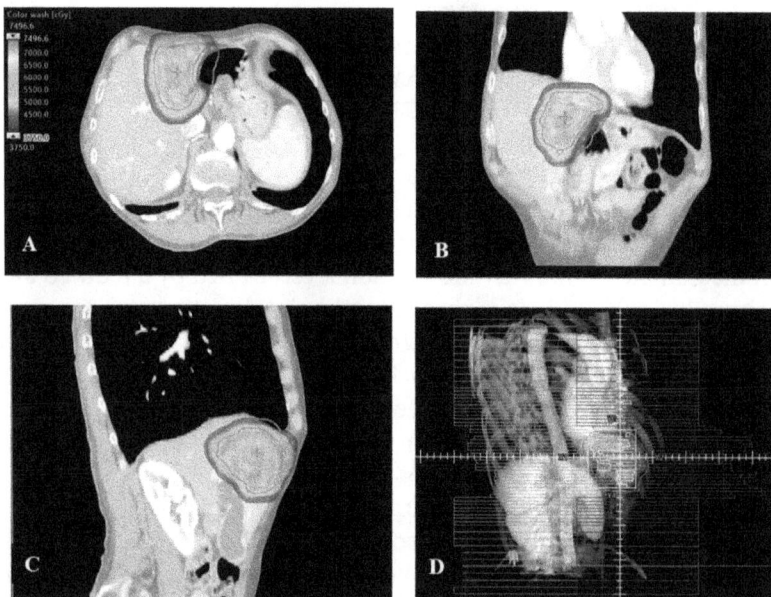

Figure 7.7. Dose coverage of the VMAT plan for the PTV3750 in three orthogonal planes: the axial plane (a), coronal plane (b), and sagittal plane (c). Panel (d) is a representative control point aperture with the dose contribution overlaid on the DRR.

optimization, whereas IMRT uses the beamlet-based inverse optimization. This is the very reason why VMAT plans have a lower total MU and are more time-efficient than IMRT plans. (3) The control point aperture focused primarily on the PTV6750. This was the higher dose level PTV. Thus, the total amount of time that the control point aperture focused on the PTV6750 was roughly 80% longer than on the PTV3750. (4) Jaws immediately followed the most trailing leaves in both directions. This was the function of the jaw tracking algorithm. Without it, the jaws would be at least 8 mm outside the PTV3750 (pink contour). Figure 7.8 shows the dose coverage of the VMAT plan for the PTV6750 in three orthogonal planes. The red contour represents PTV6750. The dose coverage appeared to be more conformal to the PTV6750 than to the PTV3750. No higher isodose lines were extending into the PTV3750. Also, there were no cold spots inside the PTV6750.

The second case was a young woman who was initially diagnosed with Stage III right breast cancer and had undergone a mastectomy. Shortly after her mastectomy, the patient developed several oligometastases involving the liver, bone, skin, and cervix. The liver MRI showed a 1.3 cm T1 hypointense lesion in segment 4b with a diffusion-weighted signal. A liver biopsy confirmed metastatic disease from the patient's primary breast cancer. The patient was enrolled in an internal treatment protocol and randomized to the SBRT cohort with a prescription dose of 900 cGy × 5. The patient was simulated with the DIBH protocol. Figure 7.9 shows the GTV (pink) and PTV (red) in the liver. The PTV was created by adding a 5 mm 3D uniform margin to the GTV with the exception at the borderline with the stomach. Relevant OARs were also contoured. These included the cord, large bowel, small bowel, liver,

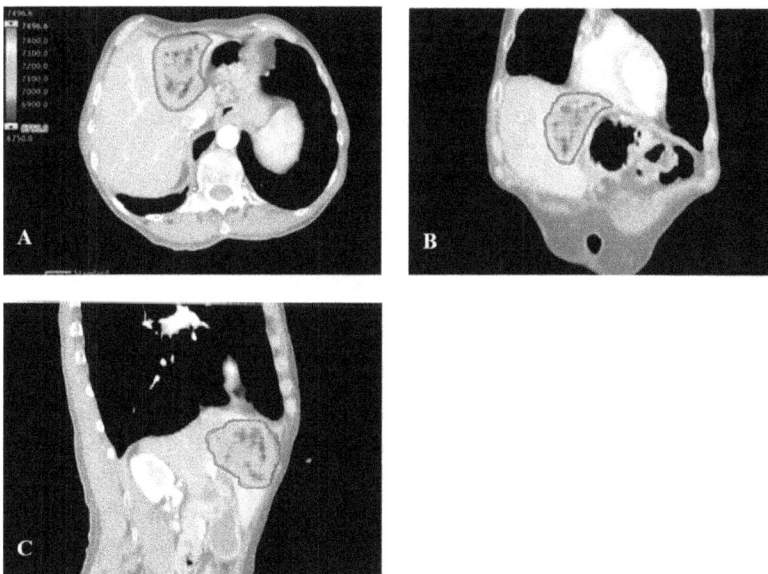

Figure 7.8. Dose coverage of the VMAT plan for the PTV6750 in three orthogonal planes: the axial plane (a), coronal plane (b), and sagittal plane (c). Target dose conformality was excellent. There were no cold spots inside the PTV6750 and hot spots outside it.

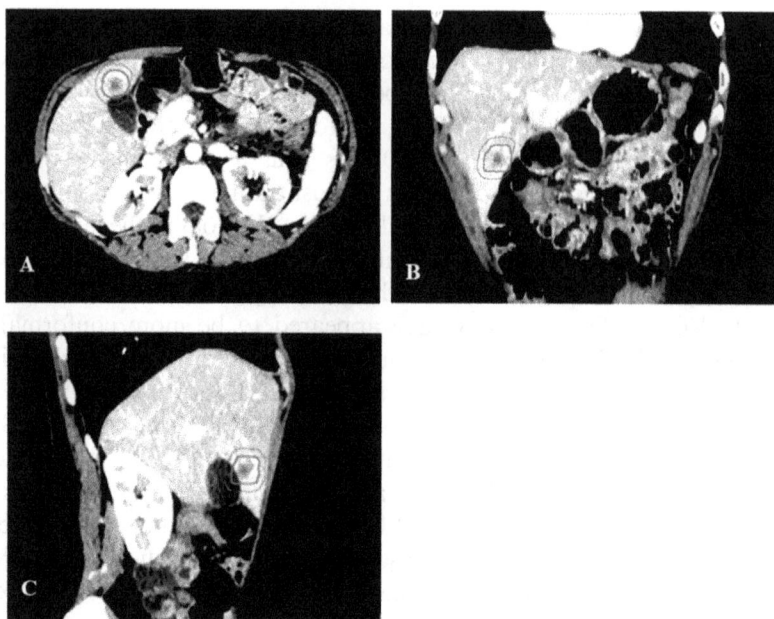

Figure 7.9. A hepatic metastasis from the primary breast cancer. The patient was scanned with the DIBH protocol. Axial plane (a), coronal plane (b), and sagittal plane (c). The hepatic metastasis measured 1.3 × 1.0 cm in the greatest dimension and was very close to the skin.

kidneys, skin, and stomach. The skin was defined as the layer of superficial normal tissue between the skin surface and 5 mm depth. Because the target was shallow, low energy photons were used. Therefore, the skin structure served a useful dose tuning tool in controlling the hot spots on the skin surface. To better spare the stomach, a protective structure was created by adding a 5 mm 3D uniform margin to the stomach (figure 7.10). Other dose tuning structures, such as rinds, PTV optimization, and normal tissue, were also created to enhance the dose gradients between the PTV and gastrointestinal (GI) structures and eliminate distant hot spots.

A VMAT plan was computed for the patient. The VMAT plan consisted of two ipsilateral partial arcs covering an arc sector of 200°. The second arc had a collimator rotation of 90° to boost intensity modulation. Each arc contained 114 control points (figure 7.10); **01** and **02** indicated Arc01 and Arc02, respectively. Short straight lines perpendicular to the arc sector represented control points. Except for the first and last control points, all other control points were evenly distributed over the arc sector, with a gantry angular interval of 1.8°. The first and the last control points had a gantry angular interval of 0.9°. The three yellow long straight lines for each arc represented the beam aperture size of the first control point. The middle line was the beam central axis. The green contour represented all GI structures, including the stomach. The blue contour represented the protective structure for the GI structures, which was created by adding a 5 mm 3D uniform margin to the GI structures. The protective structure abutted the PTV. The plan was

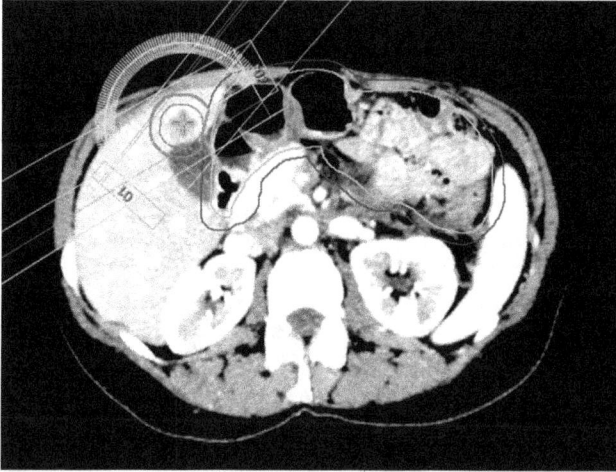

Figure 7.10. The arc arrangement of the VMAT plan. Each arc had 114 control points. The short yellow lines perpendicular to the arc sector represented the control point distribution. For isocentric planning technique, they pass through the isocenter.

optimized with 6X flattening filter–free (FFF) photons with a maximum dose rate of 1400 MU/min to improve plan delivery efficiency. The MU for each arc was 819 and 965, respectively, yielding an IMRT factor of 1.982. This was a quite efficient plan given the fact that the PTV abutted the stomach. Each arc was delivered within three DIBH sessions. During plan delivery, the dose rate varied between 854.411 MU/min and 1400 MU/min, and the gantry angular speed varied between 2.919°/s and 6.0°/s.

For Varian TrueBeam Linacs, the maximum design gantry angular speed is 6.0°/s. This is the speed limit imposed by United States regulatory bodies. For plans with conventional fractionation, the gantry always rotates at the maximum angular speed in order to achieve the highest plan delivery efficiency. However, for SBRT and stereotactic radiosurgery (SRS) plans, the gantry angular speed often has to be slowed down to achieve the highest angular dose rate (MU/°) as a result of both high MU/field and a shorter arc. In this plan, the angular dose rate varied between 2.373 MU/° and 7.993 MU/°. The lowest angular dose rate occurred when the temporal dose rate (MU/min) was at the lowest and the gantry angular speed was at the highest. The highest angular dose rate occurred when the temporal dose rate was at the highest and the gantry angular speed was at the lowest. This synergistic interplay between the MLC leaf speed, dose rate, and gantry angular speed not only improves the VMAT plan delivery efficiency but also enhances the dynamic range of intensity modulation.

Figure 7.11 shows the dose distribution of the VMAT plan in three representative orthogonal planes [figures 7.11(a)–(c)]. Figure 7.11(d) is a representative beam aperture of Arc02 projected onto the DRR. The red outline outside the dose distribution is the PTV. As the color indicates, the dose distribution within the PTV is fairly homogeneous. This feature is different from that of dose painting plans with multiple dose levels, which tend to produce a highly inhomogeneous dose

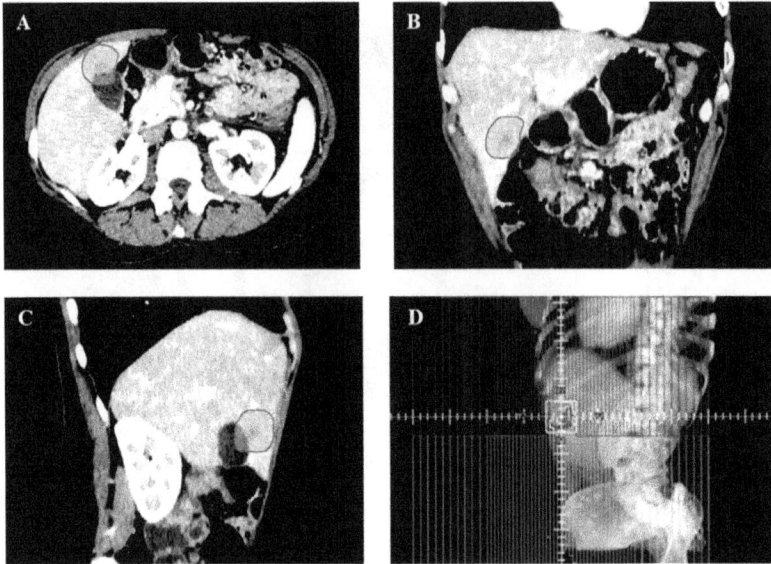

Figure 7.11. Dose distribution in three representative orthogonal planes: axial plane (a), coronal plane (b), and sagittal plane (c), and the beam aperture of the first control point of Arc02 (d).

distribution, even within the same dose level target. This is the direct consequence of discontinuity in prescribed dose levels. From a physics point of view, it is impossible to create a dose distribution like a step function. Another interesting phenomenon observed in the dose distribution is that the dose in the center is slightly cooler than the dose in the peripheral region. This is because the GTV has a lower mass density than the peripheral normal liver tissue. It is well known that the absorbed dose of a medium is related to its mass density [43]. For SBRT or ablative cases in the upper abdomen, we typically set up the patients to the implanted fiducials or stent and treat them with DIBH and intra-fractional motion review (IMR). For free-breathing cases, we typically set up the patients to a fixed bony landmark like the spine and treat the patients without IMR even if they have implanted fiducials. This is because kV imaging represents only a snapshot of the random moving position of the fiducial in a free-breathing cycle. It is not a reliable guidance structure for free-breathing cases. Free-breathing CBCT is also not very precise because it is time and volume averaged. For this case, the patient was set up to the liver with DIBH.

For SBRT or ablative cases in the upper abdomen, setting up the patient to a wrong guidance structure could result in target miss and even a serious complication. As shown in figure 7.11(d), the plan isocenter was placed in the geometric center or the mass center of the PTV. This is especially important for SBRT plans because of high MU and long treatment time. If the isocenter was placed off the geometric center of the PTV, the field sizes defined by jaws would become very large and radiation leakage throughout MLC leaves would be significantly increased. Under this circumstance, a jaw tracking technique can reduce radiation leakage through MLC leaves but cannot completely eliminate it. Table 7.9 shows the VMAT

Table 7.9. The VMAT plan dosimetry statistics.

Anatomical Structure Name	Dosimetric Parameter	Plan Value
PTV	D95%V100%Dmax	101.7%99.2%109.5%
GTV	V100%Dmax	100%109.5%
Cord	DmaxV1000 cGy	313 cGy0.0 cm^3
Kidney (right) Kidneys (left)	V1500 cGyV1500 cGy	0.0% 0.0%
Large bowel	DmaxD5 cm^3	538 cGy324 cGy
Liver-GTV	V (<1500 cGy)Dmean	1096.5 cm^{3}348.4 cGy
Small bowel	DmaxD5 cm^3	
V2000 cGy	1414 cGy213.0 cm^{3}0.0 cm^3	
Stomach	DmaxD5 cm^3	2691 cGy1579.0 cGy

plan dosimetry statistics. All dosimetric parameters met our institutional plan acceptance criteria and normal tissue constraints. The anatomical structures not listed in the table were distant from the PTV. Therefore, the doses to these structures were insignificant.

7.4 The future of Linacs in abdominal cancer treatment

For the past two decades, great progress has been made in cancer research and cancer treatment. Many novel biologic treatment modalities have emerged and implemented clinically. In addition to advances in medical therapy, there has been significant progress in the technological innovation of cancer treatment equipment and devices. Advances in the engineering and design of radiation treatment machines have played an important role in improving clinical treatments and patient outcomes for cancer patients [44]. As cancer incidence continues to rise worldwide, cost-effective cancer treatment, personalized cancer management, and eventual eradication of cancer occurrences remain a formidable challenge to biomedical engineers, cancer researchers, and clinicians. Currently, the major cancer treatment modalities include surgery, chemotherapy, targeted therapy, radiation therapy, immunotherapy, and hormonal therapy [45]. Among them, radiation therapy is still one of the most important treatment modalities. It is estimated that more than 50% of all cancer patients receive radiation therapy during their cancer treatments and management. In the next two to three decades, MR-Linacs, proton therapy, and other types of heavy-ion particle therapies will play a more important role in modern radiation oncology. Despite their advantages in physical properties, the biological effectiveness has only been marginally improved. Also, high capital investment in facility construction and maintenance and unresolved insurance reimbursement issues are other major obstacles to their widespread clinical adoption. Because of these, we believe that Linacs will continue to be the predominant modality in radiation oncology in the coming two to three decades.

References

[1] Reed A B 2011 The history of radiation use in medicine *J Vasc. Surg.* **5** 3S–5S

[2] Knutsson F 1974 Röntgen and the Nobel Prize: the discussion at the Royal Swedish Academy of Sciences in Stockholm in 1901 *Acta Radiol. Diagn.* **15** 465–73

[3] Howard-Flanders P 1954 The development of the linear accelerator as a clinical instrument *Acta Radiol.* **41** 649–55

[4] Ginzton E L, Mallory K B and Kaplan H S 1957 The Stanford medical linear accelerator: I. Design and development *Stanford Med. Bull.* **15** 123–40

[5] Brahme A 1993 Optimization of radiation therapy and the development of multileaf collimator *Int. J. Radiat. Oncol. Biol. Phys.* **25** 373–5

[6] Stein J, Bortfeld T and Dorschel B *et al* 1994 Dynamic x-ray compensation for conformal radiotherapy by means of multi-leaf collimation *Radiat. Oncol.* **32** 163–73

[7] Chui C S, LoSasso T and Spirou S 1994 Dose calculation for photon beams with intensity-modulation generated by dynamic jaw or multi-leaf collimations *Med. Phys.* **21** 1237–44

[8] Ling C C, Burman C and Chui C S *et al* 1996 Conformal radiation treatment of prostate cancer using IMRT photon beams produced with dynamic multileaf collimation *Int. J. Radiat. Oncol. Biol. Phys.* **35** 721–30

[9] Burman C, Chui C S and Kutcher G *et al* 1997 Planning, delivery and quality assurance of IMRT using dynamic multileaf collimator: a strategy for large-scale implementation for the treatment of carcinoma of the prostate *Int. J. Radiat. Oncol. Biol. Phys.* **39** 836–73

[10] Hong L, Hunt M and Chui C S *et al* 1999 Intensity modulated tangential beam irradiation of the intact breast *Int. J. Radiat. Oncol. Biol. Phys.* **44** 1155–64

[11] Hunt M A, Zelefsky M and Wolden S *et al* 2001 Treatment planning and delivery of IMRT for primary nasopharynx cancer *Int. J. Radiat. Oncol. Biol. Phys.* **48** 623–32

[12] Sidhu K, Ford E and Spirou S *et al* 2003 Optimization of conformal thoracic radiotherapy using cone-beam CT imaging for treatment verification *Int. J. Radiat. Oncol. Biol. Phys.* **55** 757–67

[13] Bucci M K, Bevan A and Roach M 2005 Advances in radiation therapy: conventional to 3D, to IMRT, to 4D, and beyond *CA Cancer J. Clin.* **55** 117–34

[14] Chui C S, Chan M F and Yorke E *et al* 2001 Delivery of intensity-modulated radiation therapy with a conventional multileaf collimator: comparison of dynamic and segmental methods *Med. Phys.* **28** 2441–9

[15] Song Y, Zhang P, Obcemea C, Mueller B, Burman C and Mychalcza B 2009 Dosimetric effects of gantry angular acceleration and deceleration in volumetric modulated radiation therapy *World Congress on Medical Physics and Biomedical Engineering* ed O Dössel and W C Schlegel 25/1 (Berlin: Springer) pp 1046–50

[16] Song Y, Zhang P, Wang P, Obcemea C, Mueller B, Burman C and Mychalczak B 2009 The development of a novel radiation treatment modality—volumetric modulated arc therapy *Annu. Int. Conf. of the IEEE Engineering in Medicine and Biology Society* 3401–4

[17] Otto K 2008 Volumetric modulated arc therapy: IMRT in a single gantry arc *Med. Phys.* **35** 310–7

[18] Zygmanski P, Hogele W and Cormack R *et al* 2008 A volumetric-modulated arc therapy using sub-conformal dynamic arc with a monotonic dynamic multileaf collimator modulation *Phys. Med. Biol.* **53** 6395–417

[19] Gladwish A, Oliver M and Craig J *et al* 2007 Segmentation and leaf sequencing for intensity modulated arc therapy *Med. Phys.* **34** 1779–88

[20] Luan S, Wang C and Cao D *et al* 2008 Leaf-sequencing for intensity-modulated arc therapy using graph algorithms *Med. Phys.* **35** 61–9

[21] Wang C, Luan S and Tang G *et al* 2008 Arc-modulated radiation therapy (AMRT): a single-arc form of intensity-modulated arc therapy *Phys. Med. Biol.* **53** 6291–303

[22] Bortfeld T and Webb S 2009 Single-arc IMRT? *Phys. Med. Biol.* **54** N9–20

[23] Li G, Citrin D, Camphausen K, Mueller B, Burman C, Mychalczak B, Miller R W and Song Y 2008 Advances in 4D medical imaging and 4D radiation therapy *Technol. Cancer Res. Treat.* **7** 67–82

[24] Sharma A 2020 Global linear accelerators for radiation market forecast 2024 by industry size and share, demand, worldwide research, prominent players, emerging trends, investment opportunities and revenue expectation https://medium.com/@anandsharma9423/global-linear-accelerators-for-radiation-market-forecast-2024-by-industry-size-and-share-demand-bcef60beed93

[25] NIH 2022 Cancer stat facts: esophageal cancer https://seer.cancer.gov/statfacts/html/esoph.html

[26] NIH 2022 Cancer stat facts: stomach cancer https://seer.cancer.gov/statfacts/html/stomach.html

[27] Song Y, Mueller B, Dow K, Saleh Z, Tang X, Zinovoy M, Gelblum D and Mychalczak B 2021 Considerations and experience in the treatment of lung cancer with VMAT SBRT +DIBH in arms-down position *Int. J. Med. Phys. Clin. Eng. Radiat. Oncol.* **10** 69-80

[28] NCCN 2022 NCCN guidelines https://nccn.org/professionals/physician_gls/default.aspx#site

[29] Unkelbach J, Bortfeld T and Craft D *et al* 2015 Optimization approaches to volumetric modulated arc therapy planning *Med. Phys.* **42** 1367–77

[30] Zhang P, Happersett L and Yang Y *et al* 2009 Optimization of collimator trajectory in volumetric modulated arc therapy: development and evaluation for paraspinal SBRT *Int. J. Radiat. Oncol. Biol. Phys.* **77** 591–9

[31] Xu D, li G, Li H and Jia F 2017 Comparison of IMRT vs. 3D-CRT in the treatment of esophagus cancer: a systematic review and meta-analysis *Medicine* **96** e76885

[32] Shi A, Liao Z and Allen P K *et al* 2017 Long-term survival and toxicity outcomes of IMRT for the treatment of esophageal cancer: a large single-institutional cohort study *Adv. Radiat. Oncol.* **2** 316–24

[33] Munch S, Aichmeier S and Hapfelmeier A *et al* 2016 Comparison of dosimetric parameters and toxicity in esophageal cancer patients undergoing 3D conformal radiotherapy or VMAT *Strahlenther. Onkol.* **192** 722–9

[34] American Cancer Society 2022 Key statistics about liver cancer https://cancer.org/cancer/liver-cancer/about/what-is-key-statistics.html#:~:text=The%20American%20Cancer%20Society's%20estimates,will%20die%20of%20these%20cancers

[35] Ravaioli M, Ercolani G and Neri F *et al* 2014 Liver transplantation for hepatic tumors: a systematic review *World. J. Gastroenterol.* **21** 5345–52

[36] Lurje I, Czigany Z and Bednarsch J *et al* 2019 Treatment strategies for hepatocellular carcinoma—a multidisciplinary approach *Int. J. Mol. Sci.* **20** 1465

[37] Schaub S K, Hartvigson P E and Lock M I *et al* 2018 Stereotactic body radiation therapy for hepatocellular carcinoma: current trends and controversies *Technol. Cancer Res. Treat.* **17** 1533033818790217

[38] Park J, Park J W and Kang M K 2019 Current status of stereotactic body radiotherapy for the treatment of hepatocellular carcinoma *Yeungnam Univ. J. Med.* **36** 192–200

[39] Que J, Lin C and Lin L *et al* 2020 Challenges of BCLC stage C hepatocellular carcinoma: results of a single-institutional experience on stereotactic body radiation therapy *Medicine (Baltimore)* **99** e21561

[40] Spieler B, Mellon E A and Jones P D *et al* 2019 Stereotactic ablative radiotherapy for hepatocellular carcinoma *Hepatoma. Res.* **5** 4

[41] Song Y 2020 *BER_TB2 AlignRT v6.2 Commissioning Summary Report* Memorial Sloan Kettering Cancer Center

[42] Song Y, Mueller B, Laguna J, Obcemea C, Saleh Z, Tang X, Both S and Mychalczak B 2016 Investigation of MLC leakage dose in VMAT head and neck treatment *ASTRO 2016 Annual Meeting*

[43] Venkatesh S K, Chandan V and Roberts L R 2014 Liver masses: a clinical, radiological and pathological perspective for: perspectives in clinical gastroenterology and hepatology *Clin. Gastroenterol. Hepatol.* **12** 1414–29

[44] Baskar R, Lee K A and Yeo R *et al* 2012 Cancer and radiation therapy: current advances and future directions *Int. J. Med. Sci.* **9** 193–9

[45] Alexander-Bryant A A, Vanden Berg-Foels W S and Wen X 2013 Bioengineering strategies for designing targeted cancer therapies *Adv. Cancer Res.* **118** 1–59

IOP Publishing

Principles and Practice of Image-Guided Abdominal Radiation Therapy

Yu Kuang

Chapter 8

Abdominal radiotherapy with tomotherapy

Yidong Yang and Xudong Xue

This chapter is dedicated to intensity-modulated radiation therapy (IMRT) of abdominal cancer using tomotherapy (TOMO). We first introduce the design and construction of the tomotherapy system. Then, we describe onboard imaging and treatment procedures with tomotherapy, and attention is especially paid to adaptive radiation therapy and motion management. Finally, we discuss the clinical indications of treating liver and pancreas cancer using tomotherapy by presenting the suitability and advantages of tomotherapy treatment and its comparison with conventional IMRT techniques on c-arm–based linear accelerator (Linac). We also briefly explain special treatment procedures, total body irradiation (TBI), total marrow irradiation (TMI) and total marrow and lymphoid irradiation (TMLI), with tomotherapy.

Tomotherapy, also called helical tomotherapy, is a unique IMRT modality developed by Mackie and colleagues [1–3]. In the first tomotherapy paper, which was published in 1992 [4], the concepts of helical radiation delivery, together with dose reconstruction and adaptive radiotherapy (ART), were proposed using the integrated megavoltage (MV) imaging and treatment system. During tomotherapy treatment, the treatment target is scanned across a modulated "slice" beam while the treatment couch is moving forward [4]. The appearance and structure of tomotherapy are similar to an MV-version helical computed tomography (CT) and are distinct from a conventional Linac. In this chapter, a conventional Linac or radiotherapy modality refers to the Linac machine or treatment with a c-armed gantry design and a two-dimensional multi-leaf collimator (MLC) field. Figure 8.1 shows the general layout of a tomotherapy unit. Since its first clinical application in 2002 at University of Wisconsin Madison clinics [2], tomotherapy has been utilized to treat various tumor types, including but not limited to prostate cancer, head and neck cancer, lung cancer, abdominal cancer, etc. In addition, its advantage of treating long targets without connecting radiation fields otherwise required by conventional radiotherapy modality makes it particularly useful for IMRT-based

doi:10.1088/978-0-7503-2468-7ch8

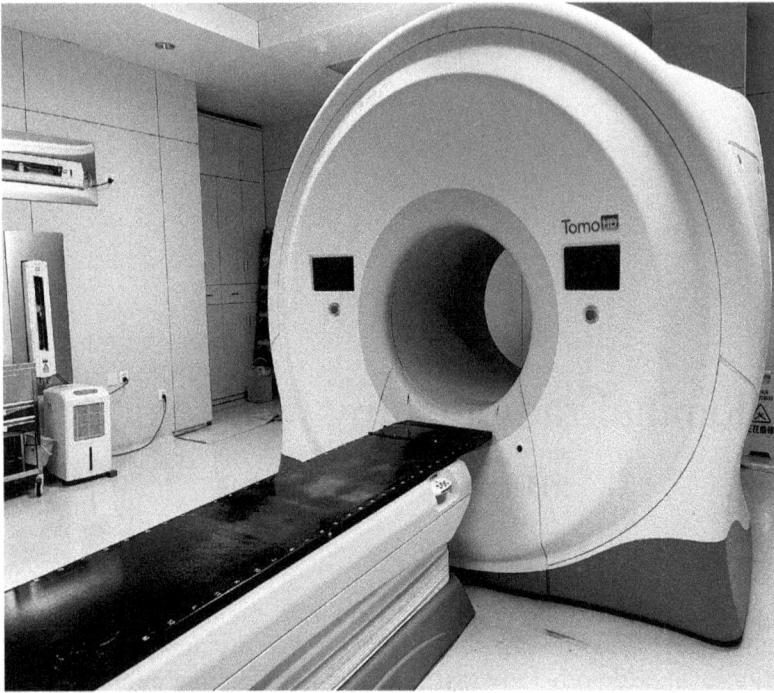

Figure 8.1. The layout of the tomotherapy system.

TBI or TMI treatment. Compared with conventional Linac, helical tomotherapy offers dosimetric advantages in terms of target dose conformity and critical structure sparing, using a continuously rotating gantry with a binary MLC. The on-board MVCT system provides precise image guidance that enables adapted treatment planning [5]. In this chapter, an overview of the tomotherapy system is provided. Treatment planning, image guidance, radiation delivery and quality assurance (QA), particularly for the treatment of abdominal tumors, are discussed.

8.1 System overview

Different than the conventional radiotherapy modalities, the linear accelerator in tomotherapy is installed on a slip ring gantry and produces a flattening-filter-free 6 MV beam. The beam passes through the primary collimator and is collimated to a fan shape by a pair of adjustable jaws. Then, it is further modulated by a row of pneumatic binary MLCs before being delivered to the treatment target. During treatment, the gantry rotates continuously, and the patient steps through the rotating beam plane via couch translation. The distance from the source to the center of rotation or radiation isocenter is 85 cm. Figure 8.2(a) shows the gantry of a typical tomotherapy system [3].

The fan-shaped radiation field has a maximum size of 40 cm in the transverse (in-plane) direction and can be collimated to three different widths, namely, 1.0, 2.5 and 5.0 cm defined by the jaw opening, along the axial (cross-plane) direction. The

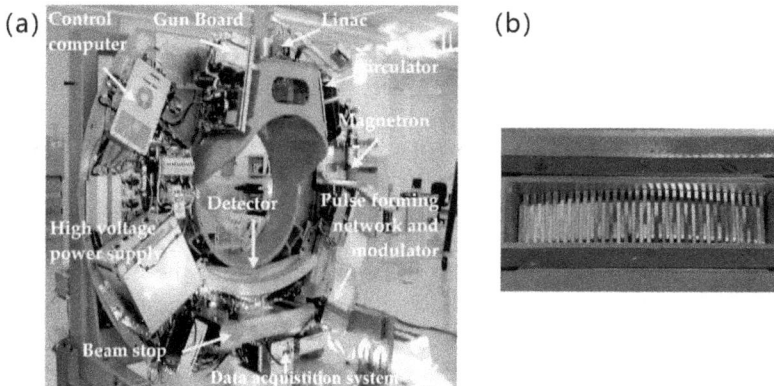

Figure 8.2. (a) The gantry of a typical tomotherapy system [3]. (b) A photo of the binary MLC. Reproduced from [3] with permission from Elsevier.

intensity modulation is achieved through programing the opening time of each of the 64 MLC leaves. Each leaf is either fully open (in "on" state) or fully closed (in "off" state). Figure 8.2(b) shows the appearance of the MLC structure. A beamlet in tomotherapy usually refers to the part of the treatment beam passed through one single MLC leaf opening. The beamlet size along the transverse direction, which is determined by the MLC leaf width, is 0.625 cm at the radiation isocenter.

In tomotherapy, each gantry rotation consists of 51 equally spaced beam projections. For one projection, each leaf is individually controlled to deliver a specific open or closed duration. The MLC moves at a speed of 250 cm s^{-1}, which is much faster than the MLC installed on Linac. Therefore, its beam intensity-modulation ability is also much greater than that of the conventional MLC.

In addition, tomotherapy overcomes some limitations of conventional radiotherapy modality. For example, the maximum length of the target volume treated on conventional accelerators is usually 40 cm at a time, but that value can be up to 160 cm in a single tomotherapy treatment [6,7]. Therefore, multiple lesions, large volume or long tumors can be treated in tomotherapy without connecting fields.

8.2 On-board imaging

Tomotherapy has a built-in on-board MVCT system with a nominal energy of 3.5 MV generated from the same accelerator source. The advantage of such a design is that the imaging and treatment beams are inherently aligned, which minimizes the laborious QA check of the imaging and radiation field registration [8]. In addition, the imaging x-ray energy is in MV range, close to that of the treatment beam, so as to provide tissue electron density suitable for dose calculation.

The image acquisition of MVCT is accomplished by the conventional xenon CT detector array on the opposite side of the radiation source. The detector array has an arc shape and is composed of 640 dual-channel ionization chambers, of which 520 chambers are used in imaging. Each ionization chamber is separated by a 0.32 mm thick tungsten spacer, and the length of each spacer in the beam direction is 2.54 cm.

The field of view has a diameter of 40 cm at the isocenter. The distance from the source to the ionization chamber is 129.2 cm, but the focal point of the arc-shaped detector is at 103.6 cm from the chamber. The on-purpose misalignment of the detector focus improves the MV signal detection efficiency by the photon interactions with the tungsten spacers [9, 10].

8.3 Treatment procedures for tomotherapy

The uniqueness of tomotherapy lies in the combination of the helical CT scanning geometry and linear accelerator–based radiation delivery. The rotational and slice-by-slice radiation delivery approach improved the dose conformity to the target and sparing to the organs at risk (OARs). To date, tomotherapy has been applied to the treatment of nearly all cancer types located in any body regions. Since its first treatment in the First Affiliated Hospital of the University of Science and Technology of China, tomotherapy has been used to treat head and neck, lung, liver, pancreatic, stomach, prostate, gynecological and rectal cancers. In addition, special clinical procedures have been developed for total central nervous system, TBI and TMI. In the following, we briefly introduce the treatment workflow and procedures of abdominal tomotherapy at our center.

8.3.1 Simulation

Patients are usually immobilized in supine position using a negative-pressure vacuum bag (as shown in Figure 8.3), with elbows folded and hands on their forehead. Crosshair reference positions are marked on both the vacuum bag and the patient's skin. Radiopaque markers are then pasted on the patient's skin, coincident with the crosshair, to indicate the positioning reference center in the CT image during treatment planning. After that, the CT scan (regular and contrast CT) is performed under free breathing (or deep expiration breath hold if there is significant tumor motion). Reconstructed CT images are transferred to the treatment planning system (TPS) for treatment planning.

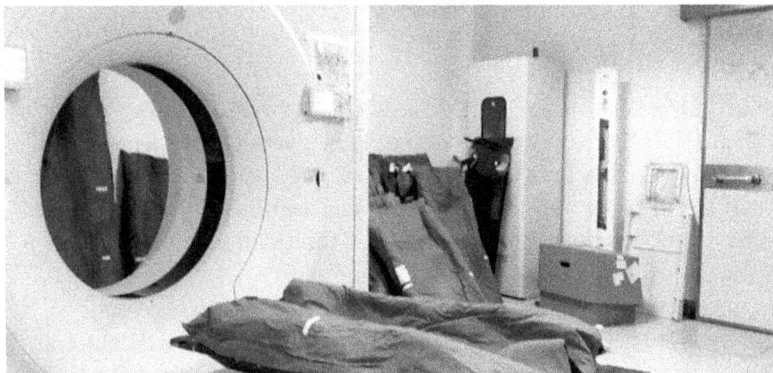

Figure 8.3. Simulation CT and Immobilization device. The blue on the CT couch is the vacuum bag used for immobilization of patients with abdominal cancer. © From the Anhui Provincial Hospital

8.3.2 Treatment planning

After patient CT images are imported into the TPS, radiation oncologists draw the contours of tumor target and OARs according to clinical protocols. Often, the contouring process is performed on some TPS, such as the Pinnacle TPS, or image process software, such as the MIM software, other than the tomotherapy TPS. If that is the case, additional data transfer is necessary to bring the patient CT and contours into the tomotherapy TPS. In tomotherapy TPS, the CT couch is first replaced by the treatment couch, and an appropriate electronic density table selected. Radiation oncologists drop the prescription and dose constraints. Dosimetrists then set the positioning reference center, set the jaw setting and pitch values, and determine the initial dose objectives for the tumor target and OARs. Subsequently, the beamlets, including their sequences and weights, are optimized to obtain an optimal dose distribution, maximizing the dose coverage to the target and minimizing irradiation of OARs and healthy tissues.

There are some planning parameters unique to tomotherapy that needed to be specifically determined when making a helical treatment plan: beam width, pitch and modulation factor. The beam width is defined by the collimating jaws moving along the longitudinal direction and selected from 1.0, 2.5 and 5.0 cm—three typical values. The pitch value is defined as the ratio of the couch travel per gantry rotation to the beam width. An empirical pitch value of $0.86/n$ is usually recommended [11], where n is a positive integer. However, as for the moving target such as tumor in the lung and abdomen, loose helical beams with pitches >1 could provide more homogeneous dose distribution [12]. The modulation factor is the ratio of the maximum leaf opening time to average leaf opening time, and its value indicates a tradeoff between dose conformality and delivery efficiency. A higher modulation factor can improve plan quality but may increase delivery time and decrease delivery efficiency.

8.3.3 Plan QA

Like any Linac-based IMRT plan, a tomotherapy plan should also be checked for its dose-delivery accuracy. A delivery QA procedure, designed to recalculate the dose in a cylindrical, virtual water phantom (also called "cheese phantom"), is integrated in the TOMO planning software package.

An appropriate module is first selected following TOMO's delivery QA procedure. One important step is to place the cheese phantom in accordance with the plan dose distribution. Particularly, the ion chamber, which is to measure the absolute dose inside the planning target volume (PTV), should be placed in a position where the dose gradient is not too large. After a dry run of dose delivery to the phantom, the QA result is analyzed by comparing the measurement and plan dose. The absolute point dose comparison is based on the ion chamber measurement, and the relative dose distribution comparison is based on the planar dose measurement (figure 8.4).

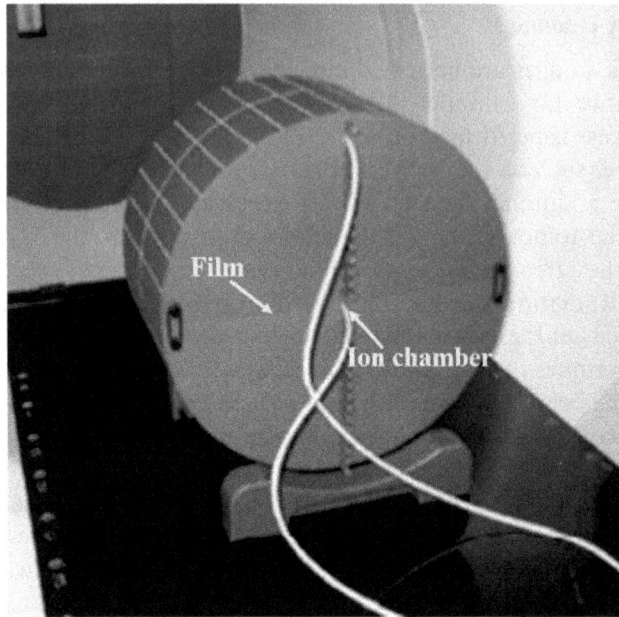

Figure 8.4. The cheese phantom used for TOMO plan QA. The film is sandwiched between the upper and lower phantom measuring the two-dimensional dose distribution. © From the Anhui Provincial Hospital.

8.3.4 Imaging guidance

As mentioned before, TOMO has a built-in on-board MVCT, which is used to verify the patient positioning prior to and potentially during radiation delivery. The MVCT imaging parameters, such as the scanning length, scan pitch value and image reconstruction interval, are determined by users. There are three imaging pitch options: fine, normal and coarse, corresponding to the 4, 8 and 12 mm/rotation couch speed, respectively. The rotation speed during image acquisition is fixed at 10 s/rotation. The reconstructed slice thickness also has three options: 2, 4 and 6 mm, respectively. A smaller reconstruction interval usually improves the longitudinal display resolution, with no impact on scan time or imaging dose. On the other hand, the imaging dose depends on the imaging pitch and slice thickness.

Following initial patient setup using positioning lasers, an on-board MVCT is acquired and rigidly registered to the planning kilovoltage CT (kVCT). The couch position is then corrected according to the registration. TOMO provides automatic registration, but the registration accuracy must be rigorously checked before proceeding to radiation delivery. It is critical to double check and to ensure the registration accuracy during the treatment of abdominal tumors, owing to the relatively large day-to-day variation in the abdominal region. Several factors can cause patient anatomical change, including but not limited to the stomach filling, weight loss and different breathing patterns.

Geometry accuracy, image uniformity, noise, spatial resolution and contrast should all be monitored in the regular MVCT QA. Figure 8.5 shows the kVCT and

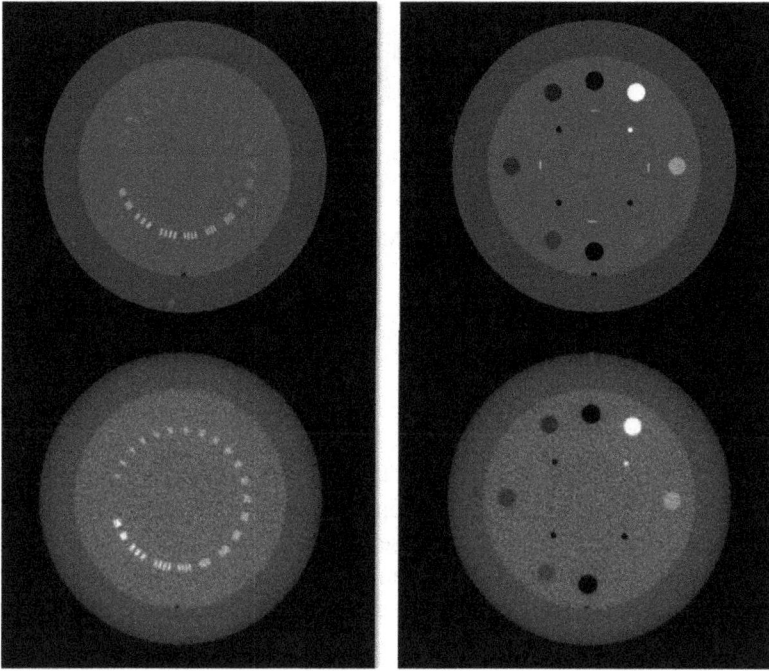

Figure 8.5. kVCT images with 2.5 mm slice thickness (row 1) and fine pitch MVCT with 1 mm slice thickness (row 2). Reproduced from [15] John Wiley & Sons.

MVCT images for the CatPhan 504 (The Phantom Laboratory, Greenwich, NY) phantom. As can be seen, the MVCT exhibits more noise, lower spatial resolution and lower contrast than kVCT at similar pitch setting and radiation dose level [13]. Following the AAPM TG-148 guidelines, Han *et al* developed a useful program to perform the MVCT QA automatically and improved the QA efficiency compared with manual analysis [14]. Velten *et al* investigated the MVCT image quality and found that, with proper scan parameters, MVCT can enable ART on the TOMO system [15]. Considering the large interfractional anatomical variation in the abdomen, adaptive therapy, which re-optimizes the treatment plan by taking the anatomical change into account, can be an attractive and promising treatment strategy in abdominal radiotherapy.

Although the overall image quality is inferior to the kVCT, the MVCT can still provide decent soft-tissue contrast and enable delineation of many soft-tissue structures. Furthermore, due to the high energy, MVCT mitigates the beam-hardening effect and significantly reduces the streak artifacts caused by high-Z material [16]. In addition, the x-ray energy in MVCT is closer to that in treatment; therefore, the CT number and electron density calculated based on the MVCT may be more accurate in terms of treatment dose calculation [17]. Figure 8.6 compares an MVCT with a kVCT. The metal artifact in the kVCT is eliminated in the MVCT, while the soft-tissue structures are still visible in the MVCT [18].

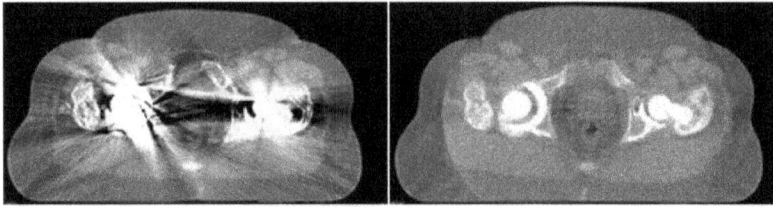

Figure 8.6. Left: A planning kVCT image. Right: An MVCT image from tomotherapy. Reproduced from [18] with permission from Elsevier.

8.3.5 Radiation delivery

Following image guidance, the tumor target is then aligned to the radiation isocenter via couch movement. The patient's size is limited by the physical dimensions of TOMO bore, that is, 85 cm in diameter. It recommends that the couch should not move more than 2 cm laterally away from the radiation isocenter. Therefore, the beam isocenter should be placed as close to the middle of the patient's body as possible during treatment planning, even if the tumor locates in the peripheral regions. This is different than the treatment planning process in the conventional Linac.

8.3.6 Adaptive radiotherapy

ART is a closed-loop radiation therapy process where the treatment response is consistently monitored and feedback is given to modify and improve the subsequent treatment procedures [19]. Tomotherapy can be a natural candidate for ART because of its initial conceptual design and availability of high-quality MVCT. With the on-board MVCT, it allows physicians to visualize the daily change of patient anatomy and thereafter to modify the treatment plan and accommodate the anatomical changes. There are two kinds of ART technologies: offline and online. Currently, TOMO is commonly used in offline ART practice.

TOMO provides an ART package called "Planned Adaptive" to check the possible change of dose distribution by re-calculating the radiation dose on the MVCT of the day. First, the MVCT is rigidly registered to the planning CT. A group of "Verification ROIs," namely, the contours of selected OARs, are mapped from the planning CT to the MVCT and then edited on the merged images. Then, the three-dimensional dose distribution is re-computed according to the planning sinogram, and the differences in isodose lines and dose-volume histogram are evaluated. Finally, the plan is re-optimized based on the MVCT image and edited contours. Wen *et al* implemented an ART program and reduced the radiation dose to the OARs surrounding the target in the subsequent treatment [20]. Martin *et al* compared treatment plans generated on standard kVCT, MVCT and hybrid MVCT/kVCT and demonstrated that the adaptive plan generated on the MVCT or hybrid kVCT/MVCT was equivalent to that on the kVCT [21].

8.3.7 Motion management in abdominal tomotherapy

Periodic motion caused by respiration, combined with the periodic gantry rotation, could lead to sinusoidal variations in radiation dose delivery. The amplitude of the variation can be severe if the two movements are in synchrony as mentioned by Mackie [2]. There are two categories of methods to offset the motion effect. (1) The dose error can be reduced by optimizing the delivery parameters. Yang *et al* studied the motion synchrony effect and found that the dose error can be reduced to an acceptable level if the gantry rotation period was much longer than the breathing period [22]. Kissick *et al* further studied the interplay between a constant scan speed and intrafraction oscillatory motion and the resulting fluence intensity modulations along the axis of motion [23]. They found that the fluence intensity modulations are only a few percent if the gantry rotation speed is slow but is sensitive to the random change of motion rhythm. The uncertainty in an accumulative dose, however, would be greatly mitigated after an adequate number of fractions. Kim *et al* investigated the potential of using "loose helical delivery" to account for respiratory tumor motion in breath-hold treatment [12]. This method divides a treatment into a set of interlaced "loose" helices, each of which starts at different gantry angles and covers the entire target length in one single gantry rotation. They found that the loose delivery keeps the dose modulation under 5% when a pitch factor <7 was selected, at the cost of doubling the total treatment time. (2) The dose error can be mitigated by implementing motion-management strategies, such as breath hold, active breathing control, respiratory gating, real-time target tracking and abdominal compression. Schnarr *et al* added a kV x-ray imaging system–mounted 90 degree offset from the MV beam to a tomotherapy system and an optical camera system above the foot of the couch for better tumor tracking and motion monitoring [24]. Hu *et al* evaluated an abdominal compression technique and found that the magnitude of liver motion was significantly reduced in hepatocellular carcinoma radiotherapy [25].

8.4 Clinical indications of abdominal tomotherapy

Abdominal radiotherapy is always risky, considering the critical structures surrounding the tumor target in the abdominal region. The helical delivery of tomotherapy can generate a high dose gradient and hence enables tumor-dose escalation and/or superior normal tissue sparing. On the other hand, the high-quality MVCT, particularly its decent soft-tissue contrast, increases the accuracy of planning-to-treatment image alignment and therefore reduces the dose-delivery uncertainty.

8.4.1 Liver cancer

Liver cancer is one of the most common malignances in the world. As a local non-invasive treatment method, radiotherapy has yielded promising results for liver cancer management. In particular, the development of image-guided radiation therapy technology has improved the precision and efficacy of liver cancer treatment [26, 27]. Baisden *et al* investigated the dose-volume parameters in liver cancer tomotherapy [28] and found that the helical tomotherapy is capable of performing stereotactic body

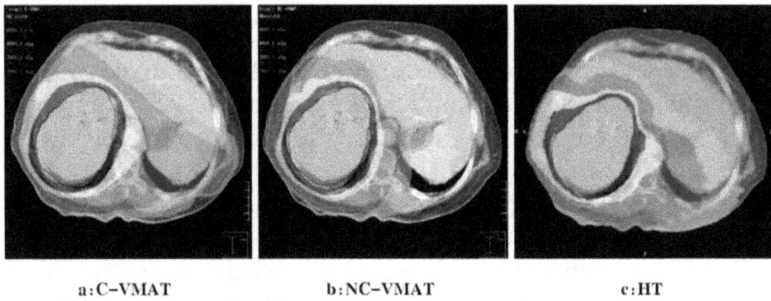

a:C–VMAT   b:NC–VMAT   c:HT

Figure 8.7. Comparison of dose distributions among coplanar VMAT, non-coplanar VMAT and Tomo plans. Reproduced from [30] with permission from *Chinese Journal of Medical Physics.*

radiotherapy for liver lesions. The data provided a framework for predicting the likely maximum tolerable dose for patients receiving liver stereotactic body radiotherapy treated by helical tomotherapy. Hsieh *et al* compared the dose-volume data among coplanar IMRT, noncoplanar IMRT and helical tomotherapy on patients with hepatocellular carcinoma and portal vein thrombosis and found that tomotherapy provided better uniformity for the PTV dose coverage, while noncoplanar IMRT reduced the V_{10Gy} (the volume percentage receiving more than 10Gy dose) to the normal liver [29]. Yan *et al* compared helical tomotherapy, coplanar volumetric modulated radiation therapy (VMAT) and non-coplanar VMAT in hepatocellular carcinoma radiotherapy and found that all three techniques achieved the prescribed dose constraints for the tumor target and OARs (figure 8.7), and tomotherapy significantly reduced the V_{30Gy} of the normal liver of normal liver and improved PTV dose uniformity [30]. Lee *et al* evaluated the tomotherapy treatment of single and multiple liver tumors and showed that tomotherapy had better dose-volume performance than step-and-shoot IMRT for multiple liver tumors at a cost of longer delivery and more monitor units [31]. The slice-by-slice dose delivery usually takes longer than conventional Linac-based IMRT or VMAT. Therefore, whether to choose tomotherapy is also determined by the patient's condition in terms of whether they can tolerate the relatively longer treatment.

8.4.2 Pancreatic cancer

Pancreatic cancer caused 432 000 deaths in the world in 2018 [32]. It ranks as the seventh leading cause of cancer-related mortality, and the 5-year survival rate was only about 9% [33]. Approximately 30% of patients with pancreatic cancer are diagnosed as locally advanced unresectable disease [34]. The role of radiotherapy for locally advanced unresectable disease is controversial owing to the high metastases rate and delicate surrounding OARs in the upper abdomen. Zschaeck *et al* demonstrated a prolonged local control and long-term survival when patients with locally advanced/recurrent pancreatic cancer were treated on tomotherapy with a dose escalation up to 66 Gy to the PTV [35]. Milandri *et al* treated patients with locally advanced pancreatic cancer using tomotherapy plus gemcitabine and oxaliplatin (GEMOX) and showed that the toxicity to both GEMOX and radiotherapy were well tolerated [36]. In another study by Passardi *et al*, patients treated with GEMOX plus tomotherapy had a median progression-free survival rate of 9.3 months and overall survival rate of 15.8 months [37]. These studies demonstrated that tomotherapy alone or chemoradiation regimen is a potential treatment strategy for unresectable locally advanced pancreatic cancer. Appropriate motion management methods should be developed to mitigate the motion effect and reduce the radiation toxicity to the critical structures surrounding the tumor target [38].

8.4.3 TBI, TMI and TMLI

TBI, TMI and TMLI are widely used in conditioning regimens for patients undergoing bone marrow transplantation and hematopoietic stem cell transplantation. The main reason for using these techniques is that removal of tumor cells and immunosuppression allows for bone marrow transplantation from the donor. From April 2000 to April 2019, a total of 1000 cord blood transplant patients were treated in a single center of the first affiliated hospital of USTC, making it the single center with the largest number of cord blood transplant cases in the world. Of them, more than 300 patients were treated with TBI technology.

A prolonged source-to-skin distance and a single large field are used in the conventional TBI technique. The patient locates in a treatment chamber. A 10 mm thick plexiglass plate was placed between the patient and the accelerator. The irradiation strategies are divided into 3~4 Gy low-intensity dose (single dose) and 12 Gy high-intensity dose (two times per day, two days in total). The semiconductor ion chambers are used to monitor patients' dose of head, left lung, right lung, umbilical, knee and ankle.

Owing to the advantage of helical tomotherapy, highly conformal dose could be delivered to large and complex target shapes while simultaneously protecting the critical normal tissues. Especially the maximum target size is 40 cm wide and 160 cm long, which allows for dose delivery in a single time. Different from TBI, TMI is defined as the skeletal bone and total marrow irradiation. TMLI is defined as the treatement of bone, major lymph node chains, liver, spleen and sanctuary site, such as brain [39, 40], as shown in figure 8.8. There are three fractionation strategies for

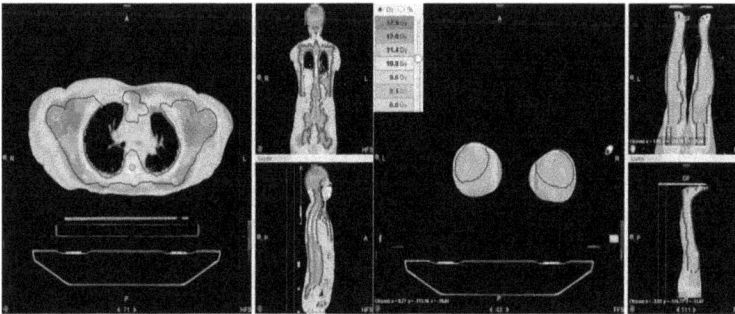

Figure 8.8. Treatment plan dose distribution from a patient treated with 12 Gy of TMLI. © From the Anhui Provincial Hospital.

TMI or TMLI patients, which are 4 Gy×3 f, 4 Gy×3 f and 5 Gy×3 f. The prescribed dose should cover at least 85% of the PTV. Helical mode is used for the upper body and TomoDirect conformal radiotherapy is used for the lower body if the length is greater than 160 cm. Figure 8.8 shows a TMLI patient dose distribution. The patient geometry is usually divided into three ranges during image guidance: the first range is the skull base and nasopharyngeal level; the second is from the upper thoracic to the level of the liver; and the third is from the middle of the knee. It should be noted that the patient should not move in longitude direction during the second range of the scan to avoid hot spots in the spinal cord.

References

[1] Welsh J S, Patel R R and Ritter M A *et al* 2002 Helical tomotherapy: an innovative technology and approach to radiation therapy *Technol. Cancer Res. Treat.* **1** 311–6

[2] Mackie T R 2006 History of tomotherapy *Phys. Med. Biol.* **51** R427

[3] Fenwick J D, Tomé W A and Soisson E T *et al* 2006 Tomotherapy and other innovative IMRT delivery systems *Semin. Radiat. Oncol.* **16** 199–208

[4] Mackie T R, Holmes T and Swerdloff S *et al* 1993 Tomotherapy: a new concept for the delivery of dynamic conformal radiotherapy *Med. Phys.* **20** 1709–19

[5] Mackie T R, Kapatoes J and Ruchala K *et al* 2003 Image guidance for precise conformal radiotherapy *Int. J. Radiat. Oncol. Biol. Phys.* **56** 89–105

[6] Langen K M, Papanikolaou N and Balog J *et al* 2010 QA for helical tomotherapy: report of the AAPM Task Group 148 *Med. Phys.* **37** 4817–53

[7] Rong Y and Welsh J S 2011 Dosimetric and clinical review of helical tomotherapy *Exp. Rev. Anticancer Ther.* **11** 309–20

[8] Bailat C J, Baechler S and Moeckli R *et al* 2011 The concept and challenges of TomoTherapy accelerators *Rep. Prog. Phys.* **74** 086701

[9] Ruchala K J, Olivera G H and Schloesser E A *et al* 1999 Megavoltage CT on a tomotherapy system *Phys. Med. Biol.* **44** 2597

[10] Keller H, Glass M and Hinderer R *et al* 2002 Monte Carlo study of a highly efficient gas ionization detector for megavoltage imaging and image-guided radiotherapy *Med. Phys.* **29** 165–75

[11] Kissick M W, Fenwick J and James J A *et al* 2005 The helical tomotherapy thread effect *Med. Phys.* **32** 1414–23

[12] Kim B, Kron T and Battista J *et al* 2005 Investigation of dose homogeneity for loose helical tomotherapy delivery in the context of breath-hold radiation therapy *Phys. Med. Biol.* **50** 2387

[13] Meeks S L, Harmon J F Jr and Langen K M *et al* 2005 Performance characterization of megavoltage computed tomography imaging on a helical tomotherapy unit *Med. Phys.* **32** 2673–81

[14] Han M C, Hong C S and Chang K H *et al* 2020 TomoMQA: automated analysis program for MVCT quality assurance of helical tomotherapy *J. Appl. Clin. Med. Phys.* **21** 151–7

[15] Velten C, Boyd R and Jeong K *et al* 2020 Recommendations of megavoltage computed tomography settings for the implementation of adaptive radiotherapy on helical tomotherapy units *J. Appl. Clin. Med. Phys.* **21** 87–92

[16] Beavis A W 2004 Is tomotherapy the future of IMRT? *Br. J. Radiol.* **77** 285–95

[17] Shaw R M and Mackie T 2006 MVCT superiority over KVCT in assessment of electron density for treatment planning *Med. Phys.* **33** 2124

[18] Paudel M R, Mackenzie M and Fallone B G *et al* 2014 Clinical evaluation of normalized metal artifact reduction in kVCT using MVCT prior images (MVCT-NMAR) for radiation therapy treatment planning *Int. J. Radiat. Oncol. Biol. Phys.* **89** 682–9

[19] Yan D, Vicini F and Wong J *et al* 1997 Adaptive radiation therapy *Phys. Med. Biol.* **42** 123

[20] Wen J, Li Z Q and Zhang J J *et al* 2012 A Preliminary application of the helical tomotherapy adaptive system *Chin. J. Med. Phys.* **29** 3515–8

[21] Martin S and Yartsev S 2010 kVCT, MVCT, and hybrid CT image studies—treatment planning and dose delivery equivalence on helical tomotherapy *Med. Phys.* **37** 2847–54

[22] Yang J N, Mackie T R and Reckwerdt P *et al* 1997 An investigation of tomotherapy beam delivery *Med. Phys.* **24** 425–36

[23] Kissick M W, Boswell S A and Jeraj R *et al* 2005 Confirmation, refinement, and extension of a study in intrafraction motion interplay with sliding jaw motion *Med. Phys.* **32** 2346–50

[24] Schnarr E, Beneke M and Casey D *et al* 2018 Feasibility of real-time motion management with helical tomotherapy *Med. Phys.* **45** 1329–37

[25] Hu Y, Zhou Y K and Chen Y X *et al* 2016 4D-CT scans reveal reduced magnitude of respiratory liver motion achieved by different abdominal compression plate positions in patients with intrahepatic tumors undergoing helical tomotherapy *Med. Phys.* **43** 4335–41

[26] Cheng J C H, Chuang V P and Cheng S H *et al* 2000 Local radiotherapy with or without transcatheter arterial chemoembolization for patients with unresectable hepatocellular carcinoma *Int. J. Radiat. Oncol. Biol. Phys.* **47** 435–42

[27] Seong J, Park H C and Han K H *et al* 2003 Clinical results and prognostic factors in radiotherapy for unresectable hepatocellular carcinoma: a retrospective study of 158 patients *Int. J. Radiat. Oncol. Biol. Phys.* **55** 329–36

[28] Baisden J M, Reish A G and Sheng K *et al* 2006 Dose as a function of liver volume and planning target volume in helical tomotherapy, intensity-modulated radiation therapy–based stereotactic body radiation therapy for hepatic metastasis *Int. J. Radiat. Oncol. Biol. Phys.* **66** 620–5

[29] Hsieh C H, Liu C Y and Shueng P W *et al* 2010 Comparison of coplanar and noncoplanar intensity-modulated radiation therapy and helical tomotherapy for hepatocellular carcinoma *Radiat. Oncol.* **5** 40

[30] Yan B, Aidong W and Zhang H B *et al* 2019 Dosimetric study on coplanar and noncoplanar volumetric modulated arc therapy and helical tomotherapy for hepatocellular carcinoma *Chin. J. Med. Phys.* **8** 877–81

[31] Lee T F, Chao P J and Fang F M *et al* 2010 Helical tomotherapy for single and multiple liver tumours *Radiat. Oncol.* **5** 58

[32] Ferlay J, Colombet M and Soerjomataram I *et al* 2019 Estimating the global cancer incidence and mortality in 2018: GLOBOCAN sources and methods *Int. J. Cancer* **144** 1941–53

[33] Siegel R L, Miller K D and Jemal A 2019 Cancer statistics, 2019 *CA: A Cancer J. Clin.* **69** 7–34

[34] Fitzmaurice C, Abate D and Abbasi N *et al* 2019 Global, regional, and national cancer incidence, mortality, years of life lost, years lived with disability, and disability-adjusted life-years for 29 cancer groups, 1990 to 2017: a systematic analysis for the global burden of disease study *JAMA Oncol.* **5** 1749–68

[35] Zschaeck S, Blümke B and Wust P *et al* 2017 Dose-escalated radiotherapy for unresectable or locally recurrent pancreatic cancer: dose volume analysis, toxicity and outcome of 28 consecutive patients *PLoS One* **12** e0186341

[36] Milandri C, Polico R and Garcea D *et al* 2011 GEMOX plus tomotherapy for unresectable locally advanced pancreatic cancer *Hepatogastroenterology-Curr. Med. Surg. Trends* **58** 599

[37] Passardi A, Scarpi E and Neri E *et al* 2019 Chemoradiotherapy (gemox plus helical tomotherapy) for unresectable locally advanced pancreatic cancer: a phase II study *Cancers* **11** 663

[38] Ferris W S, Kissick M W and Bayouth J E *et al* 2020 Evaluation of radixact motion synchrony for 3D respiratory motion: modeling accuracy and dosimetric fidelity *J. Appl. Clin. Med. Phys.* **21** 96–106

[39] Shueng P W, Lin S C and Chong N S *et al* 2009 Total marrow irradiation with helical tomotherapy for bone marrow transplantation of multiple myeloma: first experience in Asia *Technol. Cancer Res. Treat.* **8** 29–37

[40] Wong J Y C, Liu A and Schultheiss T *et al* 2006 Targeted total marrow irradiation using three-dimensional image-guided tomographic intensity-modulated radiation therapy: an alternative to standard total body irradiation *Biol. Blood Marrow Transpl.* **12** 306–15

Chapter 9

CyberKnife® in abdominal stereotactic body radiosurgery

Lei Wang, Nataliya Kovachuk and Erqi Pollom

CyberKnife® is a robotic radiosurgery system designated for stereotactic radiosurgery (SRS) and stereotactic body radiosurgery body (SBRT). With its frameless patient setup and real-time imaging guidance, the robotic system can deliver a dose over the entire body with <1 mm robotic targeting accuracy. This system was first implemented for SRS in the brain and was quickly made available for SBRT for the spine, lung, pancreas, prostate, and other sites in the body. CyberKnife pioneered imaging-guided SBRT and remains one of the most accurate SBRT delivery platforms. As the clinical approach and the efficacy of SBRT for abdominal tumors is covered in other chapters in this book, this chapter focuses on the system overview of the CyberKnife system and the treatment workflow specific to CyberKnife-based SBRT for abdominal tumors.

9.1 Introduction

CyberKnife® is a robotic radiosurgery system designated for stereotactic radiosurgery (SRS) and stereotactic body radiosurgery body (SBRT). It was developed by John Adler, a neurosurgeon at Stanford, and treated its first patient in 1994. With its frameless patient setup and real-time imaging guidance, the robotic system can deliver a dose over the entire body with <1 mm robotic targeting accuracy [1, 2]. This system was first implemented for SRS in the brain and was quickly made available for SBRT for the spine, lung, pancreas, prostate, and other sites in the body [3–7]. CyberKnife pioneered imaging-guided SBRT [4, 8] and remains one of the most accurate SBRT delivery platforms. As clinical approach and the efficacy of SBRT for abdominal tumors is covered in other chapters in this book, we focus this chapter on the system overview and treatment workflow specific to CyberKnife-based SBRT for abdominal tumors.

doi:10.1088/978-0-7503-2468-7ch9

9.2 CyberKnife® system overview

CyberKnife is composed of a light X-band linear accelerator (Linac) mounted on a six-joint robotic arm coupled with a real-time x-ray imaging guidance system and is capable of delivering high-precision radiation to any part of the body [9, 10]. The newer models of CyberKnife (G4, VSI, and M6) [11] produce a 6 megavoltage flattening filter-free beam with dose rate of 1000 cGy per minute measured at 80 cm from the radiation target at a depth of 15 mm in a water phantom for a field size of 60 mm diameter. The imaging guidance system comprises a pair of stereoscopic in-room x-ray tubes attached on the ceiling along with two flat panel detectors on the floor as shown in figure 9.1. The x-ray tubes can be synchronized to fire

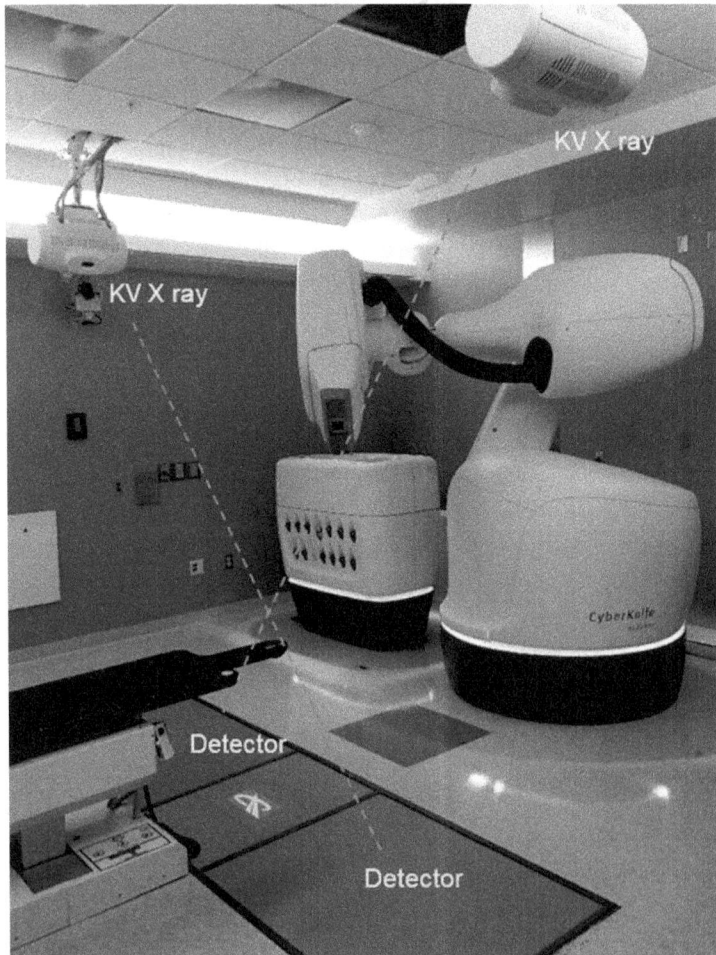

Figure 9.1. The robotic CyberKnife M6™ system with the InCise™ multi-leaf collimator (MLC) attached to the beam exit port. Target tracking is accomplished by a pair of in-room x-ray imagers on the ceiling and two flat panel detectors on the floor.

simultaneously to create a pair of near real-time x-ray images. The live images are registered to the digitally reconstructed radiographs (DRRs) generated from the planning computed tomography (CT) for patient target tracking. Automatic imaging registration algorithms are implemented to enable the precise detection and tracking of target motion for various disease sites [2]. Bony anatomy is usually used for the tracking algorithm in the treatment of intracranial and spinal targets [12, 13]. For soft tissue targets such as thoracic, abdominal, or pelvic tumors, the system typically relies on metal fiducial markers implanted inside or near the target for online tracking to guide beam deliveries.

One of the distinct features of CyberKnife SBRT is its near real-time tracking and delivery capability. The kV x-rays frequently fire to track patient motion and detect target position, and the detected shifts are used to adjust the robotic arm accordingly in near real time. For targets that move with respiration, real-time tracking is achieved through a model-based tracking mode called Synchrony™ [14, 15]. The Synchrony motion tracking system is normally used for abdominal SBRT for continuous tumor tracking. A major component of the Synchrony system is a flashpoint camera mounted on an adjustable arm attached to the ceiling near the foot end of the patient couch. This camera monitors three tracking marker LEDs attached to the patient chest for external skin motion, and an algorithm correlates the external skin movement to the internal tumor positions detected by the orthogonal kV x-ray imagers. A correlation model is first established between the external skin movement and internal tumor positions at different breathing phases before treatment, and then it is used to predict the tumor position during the treatment. Once the correlation model is established, the robotic manipulator moves correspondingly to deliver radiation beams following the guidance of the optical markers on the patient's chest [16, 17].

CyberKnife employs a spherical or '4π' working space. A typical CyberKnife plan employs 50–200 non-coplanar non-isocentric beams, which allow the creation of conformal plans with sharp dose gradients and make it an excellent platform for SRS/SBRT. The traditional CyberKnife system uses cone-based collimation systems (the fixed cone and the variable aperture collimator Iris™) with apertures ranging from 5 to 60 mm at 80 cm source to axis distance (SAD) [18]. Although complicated dosimetric shapes can be achieved with cones, delivery efficiency is not optimal, especially for irregular and larger tumors. To improve delivery efficiency, a high definition MLC (InCise™) system was introduced in the recently released CyberKnife M6 model [19, 20]. The newest version of the InCise™ MLC system includes 26 pairs of leaves with a leaf width of 0.38 cm at 80 cm SAD and maximum field size of 10 × 11.5 cm. This development has significantly improved delivery efficiency [21–23] and allowed more efficient treatment of larger target volumes.

Compared with other Linac-based systems, the largest drawback of CyberKnife is that it does not have a 3D imaging system. The requirement for fiducial markers for the treatment of extra-axial tumors introduces additional costs and risks to patients. However, the 2D planar imaging system has a distinct advantage of rapid detection, processing, and commanding of the robotic manipulator to move in response to the detected target movements, which makes real-time tracking possible. Although the use of 50–200 small non-coplanar beams results in highly conformal plans with steep dose drop off, this also contributes to the relatively longer treatment time associated with CyberKnife, which is another limitation of the system.

9.3 Clinical indication of CyberKnife® in abdominal SBRT

CyberKnife was designed for treatment on both intracranial and extracranial sites. CyberKnife's frameless image-guided treatment delivery and real-time motion tracking capability with Synchrony enables the precise treatment of tumors that move with respiration. Clinical studies of thoracic and abdominal SBRT using CyberKnife were conducted soon after Synchrony became available in the early 2000s. These studies of abdominal SBRT using CyberKnife have focused primarily on tumors in the pancreas and liver.

The first report of pancreas SBRT using CyberKnife was a phase I dose escalation trial with 15 patients, reported by Koong *et al* in 2004, and demonstrated feasibility of this treatment [4]. Subsequent phase II data from the same group reported local control of 94% in 19 patients treated with 45 Gy conventionally fractionated treatment followed by a 25 Gy in 1 fraction boost using CyberKnife [8]. Based on these experiences, a prospective multi-institutional trial was conducted of SBRT (using various treatment platforms) to 33 Gy in 5 fractions with concurrent gemcitabine for pancreatic cancer and showed a local control of 78% and acceptable rates of late gastrointestinal toxicities [24].

CyberKnife was also used in the earliest reports of SBRT for primary and metastatic tumors of the liver. A case report of CyberKnife with one patient treated to 36 Gy in 3 fractions was reported in 2006 [25]. Subsequent dose escalation series used higher doses up to 30 Gy in 1 fraction or 45 Gy in 3 fractions and demonstrated good tumor control outcomes [4, 26]. These early studies demonstrated the feasibility of using CyberKnife to deliver SBRT for moving targets using Synchrony. Since then, multiple clinical studies have been published demonstrating the efficacy of using CyberKnife for the treatment of liver tumors; for example, two recent reports showed 2-year local control of 82% in 115 patients [27] and 91% in 132 patients [28]. The clinical application of CyberKnife for liver SBRT has also been well summarized in several review papers [29, 30].

CyberKnife SBRT treatment on other abdominal tumors including adrenal and kidney tumors have also been reported [31–35], demonstrating the safety and efficacy of this approach.

9.4 Treatment workflow for abdominal SBRT on CyberKnife®

9.4.1 Fiducial marker insertion

Abdominal tumors move with respiratory motion and are usually not visible with 2D stereoscopic images. Therefore, the insertion of fiducial markers near or inside soft tissue lesions is necessary for real-time motion tracking during treatment delivery. Although fiducial placement can be accomplished by CT-guided inter-vention, laparoscopy, or laparotomy, the preferred approach is through the endoscopic route using intraluminal ultrasound to minimize tissue invasiveness and possible complications. This procedure is usually performed 1 week before obtaining the treatment planning scans to allow the seeds to stabilize following placement because any fiducial migration will degrade the accuracy of fiducial-based targeting. Three to five gold seeds are usually inserted inside or near the target [36, 37]. The relative positions of the fiducials are important for the CyberKnife tracking algorithm to accurately resolve the three translational shifts and three rotations. Recommendations for fiducial marker insertion are (1) fiducials should be placed with a minimum of 2.0 cm spacing inside or within a distance less than 5–6 cm from the lesion, (2) no three seeds should be in a line (>15° angle), and (3) no fiducials should be aligned at or near 45° axial orientation to avoid fiducial overlapping on the 2D stereoscopic tracking image [36].

9.4.2 Simulation and target delineation

Patients undergoing CyberKnife SBRT treatment are usually immobilized using a vacuum bag with arms up or on the chest. A comfortable patient position is important for reducing movement during treatment because the treatment time on CyberKnife is relatively long. The CyberKnife tracking accuracy relies on high quality DRRs generated from the patient treatment planning CT; therefore, the treatment planning CT is scanned with the patient in breath hold to reduce blurring artifacts from internal respiratory motion. To keep the best resolution, the CT field of view should be set to as small as possible to include only the patient body. Additionally, a slice thickness of 1.5 mm and thinner is required to achieve <1 mm tracking accuracy. We use 1.25 or 1 mm slice thickness in our institution. Intravenous contrast and a small amount of oral contrast can be administered for the treatment planning CT. Fluorodeoxyglucose positron emission tomography and MRI images can be fused into the treatment planning system to assist in target delineation. For pancreatic tumors, pancreas position can vary depending on the amount of gastric distension, which should be accounted for during simulation and treatment. It is recommended that patients not eat or drink 2 hours before simulation or treatment, and immediately prior to treatment, 200–250 cc of oral contrast (simulation) or water (treatment) should be consumed to ensure consistent filling.

Because CyberKnife is capable of actively tracking the target, an internal target volume, a composite volume including target movement, is not required unless the treatment is delivered without Synchrony. However, CyberKnife motion tracking accuracy is dependent on the amount of target motion, and target deformation occurs with respiration motion. For pancreatic tumors, the motion is variable, ranging from 5 to 43 mm [38]. Liver tumor motion extent is dependent on proximity to the liver dome. Therefore, it is preferred that a pair of normal expiration-hold and normal inspiration-hold CT images are acquired and fused so that the clinical target volume (CTV) can account for target deformation. If a four-dimensional CT (4DCT) is available, the expiration and inspiration phase CTs can be used. After the CTV is determined, the planning target volume (PTV) is created by adding margins for setup and delivery uncertainty. Synchrony motion tracking accuracy has been studied by multiple investigators [15, 39–41]. Although 3–5 mm margins have been used, we have used a 3 mm CTV to PTV margin (when fiducials are present and can be successfully tracked) at our institution. A non-uniform margin could be considered based on the motion study. Motion study with 4DCT is normally performed in our institution.

9.4.3 Treatment planning

SBRT dose prescription is consistent across different treatment platforms as described in the previous chapters. Treatment doses of 45–54 Gy in 3–5 fractions and 33–45 Gy in 5 fractions are typically used at our institution for liver and pancreas SBRT, respectively. Dose prescriptions are often limited by dose tolerances to the critical structures. Critical organs and risk that are considered during treatment planning for abdominal SBRT include the lungs, heart, spinal cord, liver, central hepatobiliary tract (cHBT), large and small bowel including duodenum, kidneys, esophagus, and chest wall. Typical dose constraints to the critical structures for abdominal SBRT treatment used in our institution are listed in table 9.1 [42]. The highest priority is given to spinal cord, stomach, bowel, duodenum, and livergross tumor volume (GTV) constraints.

CyberKnife treatment plans are inversely optimized based on a set of objectives and constraints. A stepwise optimization algorithm [43] has been used with the approach of optimizing multiple clinical goals in steps with built-in priority since 2008. Recently a new optimizer, VOLO, was introduced in the CyberKnife Precision planning system. With the VOLO optimizer, a plan is optimized using one single cost function based on the weighted dose volume histogram (DVH) goals and constraints. CyberKnife delivery space is composed of a group of discrete predefined delivery positions (nodes, with typically 100–200 nodes depending on treatment site) spherically distributed around the patient. Direct posterior beams are not allowed

Table 9.1. Dose constraints used in our institution for gastrointestinal SBRT. VS: volume spared.

Organ max mean DVH	Max	Mean		DVH	
Lung		<9 Gy	V5 < 24%	V20 < 10%	V30 < 5%
Heat		<12 Gy	V15 < 10%		
Cord	12 Gy				
Liver-GTV		<10 Gy	VS 15 Gy > 700 cc	VS 7 Gy > 500 cc	
Stomach, bowel, duodenum	40 Gy		V25 < 9 cc	V30 < 5 cc	V33 < 1 cc
cHBT		<19 Gy	V26 < 37 cc	V21 < 45 cc	
Kidney			V5 < 50%		
Chest wall			V30 < 30 cc		
Esophagus		<40 Gy			

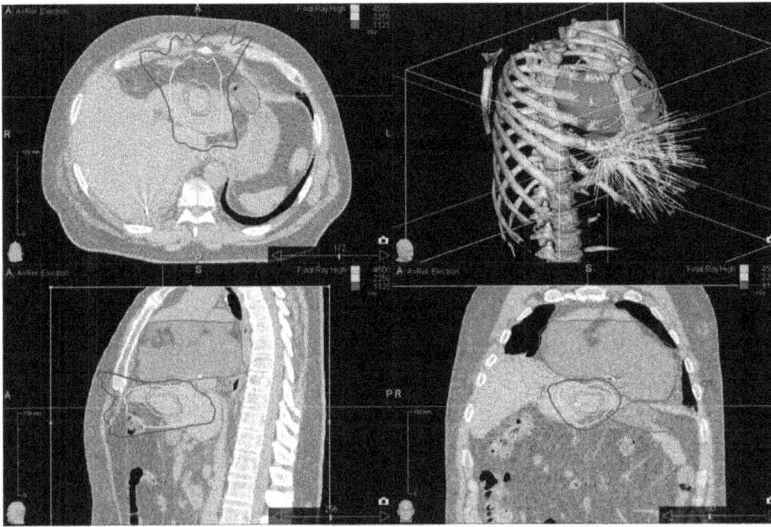

Figure 9.2. A CyberKnife treatment plan for a patient who was treated in our institution with liver SBRT. The plan delivers 45 Gy to PTV in three fractions and contains total of 125 beams with the Iris collimator.

due to collision with the ground. At each node, the planning system typically allows up to 12 beam directions for optimizations. For cone-based plans, the optimizer optimizes on the beam angle and weight from a large number of pre-generated beams in the discrete working space. For MLC-based plans, optimization starts with fluence optimization followed by segmentation, aperture adaptation, and then weight optimization. The VOLO optimizer was reported to be able to generate better quality plans with significant improvement in delivery efficiency [23]. A typical CyberKnife plan contains 50–200 non-coplanar beams. Treatment time ranges from 30 to 60 min.

Appropriate collimators should be used based on the target sizes. Because of their smaller available apertures and smaller penumbra, circular collimators (fixed cones and Iris collimators) are usually used for smaller target sizes, whereas MLC is used for relatively larger target volumes (usually >3 cm in diameter). Using MLC for larger targets could help to create superior quality plans with less MUs and shorter treatment time [21, 22] (figure 9.2).

9.4.4 Treatment delivery

Treatment delivery is quite different between CyberKnife and regular C-arm Linac systems. Therapists need to have additional training that is specific to CyberKnife treatment delivery. A physicist and a radiation oncologist are required to be present for positioning and imaging verification at the beginning of the treatment and should be available on site during the entire treatment. Abdominal SBRT treatments are typically delivered on CyberKnife with Synchrony motion tracking. Three LEDs will be positioned on the patient chest, and a correlation model will be established before the treatment can be started between the external skin motion monitored by the Synchrony camera and the internal target motion detected by the two stereo-scope x-ray detectors.

(a) LED placement:

Patient is set in the same immobilization device created during the simulation. For better signal-to-noise ratio on skin traces, the LED skin markers are recommended to be placed near the position of maximum body motion due to respiration. Attention should be paid that the position of the Synchrony camera should be directly in line with the LED tracking markers and should be sufficiently away from the patient to avoid collision with the manipulator. Continuous breathing traces should be observed on the console computer once the camera sees the LED signals.

(b) Fiducial tracking:

CyberKnife software automatically searches for the fiducials on the two stereoscopic live images. Rigid body registration algorithm is employed to estimate the couch corrections. Careful visual examination and confirmation must be performed by an examination of the live images (camera image in figure 9.3), the DRRs (synthetic image in figure 9.3), and their overlays. Fiducials that fail the tracking algorithm (often due to close distance on DRRs, such as fiducial #2 and #4 in figure 9.3) should be disabled. A fiducial with migration could introduce significant tracking errors and should be disabled. A minimum of three fiducials are required to give three translational and three rotational shifts. If fewer than three fiducials are available for tracking, treatment can be delivered without rotational correction. In that case, a setup plan should be created with tracking on the spine close to the treatment target. The patient should be set up with the spine tracking with the rotational corrections minimized before the real treatment plan is loaded.

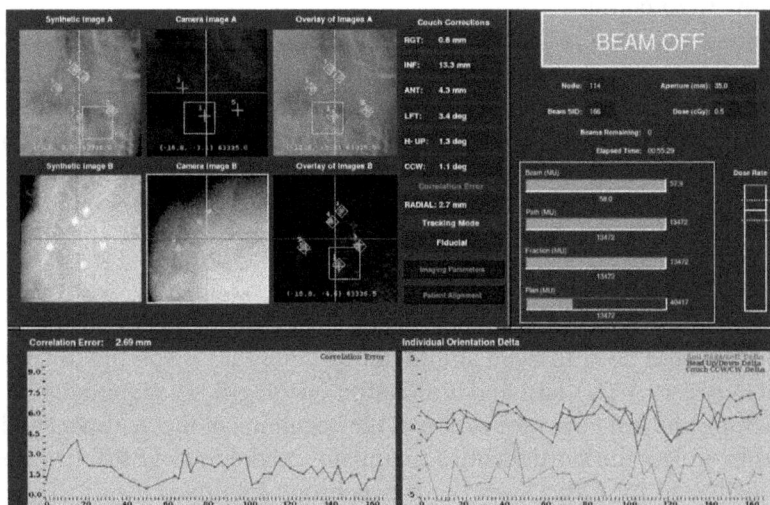

Figure 9.3. A screen capture during treatment delivery for the plan shown in figure 9.2.

(c) Synchrony modeling:

A Synchrony model stores up to 15 model points distributed over different phases and is updated in the 'first in first out approach' during treatment. Model generation is significantly improved with the automated modeling option in the new version of the software. During model generation, the system first acquires the full inspiration ('peak') and full expiration ('valley') points, prompts the user to shift the patient to middle position, and then automatically fills in all the required points. One point is added to the synchrony correlation model each time a live x-ray image is acquired. To obtain an accurate model, the model points should be distributed evenly and cover the whole range of respiratory motion. The synchrony system categorizes the correlation model as optimal if at least seven of the eight phases (the whole breathing cycle is split into eight phases) of the respiratory cycle are included in the model, including points for full inspiration ('peak') and full expiration ('valley'). An optimal model should always be used for better tracking accuracy.

(d) Treatment delivery:

During the treatment, the live images are usually taken in an interval between 30 and 90 seconds (set by the user), and the Synchrony model is updated in the 'first in first out approach.' The user should actively monitor the tracking and model status. Any unexpected movement of the Synchrony camera, LEDs, and/or the patient will invalidate the Synchrony model, which will require a model regeneration before resuming the treatment.

The correlation errors are good indicators of the model accuracy. They are displayed live during the treatment (figure 9.3). The correlation error is defined as the difference between the actual target position obtained by acquiring the live x-ray images and the position determined by the Synchrony model at that model point. An excess of 5 mm in correlation error interrupts the treatment and enables the user to update or recreate the model. In clinical practice, it is recommended that the user closely monitor the correlation errors during the treatment and keep the correlation model error as low as reasonably achievable.

9.5 Quality control and treatment safety

CyberKnife delivery for abdominal region with Synchrony active tracking is quite complex technically. Therefore, it is recommended that the physicists be actively involved in all steps of the procedures, including patient evaluation, simulation, planning, and treatment supervision. Education to the clinical team (physician, dosimetrist, and therapist) on the concept of tracking and Synchrony is essential. Policies and procedures should be in place at the beginning of establishing the program. A quality control procedure should be implemented and followed for the delivery.

Technical recommendations on CyberKnife quality assurance (QA) are well described in American Association of Physicists in Medicine Task Group 135 [44] and 135 update (currently under review). QA on Synchrony motion tracking should be conducted using a phantom on a motion platform. The Synchrony end-to-end test should be performed quarterly with all collimator assemblies used for respiratory motion tracking [45]. The tolerance on targeting accuracy should be <0.95 mm.

9.6 Summary

This chapter has been focused on the CyberKnife system overview and treatment workflow specific to CyberKnife-based SBRT for abdominal tumors. The CyberKnife system allows for the delivery of high-dose radiation in a few fractions to the tumor with millimetric precision. It is one of the most accurate SBRT platforms for abdominal SBRT and is the only FDA-approved robotic radiosurgery system currently available. CyberKnife-based abdominal SBRT has been primarily focused on tumors in the pancreas and liver. Applications on other abdominal sites such as adrenal and kidney tumors have also been reported recently. Using Synchrony, CyberKnife has the advantage of actively and accurately tracking the tumor motion with the respiration, which is preferable for abdominal treatment. However, the Synchrony tracking requires that fiducials are implanted when tumors are not visible on the 2D stereotactic imaging. Because of the technical complexity of the delivery system, active physicist involvement and education to the clinical team are recommended.

References

[1] Adler J R Jr., Chang S D, Murphy M J, Doty J, Geis P and Hancock S L 1997 The Cyberknife: a frameless robotic system for radiosurgery *Stereotact. Funct. Neurosurg.* **69** 124–8

[2] Adler J R Jr., Murphy M J, Chang S D and Hancock S L 1999 Image-guided robotic radiosurgery *Neurosurgery* **44** 1299–306 (discussion 306-7)

[3] Collins B T, Vahdat S, Erickson K, Collins S P, Suy S and Yu X *et al* 2009 Radical cyberknife radiosurgery with tumor tracking: an effective treatment for inoperable small peripheral stage I non-small cell lung cancer *J. Hematol. Oncol.* **2** 1

[4] Koong A C, Le Q T, Ho A, Fong B, Fisher G and Cho C *et al* 2004 Phase I study of stereotactic radiosurgery in patients with locally advanced pancreatic cancer *Int. J. Radiat. Oncol. Biol. Phys.* **58** 1017–21

[5] Le Q T, Tate D, Koong A, Gibbs I C, Chang S D and Adler J R *et al* 2003 Improved local control with stereotactic radiosurgical boost in patients with nasopharyngeal carcinoma *Int. J. Radiat. Oncol. Biol. Phys.* **56** 1046–54

[6] King C R, Brooks J D, Gill H, Pawlicki T, Cotrutz C and Presti J C Jr 2009 Stereotactic body radiotherapy for localized prostate cancer: interim results of a prospective phase II clinical trial *Int. J. Radiat. Oncol. Biol. Phys.* **73** 1043–8

[7] Ryu S I, Chang S D, Kim D H, Murphy M J, Le Q T and Martin D P *et al* 2001 Image-guided hypo-fractionated stereotactic radiosurgery to spinal lesions *Neurosurgery.* **49** 838–46

[8] Koong A C, Christofferson E, Le Q T, Goodman K A, Ho A and Kuo T *et al* 2005 Phase II study to assess the efficacy of conventionally fractionated radiotherapy followed by a stereotactic radiosurgery boost in patients with locally advanced pancreatic cancer *Int. J. Radiat. Oncol. Biol. Phys.* **63** 320–3

[9] Kilby J D W, Kuduvalli G, Sayeh S and Maurer C R Jr 2010 The CyberKnife® robotic radiosurgery system in 2010 *Technol. Cancer Res. Treat.* **9** 433–52

[10] Kilby M N W, Dooley J R, Maurer C R Jr and Sayeh S 2019 A technical overview of the CyberKnife System ed M Abedin-Nasab *Handbook of Robotic and Image-Guided Surgery* (Amsterdam: Elsevier) pp 15–38

[11] Fahimian P B and Wang L 2014 Introduction to CyberKnife® Technology ed D Steven and A V Chang *CyberKnife Stereotactic Radiosurgery. 1* (Hauppauge, NY: Nova Science Publishers, Inc) pp 1–12

[12] Fu D and Kuduvalli G 2008 A fast, accurate, and automatic 2D–3D image registration for image-guided cranial radiosurgery *Med. Phys.* **35** 2180–94

[13] Ho A K F D, Cotrutz C, Hancock S L, Chang S D and Gibbs I C 2007 A study of the accuracy of cyberknife spinal radiosurgery using skeletal structure tracking *Neurosurgery* **60** ONS147–56 discussion ONS56

[14] Schweikard A, Shiomi H and Adler J 2004 Respiration tracking in radiosurgery *Med. Phys.* **31** 2738–41

[15] Seppenwoolde Y, Berbeco R I, Nishioka S, Shirato H and Heijmen B 2007 Accuracy of tumor motion compensation algorithm from a robotic respiratory tracking system: a simulation study *Med. Phys.* **34** 2774–84

[16] Dieterich S, Cleary K, D'Souza W, Murphy M, Wong K H and Keall P 2008 Locating and targeting moving tumors with radiation beams *Med. Phys.* **35** 5684–94

[17] Keall P J, Mageras G S, Balter J M, Emery R S, Forster K M and Jiang S B et al 2006 The management of respiratory motion in radiation oncology report of AAPM Task Group 76 *Med. Phys.* **33** 3874–900

[18] Echner G G, Kilby W, Lee M, Earnst E, Sayeh S and Schlaefer A et al 2009 The design, physical properties and clinical utility of an iris collimator for robotic radiosurgery *Phys. Med. Biol.* **54** 5359–80

[19] Fürweger C, Prins P, Coskan H and Heijmen B J 2016 Characteristics and performance of the first commercial multileaf collimator for a robotic radiosurgery system *Med. Phys.* **43** 2063–71

[20] Asmerom G, Bourne D, Chappelow J, Goggin L, Heitz R and Jordan P et al 2016 The design and physical characterization of a multileaf collimator for robotic radiosurgery *Biomed. Phys. Eng. Exp.* **2** 017003

[21] Kathriarachchi V, Shang C, Evans G, Leventouri T and Kalantzis G 2016 Dosimetric and radiobiological comparison of CyberKnife M6 InCise multileaf collimator over IRIS variable collimator in prostate stereotactic body radiation therapy *J. Med. Phys.* **41** 135–43

[22] McGuinness C M, Gottschalk A R, Lessard E, Nakamura J L, Pinnaduwage D, Pouliot J, Sims C and Descovich M 2015 Investigating the clinical advantages of a robotic linac equipped with a multileaf collimator in the treatment of brain and prostate cancer patients *J. Appl. Clin. Med. Phys.* **16** 284–95

[23] Schuler E, Lo A, Chuang C F, Soltys S G, Pollom E L and Wang L 2020 Clinical impact of the VOLO optimizer on treatment plan quality and clinical treatment efficiency for CyberKnife *J. Appl. Clin. Med. Phys.* **21** 38–47

[24] Herman J M, Chang D T, Goodman K A, Dholakia A S, Raman S P and Hacker-Prietz A et al 2015 Phase 2 multi-institutional trial evaluating gemcitabine and stereotactic body radiotherapy for patients with locally advanced unresectable pancreatic adenocarcinoma *Cancer.* **121** 1128–37

[25] Chung Y W, Han D S, Paik C H, Kim J P, Choi J H and Sohn J H *et al* 2006 Localized esophageal ulcerations after CyberKnife treatment for metastatic hepatic tumor of colon cancer *Korean J. Gastroenterol.* **47** 449–53

[26] Vautravers-Dewas C, Dewas S, Bonodeau F, Adenis A, Lacornerie T and Penel N *et al* 2011 Image-guided robotic stereotactic body radiation therapy for liver metastases: is there a dose response relationship? *Int. J. Radiat. Oncol. Biol. Phys.* **81** e39–47

[27] Que J, Kuo H T, Lin L C, Lin K L, Lin C H and Lin Y W *et al* 2016 Clinical outcomes and prognostic factors of cyberknife stereotactic body radiation therapy for unresectable hepatocellular carcinoma *BMC Cancer* **16** 451

[28] Su T S, Liang P, Lu H Z, Liang J, Gao Y C and Zhou Y *et al* 2016 Stereotactic body radiation therapy for small primary or recurrent hepatocellular carcinoma in 132 Chinese patients *J. Surg. Oncol.* **113** 181–7

[29] Peter Ihnát E S, Tesař M and Penka I 2018 Stereotactic body radiotherapy using the CyberKnife® system in the treatment of patients with liver metastases: state of the art *Oncol. Targets Ther.* **11** 4685–91

[30] Lo C H H W, Chao H L, Lin K T and Jen Y M 2014 Novel application of stereotactic ablative radiotherapy using CyberKnife® for early-stage renal cell carcinoma in patients with pre-existing chronic kidney disease: initial clinical experiences *Oncol. Lett.* **8** 355–60

[31] Desai A, Rai H, Haas J, Witten M, Blacksburg S and Schneider J G 2015 A retrospective review of cyberknife stereotactic body radiotherapy for adrenal tumors (primary and metastatic): Winthrop University hospital experience *Front. Oncol.* **5** 185

[32] Cormier J M, Daumet P and Maire M 1972 [Complex cervico-facial arteriovenous vascular dysplasia treated under circulatory arrest] *Chirurgie* **98** 392–6

[33] Zhao X, Zhu X, Zhuang H, Guo X, Song Y and Ju X *et al* 2020 Clinical efficacy of stereotactic body radiation therapy (SBRT) for adrenal gland metastases: a multi-center retrospective study from China *Sci. Rep.* **10** 7836

[34] Senger C, Conti A, Kluge A, Pasemann D, Kufeld M and Acker G *et al* 2019 Robotic stereotactic ablative radiotherapy for renal cell carcinoma in patients with impaired renal function *BMC Urol* **19** 96

[35] Correa R J M, Louie A V, Zaorsky N G, Lehrer E J, Ellis R and Ponsky L *et al* 2019 The emerging role of stereotactic ablative radiotherapy for primary renal cell carcinoma: a systematic review and meta-analysis *Eur. Urol. Focus.* **5** 958–69

[36] Mallarajapatna G J, Susheela S P, Kallur K G, Ramanna N K, Ramachandra P G and Sudhakar *et al* 2011 Technical note: Image guided internal fiducial placement for stereotactic radiosurgery (CyberKnife) *Indian J. Radiol. Imaging* **21** 3–5

[37] Minn A Y, Koong A C and Chang D T 2011 Stereotactic body radiation therapy for gastrointestinal malignancies *Front. Radiat. Ther. Oncol.* **43** 412–27

[38] Reese A S, Lu W and Regine W F 2014 Utilization of intensity-modulated radiation therapy and image-guided radiation therapy in pancreatic cancer: is it beneficial? *Semin. Radiat. Oncol.* **24** 132–9

[39] Hoogeman M, Prevost J B, Nuyttens J, Poll J, Levendag P and Heijmen B 2009 Clinical accuracy of the respiratory tumor tracking system of the cyberknife: assessment by analysis of log files *Int. J. Radiat. Oncol. Biol. Phys.* **74** 297–303

[40] Pepin E W, Wu H, Zhang Y and Lord B 2011 Correlation and prediction uncertainties in the cyberknife synchrony respiratory tracking system *Med. Phys.* **38** 4036–44

[41] Lu X Q, Shanmugham L N, Mahadevan A, Nedea E, Stevenson M A and Kaplan I *et al* 2008 Organ deformation and dose coverage in robotic respiratory-tracking radiotherapy *Int. J. Radiat. Oncol. Biol. Phys.* **71** 281–9

[42] Pollom E L, Chin A L, Diehn M, Loo B W and Chang D T 2017 Normal tissue constraints for abdominal and thoracic stereotactic body radiotherapy *Semin. Radiat. Oncol.* **27** 197–208

[43] Schlaefer A and Schweikard A 2008 Stepwise multi-criteria optimization for robotic radiosurgery *Med. Phys.* **35** 2094–103

[44] Dieterich S, Cavedon C, Chuang C F, Cohen A B, Garrett J A and Lee C L *et al* 2011 Report of AAPM TG 135: quality assurance for robotic radiosurgery *Med. Phys.* **38** 2914–36

[45] Accuray Inc. 2017 *Physics Essentials Guide* (Sunnyvale, CA: Accuray Inc.)

Chapter 10

Varian Halcyon™

David Chamberlain, Joy Carter and Yu Kuang

This chapter consists of an introduction to the Varian Halcyon™ treatment machine as it pertains to image guided radiotherapy. We first provide a brief history of why Varian introduced this treatment machine to the radiation therapy community. We then explain differences of this compact single-photon energy treatment device compared with conventional C-arm–based linear accelerators as they relate to commissioning, workflow, and clinical applications. We specifically focus on abdomen applications and how the Halcyon™ system's imaging system is an improvement for clinicians to deliver image-guided radiotherapy treatments.

10.1 Overview of Halcyon™

At the turn of the century, Varian Medical Systems recognized a need throughout the world: to deliver high-quality radiation therapy treatment to cancer patients. Some radiation therapy departments were treating 7 days a week and 24 h a day to keep up with the patient demand. Because of time constraints, current treatment techniques were not possible. For example, intensity-modulated radiation therapy (IMRT) or volumetric-modulated arc therapy (VMAT) were not offered because of the time required to create a treatment plan, the required quality assurance (QA), the increased time to deliver the treatment, and the cost of the technology. Other factors contributing to the lack of technology in certain parts of the world included safety, staffing expertise, space requirements to build the necessary treatment vaults to house the treatment machines, expense, and the time to install the equipment.

In 2017, the Halcyon™ linear accelerator (Varian Medical Systems, Palo Alto, CA) was introduced to the radiation oncology community to address the factors listed above. Its purpose was to deliver high-quality radiation therapy treatments safely, efficiently, and at a reasonable cost.

doi:10.1088/978-0-7503-2468-7ch10

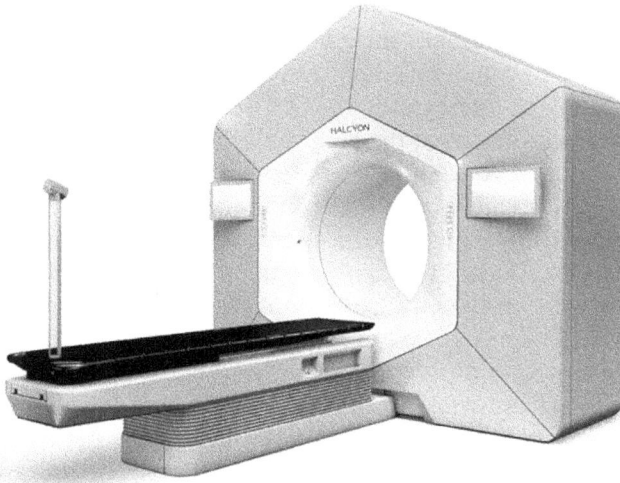

Figure 10.1. Image of the Varian Halcyon™ treatment machine (Varian Medical Systems, Palo Alto, CA). Image courtesy of Varian Medical Systems, Inc., Palo Alto, CA. Copyright 2020.

The Halcyon™ (figure 10.1) utilizes a 6-megavoltage (MV) flattening filter–free (FFF) beam on an enclosed ring-mounted gantry. The patient is no longer exposed to the moving parts of the gantry and it can rotate faster than a traditional C-arm gantry, therefore making the treatment delivery faster and safer. The ring-mounted gantry will rotate up to 4 revolutions per minute or 360° in 15 s as the patient rests within the 100 cm bore. Halcyon™ introduces a double-stacked (dual-layer) multi-leaf collimator (MLC). The maximum field size is 28 × 28 cm at the isocenter with 1-cm-wide leaves. The upper and lower banks of MLCs consist of 58 and 56 1-cm leaves, respectively. The top bank is displaced by 0.5 cm relative to the lower bank to reduce interleaf leakage. The transmission has been measured for both MLC banks (distal and proximal) and individually to be 0.008% and 0.04%, respectively [1]. The dual-layer MLC provides sufficient shielding and allows for leaf tracking rather than jaw tracking as seen on typical C-arm Linacs, thus eliminating the need for collimator jaws. Unlike the TrueBeam™ MLC with a leaf speed of 2.5 cm s^{-1}, the Halcyon™ MLC leaves travel at a maximum speed of 5.0 cm s^{-1}. The original Halcyon™ system, version 1.0, includes the dual-layer MLC; however, beam shaping is limited to the lower bank of MLCs and the upper bank was used more as a beam block or jaws. Version 2.0 utilizes both banks of the MLC to shape the beam with a 0.5 cm resolution.

The Halcyon™ system includes the Maestro control system that runs in the background. The Maestro monitors the subsystems of the treatment machine (for example, the MLC, dose rate, couch position, gantry position, temperature, collimator, etc) and continually checks the system at a frequency of 10 ms. It not only checks performance but also monitors, directs, and synchronizes the subsystems of the entire treatment delivery unit at the same rate.

Existing radiotherapy machines include certain safety features within the treatment room. Such features usually include separate audio and video devices for the staff to stay in constant communication with the patient during treatment. In many areas, regulatory bodies require the ability to have constant communication with the patient. Included in the design of the Halcyon™, a closed-circuit TV is mounted at the end of the treatment couch and a microphone and speaker are included in the bore of the treatment machine to provide constant communication and visualization of the patient. As a safety feature, if the video camera is not providing a signal, the unit will not allow treatment or imaging to proceed.

Halcyon™ delivers radiation therapy to the patient utilizing image-guided radiotherapy (IGRT) for patient setup. The design of the enclosed Halcyon™ ensures the patient is kept away from all moving parts. As a result, the radiation source is covered, and two provisions, the light field and optical distance indicator (ODI), typically available to the therapists to position the patients, are no longer available on this system. Patient positioning is accomplished using MV and kilo-voltage (kV) imaging. MV (cone-beam computed tomography) CBCT and static MV images are produced using the 6 MV FFF x-ray beam and an electronic portal image detector (EPID). There are two modes offered: a high-quality option and a low-dose option for MV imaging. Version 2.0, released in 2019, added a kV x-ray tube for kV CBCT and kV onboard imaging with an amorphous Si detector with an active area of 43 × 43 cm. The x-ray tube operates at a tube voltage of 40–150 kV. The fast rotation speed decreases the time to acquire images from 60 s on a typical C-arm gantry system, to 17–42 s on the Halcyon™ [1]. Version 2.0 also includes iterative CBCT or interactive CBCT (iCBCT). This statistical image reconstruction algorithm is designed to improve image quality, reduce noise, and improve contrast to noise [2, 3]. Imaging protocols on Halcyon™ include head, pelvis, and an image gently protocol.

Even though the single-energy Halcyon™ linear accelerator is built on fundamentals and basic principles of other linear accelerators, there are some significant differences Varian Medical Systems considered to improve efficiencies and safety. These differences can be seen in the commissioning of the machine and treatment planning system (TPS), as well as the overall treatment workflow.

10.2 Acceptance testing and commissioning of the Halcyon™ system

There are certain prerequisites for the Halcyon™ to deliver radiation treatments to patients. First, Halcyon™ operates on the backbone of Varian's electronic medical record (EMR) system, Aria. Second, the only accepted TPS for the Halcyon™ is Varian's Eclipse Version 15.x. Aria, Eclipse, and Halcyon™ are all connected to function as one unit.

Over the years, Varian has compiled and provided their representative beam data (RBD). The RBD consists of output factors, percent depth doses (PDDs), and beam profiles. Institutions are able to compare their measured data against Varian's RBD, and some have even used the RBD rather than their own data when creating models within the TPS. One of the predecessors to the Halcyon™ is the Varian TrueBeam.

Multiple institutions have reviewed and documented the TrueBeam's performance and stability with respect to the RBD [4, 5]. For institutions with multiple machines of the same model, it is often requested that the machines' beams match. This creates redundancy if one machine is inoperable so patients can be transferred to the other machine with minimal interruption to the patient's treatment.

Instead of beam matching Halcyon™ machines after they arrive in the market, Varian manufactures these machines to match the RBD when they are first assembled. Varian introduced the machine performance check (MPC) system with the TrueBeam to check multiple machine characteristics related to geometry and beam performance. Several investigators have shown the reliability and overall ability of the MPC to measure machine performance [6–8]. Prior to a Halcyon™ system being shipped to an institution for installation, Varian engineers perform MPC. It is also conducted once the Halcyon™ has been installed and the results are compared to the accepted criteria. The MPC is a good indicator that the system is operating as expected. For Halcyon™, the MPC is also a daily quality test that must be passed before patient treatments can be delivered.

As all Halcyon™ systems match the RBD, the beam models used for plan optimization and dose calculations come preconfigured within the Eclipse TPS. The models are not accessible for modification or manipulation. Unlike other treatment machines in the TPS, Halcyon™ is adjusted to match the TPS in contrast to adjusting the TPS models to match the machine. This is a paradigm shift within the medical physics community. There have been a few investigators who have reviewed the RBD with their own acquired data and have observed very good agreement [5, 9, 10]. Some say the system comes 'precommissioned.' In other words, the TPS does not require a physicist to enter the data and create the models; however, the physicist is still required to validate the existing models with their existing machine along with conducting the necessary calibration.

Prior to the Halcyon™ installation, the institution's physicist is required to specify the calibration geometry. Varian offers three different options consisting of specified calibration points, $d = 1.3$ cm, $d = 5$ cm, or $d = 10$ cm, for which 1 monitor units (MU) is equal to 1 cGy (figure 10.2) [11]. The installer of the Halcyon™ will identify the preferred calibration point of the medical physicist in system administration and the geometry will synchronize with the Eclipse TPS. The calibration point ultimately determines the dose rate. Instead of the dose rate staying constant at 800 MU min^{-1} at the calibration depth, the dose rate of the machine cannot be adjusted. If the calibration point is chosen to be at d_{max}, the dose rate will be at the maximum of 800 MU min^{-1}. If D5 or D10 is set for the calibration depth, the dose rate measured at those depths will be 740 or 600 MU min^{-1}, respectively.

Acceptance testing of the Halcyon™ system consists of the tests listed in table 10.1. The majority of the tests are performed in the presence of the medical physicist.

As the machine comes 'precommissioned' in Eclipse, the task for the medical physicist is to validate the RBD. Commissioning and QA guidelines can be found in the following references from the American Association of Physicists in Medicine (AAPM):

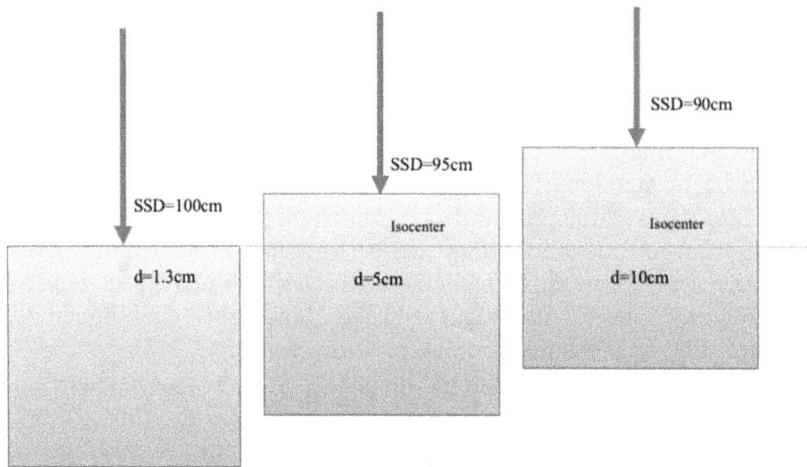

Figure 10.2. Halcyon™ calibration geometry options indicating the point of calibration where 1 MU is normalized to 1 cGy. Image courtesy of Varian Medical Systems, Inc., Palo Alto, CA. Copyright 2020.

Table 10.1. Halcyon™ 2.0 acceptance tests.

Halcyon™ acceptance tests
Interlock verification
Site radiation survey
Beam stability vs gantry rotation
Radiation isocenter and position accuracy verifications with MPC
Beam energy and profile verification
Depth of ionization
Off axis intensity
Off axis symmetry
Energy verification
Dosimetry verifications

Imaging acceptance tests	
MV imaging acquisition	kV and kV CBCT image acquisition
Dark field image	Dark field image
Noise image	Noise image
Pixel correction	Imager sensitivity
Imager sensitivity	Pixel correction
Contrast resolution	Single mode defective lines
Small object detection	CBCT image acquisitions
Portal dosimetry integration	Head mode density resolution

Table 10.1. (*Continued*)

MV CBCT imaging acquisition	Pelvis mode density resolution
Low dose density resolution	iCBCT reconstruction
High quality density resolution	Head mode spatial linearity
Low dose spatial linearity measurements	Pelvis mode spatial linearity
High quality spatial linearity measurements	Image uniformity
Image uniformity	Head mode
Low dose	Pelvis mode
High quality	High contrast resolution
High contrast resolution	Head mode
Low dose	Pelvis mode
High quality	X-ray generator verification

- AAPM TG-51 Protocol for Clinical Reference Dosimetry Amended for FFF beams
- AAPM TG-119 for IMRT Commissioning
- AAPM TG-142 Report on Quality Assurance of Medical Accelerators
- AAPM MPPG-5.a Commissioning and QA of Treatment Planning Dose Calculations

Additionally, special considerations need to be made when setting up equipment with the enclosed ring-mounted gantry as there is no ODI, light field, or manual measuring front pointer. Instead, the Halcyon™ is designed with a virtual isocenter. The system is composed of lateral and vertical lasers positioned to intersect at a point outside of the bore of the machine. The lasers' intersection is known as the 'virtual isocenter.' Once the phantom or patient is lined up at the virtual isocenter, the operator presses and holds the 'load' button to automatically advance the phantom or patient to the machine isocenter. With the absence of the ODI or front pointer, imaging is used to verify the source-to-surface distance (SSD) when setting up QA phantoms. The Varian 'Instructions for Use' includes procedures to verify proper SSD with MV imaging [11]. For phantoms set at 100 cm SSD, orthogonal imaging can be used for alignment verification. For SSDs other than 100 cm, basic trigonometry can be used to calculate the gantry angle. Figure 10.3 is an example using a 90 cm SSD setup. MV imaging is used to verify depth and SSD. The image displayed will include the electronic crosshair, representing the machine isocenter. The user is instructed to use the distance tool to measure the depth. In this example, the distance from the isocenter to the surface will be 10 cm. If the SSD is correct and the gantry is positioned correctly, the phantom surface will appear as a sharp edge. If the SSD needs to be adjusted, the depth will be incorrect or the phantom surface will appear blurred.

The ring gantry introduces additional setup challenges for using a three-dimensional (3D) scanning tank. Even though the bore is 100 cm in diameter, and most scanning systems will fit within the Halcyon™ bore, the scanning system must be positioned on the treatment couch and have the ability to level or auto level, as the

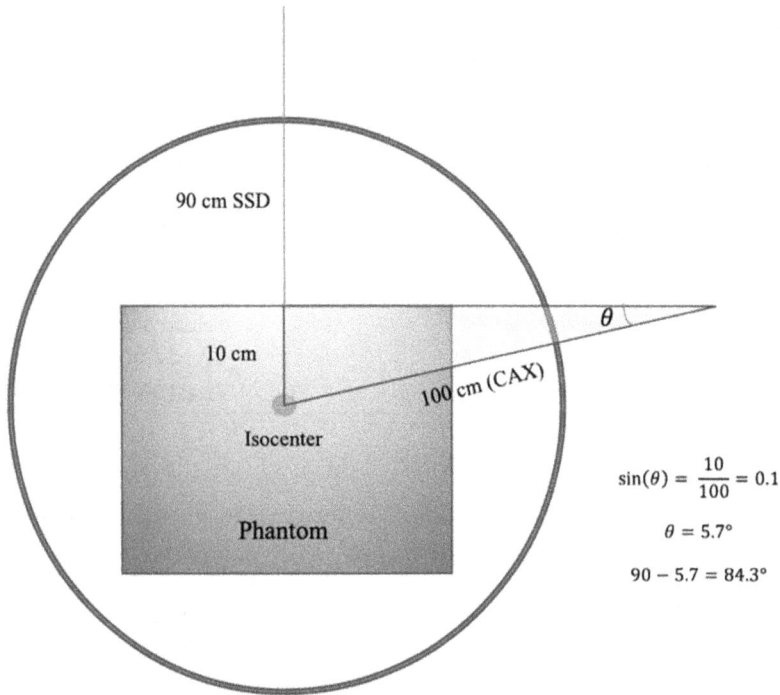

Figure 10.3. Phantom set at 90 cm geometry. Image courtesy of Varian Medical Systems, Inc., Palo Alto, CA. Copyright 2020.

couch only allows translational travel and no rotational movements. If the scanning system does not account for rotations related to leveling, then the tank must be leveled manually. Such manual manipulation can prove to be extremely tedious. There are a few 3D scanning systems, such as IBA and PTW that offer special modules to keep their system compatible with the Halcyon™ machine. Figure 10.4 depicts the PTW BeamScan™ system with the Halcyon™ package. PTW includes at transfer plate to protect the carbon fiber top of the Halcyon™ couch top, along with a longer hose to autofill the water tank once in position. This system also includes an autoleveling feature. Recently there has been work done to investigate a system for acceptance testing and verification of Halcyon™ without a 3D water-scanning system, using an ionization chamber array [10]. Results of the ionization chamber array agree very well to the 3D water system, the RBD, and appears to be a viable option for beam validation with Halcyon™.

10.3 Clinical workflow

One of the main objectives for Halcyon™ is to simplify all aspects of the treatment delivery process. This includes installation, acceptance testing, commissioning, QA, and treatment delivery.

The design and operation of the Halcyon™ focuses on safety and the patient experience during the delivery of the treatment, from minimizing patient and

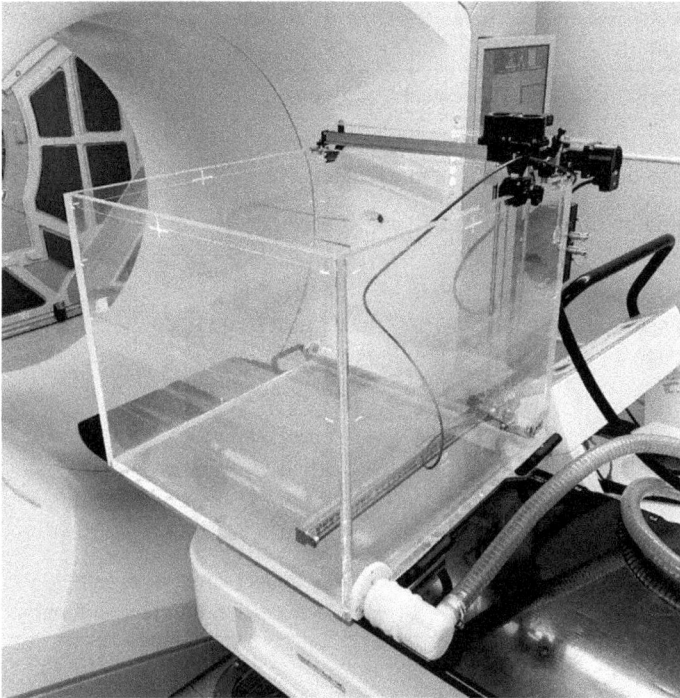

Figure 10.4. PTW BeamScan system with Halcyon™ package.

machine collisions, to lessening the noise produced by the radiation therapy machine, to providing a large 100 cm bore and lighting to reduce feelings of claustrophobia. The system also provides easy-access controls so the therapist can focus on the patient and minimize distractions by simplifying the steps leading up to treatment delivery.

The first daily task the therapist is required to perform is the MPC test. The MPC is an automated test tool to assess critical performance and geometric functions of the Halcyon™ unit and must be performed successfully daily or every day a patient is to be treated. It consists of the items listed in table 10.2. All tests must pass before clinical treatments are permitted. MPC has been reviewed extensively on Varian's TrueBeam platform and has been adopted for the Halcyon™. Varian strongly emphasizes that the MPC is not to replace the QA program established by the medical physicist; however, many authors have successfully reviewed MPC compared with other QA devices and consider it a viable review of the systems performance and geometric functions [6–8].

When the patient arrives for their radiation therapy, the therapist loads the patient's treatment plan at the console. As the patient and therapist first enter the treatment room, the patient's name, photo, and birthdate are displayed on the in-room monitors on the side panels of the Halcyon™. The patient or therapist is required to confirm the patient's name, photo, and date of birth for every treatment fraction. Setup notes, appointment notes, patient orientation, and patient setup

Table 10.2. Machine performance checks (MPC) tests for performance and geometric functions of the Halcyon™ machine.

Isocenter	Threshold	Gantry	Threshold
Size	± 0.90 mm	Absolute	± 0.50 °
MV imager projection offset	± 0.50 mm	Relative	± 0.50 °
kV imager projection offset	± 0.50 mm	**Couch**	
Beam		Lateral	± 0.50 mm
Output change	± 4.00%	Longitudinal	± 0.50 mm
Uniformity change	± 2.00%	Vertical	± 0.50 mm
MU 1 gain change	± 10.00%	Lateral (long)	± 1.00 mm
MU 2 gain change	± 10.00%	Longitudinal (long)	± 1.00 mm
Collimation		Vertical (long)	± 1.00 mm
MLC		Virtual-to-isocenter lateral	± 2.00 mm
MLC reproducibility		Virtual-to-isocenter longitudinal	± 2.00 mm
Field edges		Virtual-to-isocenter vertical	± 2.00 mm
kV field edges		**MV imager calibration gain**	
Rotation offset	0.50 °	**MV imager calibration uniformity**	

photos can be accessed from the in-room monitors for reference when setting up the patient. No external monitors are required for reference. All pertinent information is displayed on the monitors built into the Halcyon™ system. If bolus is indicated, the therapist must verify and confirm bolus and check the checkbox on the touch screen. By using the **Align** button (figure 10.5(a)), the table will automatically move to the last saved table position. If it is the patient's first treatment the Align function will move the couch to a default location (Lateral = 0.00, Longitudinal = +85.00, Vertical = −10.00). This default table location requires the therapists to align the patient to the virtual isocenter. Alignment is accomplished using the internal lasers of the Halcyon™ unit and the controls on the side of the treatment machine (figure 10.5(b)). No external lasers are required. Once the alignment is correct, the therapist presses and holds the load button (figure 10.5(c)) to move the patient to the machine isocenter.

At subsequent treatments, when the patient's plan is loaded and when the Align button is pressed, the system will move the patient and treatment couch to the last saved virtual isocenter position. This helps minimize the time it takes to position the patient. Once it is verified that the patient is lined up with the lasers, the load button (figure 10.5(c)) is used to advance the patient to the machine isocenter. The patient is now in position for image guidance. Fine tuning of the patient then occurs using the image guidance capabilities of the Halcyon™.

At the treatment console, the therapists acquire the setup images. The 'Match Assistant' will automatically try to match the acquired images with the digitally reconstructed radiographs created from the patient's computed tomography

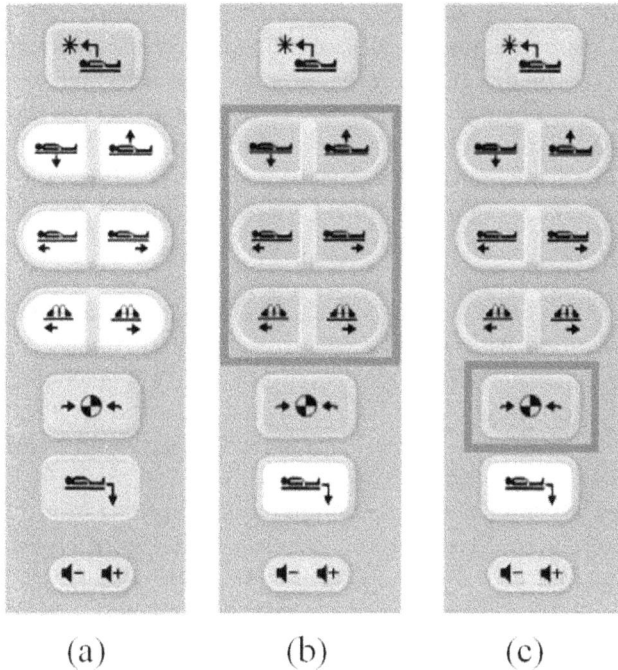

(a) (b) (c)

Figure 10.5. Side control panel of the Halcyon™. Image courtesy of Varian Medical Systems, Inc., Palo Alto, CA. Copyright 2020.

simulation. The therapists may manually adjust and match the images to determine any required shifts. When the alignment of the patient, via image guidance is acceptable, the enable button is activated and the patient is moved to the approved treatment position. If the shifts exceed 2.0 cm in any direction, the system requires the therapist to enter the treatment room and conduct the shift within the room to ensure the patient does not have contact with the bore of the machine. If the calculated shifts are less than 2.0 cm, the shifts can be performed from the console using the Enable button. Positioning is again verified with imaging and if approved, the therapists proceed to treatment delivery. Image guidance is required for every treatment delivered on this machine.

When the treatment has been delivered successfully to the patient, usually within a few minutes, the therapist presses the Enable button at the console and the couch exits the treatment bore and lowers the patient while the other therapist enters the room to help the patient off the treatment couch.

Halcyon™ is designed to deliver IMRT or VMAT treatment plans with an FFF beam. Version 2.0 includes a 3D planning feature using a dynamically flattened beam. This is achieved with the double stacked MLC [1, 12]. Another feature introduced in version 2.0 is the large field treatment option. The maximum field size of Halcyon™ is 28 × 28 cm, however, with the large field option, a treatment length of 36 cm can be achieved. Within the treatment plan, the plan is divided into two isocenters. The junction of the two fields is auto feathered to create a uniform dose distribution at the beam intersection. As all treatments must be image guided on Halcyon™, the dosimetrist may select one of the isocenters as the imaging center or an image center can be designated in between the two isocenters.

The steps for treatment planning follow the same guidelines and steps as with other IMRT/VMAT plans within Eclipse with one exception: the imaging for patient alignment must be selected before the plan can be optimized. If MV imaging is selected, the dose contribution from the image guidance will be accounted for in the optimization of the treatment plan. This has been found to be accurately modeled within the TPS [5, 13].

Halcyon™ 2.0 is designed with kV and MV imaging capabilities. kV CBCT protocols are included with the system based on quality, dose, and anatomical location. Eleven such kV CBCT protocols are included, including head, breast, thorax, and pelvis techniques. kV CBCT is accomplished using a 360° arc and has a longitudinal width of 24.3 cm. The maximum field of view is 49.1 cm in diameter [2]. The kV settings are fixed with four options: 80, 100, 125, and 140 kVp.

As MV imaging is modeled and included in the optimization of the treatment plan, if MV imaging is selected as the imaging modality of choice, and with the EPID positioned in the beam path 100% of the time, daily *in vivo* dosimetry is easily achieved. Varian offers a portal imaging dosimetry package with their TrueBeam and Halcyon™ systems. This system adds the ability to use the EPID device as a QA device for pretreatment patient-specific QA and now, on Halcyon™, efficiently measuring *in vivo* dosimetry utilizing the EPID for imaging and Eclipse for dose calculation [2, 14–16]. Not only is the treatment plan tested for deliverability, *in vivo* dosimetry closes the loop to demonstrate the actual treatment delivered to the patient. Varian has included these and other advanced artificial intelligence tools in their roadmap to the Ethos™ adaptive radiotherapy system.

10.4 Clinical applications

Halcyon™ was initially designed to provide high quality radiation treatments to patients who otherwise would not receive standard-of-care treatment on old technology. The system was built with only 6 MV x-rays. Electrons and high-energy photons, option on other radiation therapy machines, are not offered on the Halcyon™. Even though this may seem to be a limitation of the system, the single energy permits the small configuration and footprint of the treatment machine and it also makes for an efficient, reliable, and simplistic delivery system.

The Halcyon™ unit is very capable of delivering high quality IMRT and VMAT (RapidArc) treatments to all areas of the body. With the addition of the large field option on version 2.0, all IMRT or VMAT plans delivered on other treatment delivery platforms are capable of being delivered faster on Halcyon™ because of a simple workflow, fast imaging, and increased treatment speeds. Imaging, CBCT or static images, are acquired in less than a minute and treatment delivery within a few minutes. This reduces the amount of time the patient is in the treatment room.

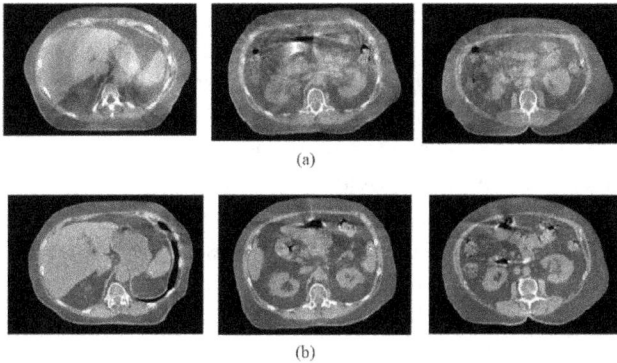

Figure 10.6. Abdominal CBCT images (a) from a Varian TrueBeam™ acquired with free breathing and (b) is an iCBCT from a Halcyon™ with breath-hold [2], reproduced with permission Cai B., et al © John Wiley & Sons

Clinically, the faster treatments improve accuracy not only for patient movement and setup but for internal movement of the target and organs at risk.

In addition to the faster treatments, Halcyon™ provides a conventional kV CBCT reconstruction Feldkamp–Davis–Kress (FDK) algorithm or the iCBCT nonlinear algorithm [3, 17]. The data acquisition for iCBCT and FDK is the same; therefore, the dose is the same. The difference is noticeably different with the iCBCT images. Soft tissue is enhanced with iCBCT as there is less noise, improved contrast to noise, and better uniformity (figure 10.6) [3]. With improved soft tissue visualization, iCBCT is superlative for volumetric imaging of the abdomen.

The features you will not find on the current Halcyon™ system include motion management, gating of the beam, six degree (6D) couch, or surface monitoring. These features might be offered on future versions; however, some may say these features are not needed. Due to the faster imaging techniques, a CBCT is completed in approximately 30 s. This allows imaging to be acquired in one breath hold, therefore, the need for gating or motion management is reduced. Surface monitoring is currently not an option offered on Halcyon™. The system is equipped with a closed-circuit video camera at the foot of the treatment couch; however, the video is only used for patient visualization and a safety feature.

Without a 6D of freedom couch, Halcyon™ is capable of delivering only coplanar beams. This limits its ability to deliver stereotactic body radiation therapy (SBRT) or stereotactic radiosurgery (SRS); however, some have reviewed spine SBRT with Halcyon™ compared to TrueBeam. They concluded that Halcyon™ delivered clinically equivalent plans with a more gradual dose fall-off and lower conformality [18]. Lung SBRT might also be an option because of the faster treatment times; however, the inability to deliver noncoplanar treatments limit the types of SBRT offered on Halcyon™.

On traditional radiation therapy machines, as imaging technology has improved, the treatment workflow has become more complex, adding more opportunities for error. The number of tasks the therapists must complete continues to grow.

The treatment console area tends to be very chaotic with multiple computers, monitors, keyboards, and notes, providing multiple distraction points. Halcyon™'s system is simplistic. Even though it has a sophisticated imaging system for IGRT, the treatment console consists of two monitors, one keyboard, one mouse, and a small control unit. Everything needed to treat the patient, imaging, patient alignment, and treatment delivery, is provided on one system. With Aria, Eclipse, and Halcyon™ all integrated, the workflow has also been simplified. From a patient safety standpoint, it has been said the majority of safety concerns are not related to the technical side of Halcyon™ but to the organizational management, procedures, and human behaviors [4]. The Halcyon™'s design and operation has simplified the workflow and made treatment delivery safer.

References

[1] Kim M M *et al* 2019 Dosimetric characterization of the dual layer MLC system for an O-ring linear accelerator *Technol. Cancer Res. Treat.* **18**

[2] Cai B *et al* 2019 Characterization of a prototype rapid kilovoltage x-ray image guidance system designed for a ring shape radiation therapy unit *Med. Phys.* **46** 1355–70

[3] Kim H *et al* 2019 Early clinical experience with Varian Halcyon V2 linear accelerator: dual-isocenter IMRT planning and delivery with portal dosimetry for gynecological cancer treatments *Int. J. Radiat. Oncol. Biol., Phys.* **105** E705

[4] Pawlicki T *et al* 2019 Clinical safety assessment of the Halcyon system *Med. Phys.* **46** 4340–5

[5] Bernard V *et al* 2018 33 Commissioning and dosimetric characteristics of new Halcyon system *Phys. Med.* **56** 54

[6] Barnes M P and Greer P B 2017 Evaluation of the TrueBeam machine performance check (MPC) beam constancy checks for flattened and flattening filter-free (FFF) photon beams *J. Appl. Clin. Med. Phys.* **18** 139–50

[7] Clivio A *et al* 2015 Evaluation of the machine performance check application for TrueBeam Linac *Radiat. Oncol. (London, England)* **10** 97

[8] Gao S *et al* 2014 Evaluation of IsoCal geometric calibration system for Varian linacs equipped with on-board imager and electronic portal imaging device imaging systems *J. Appl. Clin. Med. Phys.* **15** 164–81

[9] De Roover R *et al* 2019 Validation and IMRT/VMAT delivery quality of a preconfigured fast-rotating O-ring linac system *Med. Phys.* **46** 328–39

[10] Gao S *et al* 2019 Acceptance and verification of the Halcyon-Eclipse linear accelerator-treatment planning system without 3D water scanning system *J. Appl. Clin. Med. Phys.* **20** 111–7

[11] Systems V M 2018 *Halcyon Instructions for Use* (Palo Alto, CA: Varian Medical Systems)

[12] Bollinger D *et al* 2020 Technical Note: Dosimetric characterization of the dynamic beam flattening MLC sequence on a ring shaped, jawless linear accelerator with double stacked MLC *Med. Phys.* **47** 948–57

[13] Malajovich I *et al* 2019 Characterization of the megavoltage cone-beam computed tomography (MV-CBCT) system on Halcyon TM for IGRT: image quality benchmark, clinical performance, and organ doses *Front. Oncol.* **9** 496

[14] Bojechko C *et al* 2015 A quantification of the effectiveness of EPID dosimetry and software-based plan verification systems in detecting incidents in radiotherapy *Med. Phys.* **42** 5363–9

[15] Chuter R W *et al* 2016 Feasibility of portal dosimetry for flattening filter-free radiotherapy *J. Appl. Clin. Med. Phys.* **17** 112–20

[16] Pardo E *et al* 2016 On flattening filter-free portal dosimetry *J. Appl. Clin. Med. Phys* **17** 132–45

[17] Jarema T and Aland T 2019 Using the iterative kV CBCT reconstruction on the Varian Halcyon linear accelerator for radiation therapy planning for pelvis patients *Phys. Med.* **68** 112–6

[18] Petroccia H M *et al* 2019 Spine SBRT With Halcyon™: plan quality, modulation complexity, delivery accuracy, and speed *Front. Oncol.* **9** 319

IOP Publishing

Principles and Practice of Image-Guided Abdominal Radiation Therapy

Yu Kuang

Chapter 11

VenusX™

Yulin Song

VenusX™ is the latest cutting-edge, multifunctional, and artificial intelligence–based linear accelerator (Linac) manufactured by LinaTech (Sunnyvale, CA, USA). Founded in 2010, LinaTech soon became one of the successful high-tech radiation oncology equipment and software companies. Its signature TiGRT micro-MLC (2 mm) is among the bestselling multiple-leaf collimeters (MLCs) in the market at present time. So far, more than 500 units have been sold worldwide. Currently, the company not only manufactures modern medical Linacs and related peripheral devices but also supplies informatics systems to radiation oncology centers and medical oncology clinics worldwide. In this chapter, we discuss various new technical features of VenusX™, including the triple-layer MLC system, built-in three-dimensional optical imaging system, and on-board x-ray imaging systems. We then present abdominal cases treated with VenusX™.

11.1 Introduction

VenusX™ is the latest cutting-edge, multifunctional, and artificial intelligence (AI)–based linear accelerator (Linac) manufactured by LinaTech (Sunnyvale, CA, USA) (figure 11.1) [1]. Founded in 2010, LinaTech soon became one of the successful high-tech radiation oncology equipment and software companies. Its signature TiGRT micro-MLC (2 mm) is among the bestselling multiple-leaf collimeters (MLCs) in the market at present time. So far, more than 500 units have been sold worldwide. LinaTech is also the main supplier of micro-MLC to Accuray® for its Model M6 Cyberknife (Sunnyvale, CA, USA) [2]. LinaTech's Monte Carlo–based treatment planning system (TPS) employs graphics processing unit acceleration technology, making three-dimensional (3D) dose calculation both accurate and efficient. Currently, the company not only manufactures modern medical Linacs and related peripheral devices but also supplies informatics systems to radiation oncology

doi:10.1088/978-0-7503-2468-7ch11

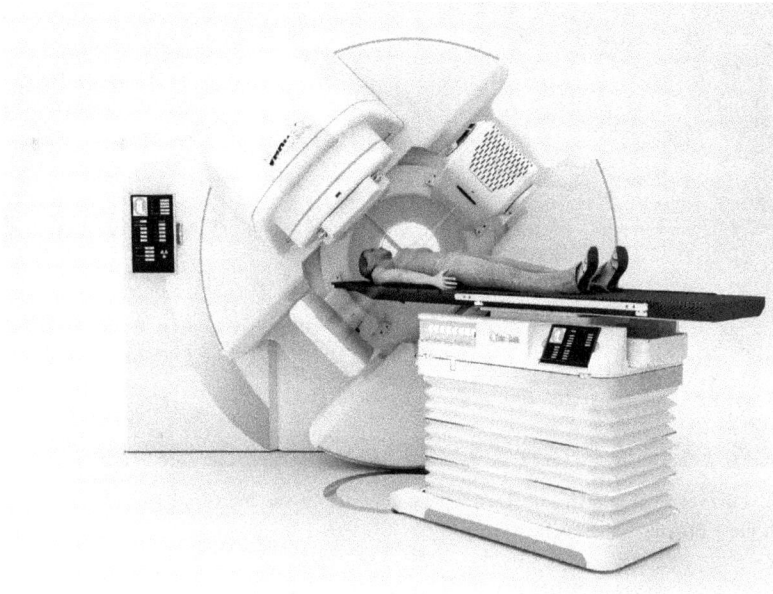

Figure 11.1. A VenusX™. Debuted in March 2020, VenusX™ features a futuristic design and many technological innovations. In this photo, the third layer MLC is mounted to the treatment head. kV imaging source and kV imaging detector are deployed in the imaging position. The side panel on the machine right displays the patient information and plan parameters. The same information is displayed on the two couch side panels as well. © LinaTech.

centers and medical oncology clinics worldwide. LinaTech has a very strong presence in the emerging markets, including China, Russia, Latin America, and Asia. The company has a comprehensive research and development center and a modern production base in the industrial park of the beautiful Chinese city of Suzhou (figure 11.2). Since June 2020, VenusX™ has been conducting large-scale clinical trials on various disease sites at Tianjin Tumor Hospital and Beijing 301 Hospital as part of its pre-market Chinese regulatory certification process.

11.2 VenusX™ technical characteristics

11.2.1 Triple-layer MLC system

VenusX™ has a futuristic design and many distinctive technical features. It incorporates AI technology into every major subsystem module. Its orthogonal triple-layer MLC system is the most noticeable one. Unlike most mainstream Linacs, VenusX™ is equipped with a triple-layer MLC system as its beam shaping and beam intensity modulation device. As shown in figure 11.3, the first layer (highest) is a 51-pair MLC. Each leaf has a projected width of 2.5 mm at the isocenter and a physical height of 70 mm. The high spatial resolution is essential to stereotactic radiosurgery (SRS) treatments for lesions less than 5 mm. The height provides enough radiation attenuation. The leaf ends are focused, offering optimal

Figure 11.2. LinaTech's Research and Development Center in the industrial park of the Chinese city of Suzhou, Zhejiang Province. © LinaTech.

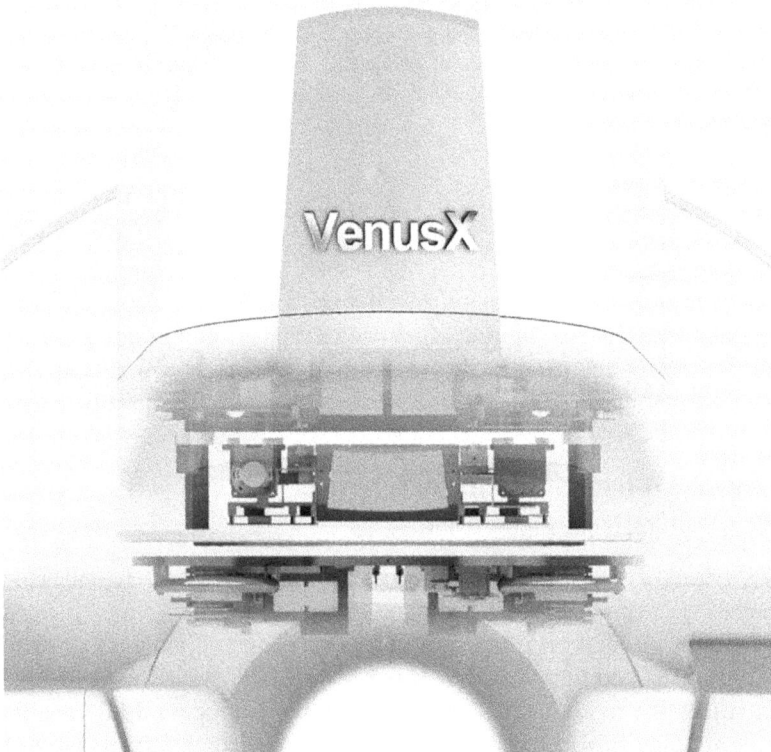

Figure 11.3. The exposed view of the orthogonal triple-layer MLC structure. The third layer MLC is an optional removable module. In total, the triple-layer MLC structure consists of 153 leaf pairs. © LinaTech.

beam characteristics. MLC leaves can travel at a maximum speed of 50 mm s^{-1} with a step precision of 0.1 mm. These technical characteristics greatly enhance the dynamic range of intensity modulation and provide a better protection to organs at risk (OARs) as well. As a comparison, the maximum MLC leaf speed of Varian TrueBeam Linacs (Palo Alto, CA, USA) is 35 mm s^{-1} [3].

The first layer MLC leaves move along a circular path centered at the treatment beam source in the y-direction. This design makes the MLC motion trajectory always perpendicular to the beam central axis during treatment delivery. Therefore, it minimizes the size of beam lateral penumbra along the y-direction. A smaller beam penumbra is one of the crucial requirements for stereotactic body radiation therapy (SBRT) and SRS treatments [4]. This is because SBRT and SRS plans consist mainly of small fields and require steep dose gradients between the planning target volume (PTV) and OARs. A larger beam penumbra makes dose distribution more diffused and less conformal to the PTV for isodose lines less than 60%. The second layer MLC (middle) has the same 51-pair MLC as the first layer but is rotated by 90°. This is the conventional location and rotation for the mainstream Linacs that only have a single-layer MLC. The MLC leaves now travel along a circular path in the x-direction, thus reducing the beam lateral penumbra in the x-direction (figure 11.4). The third layer MLC is an optional removable SBRT/SRS-dedicated MLC module with 51 pairs and

Figure 11.4. A heart-shaped field formed by the upper two-layer MLCs. The arrangement of the MLCs is such that they are always orthogonal to each other. This design aims at achieving the most optimal PTV conformity and precise intensity modulation at the boundaries of concave PTVs. The first layer MLC is oriented in the y-direction, and the middle layer MLC is oriented in the x-direction. © LinaTech.

a projected leaf width of 2 mm. There are also 10 built-in cylinder cones and one 3D optical surface imaging system at this layer. The third layer MLC is an independent micro-MLC module. It is adaptable to most conventional Linacs. The orthogonal three-layer MLC configuration has several distinctive advantages over the current one-layer MLC structure: (1) The size of beam penumbra is reduced in all directions. (2) The accuracy of beam shaping for concave PTVs is greatly improved in both conformity and continuity (figure 11.5). (3) The beam intensity is more precisely modulated. (4) The PTV dose homogeneity is improved. (5) The dose gradients between the PTV and OARs are greatly enhanced. (6) All three types of radiation leakage are reduced. The average magnitude of leakage is decreased from 1.6% for a single-layer MLC to less than 0.1% for the orthogonal triple-layer MLC configuration. (7) The island OARs inside the PTV, such as paraspinal SBRT plans, and island PTVs inside OARs, such as SRS plans with multiple targets, can now be effectively and efficiently spared (figure 11.6). (8) Multiple PTVs can be treated simultaneously, which greatly improves the treatment efficiency.

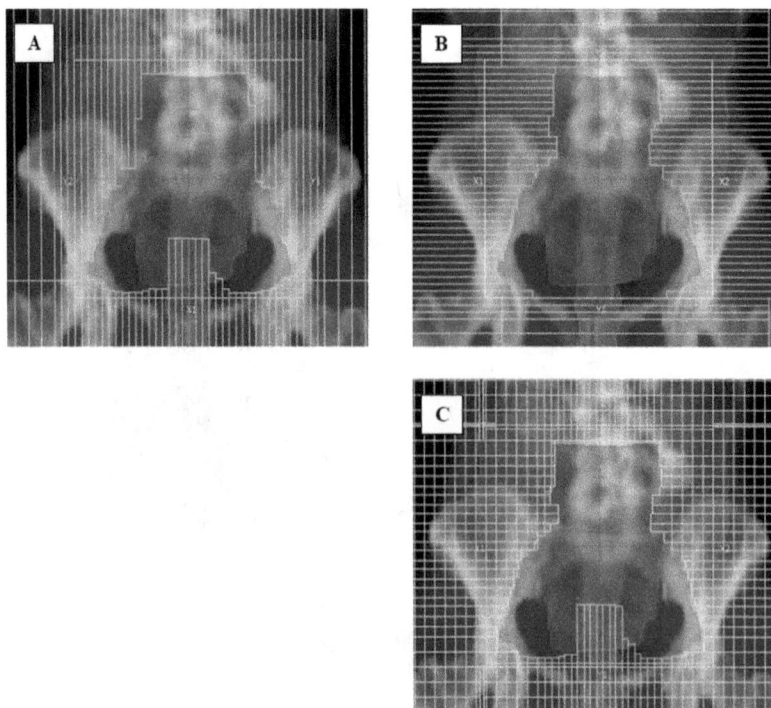

Figure 11.5. This is a pelvic case with a large concave PTV. (A) The field shaped with the collimator angle of 90°. The superior rectum (dark green) is fully blocked by the MLC, but the inferior-medial large bowel (cyan) is exposed to radiation. (B) The field shaped with the collimator angle of 0°. The inferior-medial large bowel is well spared, but the superior rectum is exposed to radiation. (C) The field shaped with the two-layer MLC. Its PTV conformity is greatly enhanced. © LinaTech.

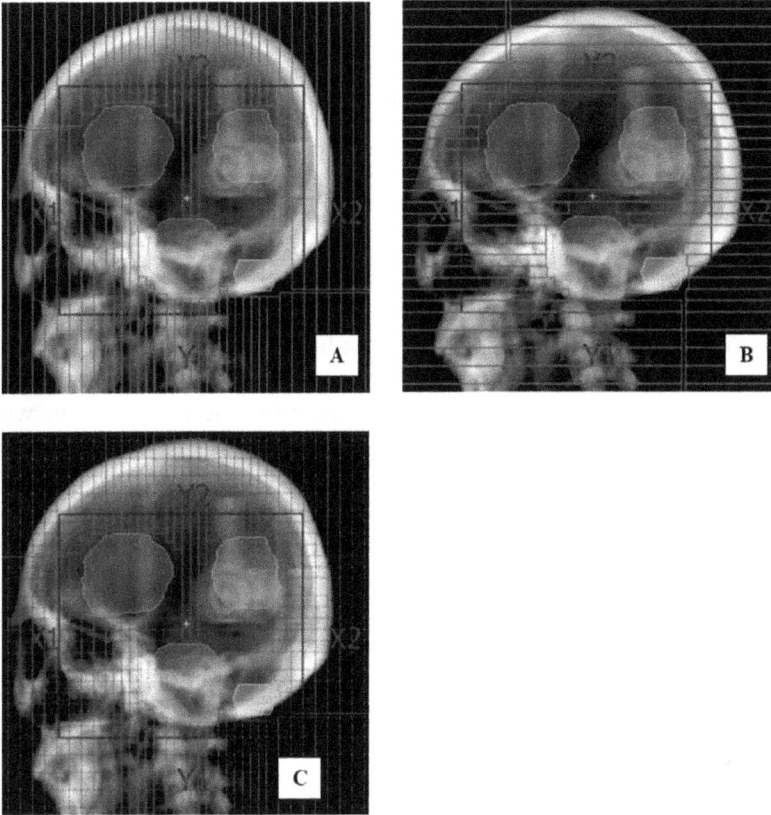

Figure 11.6. This is a case with multiple intracranial island metastasis commonly seen in patients with distant metastases. Conventional one-layer MLC has technical difficulty in shaping these isolated targets conformally. (A) The field shaped with the collimator angle of 90°. (B) The field shaped with the collimator angle of 0°. In both approaches, a large volume of normal tissue is exposed to radiation. (C) The field shaped with the two-layer MLC. The normal tissue among the targets is maximally spared. © LinaTech.

11.2.2 Built-in 3D optical surface imaging system

Another very attractive technical feature of VenusX™ is its built-in 3D optical surface imaging system, αBPS (binocular positioning system). The system consists of three sets of αBPS arranged in a triangular way separated by 120°. Each αBPS contains a binocular camera system. Unlike VisionRT's AlignRT (London, UK) [5, 6] and Varian's Optical Surface Monitoring System (OSMS), αBPS is installed inside the treatment head at a position closest to the patient skin surface, rather than on the room ceiling. This technical feature offers two distinctive advantages: (1) The resolution of the structured light spatial encoding is increased by several times as compared with the room-ceiling installation for the light source having the same spatial resolution. For example, the distance between the mount for the AlignRT center camera and isocenter is 221 cm. It forms a vertical angle of 37.65° with the couch rotation axis. The distance between the mounts for the two lateral cameras and isocenter is 219.2 cm. They form a vertical angle of 38.03° with the couch rotation axis. Depending on the room size, the distance can range from 210 to 230 cm. For αBPS, the distance between the focal point of the camera lens to the isocenter is 50 cm. This factor alone increases the surface image resolution by 4.4 times due to the reduction in light ray diversion. (2) The second advantage is that the structured light of the system and the cameras will never be blocked by the Linac gantry. This is different from AlignRT and OSMS. Their left and right sets of cameras and structured light sources can be partially and even completely blocked by the Linac gantry at some angles during a treatment delivery. This blockage forms a blind zone about an arc sector of 60° for the cameras and the structured light sources. When the gantry is rotating into the blind zone, the affected structured light source and the camera are blocked. As a result, about one third of the skin surface cannot be tracked. For this reason, these systems' tracking performance starts to deteriorate in this wide blind zone. The treatment head built-in type of 3D optical surface imaging system also has its shortcomings. One of them is that its performance could be affected by radiation as time goes by. Therefore, αBPS is heavily shielded by lead. Its design life span is 5 years. The second shortcoming is that it has a large posterior blind zone when delivering posterior static fields or rotating arcs for some disease sites, such as the abdomen and pelvis. However, this limitation will not affect breast deep inspiration breath hold (DIBH) treatments because all treatment fields are anterior.

In addition, αBPS provides a very high image data sampling frequency for real-time patient tracking and gated treatments, including respiratory gating and DIBH treatments. The highest frame rate can reach 24 frames s^{-1}. During respiratory gating and DIBH treatments, αBPS is able to trigger beam on/off based on user-selected gate window thresholds. αBPS also offers a wide field of view (FOV). The maximum FOV in the lateral and longitudinal directions is 30 × 40 cm, which is sufficient for all types of clinical applications. αBPS also incorporates AI technology into its tracking software for automatic patient recognition and source-to-skin distance verification. Figure 11.7 is a 3D chest surface image of a volunteer captured by αBPS. The cross marks on the surface are clearly visible, making patient setup easy and efficient. Figure 11.8 is a 3D head surface image of another volunteer. The hair is differentiable despite its small size.

11.2.3 On-board x-ray imaging systems

VenusX™ provides multifunctional on-board x-ray imaging systems. These include two-dimensional (2D) megavoltage (MV), 3D MV cone-beam computed tomography (CBCT), 2D kilovoltage (kV), 3D kV CBCT, and four-dimensional (4D) kV CBCT imaging. Its kV CBCT, called αCBCT, adopts the dual-ring technology, as shown in figure 11.9. VenusX™ employs a hybrid C-arm and an open dual-ring gantry system. The two rings rotate independently. The hybrid dual-ring gantry system has several advantages over the traditional C-arm gantry, including enabling whole-body CBCT scan, faster CBCT scan, non-coplanar treatment delivery, and

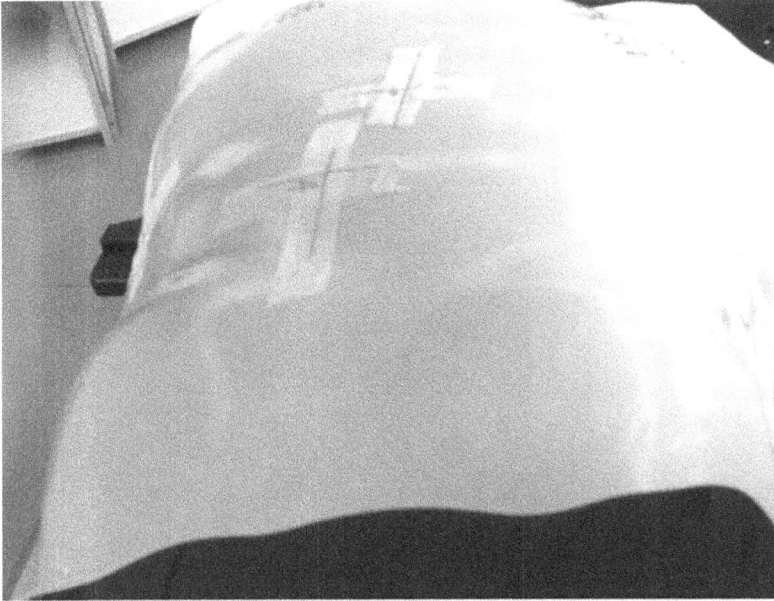

Figure 11.7. This is a 3D chest surface image of a volunteer captured by αBPS. The image shows detailed surface texture. The cross marks and numbers are clearly visible. The FOV is large enough for all clinical applications. The details of the surface are not overshadowed by the structured light pattern, as observed on AlignRT v6.2.

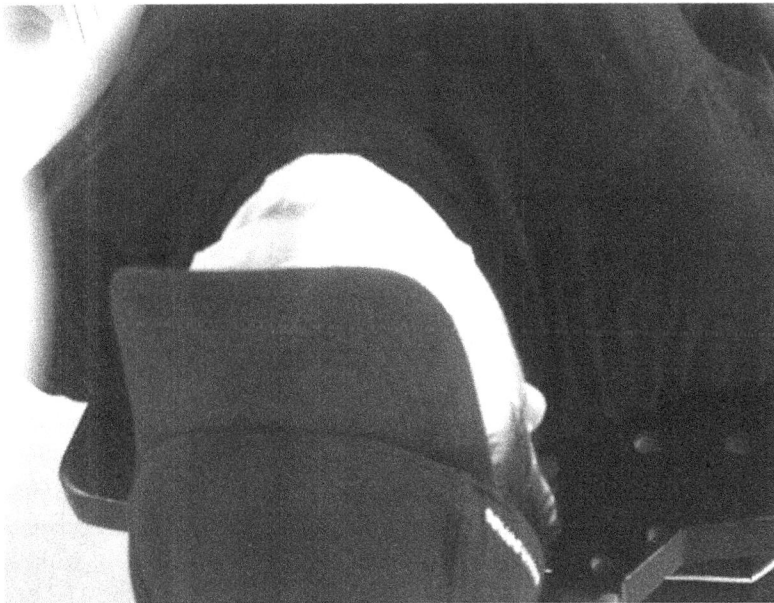

Figure 11.8. This is a 3D head surface image of another volunteer captured by αBPS. Even though there is a high color contrast among the face, hat, and T-shirt, the details of the surface were well captured, as illustrated by the hair on the right side of the head.

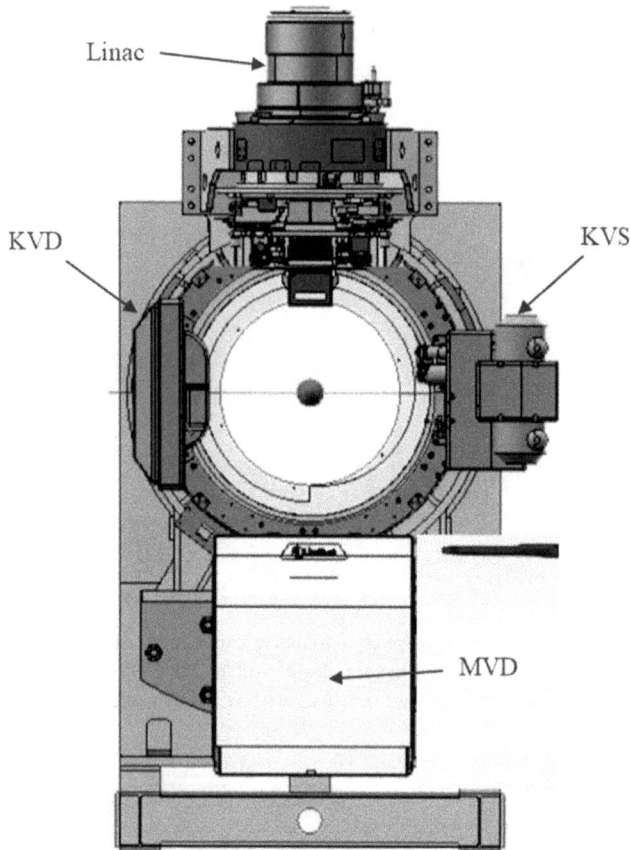

Figure 11.9. A schematic engineering drawing of the hybrid C-arm and dual-ring gantry system. αCBCT rotates independently on the inner ring at an angular speed of 6° s⁻¹, while the Linac rotates on the outer ring. MVD: MV imaging detector, KVS: kV imaging source, KVD: kV imaging detector. © LinaTech.

whole-body radiation treatment. αCBCT is mounted on the inner ring of a diameter of 85 cm. This bore size is larger than those of most computed tomography simulators and can accommodate various types of immobilization devices, including those for the breast and lung treatment. αCBCT consists of a paired kV source and kV detector. Their positions are adjustable, providing flexibility for scanning off-center patients. The energy of kV source is continuously adjustable. The maximum kV peak is 120 kV. Both kV and MV detectors have a physical size of 40×40 cm and pixel matrix size of 2688×2688, which provides a pixel resolution of 0.15 mm (figure 11.10). The shortest αCBCT scan time is 15 s with a 90° fan projection. The longest αCBCT scan range in the caudal-cranial direction is 140 cm with the couch moving through the bore, as compared with about 20 cm for most C-arm–based Linacs. In addition, 4D kV CBCT provides maximum intensity projection and minimum intensity projection options for soft tissue matching and can be directly used for adaptive treatment planning.

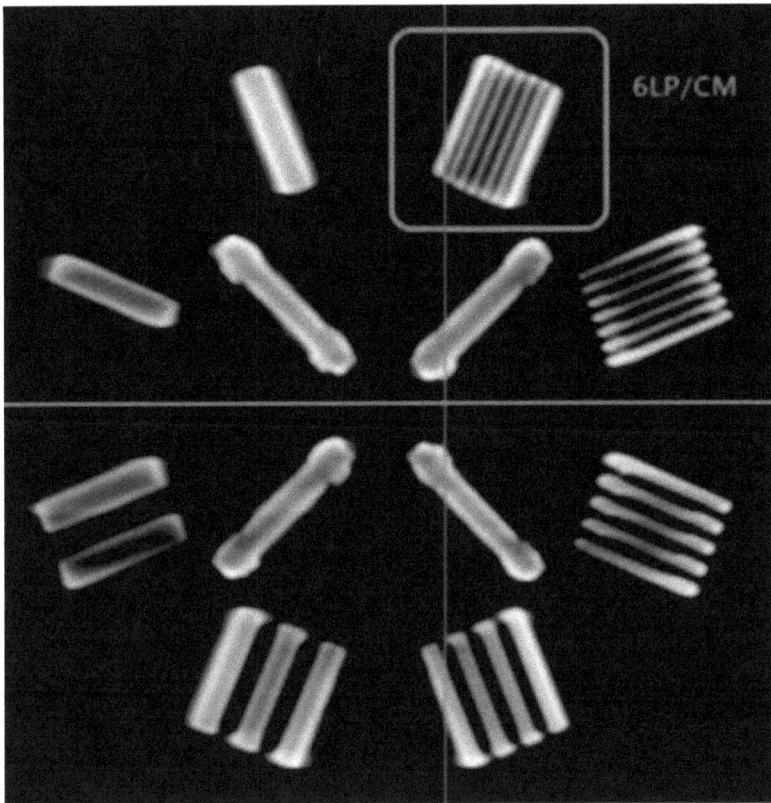

Figure 11.10. A CBCT image illustrating the spatial resolution test of αCBCT with a CBCT phantom. In this example, the measured spatial resolution was 6 lp cm^{-1}.

11.3 Abdominal cancers treated with VenusX™

11.3.1 Metastasis of liver cancer to the iliac bone and T-spine

A 44-year-old male patient had a prior liver transplant. Recent imaging studies revealed multiple metastases of liver cancer to the iliac bone and T-spine. The patient was prescribed 300 cGy ×20 to the iliac bone and 300 cGy ×10 to the T-spine, respectively. Two intensity-modulated radiation therapy (IMRT) plans were computed using the sliding window technique and delivered with multilayer MLC. The plan for the iliac bone consisted of seven equally spaced ipsilateral fields with the gantry angles of 150°, 190°, 230°, 270°, 310°, and 350°, respectively. Figure 11.11 shows the beam arrangement and dose distribution in absolute dose. The blue solid line represents the gross tumor volume (GTV). The solid red dot represents the plan isocenter. The pink isodose line is the prescription dose line (6000 cGy, 100%). The maximum target dose was 6664.7 cGy (111.08%). The minimum target dose was 5731.6 cGy (95.53%). The mean target dose was 6300.7 cGy (105.01%). Figure 11.12 shows the digitally reconstructed radiographs (DRRs) computed at the gantry angles of 0° (A) and 90° (B), respectively. Figure 11.13 shows the MV and kV image

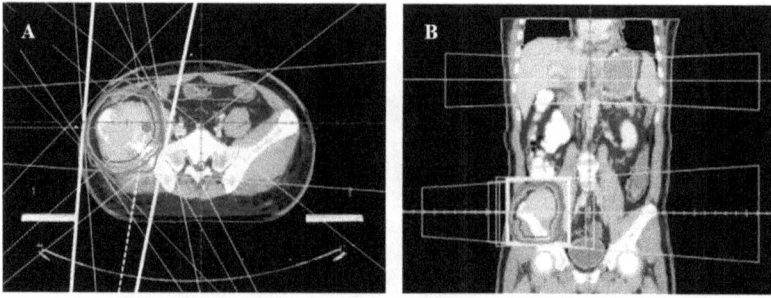

Figure 11.11. The beam arrangement and dose distribution of the plan for the iliac bone. The blue contour represents the GTV. The pink solid line represents the 100% isodose line. The red solid line represents the 80% isodose line. (A) Axial plane. (B) Coronal plane.

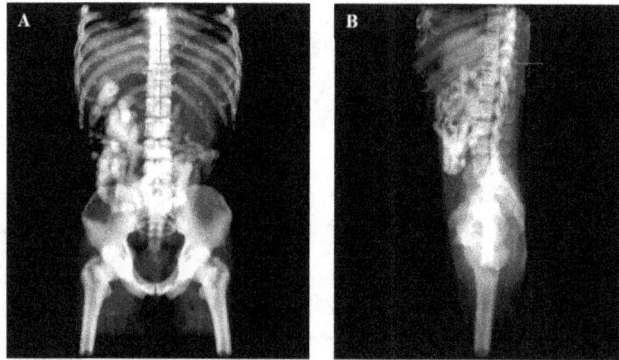

Figure 11.12. Two DRRs computed at the gantry angles of 0° (A) and 90° (B). The green crosshair represents the isocenter of the T-spine plan.

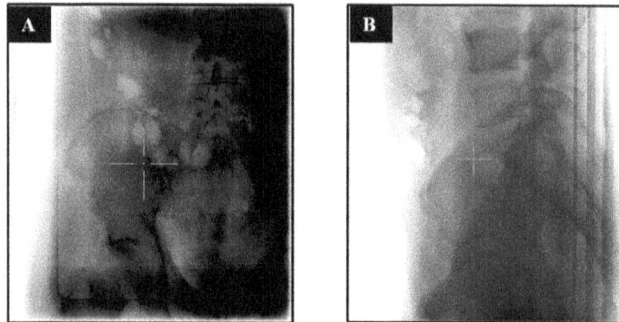

Figure 11.13. An MV/kV image pair acquired by electronic portal imaging detector (EPID) and αCBCT at the gantry angles of 0° (A) and 90° (B). The green crosshair represents the isocenter of the plan for the iliac bone.

pair acquired at the gantry angles of 0° (A) and 90° (B), respectively. Figure 11.14 shows image registration between two image pairs. The plan for the T-spine consisted of four posterior fields with the gantry angles of 115°, 155°, 205°, and 250°, respectively. Figure 11.15 shows the beam arrangement for the T-spine plan.

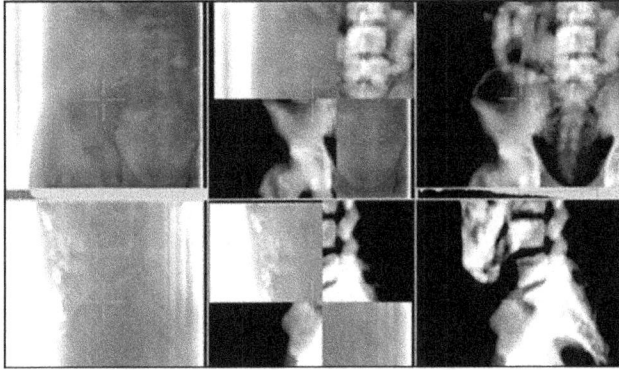

Figure 11.14. MV/kV image pair registration using split window technique. The images at upper row were DRR/MV at the gantry angle of 0°. The images at lower row were DRR/kV at the gantry angle of 90°. The split window technique is a convenient way to examine the goodness of image registration. VenusX™ offers several different image registration tools, including rigid body, mutual information, and deformable body.

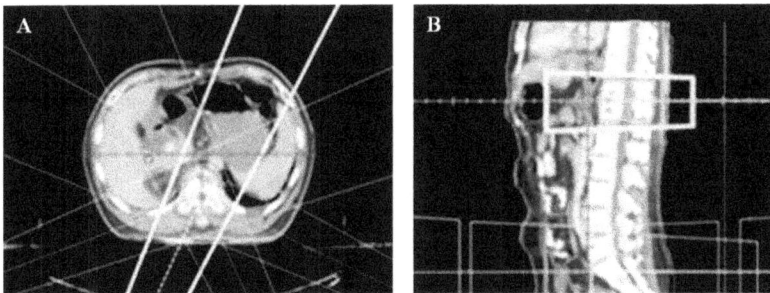

Figure 11.15. The beam arrangement for the T-spine plan. Four posterior beams were used in the plan. The green contour represents the GTV. The red contour represents the PTV. The red line represents the 80% isodose line. (A) Axial plane. (B) Sagittal plane.

The isocenter was placed inside the spinal cord. The green contour represents the GTV, and the red contour represents the PTV. The maximum target dose was 3356.6 cGy (111.89%). The minimum target dose was 2876.8 cGy (95.89%). The mean target dose was 3172.0 cGy (105.73%). Figure 11.16 shows the composite dose volume histograms of the two plans. The doses to the OARs met the institutional dose constraints.

11.4 The competitive market landscape for VenusX™

Currently, VenusX™ has two major competitors in the market, especially in the emerging markets and developing countries. They are Varian Ethos™ [7] and Accuray Radixact™ [8]. Using the latest technologies and patient-centric concepts, both systems are designed to streamline radiation therapy and improve treatment

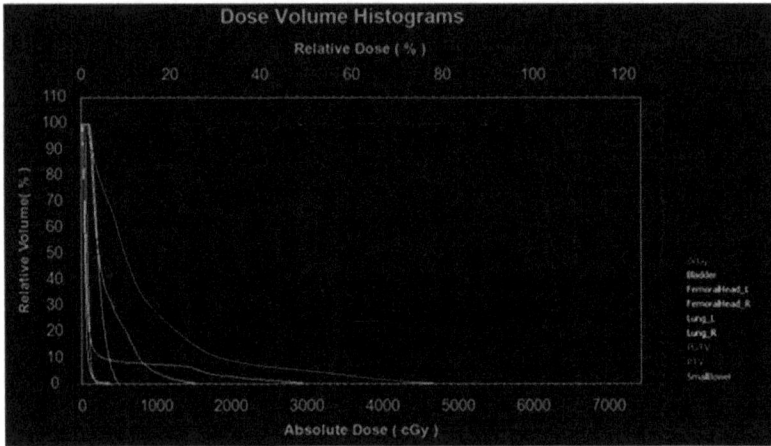

Figure 11.16. The composite dose volume histograms of the two plans. The red line on the right represents the PTV of the left iliac bone plan. The red line on the left represents the PTV of the T-spine plan. The dose volume constraints for the OARs met the institutional dose constraints.

accuracy while enhancing treatment efficiency. Both systems have their respective unique advantages and disadvantages. For example, the foundation of Ethos™ therapy is adaptive intelligence. It streamlines adaptive workflow by integrating visualizing today's anatomy, accessing diagnostic computed tomography, magnetic resonance imaging, and positron emission tomography images at the treatment console, creating AI-aided contours online, calculating on-couch adaptive plan based on today's anatomy, and performing patient-specific quality assurance (QA). All these tasks can be accomplished within a typical 15-min treatment timeslot. Another attractive feature of Ethos™ therapy is that no additional staff or training is required for machine commissioning or QA. Thus, Ethos™ therapy is highly suitable to radiation oncology centers with very limited physics resources. One of the disadvantages of Ethos™ is that it cannot deliver non-coplanar treatment, which is required for intracranial SRS treatment. In addition, limited by the ring-based gantry structure, Ethos™ is equipped with a 3D couch. As to Radixact™, its unique feature among all radiation therapy systems is its continuous 360° helical delivery using its ultrafast binary MLC when the couch moves through the bore. In addition, Radixact™ also has another unique technical feature of real-time target motion synchronization with Synchrony®, which corrects for target motion as a result of respiration, digestion, and patient body movement. The disadvantages of Radixact™ include relatively long treatment time and coplanar-only delivery. Table 11.1 compares the major technical specifications of VenusX™, Ethos™, and Radixact™. Through healthy and constructive competition, the quality and functions of all treatment systems will be improved. The real beneficiaries from this benign, open, and fair competition are the patients.

Table 11.1. Major technical specifications of VenusX™, Ethos™, and Radixact™.

Parameter	VenusX™	Ethos™	Radixact™
Energy (MV-FFF)	6	6	6
Dose rate (MU min^{-1})	1200, 1000, 100	800	1000 or 850
Max field size (cm × cm)	40×40	28×28	5×20 (up to 135 cm in field length)
MLC: no of pairs (leaves)	Orthogonal triple layers, 51×3 (306)	Parallel double layers, 29/28 (114)	Ultrafast single layer, 32 (64)
Built-in SRS cylinder cones	10	None	None
CBCT	Dual rings kV/MV CBCT	kV/MV CBCT	kV/MV CBCT
3D optical surface imaging	Built-in three sets of BPS	None	None
MV EPID no of pixels (pixel resolution in mm)	2688×2688 (0.15)	1280×1280 (0.34)	512×512 (0.78)
Couch type	6D	3D	3D
Auto QA	αQC (TG142, TG106, TG100, TG51)	MPC, DoseLab	None
Non-coplanar delivery	Yes	None	None
Treatment mode	3D-CRT, IMRT, vART (VMAT), sARC (4Pi)	IMRT, VMAT	IMRT, VMAT
Real-time adaptive delivery	Yes	None	Yes, Synchrony®
AI availability	Yes	Yes	Yes
Typical treatment time (min)	4–8	5–10	15–30

MU = monitor unit; MV-FFF = megavoltage flattening filter free; sARC = sphere arc; VMAT = volumetric-modulated arc therapy.

References

[1] LinaTech http://linatech.com/
[2] Accuray https://accuray.com/
[3] Varian https://varian.com/
[4] Padilla L and Palta J R 2020 *Overview of Technologies for SRS and SBRT Delivery* https://radiologykey.com/overview-of-technologies-for-srs-and-sbrt-delivery-2/
[5] VisionRT https://visionrt.com/
[6] Song Y 2020 *BER_TB2 AlignRT v6.2 Commissioning Summary Report* Memorial Sloan Kettering Cancer Center
[7] Varian Ethos https://varian.com/products/adaptive-therapy/ethos
[8] Accuray Radixact https://accuray.com/radixact

Chapter 12

Imaging-guided proton therapy for gastrointestinal tumors

Yao Hao and Tianyu Zhao

This chapter is dedicated to image-guided proton therapy in gastrointestinal (GI) cancers. We first introduce the physics governing protons interacting with matter, followed by discussing the advantages and disadvantages of proton therapy compared to conventional external beam radiotherapy. We also present the clinical indications and manifestation of proton therapy in some predominant forms of GI cancers. We then elaborate on the imaging and motion management modalities used for patient setup and plan adaptation in proton therapy. Ultimately, the planning and delivery of proton radiotherapy is briefly summarized.

12.1 Introduction to proton beam therapy

Protons interact with atoms along their paths, leaving behind cascades of ionization and atomic excitations. The mean kinetic energy loss of protons per unit length of travel is well described by the Bethe–Bloch equation [1]. The kinetic energy lost is inversely proportional to the square of velocity. Energy deposits by protons are small at the entrance but increase as protons slow down and eventually rise suddenly when all protons ultimately cease to move. The pronounced peak at the end of the proton range was observed by and named after Bragg [2]. The finite range of protons beyond the Bragg peak was explored and suggested for medical application utilizing fast protons from a cyclotron by Wilson [3]. The first patient was treated with proton beam therapy (pencil beam scanning [PBS]) at the Lawrence Berkeley National Laboratory, Berkeley, California, in 1954. The first hospital-based proton center was opened at the Loma Linda University Medical Center in 1990, jointly developed by Fermilab and the Harvard Cyclotron Laboratory. It is the first particle accelerator explicitly built for clinical use.

The first commercial cyclotron for medical use was manufactured by Ion Beam Applications SA and was installed at Massachusetts General Hospital in 2001. As of 2020, 109 proton/heave ion therapy institutes are in clinical operation, and have treated over 250 000 patients globally [4].

Monoenergetic protons are produced in short pulses of a few millimeters in diameter inside a cyclotron or synchrotron. Three generations of technologies have been employed in the proton therapy's history to broaden the beam for large tumors. The first generation of technology utilized a single scattering foil, which produces a Gaussian-like dose profile that only covers a small tumor with acceptable dose homogeneity. This early technology has been discontinued for clinical use, except for some special applications such as eye treatment, due to significant waste of protons. The second generation of technology added a second scatter in the beamline that flattens the dose distribution to a double Gaussian-like profile [5]. This type of technology is called double scattering (DS), and it was used for larger tumors, usually up to 25 cm in diameter. The first and second generations of this technology are called passive scattering, with DS explicitly referring to the second generation. The third generation of delivering technology, pencil beam scanning (PBS), replaces the scatters with a sweeping magnet to fill a tumor by manipulating proton spots in the horizontal and vertical directions [6]. The intensity of an individual proton beamlet delivered in PBS mode can be optimized to achieve modulated dose distribution inside a tumor in addition to the drastically reduced neutron dose when compared to DS [7]. Due to the clear dosimetric benefit of intensity-modulated proton therapy, proton machines ordered and installed after 2014 were made exclusively carrying PBS.

12.1.1 Advantages

PBS's main dosimetric advantages over conventional photon radiotherapy are the sparing of normal tissues distal to the tumor due to the finite and well-controlled range of proton beam and the reduced total radiation dose deposited in the patient ("integral dose"). Unlike a photon beam that decays exponentially along the beam path, quasi-monoenergetic protons come to a full stop at nearly the same depth, dominantly determined by the initial kinetic energies of protons extracted from an accelerator and the proton stopping power of the media along the beam path.

A proton interacts with atoms of the media along the beam path in three distinct ways. First, nuclear interaction is a much less frequent event, during which the nucleus inside an atom captures a direct hit proton. This interaction usually results in releasing but not cascading one or more protons, neutrons and other lightly charged particles, such as alpha particles. Secondly, scattering occurs more frequently when a proton is deflected by orbital electrons (most likely) or the nucleus (rare). A series of such deflections, called multiple Coulomb scattering, contributes to the angular spread of a proton beam. Lastly, ionization of orbital electrons through electromagnetic interaction is the dominant process by which a proton loses its kinetic energy. The mean rate of energy loss by protons in the therapeutic energy range is well described by the Bethe–Bloch equation [8].

The proton stopping power, which is defined as energy loss per unit distance that protons travel, increases as protons slow down, giving rise to a narrow but prominent peak where the majority of protons deplete all their residual energy. The peak is called the Bragg peak, named after William Bragg, who observed it in 1903. As very few protons still carry kinetic energy after the Bragg peak, PBS offers significant dosimetric advantages over photons concerning the entrance dose, the dose around the Bragg peak and the exiting dose beyond the Bragg peak.

12.1.1.1 Entrance dose

According to the Bethe–Bloch equation, the transfer of energy from a proton to the surrounding media it passes by is mostly inversely proportional to the square of proton velocity. As protons move deeper into the media, they start to slow down due to the loss of kinetic energy by interacting with orbital electrons and nuclei of atoms along the beam paths. Therefore, proton stopping power is lowest at the entrance and decreases with depths following the beam path.

Similar to photons, there is also a buildup region called nuclear buildup at the entrance, mainly caused by the inequilibrium of secondary protons generated from nuclear reaction. However, the magnitude of nuclear buildup, typically around a few percentages of the peak dose and spanning up to 4 cm, is significantly smaller than the buildup observed in photons. The skin-sparing effect achieved in high-energy photon beams is much less prominent in proton beams.

12.1.1.2 Bragg peak

As protons slow down, the cross-section of a moving protons interacting with orbital electrons and nuclei grows quadratically with the reciprocal of the proton's kinetic energy. A prominent peak is observed near the end of the beam range as all protons release most of their energy at almost the same depth. However, not all protons with the same initial kinetic energy stop at exactly the same depth for two reasons. First, the number of interactions and the energy transfer between individual protons and atoms in the media, varying in electron binding energy and atomic shell structure, are subject to statistical fluctuations. This fluctuation broadens the energy spectrum of incident protons and is governed by Landau distribution as the protons travel deeper into the media. Secondly, individual protons travel in slightly different trajectories due to multiple Coulomb scattering, which continuously pushes protons to a slight deflection angle through electrostatic force when protons approach the orbital electrons or nuclei of atoms along the beam path. Protons deflected further from their initial incident direction suffer more energy loss along the beam path and stop earlier than those that remain on course. The two statistical disturbances increase the width of the Bragg peak. They are the significant contributors to the range staggering in PBS

12.1.1.3 Exit dose

The number of orbital electrons knocked out by protons increases rapidly as protons approach the Bragg peak and completely stops after all protons' kinetic energy is dissipated. A sharp distal edge for a single proton follows the depth dose curve down from the location where the proton loses all the kinetic energy and stops moving. However, since not all protons stop at the same depth due to energy straggling, the fluctuation in the rate of energy loss along the beam path results in a slight spreading out of the distal edge, which is called distal fall-off of the Bragg peak. The distal fall-off is usually 1.1% of the residual range that protons travel in the media, in addition to the energy spectrum of protons extracted from the cyclotron and accumulated energy staggering through beamline components.

12.1.2 Disadvantages

Unlike photons that only attenuate fluence with penetration depth, coverage of tumor or sparing of normal organs could be compromised by the uncertainty in beam range that is sensitive to a variety of factors, including the initial kinetic energy, energy spectrum of protons out of the beamline, and the thickness and composition of tissues along the beam path.

12.1.2.1 Range uncertainty

Proton range is clinically defined as the penetration depth of the proton beam in water. However, not all protons stop at the same depth due to a slight spread of energies among incident protons. There are multiple practical ways to define proton range statistically. The range of a proton beam could be either at a depth where 50% of proton fluence stop or at a point on the distal fall-off of the percentage depth curve that corresponds to a predefined percentage of the dose at the Bragg peak or middle of spread-out Bragg peak (SOBP). In clinical practice, the most appropriate definition of range is the R80 (i.e., the position of the 80% dose in the distal fall-off). For quasi-monoenergetic proton beams with clinically acceptable energy spectrums, the R80 corresponds to the mean projected range of protons (i.e., the range at which 50% of the protons have stopped). Thus, the R80 is a good surrogate of nominal proton energy and relatively insensitive to the initial energy spread of the proton beam. Another popular definition of proton range, especially popular in clinical practice, is the R90 (90% dose in the distal fall-off) of a pristine beam or SOBP. The R90 is used mainly for the clinical purpose of ensuring proper energy is used so that a tumor is covered by at least 90% of the isodose line.

Even the initial energies of incident protons are exactly the same, they don't stop at exactly the same depth due to the statistical fluctuation in the interactions between protons and atoms along the beam path. Some protons travel deeper and some shallower, widening the Bragg peak's width and elongating the distal fall-off. This is called range straggling. Studies in homogenous media showed that the extent of range straggling varies with materials, lower for lighter materials and higher for denser materials [9]. For heterogeneous media, the modeling of range straggling is further complicated by the numerous combinations of proton trajectories continuously deflected through multiple Coulomb scattering [10].

Characterization of fundamental physical properties, such as the mass density or proton stopping power of tissues in patients, plays a vital role in the accurate estimation of proton range in PBS. This is no easy job. Single-energy computed tomography (CT) is widely used for patient simulation in radiotherapy. It provides Hounsfield units (HU), a linear transformation of the measured attenuation coefficient relative to that of pure water. To convert from HU to mass density or proton stopping power, a well-established stoichiometric calibration technique has been used as the current standard of practice [11]. However, uncertainty regarding proton stopping power converted from HU was estimated between 1.6% and 5% for different tissue groups [12]. This uncertainty is mainly caused by the lack of linear correlation between HU and tissue composition, the latter of which has a strong impact on proton stopping power. Dual-energy CT (DECT) has been proposed to further improve the accuracy of tissue characterization [12, 13]. However, a key parameter, the mean excitation energy, cannot be determined through DECT, which brings a 1.5% uncertainty to the proton range defining the limit achievable with current technology [14].

An additional margin is typically applied both proximally and distally to the tumor in treatment planning to ensure the maximum coverage of a tumor. A widely accepted recipe of the uncertainty margin is 3.5% of the nominal proton range in water plus an additional 1 mm that accounts for uncertainties in beam commissioning, beam reproducibility and beamline variation [14].

12.1.2.2 Sensitivity on daily anatomic change

A plan is calculated and approved for treatment on the assumption that the patient's anatomy and positioning would be maintained through the entire course of radiotherapy. However, this premise never strictly holds. The impact of interfractional anatomic change poses greater challenges to proton beam therapy than intensity-modulated radiation therapy (IMRT) because of the sensitivity of the proton range. They could cause significant loss of coverage or excessive toxicity to distal normal organs as proton range varies within patients. A retrospective study with a cohort of 730 patients identified some important anatomic changes, including tumor growth or shrinkage, weight change, deformation of bowel and stomach in the abdomen, pelvic cavity and paranasal sinus filling [15]. Anatomic changes were observed in 244 (33.5%) patients, of which 16% were replanned at least once during the course of radiotherapy. A 9% reduction in target coverage was observed in one out of eight lung patients [16], and range changes between +8 and −13 mm were reported in patients receiving proton dose to paraspinal tumors [17] due to weight loss. Another study showed a significant correlation between gas contained in the bowel and target coverage in six out of seven subjects investigated [18].

12.1.2.3 Lack of understanding of biological effect

Radiotherapy treatments are prescribed based on clinical experiences acquired dominantly on the physical dose levels delivered with high-energy photons. The mechanisms of DNA damage by photons and protons are different. Photons are low-LET (linear energy transfer) radiation. The δ-electron releases a photon along its path, creating free radicals that are only lethal to DNA within reach of the radicals before they are neutralized. Protons, on the other hand, are high-LET radiation. A proton is capable of producing δ-electrons much higher in energy, albeit less in number along a unit length the proton travels. The high-energy δ-electron leaves behind a cluster of ionizations, increasing the chance of a direct hit to the DNA, and hence a larger number of complex chromosome aberrations.

Protons are more biologically effective than photons. A constant value of 1.1 is added to the physical dose in proton therapy for the equivalent biological effectiveness dose that would be required to achieve the same biological outcome if delivered with photons. The generic value, which is also called the relative biological effectiveness (RBE) of protons, regardless of tissue types, biological endpoints, fractionations and variations of LET along the beam path is an empirical number without support from solid experimental or clinical data. The uncertainty in the RBE value poses a significant challenge to proton beam therapy.

12.2 Clinical indications of proton therapy in managing gastrointestinal cancers

Adjuvant radiotherapy has been widely used for treating gastrointestinal (GI) cancers. Despite the fact that the strong and persistent benefits from adjuvant radiochemotherapy have been shown in several multicenter clinical trials [19], severe toxicities were reported [20, 21]. It was reported that only 65% of gastric cancer patients enrolled in the Intergroup 0116 trial completed chemoradiotherapy after gastrectomy [19, 22]. Studies showed that toxicities could be reduced significantly by minimizing radiation exposure of the bowel during the dose delivery [23]. The rationale of utilizing proton beam therapy in GI cancers is that further sparing of a critical organs would lower acute and late toxicity while maintaining or even improving treatment efficacy.

12.2.1 Esophageal cancer

The esophagus runs in close vicinity to heart and both lungs. Radiation can cause inflammation in the lining of the esophagus and mouth as well as induce pulmonary and cardiac complications. Common side effects include esophagitis, mucositis, cardiac, pneumonitis and cardiac toxicities such as effusion, pericarditis, etc. The clinical outcomes of patients receiving proton beam therapy alone or with concurrent chemotherapy were reported in a multicenter retrospective review study [24]. The study included 202 patients from four proton centers in Japan. A total of 100 (49.5%) patients had stage III/IV cancer, and 90 were inoperable. The five year local control and overall survivals (OSs) were 64.4% and 56.3%. No grade IV or higher cardiopulmonary toxicities were observed, although two patients were reported with grade III pericardial effusion and one patient with grade III pneumonia. The dosimetric advantage of proton over x-ray was reported in multiple studies. A multi-institutional study comparing outcomes in patients treated with conventional three-dimensional conformal radiotherapy, IMRT and proton beam therapy found radiation therapy (RT) modality was a good indicator of the incidence of pulmonary, cardiac and wound-healing complications [25]. Outcome comparison between IMRT and PBS was reported in a phase IIB trial in which 145 patients were randomized to PBS and IMRT arms by the same group. Total toxicity burden, a composite score of 11 distinct adverse events, was reported to be 2.3 times lower in the patients randomized to the PBS arm. The trial ended early due to strong evidence showing that PBS reduces the risk and severity of adverse events while maintaining similar progression-free survival [21].

12.2.2 Liver cancer

Hepatocellular carcinoma (HCC) is the most common type of liver cancer and is a major contributor to the rising death toll of 700 000 people globally. The role of conventional x-ray therapy has been limited due to the excessive risk of radiation-induced liver disease (RILD) in patients with HCC. High-dose proton beam therapy for HCC was pioneered at the University of Tsukuba in the early 1990s with excellent outcomes [26–29]. A systematic multi-institutional review of 900 patients treated with proton beam therapy between 1990 and 2012 found local control was approximately 80% at 3–5 years, and average OS was 32% at 5 years with low GI toxicities [30]. It was reported that the ratio between unirradiated liver volume and the standard liver volume was one of the significant predictors associated with RILD occurrence [31]. This finding suggests that PBS is superior to conventional x-ray therapy in treatment efficacy, which can largely be attributed to its abrupt dose fall-off beyond the end of the beam range. A single-institution retrospective study followed up on 113 patients with nonmetastatic and unresectable HCC treated with radiotherapy. Patients receiving proton beam therapy were associated with a decreased risk of RILD. The median OS, which was found to be significantly associated with treatment modality in multivariable analysis, was 31 and 15 months for proton and photon patients, respectively [32]. A phase III randomized trial (NRG-GI003) comparing the treatment efficacy and toxicity of protons versus photons for unresectable or locally recurrent HCC is currently accruing patients.

12.2.3 Pancreatic cancer

Pancreatic cancer is the second most common GI cancer in the USA, trailing colon cancer with 53 070 new incidences and 41 780 deaths in 2016 [33]. The prognosis of pancreatic cancer is very poor, with a reported five year OS rate of less than 5% [34]. The role of radiotherapy in the management of pancreatic cancer is controversial, as 70% of patients die with distant metastatic disease [35]. Preoperative radiotherapy has been explored as an effective way to convert borderline resectable tumors to resectable [36]. For locally advanced pancreatic cancer, a viable option is to escalate the radiation dose to the tumor while sparing surrounding normal tissues. Krishnan *et al* reported superior two year and three year OS in patients receiving a biologically effective dose higher than 70 Gy (36% versus 19% and 31% versus 9%, respectively) [37]. The key to the success of preoperative RT and dose escalation is respecting the dose constraints in the proximal GI mucosa. A phase III randomized trial found that radiation dose escalation is potentially deleterious if a large treatment margin has to be used without proper imaging guidance [38]. Proton beam therapy has been reported by multiple groups to reduce the volume of GI mucosa in the low-dose region and significantly cut the clinically relevant dose to kidneys and spinal cord compared to photons [39, 40]. Clinical evidence for proton beam therapy is limited. Some single-institutional and single-arm studies with small cohorts were reported with promising results [41, 42]. Patients involved in those studies were treated with passive scatter protons and set up without daily volumetric imaging. Further investigations with advances in proton planning, delivery and imaging guidance are needed to further explore the benefits of proton beam therapy for pancreatic cancer.

12.3 Imaging and motion management

Similar to conventional x-ray radiotherapy, proton beam therapy uses a variety of imaging modalities for treatment planning, patient setup, plan verification and adaption. Special attention is paid to the images used for planning and adaption due to the higher sensitivity of proton range to subtle variation along the planned beam path, such as daily anatomic changes that can be easily accommodated by x-ray radiotherapy.

12.3.1 Imaging for simulation

Traditionally, patients were simulated with a single energy. A respiratory sensor can be used to guide simulation. Generally, multiple scans may be acquired and fused for treatment planning. Depending on patient condition and tumor motion, simulation procedures can be different from patient to patient. Motion management is performed when there is a need to account for the motion of the abdominal target. For a patient with regular breathing patterns and small target motion (such as motion less than 5 mm), the motion management device may not be necessary. A variety of motion management approaches and devices are available. An abdominal compression device can limit motion for some patients. Interactive breathing management also plays a very important role both during simulation and treatment. A reproducible breathing pattern is needed in order to acquire appropriate images. A patient can be coached in multiple ways, such as verbal coaching, using a respiratory sensor as a guide, or coaching patients to interactively control their breathing pattern. Figure 12.1 shows an SDX™ Spirometric Motion Management System. Such devices may help with scanning by reducing tumor motion during imaging, improving image quality, generating reproducible motion and so on. Some patients are capable of holding their breath at the exhale or inhale phase for the entire scan, while others have to take their free-breathing scan. The patient may or may not receive an additional four-dimensional (4D) CT scan. 4D CT captures the location and movement of organs and target over time. This information will allow us to design a more precise treatment approach and reduce the risk of treatment-related side effects.

Target visualization can be poor using conventional single-energy CT for the abdominal region. Many studies have investigated the benefits of DECT. Two very important advantages for particle therapy include more accurate prediction of the stopping power ratio and improving tumor contrast and contrast-to-noise ratio. Figure 12.2 is one type of dual-source photon-counting CT scanner, a NAEOTOM

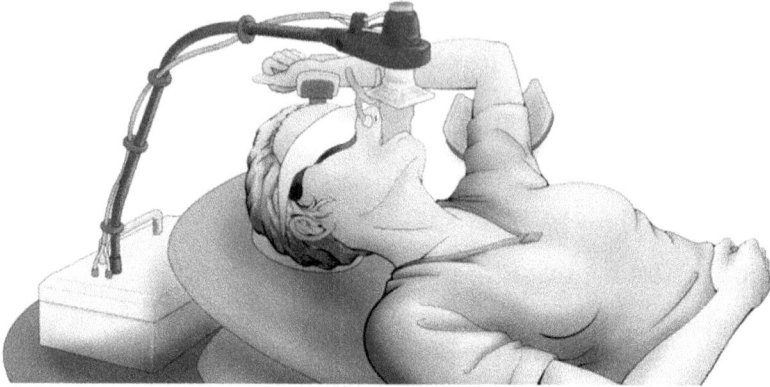

Figure 12.1. A spirometric motion management system. Courtesy of DYN'R Medical Systems.

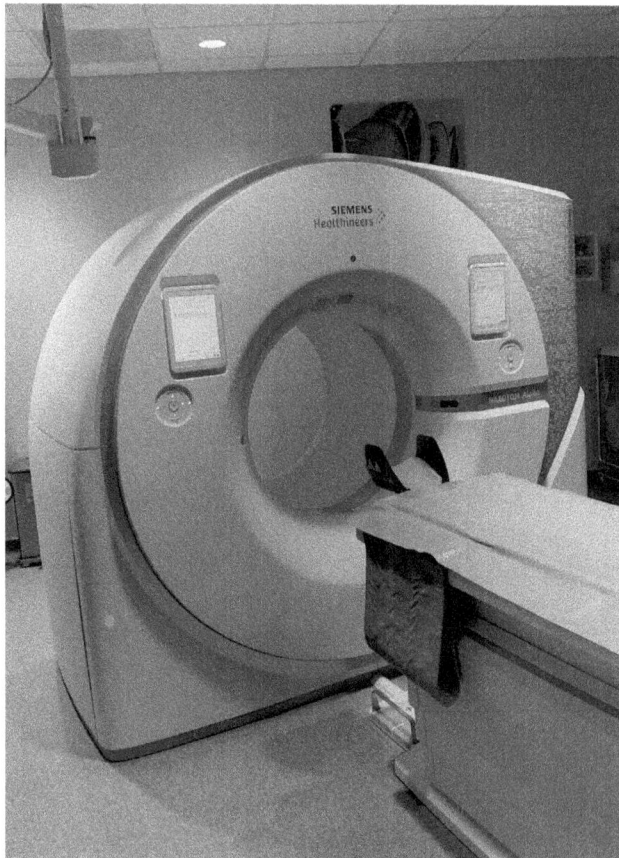

Figure 12.2. A DECT scanner. Courtesy of Dr Zhongwei Zhang. Photo was taken at St. Louis Children's Hospital.

Alpha® with Quantum Technology. This scanner is equipped with two VectronTM x-ray tubes and two QuantaMax photon-counting CT detectors.

12.3.2 Imaging for treatment setup

Usually, particle therapy in-room imaging techniques include orthogonal imaging, cone beam CT (CBCT), CT-on-rail, proton radiography, and proton tomography. The orthogonal imaging system was widely used for patient positioning with the lowest image dose. The CBCT dose can be 10 times higher than radiographic imaging while providing additional data regarding soft tissue. Some centers are also equipped with an in-room CT on rails. Figure 12.3 shows a Siemens SOMATOM Definition Edge inside of a proton vault. CT-equivalent images for treatment planning can be acquired using an in-room CT such as this. Ideally, the patient should be shifted based on one of the imaging modalities and then verified by another modality. Some patients may have internal fiducial markers that can help with abdominal target alignment. Repeat imaging should be avoided, especially for in-room CBCT and CT.

Besides the above imaging modalities, a patient may also use a motion management device for treatment to reproduce the breathing pattern acquired during the simulation, such as the breath-hold treatment illustrated in figure 12.4. The patient can be coached to hold their breath, and a radiation beam can be turned on/off automatically or manually. The breathing pattern window in the figure is what the patient will see during their treatment. Some centers also implemented respiratory-gated spot-scanning proton beam therapy [43]. A real-time position management

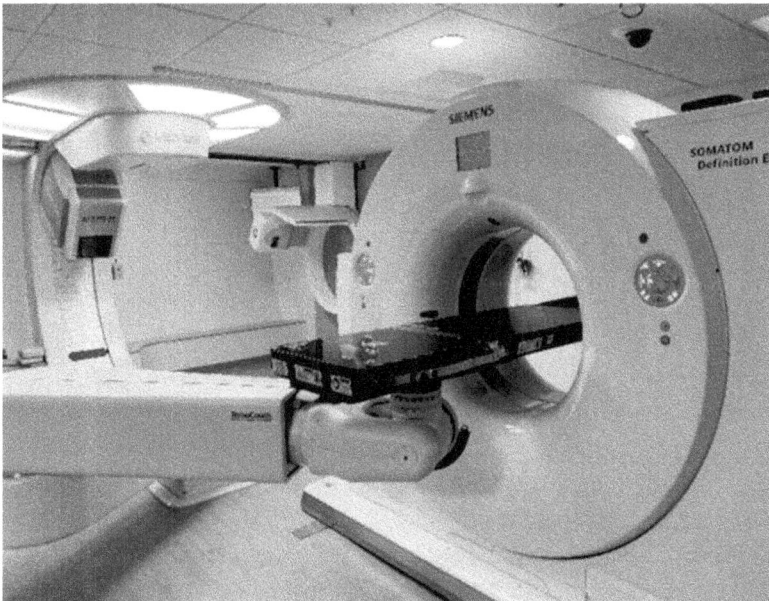

Figure 12.3. A DECT-on-rails for daily setup. Courtesy of Washington University School of Medicine.

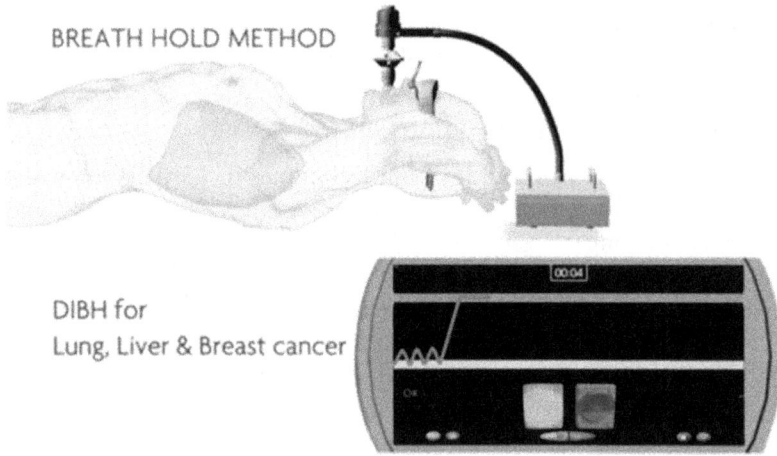

Figure 12.4. A breath-hold system. Courtesy of DYN'R Medical Systems.

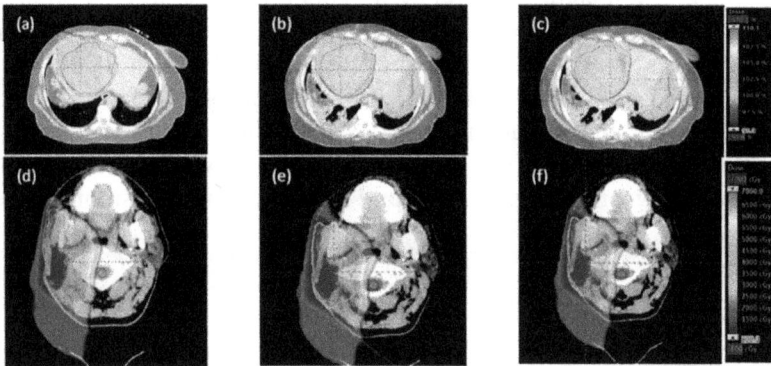

Figure 12.5. Dose distribution of the original treatment plan in a liver and head and neck case (a and d), recalculated dose on daily CT (b and e) and replanned dose on daily CT (c and f). Reprinted from [46], Copyright (2018), with permission from Elsevier.

system (Varian Medical Systems, Inc., Palo Alto, CA, USA) was used for the synchrotron-based delivery system.

12.3.3 Imaging for plan verification and adaption

Patient anatomy can change during the course of treatment. Particle therapy is less forgiving in changes such as weight loss/gain, tumor motion or shrinkage/swell and organ-at-risk motion. In-room CBCT or CT not only serves for daily positioning but also as a "trigger" for replanning. In-room CT can be directly used for replanning if the calibration curve is ready. Several studies investigated the feasibility of using CBCT for particle adaptive therapy [44, 45]. Figure 12.5 shows dose distribution of the original treatment plan on planning CT ((a) and (d)) and daily CT ((b) and (e)) and the adaptive plan on daily CT ((c) and (f)) [46]. Recent advancements on volumetric imaging, registration and dose accumulation allow a patient to be treated with online adaptive replan or offline adaptive replan.

12.4 Treatment planning and delivery

Treatment plans of PBS are tailored based on how proton beams are delivered. Proton beams out of a cyclotron or synchrotron are pencil beams of a few millimeters in diameter. To reach the large radiation field size for large tumors, either scatters or a steering magnet is used in the beamline. A proton beam delivered with one or more scatters in the beamline is called passive scattering. A field shaped by a steering magnet is called PBS. As PBS exhibits more flexibility in field modulation and superior dose distribution than passive scattering, it becomes popular and dominant in newly constructed proton centers.

12.4.1 Passive scattering

Passive scattering features one or two scatters that broaden a narrow beamlet from the accelerator to a large field. Every scatter adds a Gaussian convolution to the beam profiles. In general, two well-designed scatters add a double-Gaussian convolution to the beam profiles and are able to maintain a ±5% uniformity in a field 25 cm in diameter. Several scatters are placed in a carousel or rail for various field sizes needed for different disease sites. Brass apertures are used for further trimming to fit for specific tumor shapes. It was estimated that 60%–80% of protons are wasted in the double-scatter configuration [47]. Modulation of longitudinal profiles is achieved with either a modulation wheel, installed predominantly in most clinical machines, or a ridge filter. A modulation wheel is a premanufactured wheel with steps of various thicknesses arranged in circular order. A stack of Bragg peaks with various ranges is produced and forms an SOBP when the proton beam hits the steps of a rotating modulation wheel. The widths of the steps are optimized to carry proper weights for uniform SOBP. A beam modulation system that synchronizes the beam delivery and modulation position is equipped for further tuning of SOBP uniformity in some modern passive-scattering machines.

The combinations of scatters and modulation wheels are limited on a passive-scattering machine. Planning for such a machine requires finding the optimal combinations of beamline components to maximize dose to the target while sparing normal tissues as much as possible. Brass apertures are used to shape the field laterally. Margins that account for possible intra- and interfractional motion, setup uncertainty and lateral penumbra are used in the design of the brass apertures. A plastic compensator is used to fine-tune the residual energy of the proton beam so that the delivery is distally conformal to the target.

12.4.2 Pencil beam scanning

PBS has a relatively simpler beamline, where a steering magnet replaces all the scatters to manipulate charged protons for lateral spreading. Unlike passive scattering, during which the whole tumor receives radiation simultaneously, radiation scans through the tumor volume layer by layer and spot by spot in PBS. Longitudinal modulation is done by optimizing the weights (or monitor units) of all the proton beamlets. The optimization is an inverse procedure performed in the treatment planning system [48].

The optimization starts with a uniform distribution of spots in the treatment target. The weights and energies of the spots are optimized to reduce the quadratic penalty function that evaluates the difference between objectives and the current iteration. Some spots gain more weight, while some are reduced or even eliminated from the optimization if the weights consistently drop below a preset threshold for the purposes of delivery efficiency and accuracy. Beams can be optimized separately or in combination. If optimized separately, each beam is designed to deliver radiation uniformly to the target. This is called single-field optimization [49]. If all beams are optimized in a group to deliver radiation to the target, each beam only contributes part of the radiation dose to part of the target. This is called multiple-field optimization (MFO) [50]. Although MFO improves dose conformality and sparing of normal organs, it depends on the distal and lateral matching of spots from beams incident from different angles and hence is more vulnerable to anatomic change, setup error and range uncertainty. Robustness optimization with setup error, range uncertainty and organ motions should be used in plan optimization to reduce the variation in the delivered dose [51].

References

[1] Bethe H and Ashkin J 1953 *Experimental Nuclear Physics* (New York: Wiley)
[2] Bragg W H and Kleeman R 1904 LXXIV. On the ionization curves of radium *Lond. Edinb., Dublin Philos. Mag. J. Sci.* **8** 726–38
[3] Wilson R R 1946 Radiological use of fast protons *Radiology* **47** 487–91
[4] Particle Therapy Co-Operative Group 2020 https://ptcog.ch/index.php/facilities-in-operation-restricted
[5] Koehler A M, Schneider R J and Sisterson J M 1977 Flattening of proton dose distributions for large-field radiotherapy *Med. Phys.* **4** 297–301
[6] Goitein M and Chen G T 1983 Beam scanning for heavy charged particle radiotherapy *Med. Phys.* **10** 831–40
[7] Schneider U, Agosteo S, Pedroni E and Besserer J 2002 Secondary neutron dose during proton therapy using spot scanning *Int. J. Radiat. Oncol., Biol., Phys.* **53** 244–51
[8] Groom D E and Klein S R 2000 Passage of particles through matter *Eur. Phys. J. C – Part. Field.* **15** 163–73
[9] Janni J F 1982 Energy loss, range, path length, time-of-flight, straggling, multiple scattering, and nuclear interaction probability: in two parts. Part 1. For 63 compounds: part 2. For elements $1 \leqslant Z \leqslant 92$ *At. Data Nucl. Data Tables* **27** 147–339
[10] Jia S B, Mowlavi A A, Hadizadeh M H, Ebrahimi and Loushab M 2014 Impact of range straggling and multiple scattering on proton therapy of brain, using a slab head phantom *Int. J. Radiat. Res* **12** 161–7
[11] Schneider U, Pedroni E and Lomax A 1996 The calibration of CT Hounsfield units for radiotherapy treatment planning *Phys. Med. Biol.* **41** 111–24

[12] Yang M, Zhu X R and Park P C *et al* 2012 Comprehensive analysis of proton range uncertainties related to patient stopping-power-ratio estimation using the stoichiometric calibration *Phys. Med. Biol.* **57** 4095–115

[13] Williamson J F, Li S, Devic S, Whiting B R and Lerma F A 2006 On two-parameter models of photon cross sections: application to dual-energy CT imaging *Med. Phys.* **33** 4115–29

[14] Paganetti H 2012 Range uncertainties in proton therapy and the role of Monte Carlo simulations *Phys. Med. Biol.* **57** R99–117

[15] Placidi L, Bolsi A and Lomax A J *et al* 2017 Effect of anatomic changes on pencil beam scanned proton dose distributions for cranial and extracranial tumors *Int. J. Radiat. Oncol., Biol., Phys.* **97** 616–23

[16] Hui Z, Zhang X and Starkschall G *et al* 2008 Effects of interfractional motion and anatomic changes on proton therapy dose distribution in lung cancer *Int. J. Radiat. Oncol., Biol., Phys.* **72** 1385–95

[17] Albertini F, Bolsi A, Lomax A J, Rutz H P, Timmerman B and Goitein G 2008 Sensitivity of intensity modulated proton therapy plans to changes in patient weight *Radiother. Oncol.* **86** 187–94

[18] Berger T, Petersen J B B, Lindegaard J C, Fokdal L U and Tanderup K 2017 Impact of bowel gas and body outline variations on total accumulated dose with intensity-modulated proton therapy in locally advanced cervical cancer patients *Acta Oncol.* **56** 1472–8

[19] Smalley S R, Benedetti J K and Haller D G *et al* 2012 Updated analysis of SWOG-directed intergroup study 0116: a phase III trial of adjuvant radiochemotherapy versus observation after curative gastric cancer resection *J. Clin. Oncol.* **30** 2327–33

[20] Lee H J Jr., Macomber M W and Spraker M B *et al* 2019 Analysis of gastrointestinal toxicity in patients receiving proton beam therapy for prostate cancer: a single-institution experience *Adv. Radiat. Oncol.* **4** 70–8

[21] Lin S H, Hobbs B P and Verma V *et al* 2020 Randomized phase IIB trial of proton beam therapy versus intensity-modulated radiation therapy for locally advanced esophageal cancer *J. Clin. Oncol.* **38** 1569–79

[22] Macdonald J S, Smalley S R and Benedetti J *et al* 2001 Chemoradiotherapy after surgery compared with surgery alone for adenocarcinoma of the stomach or gastroesophageal junction *New Engl. J. Med.* **345** 725–30

[23] Parekh A, Truong M T and Pashtan I *et al* 2013 Acute gastrointestinal toxicity and tumor response with preoperative intensity modulated radiation therapy for rectal cancer *Gastrointest. Cancer Res.: GCR* **6** 137–43

[24] Ono T, Wada H, Ishikawa H, Tamamura H and Tokumaru S 2019 Clinical results of proton beam therapy for esophageal cancer: multicenter retrospective study in Japan *Cancers* **11** 993

[25] Lin S H, Merrell K W and Shen J *et al* 2017 Multi-institutional analysis of radiation modality use and postoperative outcomes of neoadjuvant chemoradiation for esophageal cancer *Radiother. Oncol.* **123** 376–81

[26] Wang X, Krishnan S and Zhang X *et al* 2008 Proton radiotherapy for liver tumors: dosimetric advantages over photon plans *Med. Dosim.* **33** 259–67

[27] Tsujii H, Tsuji H and Inada T *et al* 1993 Clinical results of fractionated proton therapy *Int. J. Radiat. Oncol., Biol., Phys.* **25** 49–60

[28] Matsuzaki Y, Osuga T and Saito Y *et al* 1994 A new, effective, and safe therapeutic option using proton irradiation for hepatocellular carcinoma *Gastroenterology* **106** 1032–41

[29] Mizumoto M, Okumura T and Hashimoto T *et al* 2011 Proton beam therapy for hepatocellular carcinoma: a comparison of three treatment protocols *Int. J. Radiat. Oncol., Biol., Phys.* **81** 1039–45

[30] Dionisi F, Widesott L, Lorentini S and Amichetti M 2014 Is there a role for proton therapy in the treatment of hepatocellular carcinoma? A systematic review *Radiother. Oncol.* **111** 1–10

[31] Hsieh C E, Venkatesulu B P and Lee C H *et al* 2019 Predictors of radiation-induced liver disease in eastern and western patients with hepatocellular carcinoma undergoing proton beam therapy *Int. J. Radiat. Oncol., Biol., Phys.* **105** 73–86

[32] Sanford N N, Pursley J and Noe B *et al* 2019 Protons versus photons for unresectable hepatocellular carcinoma: liver decompensation and overall survival *Int. J. Radiat. Oncol., Biol., Phys.* **105** 64–72

[33] Siegel R L, Miller K D and Jemal A 2016 Cancer statistics, 2016 *CA: A Cancer J. Clin.* **66** 7–30

[34] Hidalgo M 2010 Pancreatic cancer *New Engl. J. Med.* **362** 1605–17

[35] Iacobuzio-Donahue C A, Fu B and Yachida S *et al* 2009 DPC4 gene status of the primary carcinoma correlates with patterns of failure in patients with pancreatic cancer *J. Clin. Oncol.* **27** 1806–13

[36] Gillen S, Schuster T, Meyer Zum Büschenfelde C, Friess H and Kleeff J 2010 Preoperative/neoadjuvant therapy in pancreatic cancer: a systematic review and meta-analysis of response and resection percentages *PLoS Med.* **7** e1000267

[37] Krishnan S, Chadha A S and Suh Y *et al* 2016 Focal radiation therapy dose escalation improves overall survival in locally advanced pancreatic cancer patients receiving induction chemotherapy and consolidative chemoradiation *Int. J. Radiat. Oncol., Biol., Phys.* **94** 755–65

[38] Chauffert B, Mornex F and Bonnetain F *et al* 2008 Phase III trial comparing intensive induction chemoradiotherapy (60 Gy, infusional 5-FU and intermittent cisplatin) followed by maintenance gemcitabine with gemcitabine alone for locally advanced unresectable pancreatic cancer. Definitive results of the 2000-1 FFCD/SFRO study *Ann. Oncol.* **19** 1592–9

[39] Thompson R F, Mayekar S U and Zhai H *et al* 2014 A dosimetric comparison of proton and photon therapy in unresectable cancers of the head of pancreas *Med. Phys.* **41** 081711

[40] Zurlo A, Lomax A and Hoess A *et al* 2000 The role of proton therapy in the treatment of large irradiation volumes: a comparative planning study of pancreatic and biliary tumors *Int. J. Radiat. Oncol., Biol., Phys.* **48** 277–88

[41] Kim T H, Lee W J and Woo S M *et al* 2018 Effectiveness and safety of simultaneous integrated boost-proton beam therapy for localized pancreatic cancer *Technol. Cancer Res. Treat.* **17** 1533033818783879

[42] Sachsman S, Nichols R C Jr. and Morris C G *et al* 2014 Proton therapy and concomitant capecitabine for non-metastatic unresectable pancreatic adenocarcinoma *Int. J. Part. Ther.* **1** 692–701

[43] Gelover E, Deisher A J, Herman M G, Johnson J E, Kruse J J and Tryggestad E J 2019 Clinical implementation of respiratory-gated spot-scanning proton therapy: an efficiency analysis of active motion management *J. Appl. Clin. Med. Phys.* **20** 99–108

[44] Yao W, Krasin M J, Farr J B and Merchant T E 2018 Feasibility study of range-based registration using daily cone beam CT for intensity-modulated proton therapy *Med. Phys.* **45** 1191–203

[45] Thummerer A, Zaffino P and Meijers A *et al* 2020 Comparison of CBCT based synthetic CT methods suitable for proton dose calculations in adaptive proton therapy *Phys. Med. Biol.* **65** 095002

[46] Sun B, Yang D and Lam D *et al* 2018 Toward adaptive proton therapy guided with a mobile helical CT scanner *Radiother. Oncol.* **129** 479–85

[47] Liu H and Chang J Y 2011 Proton therapy in clinical practice *Chin. J. Cancer* **30** 315–26

[48] Lomax A J, Böhringer T and Bolsi A *et al* 2004 Treatment planning and verification of proton therapy using spot scanning: initial experiences *Med. Phys.* **31** 3150–7

[49] Meyer J, Bluett J and Amos R *et al* 2010 Spot scanning proton beam therapy for prostate cancer: treatment planning technique and analysis of consequences of rotational and translational alignment errors *Int. J. Radiat. Oncol., Biol., Phys.* **78** 428–34

[50] Pugh T J, Amos R A and John Baptiste S *et al* 2013 Multifield optimization intensity-modulated proton therapy (MFO-IMPT) for prostate cancer: robustness analysis through simulation of rotational and translational alignment errors *Med. Dosim.* **38** 344–50

[51] Liu W, Zhang X, Li Y and Mohan R 2012 Robust optimization of intensity modulated proton therapy *Med. Phys.* **39** 1079–91

IOP Publishing

Principles and Practice of Image-Guided Abdominal
Radiation Therapy

Yu Kuang

Chapter 13

Application of image-guided radiation therapy for abdominal stereotactic body radiotherapy

Hui Zhao and Adam Paxton

This chapter is dedicated to the application of image-guided radiation therapy for abdominal stereotactic body radiotherapy (SBRT). General challenges for abdominal SBRT are briefly described, including patient positioning and immobilization, organ-at-risk considerations, and patient inter- and intrafractional motion management. Commonly applied image-guided radiation therapy modalities for abdominal SBRT are then introduced, including CT-on-rails, cone-beam computed tomography, ultrasound, implanted body global positioning system, radiographic imaging with implanted fiducials, and magnetic resonance imaging. The advantages and limitations for each modality are discussed. Lastly, common technologies available for patient intrafractional motion management and monitoring for abdominal SBRT, such as direct treatment target motion monitoring, patient surface guidance, gated treatment delivery, and breath hold techniques, are described, and their advantages and limitations are compared.

13.1 Introduction

Stereotactic body radiation therapy (SBRT) has become more popular in the last decade. This hypofractionated technique delivers high doses in a few numbers of fractions to shorten the whole course of treatment, which requires a high degree of dose conformity on targets to spare normal tissues but results in improved treatment response [1–16]. SBRT treatment sites in the abdominal region have included liver, pancreas, and adrenal glands. SBRT to these sites has shown better tumor control and lower normal tissue toxicity compared with normal fractionated radiation therapy [1–14].

doi:10.1088/978-0-7503-2468-7ch13

Delivering high doses in a few number of fractions creates technical challenges for SBRT. Target inter- and intrafractional motions are common challenges for SBRT [17–30]. For abdominal targets, these motions are often due to organ filling and emptying and patient respiration. Target proximity to organs at risk (OARs) is another challenge for SBRT, and the OARs in abdominal SBRT typically include radiosensitive gastrointestinal organs. Patient interfractional motion can also cause variations in the relative location between OARs and targets. This creates a need for three-dimensional (3D) image-guided radiation therapy (IGRT) for the treatment of these abdominal sites. However, accurate IGRT for abdominal SBRT is also challenged by the relatively homogeneous tissues in this region, which may not provide enough contrast to clearly visualize targets and OARs, and the intrafractional motion due to respiration, which can further degrade imaging.

All SBRT challenges need to be anticipated and addressed during the entire radiation therapy process, from computed tomography (CT) simulation and treatment planning to treatment day image guidance and treatment delivery. During CT simulation, patient positioning and immobilization may limit patient intrafractional motion and help accommodate the designing of a more conformal treatment plan with less OAR doses. Application of four-dimensional CT (4DCT) and the definition of an internal target volume (ITV) may help better account for the patient intrafractional motion. The creation of planning organ-at-risk volumes (PRVs) for OARs may help to design a treatment plan for better managing OAR doses and attempt to account for changes in their position relative to the target. The choice of IGRT modality can contribute to more effective treatment through target visibility and localization; however, multiple IGRT modalities may be needed to ensure treatment accuracy of abdominal SBRT sites. Patient intrafractional motion management and monitoring during imaging for localization and radiation treatment delivery may also help to ensure the overall accuracy in abdominal SBRT.

This chapter discusses the general IGRT modalities utilized for abdominal SBRT, including their advantages and limitations, and also discusses techniques for addressing abdominal SBRT challenges.

13.2 Abdominal SBRT simulation and treatment planning

CT simulation and treatment planning procedures for abdominal SBRT should take specific considerations of addressing SBRT challenges, such as: patient positioning, patient immobilization, patient setup reproducibility and stability during treatment, patient intra- and interfractional motions, treatment target and OAR margins, treatment beam management, and treatment plan dose conformity and quality.

13.2.1 Immobilization/patient position

Patient positioning and immobilization design for abdominal SBRT should consider the positional setup reproducibility, patient mobility and stability during imaging and radiation treatment delivery, and the ease of treatment planning beam arrangements. Patient immobilization may also need to consider options for limiting patient intrafractional motion. Additionally, patient comfort should be considered as SBRT can take longer than regular fractionated treatments, and patients' comfort levels will correlate with their ability to remain still and compliant with instructions.

Patients may be in the supine or prone position for abdominal SBRT. The supine position is more common due to its reproducibility and patient stability during the entire SBRT procedure, including imaging and treatment delivery [32, 33]. Patients may be in an arms-up or arms-down position, depending on the treatment site and target location. An arms-down position may be more comfortable for patients, allowing them to remain still during the entire SBRT delivery, but may create some treatment planning challenges [38].

Patient immobilization systems designed specifically for SBRT are often designed to immobilize the entire body of the patient, rather than just a particular body site. These systems also typically accommodate the use of abdominal compression devices. Any immobilization devices used for SBRT need to be able to hold the patient in a relatively stable position with the least possible motion, and to be able to reproduce the patient setup position at treatment vaults. Abdominal compression devices may be used to force the patient to breathe shallowly, thereby limiting the patient's respiratory motion and further decreasing the ITV volume. The two most common types of abdominal compression systems use either a screw-adjustable paddle or a pneumatic belt. The paddle system is held in place with a bridge that goes above the patient and attaches to the couch. The paddle is placed in the patient's abdomen and the screw can be used to increase or decrease the level of compression. The belt system wraps around the patient's abdomen and the air pressure is adjusted to vary the level of compression [34, 35]. It is worth mentioning that sometimes abdominal compressions may be harsh on some patients and cause discomfort [36, 37]. The success of these devices can vary from patient to patient, and some patients will not be able to tolerate the needed amount of compression to achieve a significant reduction in motion.

13.2.2 Four-dimensional CT/ITV

Four-dimensional CT is a way to evaluate target intrafractional motion due to respiration, especially for patients with lung, liver, or pancreatic cancer. The consistency of the patient respiratory pattern needs to be evaluated before 4DCT acquisition because an accurate 4DCT reconstruction depends on consistent respiratory cycles. If the patient respiratory pattern is irregular and random, the reconstructed 4DCT may have blurred images and other artifacts. For targets in the abdomen, intravenous contrast may be helpful to aid in visualizing targets that may not otherwise be visible from the surrounding tissues. Other imaging modalities, such as magnetic resonance imaging (MRI) or positron emission tomography (PET), can also be helpful for visualizing targets that may not contrast with the surrounding tissue. However, these other imaging modalities may not account for respiratory motion and so should be used with caution.

An ITV may be delineated to include the target motion shown in all ten phases of the 4DCT. A planning target volume (PTV) may be expanded from ITV with certain margins according to the location of the target and the surrounding OARs. The treatment plan and dose calculation may be performed on an averaged 4DCT. It is important to remember that the ITV is only a representation of the motion of the target on that day. There may be day-to-day variations in the motion [31].

13.2.3 PRV for OARs

PRVs may be obtained by expanding OARs with estimated margins of patient intra- and interfractional motion. The intrafractional motion may be evaluated via 4DCT. The concept of PRV is important for SBRT because the treatment plan is highly conformal, and any variation of OAR relative position to the target may cause a discrepancy in OAR dose expected from the treatment plan. OAR dose evaluation should include the PRV dose evaluation for counting both inter- and intrafractional patient motion.

13.3 IGRT for SBRT (target localization)

There are a wide variety of IGRT modalities commercially available for abdominal SBRT target localization, and some are configured for specialized treatment machines. The IGRT modalities for abdominal SBRT include, but are not limited to, CT-on-rails (CTOR), cone-beam CT (CBCT), ultrasound, implanted body global positioning system (GPS), radiographic imaging with implanted fiducials, and MRI. Each imaging modality has its own benefits and limitations.

13.3.1 CTOR

CTOR has been utilized regularly as an IGRT modality over the last decade [32, 33, 39]. However, its use is much less common than CBCT. The most significant advantage of CTOR over CBCT is its image quality, particularly in the low contrast setting of the abdomen. CTOR has the same image quality as a simulation CT. This makes the image registration between treatment day and simulation CT relatively more straightforward. Figure 13.1 shows a comparison between a simulation CT and CTOR scan on a treatment day for a liver SBRT patient, and the almost identical image quality may be seen from the comparison. Another important advantage of CTOR is that a 4DCT workflow may be available. With the availability of 4DCT pre-treatment, the intra- and interfractional motion for both target and OARs may be evaluated and compared with the simulation 4DCT. This verification is extremely important for liver and pancreas SBRT because the intra- and interfractional motion variations for patients with liver and pancreatic cancer may be significant. However, CTOR has some limitations. The main limitation of non-4D CTOR is that the CTOR scan only captures a moment in the respiratory cycle. It may not be known which respiratory phase was captured during the scan, which can make matching with the simulation 4DCT scan difficult. The other limitation of CTOR is that it is a relatively more complicated workflow compared

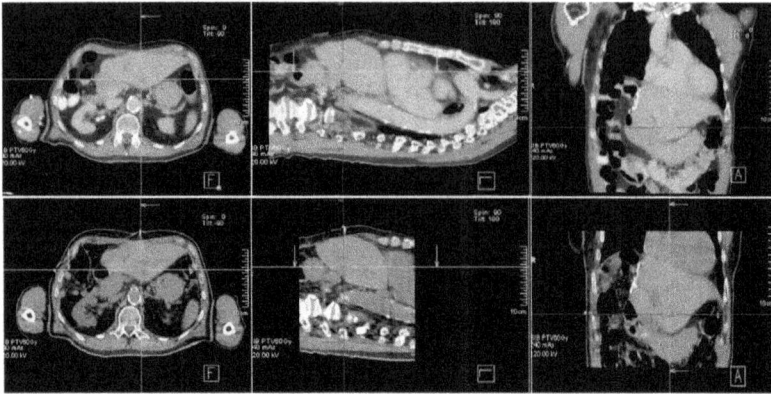

Figure 13.1. Comparison between a simulation CT and CTOR for a liver SBRT patient. Images shown from left to right are the axial, sagittal, and coronal planes for simulation CT (top), and CTOR for IGRT (bottom).

with CBCT in that the patient must be moved away from treatment isocenter in order to acquire the images. This may be more time consuming and require verifying that the patient has not moved away from the setup position they were in prior to CTOR acquisition.

Currently, there are two widely used room designs for CTOR in treatment vaults: one is the CTOR and the treatment machine 180 degrees apart, and the other is 90 degrees apart. For both designs, the patient is moved away from the setup/treatment position for the CTOR scan, and it is critical that the patient be returned to the same couch position after the CTOR scan. The IGRT-indicated shifts can then be applied.

13.3.2 CBCT

CBCT, including 4D CBCT, is a very commonly used IGRT image modality [40–43], and it is part of the on-board imaging for most modern treatment machines. The advantages of CBCT include its ease of use, relatively fast acquisition, well-established image registration process, the potential for easy and fast verification scans after IGRT shifts, and integration with linear accelerator (LINAC) treatment machines. However, CBCT suffers from image quality issues compared with simulation CT, especially for large patients. This can result in difficulty identifying the target on CBCT, especially for targets in the abdomen. Figure 13.2 shows a comparison of simulation CT and CBCT for a pancreas SBRT treatment. In this example, it is extremely hard to localize the target from the CBCT scan.

Most CBCTs take about one minute to acquire. This time frame results in an approximate averaging of all phases of the respiratory motion. If the patient had a 4DCT simulation scan, the CBCT scan is comparable to the average 4DCT image. It is very important that the patient remain still during the CBCT acquisition. If the patient moves during the CBCT scan, the CBCT image may be blurred or have artifacts. 4D CBCT has limited commercial availability; it may help identify the

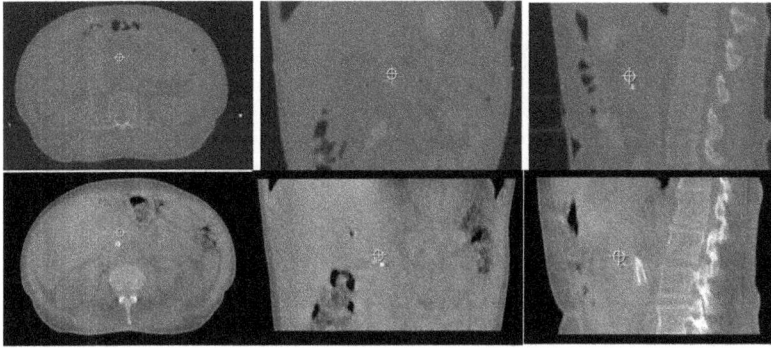

Figure 13.2. From left to right, the axial, coronal, and sagittal images from a simulation CT (top) and a treatment day CBCT (bottom) from a pancreas SBRT treatment. Note the better image quality of the simulation CT compared with the CBCT. The target and OARs are more easily identified on the simulation CT.

target and OARs intrafractional motion, but it still lacks the image quality of 4DCTs acquired on a simulation CT.

13.3.3 Ultrasound

Ultrasound image guidance in radiation therapy is most commonly used for prostate cancer, but also has applications for liver and pancreas treatments [44–48]. There are mainly two ultrasound systems commercially available for IGRT: BATCAM (Best NOMOS, Pittsburg, PA) and Clarity (Elekta, Stockholm, Sweden). BATCAM is able to scan two-dimensional (2D) ultrasound images, and Clarity is able to scan both 2D and 3D ultrasound images. Figure 13.3 shows a comparison of a simulation CT and BATCAM ultrasound images for a pancreas treatment. In this example, the aorta can be clearly distinguished from the ultrasound images and can be used to align the patient for treatment.

Clarity also has the ability to provide real-time target tracking, and this function is extremely useful for SBRT treatments. When the target is moved out of the preset tolerance, the treatment beam may be turned off. This way the accuracy of the treatment delivery is maintained.

One limitation of ultrasound for SBRT is the lack of familiarity most radiation oncology staff members have in the interpretation of ultrasound images. This creates a challenge in correctly aligning targets and assessing the alignment of OARs. Staff training for ultrasound is critical to ensure target alignment and accurate treatment delivery.

The introduction of ultrasound to abdominal SBRT does require some attention to detail, such as the verification of the fusion between simulation CT and simulation ultrasound images, and the isocenter coincidence of treatment plan and ultrasound. However, the imaging modality can provide better target visualization compared with radiographic methods and the ability to track the target position during treatment while not delivering any additional imaging dose to the patient.

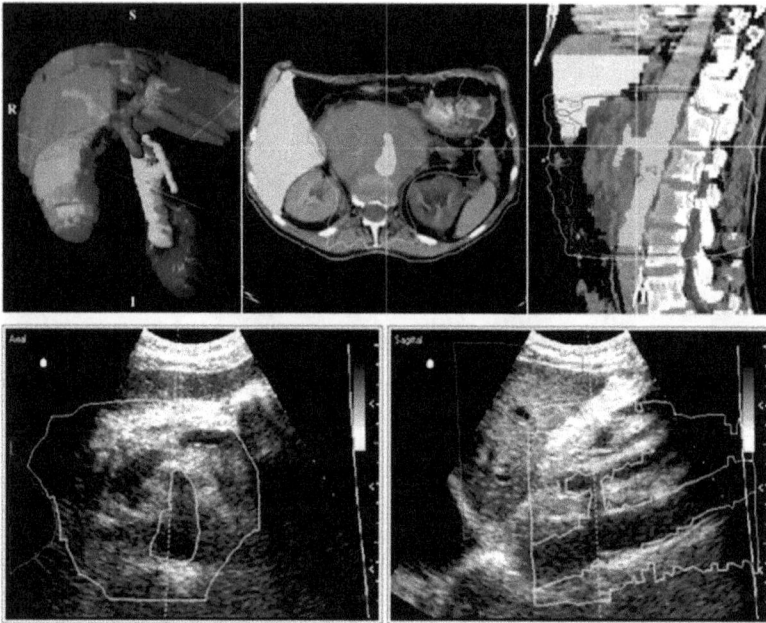

Figure 13.3. Simulation CT (top) and BATCAM ultrasound images (bottom) from a patient whose pancreas was being treated. The CT images, from left to right, are the 3D renderings of the contours, an axial slice, and a sagittal slice. On the CT simulation images, the pancreas target contour is red, the aorta is green, and the liver is yellow. The BATCAM ultrasound images from the time of treatment are of two orthogonal planes. On the ultrasound images, the pancreas target is yellow, the aorta is cyan, and the liver is red.

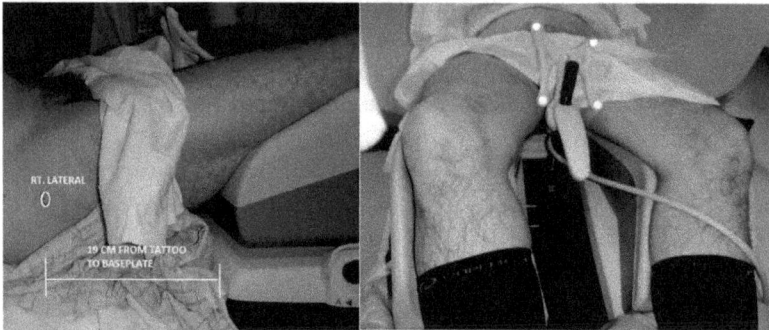

Figure 13.4. A typical transperineal Clarity prostate SBRT setup. The patient's lower body was immobilized in an Alpha Cradle (Smithers Medical Products, Inc., North Canton, OH), and his legs were positioned on the knee cushion. The ultrasound probe was placed against the patient's perineum.

13.3.3.1 Ultrasound-guided prostate SBRT

Prostate SBRT is beyond the scope of this book, but it is valuable to demonstrate ultrasound-guided radiation therapy. For prostate treatments guided with the Clarity system, 2D and 3D scans are acquired transabdominally and transperineally, respectively. Figure 13.4 shows a typical transperineal Clarity setup. The patient was

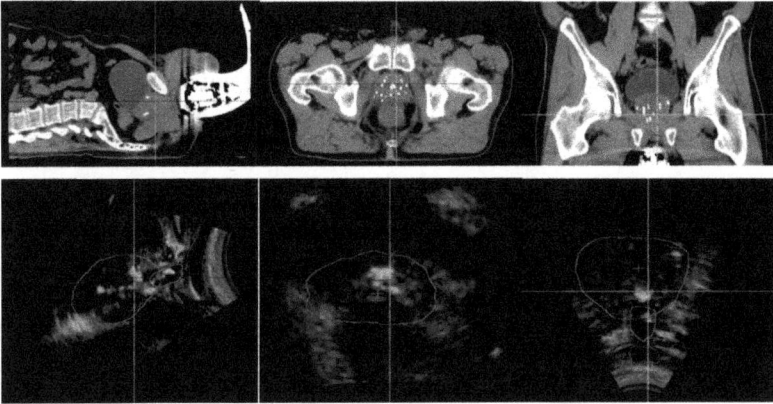

Figure 13.5. Simulation CT and Clarity ultrasound images. From left to right, the sagittal, axial, and coronal image planes of simulation CT (top) and the Clarity ultrasound (bottom) are shown. The displayed contours are the prostate (pink), sphincter (yellow), and penile bulb (purple).

immobilized in an Alpha Cradle, and his legs were positioned on the knee cushion. The ultrasound probe was placed against his perineum and good contact was aided with ultrasound gel. The white dots on the probe frame are localization fiducial markers which may be detected by cameras mounted on the ceiling. During patient simulation, both CT and Clarity scans are acquired and image fusion may be automatically performed if the Clarity scan is acquired when the patient is positioned by the CT origin lasers. On the treatment day, a Clarity scan is acquired, and the image registration is performed between the treatment Clarity scan and the simulation Clarity scan, ultrasound to ultrasound. Figure 13.5 shows a simulation CT and simulation transperineal Clarity ultrasound images. The contours shown in the images are the prostate, sphincter, and penile bulb.

13.3.4 Body GPS

Calypso (Varian Medical Systems, Inc., Palo Alto, CA) is a body GPS that may be used for target localization and intrafractional motion monitoring for abdominal SBRT [24, 49, 50]. Three Calypso beacon transponders are typically implanted into the patient's body, and the locations of the transponders are detected by the receiver via radiofrequency waves. The offsets of the transponders compared with simulation positions reflect the target offsets. Figure 13.6 shows a simulation CT for Calypso beacon transponders in the liver. The advantages of Calypso include easy setup, automatic registration, fast offset calculation, and automatic real-time radiation beam control if the offsets are out of a preset tolerance. The disadvantages of Calypso are the invasive beacon transponder implantation and the limitations for tracking if the transponders are not close enough to the receiver—that is, tracking may not be possible for deep target locations or if the patient is large.

The accuracy of Calypso on target localization is 2 mm based on manufactural specification. During beacon transponder coordination setup, special attention is needed to make sure that the three transponder coordinates are not swapped.

13.3.5 Implanted fiducials

Implanted fiducials in combination with radiographic imaging may be used for abdominal SBRT. Fiducials can include surgical markers or clips, gold seeds, and other markers easily identified with imaging. Implanted fiducials may be used for localization by aligning with simulation images via CBCT, kV 2D–2D, or MV 2D–2D. Additionally, the motion of the target can be assessed with fluoroscopy. The advantage of the application of fiducials is adding an accurate surrogate of the target localization to the existing IGRT workflow [50]. The disadvantages of applying implanted fiducials are the invasive procedure and the possibility of the fiducial migration after CT simulation. It will be critical to make sure that the fiducials are in the same positions as simulation CT during the SBRT IGRT procedure.

13.3.6 MRI

The MRI LINAC has recently become more popular [51–54]. The advantage of MRI-guided radiation therapy is the superb soft tissue contrast. Figure 13.7 shows the comparison of simulation CT and MRI images for a patient with liver cancer. The soft tissue contrast and the target visibility on the MRI are significantly superior to the simulation CT.

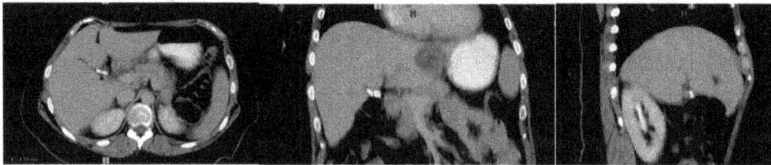

Figure 13.6. From left to right, simulation CT axial, coronal, and sagittal images of Calypso beacon transponders in a liver.

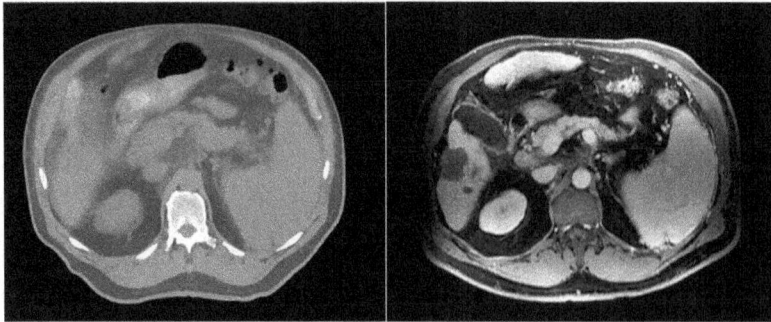

Figure 13.7. Simulation CT (left) and MRI (right) images of a liver. The target visibility is superior on the MRI.

13.4 Intrafractional motion management/monitoring

Patient intrafractional motion is a common challenge for abdominal SBRT, and it usually includes two types of motions—patient body motion and respiratory motion. Patient intrafractional motion may cause the motion of the target and OARs. This motion may happen during IGRT imaging and/or during radiation treatment delivery. Therefore, patient intrafractional motion monitoring during IGRT imaging and treatment delivery becomes critical for abdominal SBRT [55–57].

There are several IGRT imaging modules commercially available and commonly used for patient intrafractional motion monitoring, including (but not limited to) ultrasound, body GPS, implanted fiducials with fluoroscopy, and surface guidance.

13.4.1 Direct target motion monitoring

Direct target motion monitoring is ideal for SBRT treatment delivery. There are some imaging modalities with the function of direct target monitoring, such as ultrasound or MRI. Additionally, some modalities offer nearly direct monitoring, such as Calypso and fiducial marker tracking via imaging.

The Clarity ultrasound system has the function of target tracking during treatment delivery and automatically turning off the treatment beam if the target is moved out of the preset tolerances. During the SBRT procedure, after image registration and the patient being shifted, the residue of patient offset is calculated in real-time based on the preset target, with the frame rate of ten frames per second. The continuous tracking of the target position ensures the accurate treatment delivery.

Treatment machines with on-board MRI also have the ability to track targets during treatment and gate the beam if the target moves outside of a predefined threshold. The cine MR images typically include one sagittal plane and are acquired at approximately four frames per second [60]. The sagittal plane to be imaged during treatment is typically selected prior to beam delivery.

Calypso has a tracking function during radiation treatment delivery and can automatically hold the treatment beam if the target moves out of the preset tolerances. All three implanted beacon transponder locations are detected in real-time by the receiver, and then target offsets are calculated. The continuous tracking of the target position ensures accurate treatment delivery, especially for liver SBRT. The reported liver target motion amplitude due to respiration may be up to 2.3 cm, depending on the target location and the patient respiratory cycles [29, 64].

Implanted fiducials may be tracked by fluoroscopy, planar kV imaging, and MV imaging during treatment delivery. This direct target motion tracking ensures accurate treatment delivery.

Continuous imaging during treatment delivery via kV, MV, and fluoroscopy is commercially available for some LINACs, and the direct target motion tracking is available via the continuous imaging capture.

13.4.2 Surface guidance

Surface guided radiation therapy (SGRT) has become more and more popular in the last decade [32, 33, 58, 59]. Patient surface tracking may detect patient intrafractional motion, including body and respiratory motion. Even though the patient surface motion amplitude does not represent the exact target motion amplitude, there is still often a correlation between surface motion and target motion. Patient surface motion may be up to 1.6 cm during radiation therapy [32, 33]; therefore, effective surface motion monitoring during radiation therapy is significant. The tolerance of patient surface motion may be preset to numbers correlated to the SBRT treatment planning target margins and the relative positions of OARs.

There are multiple SGRT products commercially available, including AlignRT (Vision RT, London, United Kingdom), Catalyst (C-RAD, Uppsala, Sweden), and IDENTIFY (Varian Medical Systems, Inc., Palo Alto, CA). All SGRT products are able to warn the user when the patient surface is out of a preset position tolerance, and then the treatment beams may be turned off automatically or manually depending on the setup between the product and the LINAC treatment machine.

13.4.3 Gated treatment delivery

Gated abdominal SBRT treatments are often applied to liver and pancreas cancer [61–64]. IGRT image modalities for gating include fluoroscopy and surface guidance. Fiducials may be used to guide the gated procedure. The respiratory phase for treatment delivery may be any part of the respiratory cycle, depending on the goals of the gating. The goals of gating include optimal dosimetric treatment planning, best patient stability, and easiest patient position reproducibility. Near the end of exhale phase is commonly used for gating because it is the most consistent and reliable respiratory phase. Direct target monitoring during SBRT gating is critical because the correlation between respiratory cycle and target location is not always consistent.

13.4.4 Breath hold for SBRT

For liver and pancreas SBRT, target motion amplitude due to respiration may be up to 2.3 cm, depending on the target location and the patient respiratory cycles [65]. Breath hold techniques may help eliminate the ITV, minimize the overall PTV size, and lower normal tissue dose. Patients may perform a deep inspiration breath hold or just simply holding their breath comfortably [66, 67], depending on dosimetric benefits. Generally, as long as the target does not have intrafractional motion, the PTV may be smaller than ITV plus margins, and more OARs may be spared. Consistent deep inspiration breath holds are hard to achieve for SBRT patients because of patient fatigue over a long period procedure of SBRT.

SGRT imaging is generally used to track the patient surface during breath hold to verify patient position consistency during each breath hold. All SGRT modalities have the workflow for breath hold. Calypso has a tracking function that places surface markers on the patient skin to detect the marker position, and may be used for breath hold treatment delivery. Some LINAC treatment machines have integrated the workflow for breath hold functionality.

There is a common issue in the workflow of CBCT with SGRT during breath hold. The gantry rotation during CBCT may block different SGRT cameras and cause variations of detected patient offsets, and in some gantry positions, it is possible that the patient surface is not able to be tracked at all. One abdominal CBCT scan may take up to one minute, and it might take two to three patient's breath holds to finish the scan. Careful design of when and where to start and stop the gantry for each breath hold is significant and critical for an accurate and high-quality reconstructed CBCT image.

13.5 Quality assurance of IGRT for SBRT

Quality assurance of IGRT for abdominal SBRT is extremely important to ensure accurate target localization and precise radiation treatment delivery. IGRT image quality, localization accuracy, and geometric accuracy checks are needed. Isocenter consistency between treatment plan and IGRT image modality is needed to ensure the target is localized correctly, particularly if the IGRT modality is not integrated with the treatment machine. Additionally, careful workflow design and optimization will ensure the integrity of the entire SBRT procedure.

Initial workflow design for a specific IGRT imaging modality application for SBRT should consider all aspects of the SBRT procedure, from simulation CT to the delivery of radiation treatment. An end-to-end test is the best way to evaluate the workflow and find out the latent issues. Ongoing workflow optimization is needed to improve the current procedure efficiency while maintaining quality.

References

[1] Jackson W C, Silva J and Hartman H E *et al* 2019 Stereotactic body radiation therapy for localized prostate cancer: a systematic review and meta-analysis of over 6,000 patients treated on prospective studies *Int. J. Radiat. Oncol. Biol. Phys.* **104** 778–89

[2] Cushman T R, Verma V, Khairnar R, Levy J, Simone C B II and Mishra M V 2019 Stereotactic body radiation therapy for prostate cancer: systematic review and meta-analysis of prospective trials *Oncotarget* **10** 5660–8

[3] Jiang N Y, Dang A T and Yuan Y *et al* 2019 Multi-institutional analysis of prostate-specific antigen kinetics after stereotactic body radiation therapy *Int. J. Radiat. Oncol. Biol. Phys.* **105** 628–36

[4] Kellock T, Liang T and Harris A *et al* 2018 Stereotactic body radiation therapy (SBRT) for hepatocellular carcinoma: imaging evaluation post treatment *Br. J. Radiol.* **91** 20170118

[5] Baumann B C, Wei J and Plastaras J P *et al* 2018 Stereotactic body radiation therapy (SBRT) for hepatocellular carcinoma: high rates of local control with low toxicity *Am. J. Clin. Oncol.* **41** 1118–124

[6] Tétreau R, Llacer C, Riou O and Deshayes E 2017 Evaluation of response after SBRT for liver tumors *Rep. Pract. Oncol. Radiother.* **22** 170–5

[7] Wang H, Jin C, Fang L, Sun H, Cheng W and Hu S 2020 Health economic evaluation of stereotactic body radiotherapy (SBRT) for hepatocellular carcinoma: a systematic review *Cost Eff. Resour. Alloc.* **18** 1

[8] Qiu H, Moravan M J, Milano M T, Usuki K Y and Katz A W 2018 SBRT for hepatocellular carcinoma: 8-year experience from a regional transplant center *J. Gastrointest. Cancer* **49** 463–9

[9] de Geus S W L, Eskander M F and Kasumova G G *et al* 2017 Stereotactic body radiotherapy for unresected pancreatic cancer: a nationwide review *Cancer* **123** 4158–67

[10] Herman J M, Chang D T and Goodman K A *et al* 2015 Phase 2 multi-institutional trial evaluating gemcitabine and stereotactic body radiotherapy for patients with locally advanced unresectable pancreatic adenocarcinoma *Cancer* **121** 1128–37

[11] Lischalk J W, Burke A and Chew J *et al* 2018 Five-fraction stereotactic body radiation therapy (SBRT) and chemotherapy for the local management of metastatic pancreatic cancer *J Gastrointest. Cancer* **49** 116–23

[12] Francolini G, Detti B and Ingrosso G *et al* 2018 Stereotactic body radiation therapy (SBRT) on renal cell carcinoma, an overview of technical aspects, biological rationale and current literature *Crit. Rev. Oncol. Hematol.* **131** 24–9

[13] Plichta K, Camden N and Furqan M *et al* 2017 SBRT to adrenal metastases provides high local control with minimal toxicity *Adv. Radiat. Oncol.* **2** 581–7

[14] Ippolito E, D'Angelillo R M, Fiore M, Molfese E, Trodella L and Ramella S 2015 SBRT: a viable option for treating adrenal gland metastases *Rep. Pract. Oncol. Radiother.* **20** 484–90

[15] Yanez L, Ciudad A M, Mehta M P and Marsiglia H 2018 What is the evidence for the clinical value of SBRT in cancer of the cervix? *Rep. Pract. Oncol. Radiother.* **23** 574–9

[16] Snyder J E, Willett A B, Sun W and Kim Y 2019 Is SBRT boost feasible for PET positive lymph nodes for cervical cancer? Evaluation using tumor control probability and QUANTEC criteria *Pract. Radiat. Oncol.* **9** e156–63

[17] Shimohigashi Y, Toya R and Saito T *et al* 2017 Tumor motion changes in stereotactic body radiotherapy for liver tumors: an evaluation based on four-dimensional cone-beam computed tomography and fiducial markers *Radiat. Oncol.* **12** 61

[18] Yang W, Van Ausdal R, Read P, Larner J, Benedict S and Sheng K 2009 The implication of non-cyclic intrafractional longitudinal motion in SBRT by TomoTherapy *Phys. Med. Biol.* **54** 2875

[19] Dieterich S, Green O and Booth J 2018 SBRT targets that move with respiration *Phys. Med* **56** 19–24

[20] Ge J, Santanam L, Noel C and Parikh P J 2013 Planning 4-dimensional computed tomography (4DCT) cannot adequately represent daily intrafractional motion of abdominal tumors *Int. J. Radiat. Oncol. Biol. Phys.* **85** 999–1005

[21] Sonier M, Chu W, Lalani N, Erler D, Cheung P and Korol R 2016 Evaluation of kidney motion and target localization in abdominal SBRT patients *J. Appl. Clin. Med. Phys.* **17** 429–33

[22] Habermehl D, Naumann P and Bendl R *et al* 2015 Evaluation of inter- and intrafractional motion of liver tumors using interstitial markers and implantable electromagnetic radio-transmitters in the context of image-guided radiotherapy (IGRT) – the ESMERALDA trial *Radiat. Oncol.* **10** 143

[23] Prins F M, Stemkens B and Kerkmeijer L G W *et al* 2019 Intrafraction motion management of renal cell carcinoma with magnetic resonance imaging-guided stereotactic body radiation therapy *Pract. Radiat. Oncol.* **9** e55–61

[24] Bertholet J, Toftegaard J and Hansen R *et al* 2018 Automatic online and real-time tumour motion monitoring during stereotactic liver treatments on a conventional linac by combined optical and sparse monoscopic imaging with kilovoltage x-rays (COSMIK) *Phys. Med. Biol.* **63** 055012

[25] de Muinck Keizer D M, Pathmanathan A U and Andreychenko A *et al* 2019 Fiducial marker based intra-fraction motion assessment on cine-MR for MR-linac treatment of prostate cancer *Phys. Med. Biol.* **64** 07NT02

[26] Liang Z, Liu H and Xue J *et al* 2018 Evaluation of the intra- and interfractional tumor motion and variability by fiducial-based real-time tracking in liver stereotactic body radiation therapy *J. Appl. Clin. Med. Phys.* **19** 94–100

[27] Loi M, Magallon-Baro A and Suker M *et al* 2019 Pancreatic cancer treated with SBRT: effect of anatomical interfraction variations on dose to organs at risk *Radiother. Oncol.* **134** 67–73

[28] Pollock S, Tse R and Martin D *et al* 2018 Impact of audiovisual biofeedback on interfraction respiratory motion reproducibility in liver cancer stereotactic body radiotherapy *J. Med. Imaging Radiat. Oncol.* **62** 133–9

[29] Park J C, Park S H and Kim J H *et al* 2012 Liver motion during cone beam computed tomography guided stereotactic body radiation therapy *Med. Phys.* **39** 6431–442

[30] Case R B, Moseley D J and Sonke J J *et al* 2010 Interfraction and intrafraction changes in amplitude of breathing motion in stereotactic liver radiotherapy *Int. J. Radiat. Oncol. Biol. Phys.* **77** 918–25

[31] Sarkar V, Lloyd S and Paxton A *et al* 2018 Daily breathing inconsistency in pancreas SBRT: a 4DCT study *J. Gastrointest. Oncol.* **9** 989–95

[32] Zhao H, Wang B and Sarkar V *et al* 2016 Comparison of surface matching and target matching for image-guided pelvic radiation therapy for both supine and prone patient positions *J. Appl. Clin. Med. Phys.* **17** 14–24

[33] Zhao H, Sarkar V, Huang L, Wang B, Rassiah-Szegedi P, Huang Y J, Szegedi M, Gonzalez V and Salter B 2017 Preferred treatment position between supine and prone for pelvic radiation therapy; quantification of the intrafractional body motion component by 3D surface imaging system *Int. J. Cancer Ther. Oncol.* **5** 8

[34] Eccles C L, Patel R, Simeonov A K, Lockwood G, Haider M and Dawson L A 2011 Comparison of liver tumor motion with and without abdominal compression using cine-magnetic resonance imaging *Int. J. Radiat. Oncol. Biol. Phys.* **79** 602–8

[35] Lovelock D M, Zatcky J, Goodman K and Yamada Y 2014 The effectiveness of a pneumatic compression belt in reducing respiratory motion of abdominal tumors in patients undergoing stereotactic body radiotherapy *Technol. Cancer Res. Treat.* **13** 259–67

[36] Piippo-Huotari O, Funk E, Geijer H and Anderzén-Carlsson A 2020 Patients and radiographers experiences of dose reducing abdominal compression in radiographic examinations-a qualitative study *Nurs. Open* **7** 680–9

[37] Dagia C, Ditchfield M, Kean M and Catto-Smith T 2008 Imaging for Crohn disease: use of 3-T MRI in a paediatric setting *J. Med. Imaging Radiat. Oncol.* **52** 480–8

[38] van den Wollenberg W, de Ruiter P, Nowee M E, Jansen E P M, Sonke J J and Fast M F 2019 Investigating the impact of patient arm position in an MR-linac on liver SBRT treatment plans [published correction appears in Med Phys. 2020;47:1411] *Med. Phys.* **46** 5144–51

[39] Zhao H, Williams N and Poppe M *et al* 2019 Comparison of surface guidance and target matching for image-guided accelerated partial breast irradiation (APBI) *Med. Phys.* **46** 4717–24

[40] Zhou D, Quan H and Yan D *et al* 2018 A feasibility study of intrafractional tumor motion estimation based on 4D-CBCT using diaphragm as surrogate *J. Appl. Clin. Med. Phys.* **19** 525–31

[41] Nakamura M, Ishihara Y and Matsuo Y *et al* 2018 Quantification of the kV x-ray imaging dose during real-time tumor tracking and from three- and four-dimensional cone-beam computed tomography in lung cancer patients using a Monte Carlo simulation *J. Radiat. Res.* **59** 173–81

[42] Chan M, Chiang C L and Lee V *et al* 2017 Target localization of 3D versus 4D cone beam computed tomography in lipiodol-guided stereotactic radiotherapy of hepatocellular carcinomas *PLoS One* **12** e0174929

[43] Jmour O, Benna M and Champagnol P *et al* 2020 CBCT evaluation of inter- and intra-fraction motions during prostate stereotactic body radiotherapy: a technical note *Radiat. Oncol.* **15** 85

[44] Grimwood A, McNair H A and O'Shea T P *et al* 2018 In vivo validation of Elekta's clarity autoscan for ultrasound-based intrafraction motion estimation of the prostate during radiation therapy *Int. J. Radiat. Oncol. Biol. Phys.* **102** 912–21

[45] Omari E A, Erickson B and Ehlers C *et al* 2016 Preliminary results on the feasibility of using ultrasound to monitor intrafractional motion during radiation therapy for pancreatic cancer *Med. Phys.* **43** 5252–60

[46] O'Shea T P, Bamber J C and Harris E J 2016 Temporal regularization of ultrasound-based liver motion estimation for image-guided radiation therapy *Med. Phys.* **43** 455–64

[47] Han B, Najafi M and Cooper D T *et al* 2018 Evaluation of transperineal ultrasound imaging as a potential solution for target tracking during hypofractionated radiotherapy for prostate cancer *Radiat. Oncol.* **13** 151

[48] Richardson A K and Jacobs P 2017 Intrafraction monitoring of prostate motion during radiotherapy using the Clarity® Autoscan Transperineal Ultrasound (TPUS) system *Radiography (Lond)* **23** 310–3

[49] Petasecca M, Newall M K and Booth J T *et al* 2015 MagicPlate-512: a 2D silicon detector array for quality assurance of stereotactic motion adaptive radiotherapy *Med. Phys.* **42** 2992–3004

[50] Heinz C, Gerum S and Freislederer P *et al* 2016 Feasibility study on image guided patient positioning for stereotactic body radiation therapy of liver malignancies guided by liver motion *Radiat. Oncol.* **11** 88

[51] Harris W, Wang C, Yin F F, Cai J and Ren L 2018 A novel method to generate on-board 4D MRI using prior 4D MRI and on-board kV projections from a conventional LINAC for target localization in liver SBRT *Med. Phys.* **45** 3238–45

[52] Fast M, van de Schoot A, van de Lindt T, Carbaat C, van der Heide U and Sonke J J 2019 Tumor trailing for liver SBRT on the MR-Linac *Int. J. Radiat. Oncol. Biol. Phys.* **103** 468–78

[53] Al-Ward S, Wronski M and Ahmad S B *et al* 2018 The radiobiological impact of motion tracking of liver, pancreas and kidney SBRT tumors in a MR-linac *Phys. Med. Biol.* **63** 215022

[54] Feldman A M, Modh A, Glide-Hurst C, Chetty I J and Movsas B 2019 Real-time magnetic resonance-guided liver stereotactic body radiation therapy: an institutional report using a magnetic resonance-linac system *Cureus* **11** e5774

[55] Tyagi N, Hipp E and Cloutier M *et al* 2019 Impact of daily soft-tissue image guidance to prostate on pelvic lymph node (PLN) irradiation for prostate patients receiving SBRT *J. Appl. Clin. Med. Phys.* **20** 121–7

[56] Jackson W C, Dess R T and Litzenberg D W *et al* 2018 A multi-institutional phase 2 trial of prostate stereotactic body radiation therapy (SBRT) using continuous real-time evaluation of prostate motion with patient-reported quality of life *Pract. Radiat. Oncol.* **8** 40–7

[57] Vinogradskiy Y, Goodman K A, Schefter T, Miften M and Jones B L 2019 The clinical and dosimetric impact of real-time target tracking in pancreatic SBRT *Int. J. Radiat. Oncol. Biol. Phys.* **103** 268–75

[58] Heinzerling J H, Hampton C J and Robinson M *et al* 2020 Use of surface-guided radiation therapy in combination with IGRT for setup and intrafraction motion monitoring during stereotactic body radiation therapy treatments of the lung and abdomen *J. Appl. Clin. Med. Phys.* **21** 48–55

[59] Leong B and Padilla L 2019 Impact of use of optical surface imaging on initial patient setup for stereotactic body radiotherapy treatments *J. Appl. Clin. Med. Phys.* **20** 149–58

[60] Fischer-Valuck B W, Henke L and Green O *et al* 2017 Two-and-a-half-year clinical experience with the world's first magnetic resonance image guided radiation therapy system *Adv. Radiat. Oncol.* **2** 485–93

[61] Worm E S, Høyer M and Hansen R *et al* 2018 A prospective cohort study of gated stereotactic liver radiation therapy using continuous internal electromagnetic motion monitoring *Int. J. Radiat. Oncol. Biol. Phys.* **101** 366–75

[62] Riou O, Serrano B and Azria D *et al* 2014 Integrating respiratory-gated PET-based target volume delineation in liver SBRT planning, a pilot study *Radiat. Oncol.* **9** 127

[63] Jeong Y, Jung J and Cho B *et al* 2018 Stereotactic body radiation therapy using a respiratory-gated volumetric-modulated arc therapy technique for small hepatocellular carcinoma *BMC Cancer* **18** 416

[64] Chung H, Jung J and Jeong C *et al* 2018 Evaluation of delivered dose to a moving target by 4D dose reconstruction in gated volumetric modulated arc therapy *PLoS One* **13** e0202765

[65] Li W Z, Liang Z W and Cao Y *et al* 2019 Estimating intrafraction tumor motion during fiducial-based liver stereotactic radiotherapy via an iterative closest point (ICP) algorithm *Radiat. Oncol.* **14** 185

[66] Lee S, Zheng Y and Podder T *et al* 2019 Tumor localization accuracy for high-precision radiotherapy during active breath-hold *Radiother. Oncol.* **137** 145–52

[67] Boda-Heggemann J, Sihono D S K and Streb L *et al* 2019 Ultrasound-based repositioning and real-time monitoring for abdominal SBRT in DIBH *Phys. Med.* **65** 46–52

IOP Publishing

Principles and Practice of Image-Guided Abdominal
Radiation Therapy

Yu Kuang

Chapter 14

Uncertainties of image-guided radiotherapy for abdominal cancer radiotherapy

You Zhang, Jing Wang and Xun Jia

Uncertainties are inevitable in the image guidance process for abdominal cancer radiotherapy due to a number of factors, such as imperfect imaging devices, human performance in the operation workflow, and intra-fractional organ motions. This chapter focuses on the discussions on uncertainties of image-guided radiotherapy (IGRT). We will first introduce the concept of uncertainty and present several key topics related to it. We will then present an overview of uncertainties that occurred in key steps of the abdominal IGRT workflow. We will then discuss in detail several major uncertainties. For each one, we will present the causes of the uncertainties, their magnitudes, and their consequences. After that, we will also present the margin approach to effectively address uncertainties in clinical practice.

The goal of radiation therapy is to deliver a potent dose to the cancerous target while effectively sparing doses to nearby normal organs. In IGRT, a certain image guidance technology is used to help us see and target the tumor, thereby ensuring the achievement of this goal. Nonetheless, the image guidance system is not perfect in reality, and the delivered treatment under image guidance is associated with uncertainties of different types and to different degrees. Understanding the causes and magnitude of the uncertainties is of critical importance for the development of corresponding approaches to manage the uncertainty and hence ensure the success of treatments.

In this chapter, we discuss uncertainties of IGRT for the context of abdominal cancer radiotherapy. As the main goal of IGRT is to ensure geometrical alignment of the tumor to treatment beams, our discussions on uncertainties will be specifically focused on geometrical uncertainties. We will cover uncertainties that occurred throughout the IGRT workflow, starting from the acquisition of the offline images for tumor localization, to the acquisition of the online images for treatment

doi:10.1088/978-0-7503-2468-7ch14　　　　14-1

guidance, and then to the treatment delivery. Uncertainties occurring outside this range are beyond the scope of this chapter and hence will not be discussed. In addition, while we are trying to comprehensively cover different aspects within the topic, due to space limitations, it is inevitable that certain issues will be missed. Readers are encouraged to find more details in the literature.

The rest of this chapter is organized as follows. Section 14.1 introduces the concept of uncertainty and presents an overview of uncertainties occurring in the workflow of abdominal IGRT. Sections 14.2–14.5 will discuss in detail each type of uncertainty, the reasons for the uncertainties, and their magnitudes. Again, as it is impossible to enumerate all possible types of uncertainties for all cancer sites in the abdominal area, the discussions here will only focus on a few representative ones. In section 14.6, we will discuss the most effective approach, margin, to address uncertainties in clinical practice. Finally, section 14.7 will conclude this chapter.

14.1 Uncertainty overview

Generally speaking, uncertainty is associated with every measurement: for example, determining the couch shift by matching a cone-beam computed tomography (CBCT) image with the treatment planning computed tomography (CT) image. For a measurement, the difference between the obtained measurement value and the ground truth value is, in general, unknowable, since otherwise it would be possible to derive the ground truth by subtracting the difference from the measurement and hence reduce the uncertainty to zero. This difference is termed as an error of a measurement. Psychologically, the term may make people think of mistakes. However, we would like to emphasize that this difference, or measurement error, exists in nature all the time, no matter how carefully a measurement is performed, due to certain practical limitations such as finite resolution of a device.

The concept of uncertainty is introduced to allow us to characterize the distribution of this difference if we were to make measurements repeatedly. In other words, uncertainty expresses the chance that the differences fall within a certain range. It is important to keep in mind that uncertainty depicts the behaviors of repeated measurements, but we probably have nothing to say about any given measurement. A measurement technology may have a large uncertainty, but the measured value could be surprisingly close to the ground truth in a given measurement just by chance.

14.1.1 Random and systematic errors and uncertainties

There are two broad classes of observational errors: random error and systematic error. Random error varies unpredictably from one measurement to another. It is often caused by the difficulty of taking measurements. Suppose, again, that we try to determine the couch shift by matching a CBCT image with the treatment planning CT image. If we were able to repeatedly perform this matching task many times, each with a new CBCT image acquired, the resulting couch shift values for this series of CBCT images are likely close to each other but are not identical. The reasons accounting for the variations among these values are multifactorial, such as the noise signal in the CBCT images that affect the registration results between the CBCTs and the planning CT, or human factors when manually matching the CBCTs with CT, etc.

Systematic error, on the other hand, causes predictable and consistent deviations from the ground truth value due to problems such as the calibration of equipment. In the couch shift example, if the isocenter of the CBCT image is off from the therapeutic beam isocenter, the resulting couch shift would position the patient such that the therapeutic beams would be delivered slightly off from the targeted position. Note that this error occurs in the same way for all the patients in every couch shift task.

Correspondingly, uncertainties are categorized into random uncertainties and systematic uncertainties. Nowadays, it has been recommended that we term these as type A and type B uncertainties. Specifically, the type A uncertainty is evaluated based on a valid statistical method of a series of observations, whereas the type B uncertainty is evaluated based on scientific judgment using all of the relevant information available.

Another two related concepts worth clarifying are accuracy and precision, which have often incorrectly been used interchangeably. In fact, precision is related to random uncertainty/error, and accuracy to systematic uncertainty/error. These concepts are illustrated in different scenarios when playing darts in figure 14.1. A series of measurements with a low systematic uncertainty are called accurate, or achieve a high level of accuracy. In contrast, the measurements with a low random uncertainty are precise, or achieve a high level of precision. It is important to keep in mind that these two concepts are independent of each other, and hence, depending on whether they are high or low, there are four difference scenarios as illustrated in figure 14.1. Achieving a high or low level in one concept does not indicate the other. For instance, when repeated measurements give results very close to each other, it is apparent that the precision is high. However, this does not mean the measurement results are accurate and trustworthy: the results may be consistently off from the ground truth due to systematic uncertainty, as shown in the third scenario of figure 14.1.

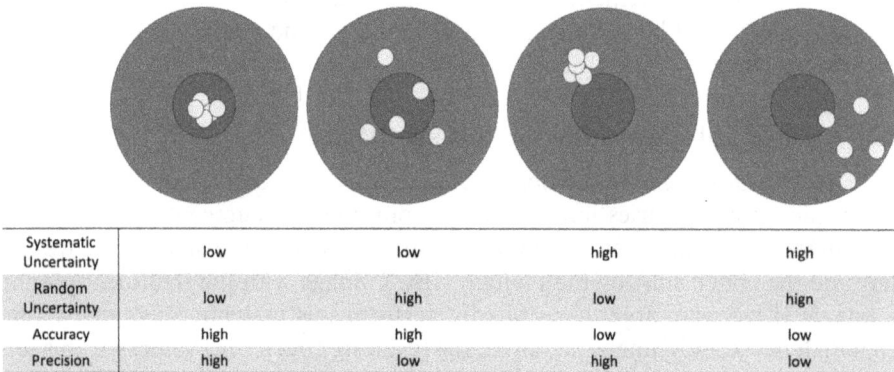

Systematic Uncertainty	low	low	high	high
Random Uncertainty	low	high	low	hign
Accuracy	high	high	low	low
Precision	high	low	high	low

Figure 14.1. Illustration of the concepts of systematic and random uncertainties and accuracy and precision.

14.1.2 IGRT to reduce geometric uncertainty

To achieve the general goal of delivering the prescribed dose to the target while sparing dose to normal organs, IGRT has indeed clearly demonstrated its effectiveness over the years. It allows the visualization of tumor target at the treatment time and makes corresponding adjustments, e.g. couch shifts, to accurately and precisely align the target with treatment beams. This effect can be seen in figure 14.2. Without using any types of IGRT procedures, daily positions deviate from the targeted position with both large systematic and random uncertainties. Studies in early days have attempted to address this patient position uncertainty by measuring patient position deviations in a series of treatment fractions via comparing portal images to simulation images. This allowed derivation of the systematic component of the uncertainty caused by tattoo laser–based localization, for example, and compensating for it in the patient positioning stage [1, 2]. As shown in figure 14.2(B), this approach was effective in terms of reducing the systematic uncertainty. However, it was not able to address the large random uncertainties, making the daily positions vary a lot around the targeted position. Later on, the introduction of online IGRT, i.e. to image the patient position on the treatment couch and adjust the patient position accordingly, can correct both the random and systematic components of the uncertainty, yielding patient positions that are both accurate and precise.

The advantage of IGRT in terms of reducing geometric variation and increasing the chance of accurately and precisely delivering the planned dose to the patient has translated to clinical benefits. In fact, the clinical effectiveness of IGRT has been demonstrated via both computer simulations [4, 5] and real clinical studies [6]. For instance, a study at Memorial Sloan Kettering Cancer Center compared toxicity profiles and biochemical tumor control outcomes between patients treated with IGRT (using orthogonal kilovoltage radiographs to match fiducial markers) and those without IGRT for clinically localized prostate cancer. It was found that IGRT is associated with an improvement in biochemical tumor control among high-risk patients and a lower rate of late urinary toxicity.

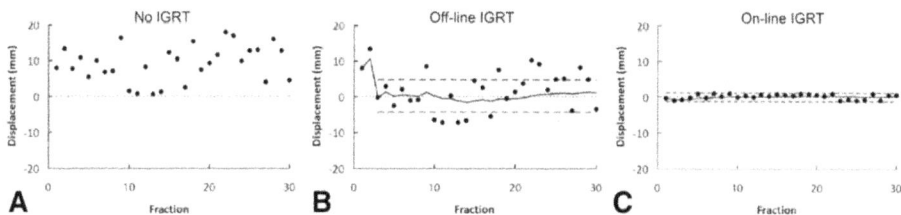

Figure 14.2. Position of a single patient showing daily positions (circles), running mean (solid line), and control limits (dashed lines). (A) The daily position without image guidance shows a large systematic error (10 mm) and a large random uncertainty. (B) Using an offline correction protocol results in small systematic errors but does not reduce large random uncertainties. (C) Using daily online corrections results in small systematic and random uncertainties. Reprinted from [3], Copyright (2012), with permission from Elsevier.

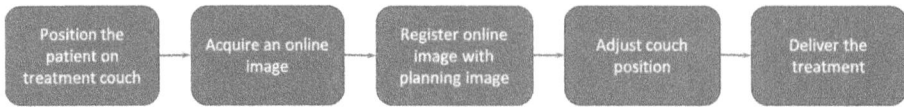

Figure 14.3. Workflow of a typical IGRT treatment.

14.1.3 Uncertainty in abdominal IGRT

Nonetheless, no approach is perfect, and uncertainty is inevitable. IGRT reduces geometric uncertainty compared to non-IGRT approaches, but there are still uncertainties in the IGRT framework due to a number of issues. Looking at the standard workflow of an IGRT treatment (figure 14.3), after positioning the patient on the treatment couch, a number of subsequent steps take place to adjust the tumor position and align it to the treatment beams. Due to technical or human factors, each of the subsequent steps bears with uncertainty that affects the final tumor position accuracy and precision. In the acquisition of an online image, what is most important is to visualize the tumor. Yet depending on the imaging modality and specific imaging systems, uncertainty exists in determining the tumor position in the acquired image. Subsequently, a registration step is needed to align the tumor in the online image with that appearing in the planning image, which is associated with uncertainty caused by the registration algorithm or observers. In the step of adjusting the couch position, the uncertainty is caused by the accuracy of the couch to execute the determined couch motion, as well as the capability of compensating a non-rigid anatomy change with the rigid couch motion. Finally, during treatment delivery, intra-fractional organ motions also generate uncertainty of the tumor position that was determined at the pre-treatment delivery stage.

In the following sections, we will discuss in detail these uncertainties caused by different reasons.

14.2 Uncertainty in image acquisition

Different imaging modalities have been used in IGRT for cancer treatment in the abdominal region. Due to intrinsic characteristics, such as low-contrast soft-tissue detectability in CBCT and image distortions in magnetic resonance imaging (MRI), the degree of uncertainty will be different in IGRT of abdominal tumors for different imaging modalities. In this section, we will discuss the uncertainties associated with a few major imaging modalities, including CBCT, ultrasound (US), and MRI.

14.2.1 CBCT

CBCT employs cone-shaped x-ray beams and a flat panel detector and acquires a three-dimensional (3D) volumetric image within the field of view. With the integration of CBCT to the linear accelerator (Linac), 3D CBCT volumetric imaging can reveal patient anatomy right before the beam delivery. The on-board CBCT can then be used to align patients by comparing it with the reference planning CT.

Ideally, such an alignment should be based on the locations of the tumors or treatment targets (e.g., gross tumor volume [GTV]). However, without using intravenous contrast agents, the contrast of abdominal tumors is typically low in CBCT and the identification of tumor location is extremely challenging. Therefore, some surrogates, such as bony anatomy, fiducial markers, or nearby soft tissues with sharp boundaries, are used to determine couch shifts for the patient alignment. When bony anatomy (e.g., vertebral bodies) is used for setup, the tumor position relative to the bony anatomy (also known as baseline shift) will not be corrected; thus, uncertainty of the setup is relatively high, and a large margin to generate the planning target volume (PTV) from the clinical target volume (CTV) is needed to account for the baseline shift. By analyzing CBCTs acquired before the treatment for 13 patients with pancreatic cancer with implanted Visicoil markers (IZI Medical Products, Owings Mills, MD, USA), large systematic displacements of the fiducial markers relative to bony anatomy were observed, with group systematic errors of 3.9, 5.5, and 3.6 mm and group random errors of 3.6, 4.7, and 2.4 mm in left–right (LR), superior–inferior (SI), and anterior–posterior (AP) directions, respectively [7]. Therefore, fiducial markers for daily online position verification are preferred over bony anatomy–based setup for patients with pancreatic cancer.

Similarly, in an analysis of 12 patients with liver cancer with 70 paired CBCTs (right before and after the treatment) [8], when fiducial markers placed in the liver tumor bed were used as surrogates for actual target positions, margins of 1.8, 3.8, and 1.4 mm in LR, SI, and AP directions were needed to compensate intra-fractional uncertainties when daily online CBCT correction was used. However, when vertebral body match was used, which cannot correct the baseline shifts, the margins needed to account for intra-fractional uncertainties became 5.5, 14.6, and 7.2 mm in LR, SI, and AP directions, respectively [8]. In another study on 11 patients with 13 liver tumors [9], when the liver contour was used as a surrogate for the actual target position, the margins for compensation of intra-fractional uncertainties were 6 and 4 mm in the SI direction and LR direction, respectively.

14.2.2 US

US provides non-ionizing, real-time volumetric imaging capability with excellent soft-tissue contrast. US is considered an operator-dependent imaging modality. Thus, the uncertainty associated with US can potentially be high, although it provides high soft-tissue contrast. Several robotic-arm–based US systems have been under development for intra-fractional guidance of external beam radiation therapy [10, 11]. Commercially, the Clarity Autoscan system (Elekta AB, Stockholm, Sweden) using transperineal US (TPUS) is available for intra-fractional prostate motion tracking. By positioning the US probe on the perineum, the probe will not interfere with the radiation beam, allowing for real-time volumetric tracking during the beam delivery. Studies based on a male pelvis phantom showed that the accuracy of TPUS-based Clarity Autoscan system for prostate tracking was within 1 mm [12]. However, for abdominal sites, there are currently no commercially available US-based real-time tracking systems.

Nevertheless, there have been active investigations into the potential use of US-based tracking system for abdominal sites. One study used the research version 'Anticosti' of the Clarity Autoscan (Elekta AB, Stockholm, Sweden) on a total of 14 patients with 16 abdominal targets (15 liver metastasis and one lymph node metastasis near the pancreatic head) to evaluate the accuracy of US-based daily repositioning [13]. The workflow of US guidance is shown in figure 14.4. To avoid pressure-induced deformations, the probe was placed with a multi-joint fixation arm along the right mid-axillary line instead of transabdominally. At each treatment fraction, 3D US was first used to correct patient positioning based on GTV or portal vein branch. CBCT was then acquired for final repositioning based on either fiducial markers or liver contours. Residual error after US-based positioning measured by CBCT was 0.4 ± 3.3 mm in LR, 0.2 ± 4.3 mm in SI and 1.0 ± 3.0 mm in AP directions, respectively. Although the systematic errors are relatively small after US-based corrections (<0.5 mm in the LR and SI directions and 1.2 mm in the AP

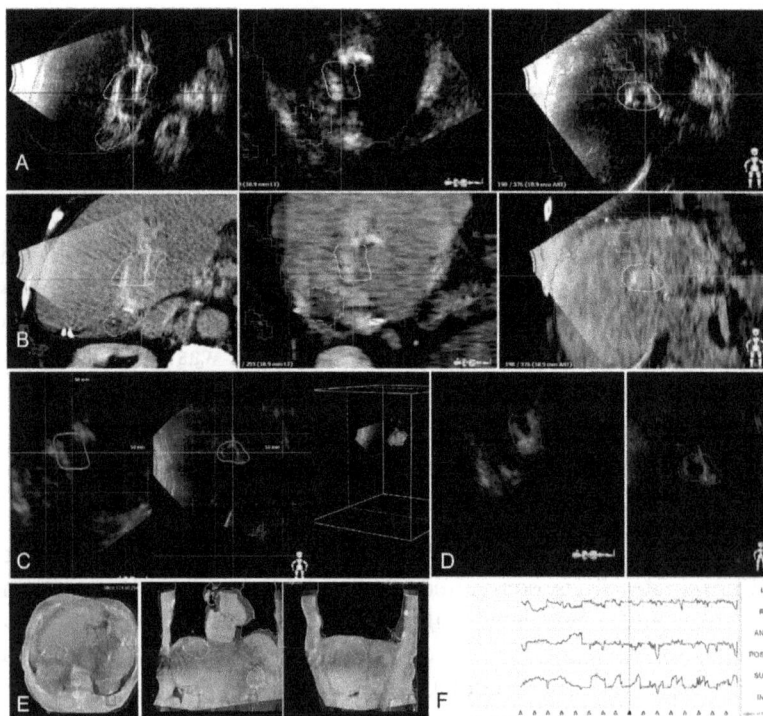

Figure 14.4. US setup, positioning reference and repositioning workflow. (A) US reference in transversal, coronal and sagittal reconstruction (portal vein branch in yellow in the vicinity of the target volume, marked orange) acquired directly after the planning CT. (B) US reference matched to the planning CT. (C) US-based table position correction before CBCT imaging. (D) Tracking of the surrogate structure during deep-inspiration breath-hold CBCT. (E) Marker-based matching in the CBCT after US-based positioning. (F) Tracking during repeated deep-inspiration breath holds. Reprinted from [13], Copyright (2019), with permission from Elsevier.

direction), CBCT was still recommended for daily patient positioning, as US imaging can only acquire a small portion of liver and relevant organs at risk [13]. On the other hand, US is essential to track intra-fractional motion.

14.2.3 MRI

MRI offers a superior soft-tissue contrast without using ionization radiation. MRI has recently been integrated into Linacs, which can acquire 3D volumetric imaging for initial patient setup and real-time imaging for intra-fractional target position tracking during beam delivery. MRI-guided radiation therapy is particularly useful for abdominal sites, as many targets and organs at risk in the abdominal region cannot be visualized on CBCT directly. Currently, there are two commercially available MRI-Linac systems: the ViewRay MRIdian system (ViewRay, Inc., Oakwood Village, Ohio, USA) and the Elekta Unity system (Elekta AB., Stockholm, Sweden). The MRIdian MRI-Linac combines a 0.35 T field-strength split-bore magnet MRI with a 6 megavoltage (MV) flattening filter–free Linac. The Unity MRI-Linac integrates a high-field (1.5 T) closed-bore MRI with a 7 MV flattening filter–free Linac.

While MRI provides superior soft-tissue contrast compared with CBCT, MRI geometric distortion introduces additional uncertainties for target localization and tracking. One study investigated the geometric distortion for an MRIdian Linac using a phantom consisting of ~6300 landmarks and found that all landmarks within 10 cm from the isocenter had a distortion <1 mm in the AP, LR, and SI directions [14]. The geometric distortion increased in the regions further away from the isocenter. In the 10–20 cm range, less than 1.5% and 4% of the distortions exceeded 2 mm along the AP and LR axes, respectively. In the regions 20–25 cm away from the isocenter, absolute displacements in the LR and AP directions were up to 5 and 7 mm, respectively [14]. Similar trends were observed on the Unity MRI-Linacs, where the maximum (99th percentile) displacement values within 7.5 and 17.5 cm from the isocenter were 0.7 and 2.0 mm, respectively [15].

14.3 Uncertainty in image matching

After image acquisition, the next step in the workflow of IGRT is to match the online image acquired with the patient lying on the treatment couch with the treatment planning image to determine the couch motion to put the tumor into the planned position. This matching task can be achieved either via an automatic matching algorithm, e.g. image registration, or manually by experienced personnel.

On the automatic matching side, rigid registration algorithms with translational degrees of freedom or translational and rotational degrees of freedom are used depending on the available motion degrees of freedom of the treatment couch. The residual uncertainty depends on multiple facts, such as the inherent uncertainty of the registration algorithms, image resolution, contrast, cost function of the registration algorithm, etc. A number of studies have been conducted to evaluate the residual uncertainty after automatic matching. End-to-end tests using phantom studies have revealed that the residual uncertainty is of the order of 0.3 mm and 0.3°

in translational and rotational directions when using the gray value registration with high-resolution CBCT images on the Elekta XVI (Elekta AB., Stockholm, Sweden) registration algorithm together with the HexaPOD couch (Elekta AB., Stockholm, Sweden) with six degrees of freedom. However, the residual error may increase when low-resolution images are used, especially in rotational directions [16]. Another study compared performance for three IGRT imaging systems: Elekta XVI CBCT (Elekta AB., Stockholm, Sweden), Varian TrueBeam CBCT (Varian Medical System, Palo Alto, CA, USA), and TomoTherapy MV CT (Accuray Inc., Sunnyvale, CA, USA) using a CIRS virtual human male pelvis phantom (CIRS Inc., Norfolk, VA, USA). Residual translation errors were similar for all systems and less than 0.5 mm [17].

Nonetheless, the uncertainty could be larger in real patient cases. A recent study collected data of patient positioning for hypofractionated liver cancer IGRT [18]. For each treatment, the online CT image was initially aligned to the planning CT based on the shape of the liver automatically using a registration software. Manual adjustments were applied after that. Considering the final position as the ground truth, uncertainty in the automatic matching step was evaluated. The median discrepancy between the final and automatic registration was 1.1 mm (0–24.3 mm), and 38% of treated fractions required manual corrections of $\geqslant 3$ mm. The systematic uncertainty was 1.5 mm in the AP direction, 1.1 mm in the LR direction, and 2.4 mm in the SI direction. The random uncertainty was 2.2 mm in the AP, 1.9 mm in the LR, and 2.2 mm in the SI direction.

Although manual matching was considered as the ground truth in the afore-mentioned study, keep in mind that it is not perfect either. When manual matching is performed, uncertainty arises due to inter- and intra-observer variability. This is particularly a concern for low-contrast soft tissue–based patient positioning, as it is hard for a human to perceive boundaries in the images to guide the matching task. In a study evaluating observer uncertainties of soft tissue–based patient positioning in prostate cancer IGRT [19], the authors compared positioning error and reference error that was calculated by matching the centroids of contours in the planning CT and the pre-treatment CBCT. Contours in the latter were drawn retrospectively by oncologists for the study purpose. They further computed systematic and random components of the errors. Inter-observer variations in AP, SI, and LR directions were 0.9, 0.9, and 0.5 mm, respectively, for the systematic error, and 1.8, 2.2, and 1.1 mm, respectively, for random error. Intra-observer variations were relatively small, being <0.2 mm in all directions.

The performance of manual matching could be improved when visually obvious features exist in the images to aid the matching task. For instance, when matching fiducial markers, even using a two-dimensional (2D) image modality of electronic portal imaging device, the mean and median intra-observer variability ranged from 0.4 to 0.7 mm and 0.3 to 0.6 mm, respectively, with a standard deviation of 0.4–1.0 mm for prostate cancer radiotherapy. Inter-observer results were similar with a mean variability of 0.9 mm, a median of 0.6 mm, and a standard deviation of 0.7 mm [20]. Compared with the numbers reported for soft tissue–based manual matching, the effectiveness of using fiducial markers mainly existed in terms of reducing inter-observer variability.

14.4 Uncertainty in couch motion

After determining the couch motion by registering the planning and the online images, the next step is to execute this motion physically by translating and/or rotating the couch. For the uncertainties related to hardware, there are two factors to consider. The first one is the agreement between imaging isocenter and Linac radiation isocenter. The online image is acquired with the information that the radiation isocenter is defined at a certain place of the imaging coordinate. The uncertainty of this coordinate contributes to the uncertainty of the final tumor position relative to the radiation beam. Verification of the coincidence between the two isocenters is recommended to be carried out through quality assurance tests on a regular basis depending on the Linac used [21]. Technologies have been developed by users or vendors for these tests [22, 23]. Recent advancements have also permitted direct visualization of the delivered dose in the imaging coordinate using N-isopropylacrylamide 3D dosimeters [24, 25]. The agreement between the imaging and radiation isocenters was found generally in the submillimeter range.

The source of uncertainty in the hardware side is couch motion accuracy. Again, this is often verified in periodic quality assurance tests as recommended by the radiotherapy community [21]. Studies have shown that the mean ± standard deviations of couch movement based on phantom measurements with an optical tracking system were 0.16 ± 0.48 mm, 0.32 ± 0.30 mm, and 0.11 ± 0.12 mm in the LR, AP, and SI couch directions, respectively [26].

A more prominent factor in the abdominal region is probably that anatomical change is usually in the form of deformation due to physiological processes such as bowel or bladder filling. This cannot be fully corrected by the rigid motion provided by couch translations or rotations. Previous studies on radical radiotherapy of urinary bladder cancer revealed a substantial patient setup variation as a result of bladder filling [27]. Repeat weekly CT scans found that bladder volumes extended outside the planning scan bladder contours in 89% of the scans, on average with 9% of the volume (range: 0%–47%). The rectum expanded outside the planning contours in all repeat scans, on average with 24% of the volume (range: 2%–69%). The ultimate solution to this problem is probably online adaptive replanning, such as that enabled by recent MRI-Linacs. If the clinical practice is constrained by the rigid couch motion, it is usually the decision of the physician regarding the best couch motion to compensate the deformation using a rigid motion.

14.5 Uncertainty due to intra-fractional motion

In addition to inter-fractional motion and deformation of abdominal structures, which are mostly induced by the organ fixation or organ filling variations from fraction to fraction, the intra-fractional motion and deformation of abdominal structures constitute another major source of uncertainties that would degrade the radiotherapy treatment quality if without any mitigation efforts. For abdominal organs that are downstream of the diaphragm, the most significant source of intra-fractional motion is respiration-induced periodical motion [28, 29]. Structures including the liver, pancreas, stomach and kidneys are all affected by respiration-induced motion, with varying degrees of amplitudes. The variation of organ fixation, as well as organ filling/peristaltic motion, makes other sources of intra-fractional motion of abdominal sites [29, 30]. Due to the compactness of organs/structures within the abdominal region [31], it is pivotal to gain knowledge of the intra-fractional motion to design corresponding motion management strategies to focus the radiotherapy dose towards the diseased site, while avoiding the surrounding normal tissues, and reduce the dose delivery uncertainties. In general, the uncertainties from intra-fractional motion can be linked to the different motion management techniques used and categorized into two general types: free breathing–based motion uncertainties and breath hold–based motion uncertainties, which will be introduced in the following section with the corresponding intra-fractional motion features, uncertainties and solutions.

14.5.1 Motion and uncertainties in free breathing–based treatment

Compared with other types of intra-fractional motions, respiration-induced motion in the abdominal region is generally periodical, stable, and easy to capture. It is often also the most prominent source of intra-fractional motion. A straightforward motion management technique is to allow the patient to breathe freely during the treatment and to use a large radiation treatment field to cover the whole region traversed by the tumor motion [32]. An internal target volume (ITV) is defined as an envelope to cover the full tumor motion range. Subsequently the treatment fields are designed to cover this ITV, usually by an additional margin to account for patient setup errors of each treatment fraction. The use of an all-encompassing ITV significantly reduces the chance of target miss during radiotherapy, which, however, comes at the expense of increased normal tissue dose.

For free breathing–based motion and motion management, there are two major sources of uncertainties: intra-fractional motion estimation from treatment simulation and intra-fractional motion during treatment.

14.5.1.1 Uncertainties of intra-fractional motion estimation from treatment simulation

Since radiotherapy is a multi-fractional process that can extend up to 5–6 weeks, a treatment plan is usually designed prior to the overall treatment course. To design a proper plan to account for the intra-fractional motion, it is crucial to estimate the intra-fractional motion range during the plan simulation stage via imaging techniques. The estimation, however, contains inherent uncertainties from the imaging technology used. Four-dimensional CT (4D-CT) is usually used as the standard imaging tool to estimate the intra-fractional respiration-induced motion due to its spatial-temporal imaging capability and wide availability [33]. Mori *et al* used a large-field 4D-CT to acquire pancreatic motion and found that the center of mass of the GTV moved by an average of 8.9 mm in the SI direction, 2.6 mm in the AP

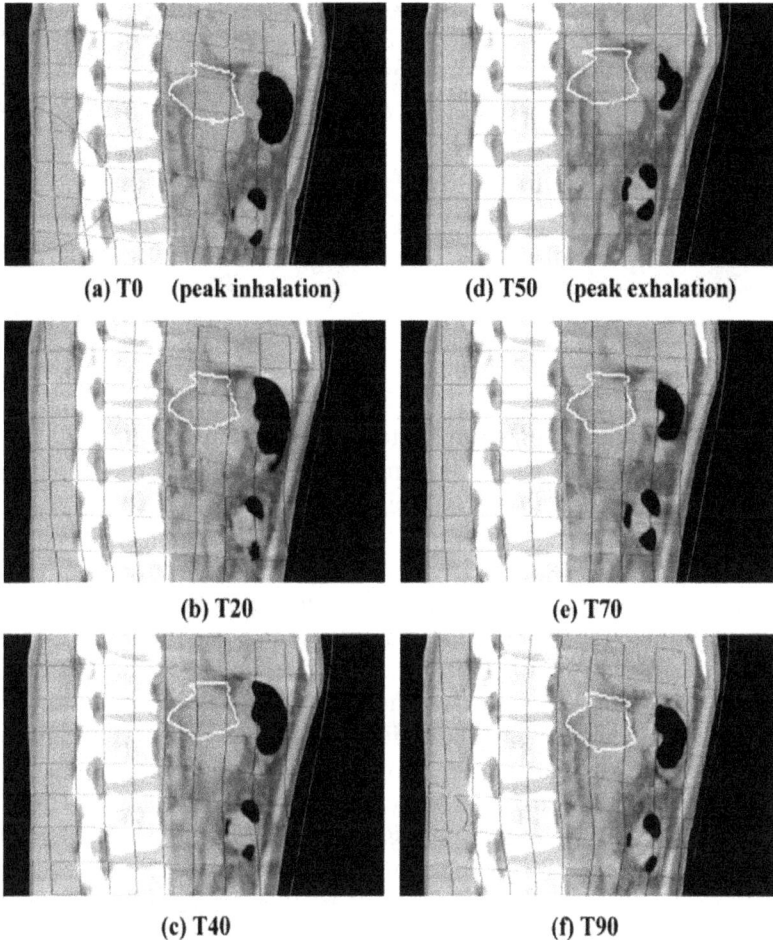

(a) T0   (peak inhalation)   **(d) T50   (peak exhalation)**

(b) T20   **(e) T70**

(c) T40   **(f) T90**

Figure 14.5. Four-dimensional-CT images at the sagittal view for one of the patient cases. The notations under subfigures (a)–(f), from T0 to T90, indicate the respiratory phases, with T0 being the peak-inspiration phase and T50 the peak-expiration phase. The yellow contour shows the GTV at each phase, whereas the blue mesh grid shows the deformation field from the T50 phase for each phase image. Reprinted from [34], Copyright (2009), with permission from Elsevier.

direction, and 1.4 mm in the LR direction (figure 14.5) [34]. In a different study, Watanabe *et al* used CT scans at the mild inspiration and expiration phases (as shallow free breathing) to evaluate the intra-fractional stomach motion and found that the average ± standard deviation vector length of the stomach center movement was 17.5 ± 11.4 mm [35].

However, as an x-ray–based imaging technique, the soft-tissue contrast in the abdominal region for 4D-CT is usually poor due to similar tissue densities between tumor and normal structure, as well as between different normal structures, introducing uncertainties to the motion characterization accuracy [36]. For assistance, radiopaque high-density markers can be implanted near the tumor prior to

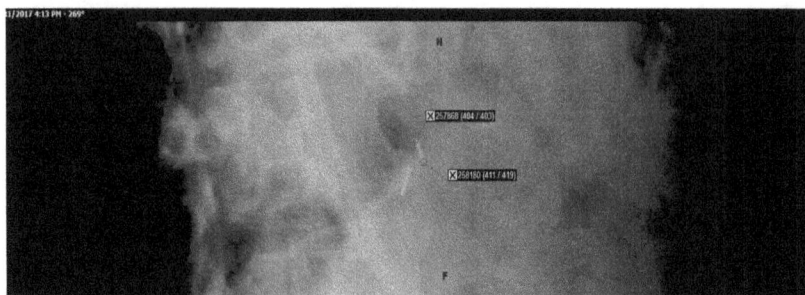

Figure 14.6. X-ray projection image of abdominal anatomy with implanted fiducial markers. [39] John Wiley & Sons.

the simulation stage to serve as surrogates of tumor motion (figure 14.6) [37]. A study by Park *et al* used 4D-CT plus implanted fiducial markers for liver tumor motion tracking and found that the respiration-induced motion has an average ± standard deviation amplitude of 17.9 ± 5.1 mm along the SI direction, 5.1 ± 3.1 mm along the AP direction, and 3.0 ± 2.0 mm along the LR direction [38]. The fiducial markers are also frequently used as a motion surrogate for the pancreatic cancer [39, 40]. In addition to the fiducial markers, the iodinated contrast agents can also be administered to enhance the tumor contrast. Figure 14.7 shows contrast-enhanced, hypo-dense liver tumors in CT images after contrast agent injection [41].

Although the fiducial markers and iodinated contrast agents can help to reduce the uncertainties by facilitating respiration-induced motion measurement, they do harbor their own limitations. The fiducial marker implantation is an invasive process, with potential side effects to organs already fragile with cancer [40]. The implanted fiducial markers may also migrate from place to place and fail to represent the actual tumor motion. They cannot capture the respiration-induced tumor deformation, either [36]. The contrast agent has a potential toxicity issue, and it remains challenging to control the timing of contrast agent administration due to fast wash-in and wash-out [41]. The contrast enhancement induced by the contrast agent is also often not ideal.

In contrast to CT imaging, MRI presents significantly better soft-tissue contrast and can be used to co-register with the CT images or used alone to facilitate motion characterization. One study applying high-speed MRI for liver motion tracking has quantified a magnitude of 21 mm in the SI direction, 8 mm in the AP direction, and 9 mm in the LR direction [42]. Dodle *et al* used 4D-MRI to assess the motion range of the pancreas and found that the mean SI motion amplitude can be up to 28.5 mm (figure 14.8) [43]. The motion amplitudes measured in 4D-MRI tend to be larger than those measured in 4D-CT, another uncertainty of intra-fractional motion estimation remaining to be investigated [29]. Such a discrepancy could be due to the contrast differences between CT and MRI. It is also potentially caused by the relatively poor spatial resolution of high-speed MRI/4D-MRI, which has to be used as a trade-off to maintain an acceptable imaging time.

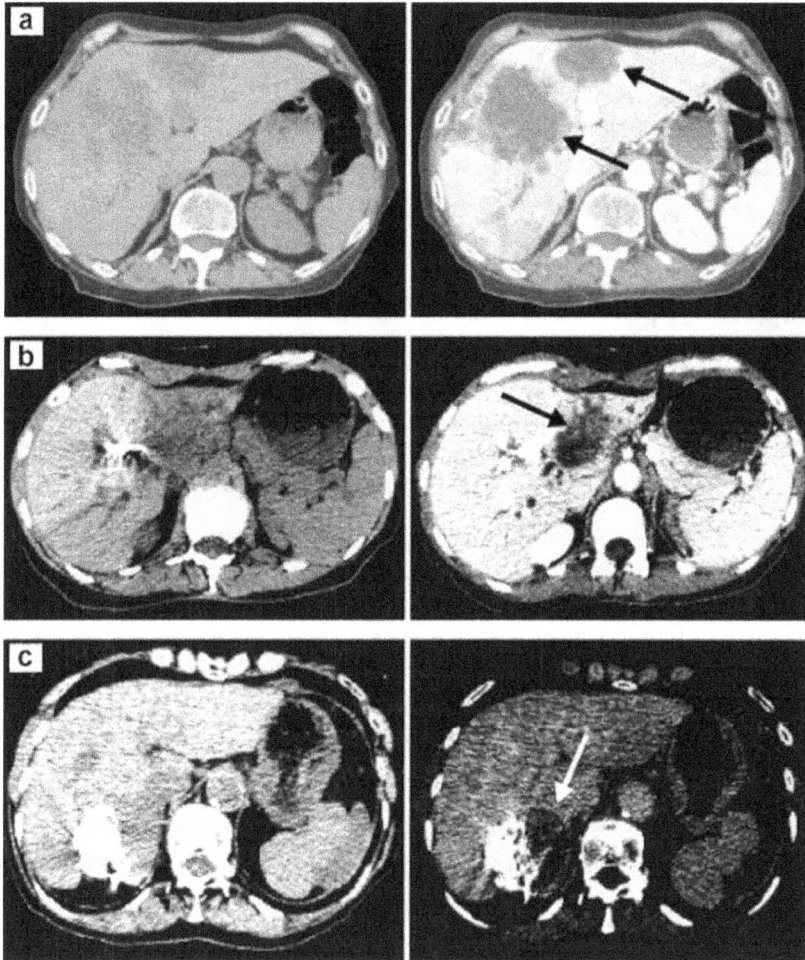

Figure 14.7. Comparison between non-contrast (left column) and contrast-enhanced (right column) liver CT images for three patient cases ((a)–(c)). The arrows on the right column are pointing to the contrast-enhanced tumor regions. Reprinted from [41], Copyright (2012), with permission from Elsevier.

In general, many studies reported the respiration-induced motion to be a major source of intra-fractional motion in abdominal sites, with the largest extent of motion along the SI direction, followed by AP, with LR motion usually the smallest [29]. Four-dimensional-CT is currently the mainstay for intra-fractional motion estimation of abdominal sites for treatment planning, which, however, is limited by uncertainties from poor soft-tissue contrast. The fiducial markers and iodinated contrast agents are frequently applied during 4D-CT simulation to assist motion characterization, but with their own uncertainties. High-speed MRI and 4D-MRI are promising alternatives to 4D-CT while presenting prominent motion amplitude variations from 4D-CT according to a few studies [29], another source of uncertainty for intra-fractional motion estimation that needs to be further investigated.

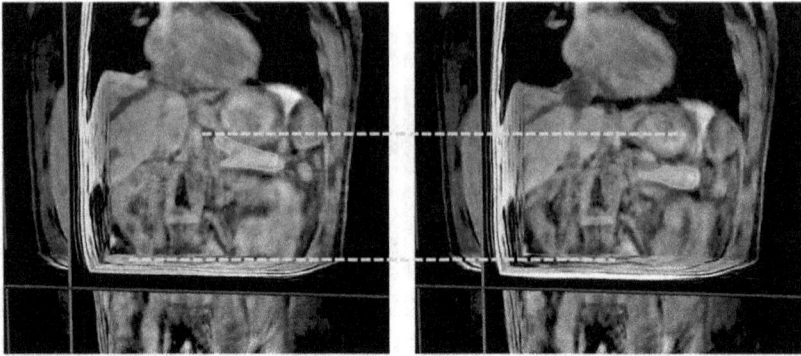

Figure 14.8. Four-dimensional-MRI at the coronal view for (left) the end-expiration phase and (right) the end-inspiration phase. The pancreas at the shown slice was contoured in solid green, with the adjacent green cloud showing the 3D pancreas segmentation anterior to the visible coronal slice. The dashed yellow lines mark the motion range of pancreas in the SI direction.

14.5.1.2 Uncertainties of intra-fractional motion during the actual treatment

Although the intra-fractional respiratory motion can be fairly well estimated during plan simulation with image guidance, at each fractional radiotherapy treatment the motion of the abdominal structures may still vary and deviate, constituting another major source of uncertainties for free breathing–based treatments. These types of uncertainties can be further divided into two sub-categories: (1) uncertainties from the difference between the motion estimated during plan simulation and the actual intra-fractional motion and (2) uncertainties from the variation of the intra-fractional motion within each treatment. With advanced on-board imaging techniques, such variations and corresponding uncertainties can be captured and evaluated. In a study from Shimohigashi *et al* [44], 4D-CBCT was applied to measure the liver tumor's respiratory motion variation between simulation and treatment (type 1) and within each treatment (type 2). Substantial motion amplitude changes (>3 mm) of type 1 were observed for 10% of treatment fractions. When evaluated within each fraction, 2% of treatment fractions yielded a >3 mm tumor motion amplitude change of type 2, which was measured through two sequential 4D-CBCTs of each fractional treatment, one acquired before treatment delivery and the other immediately after.

In addition to motion amplitude variations, tumor baseline drift makes another major source of type 2 variation. One study by Liang *et al* used planar x-ray imaging plus implanted fiducial markers, as well as external markers, to track the liver tumor motion in real time during liver stereotactic body radiation therapy (figure 14.9) and reported that intra-fractional baseline shifts that exceeded 2, 3, and 5 mm were observed in respectively 66.7%, 38.1%, and 19% of all fractions for a measurement duration of 30 min [45]. The same study also reported median tumor motion amplitude variations of 11.9 mm (SI), 3.8 mm (AP), and 1.3 mm (LR) from simulation to treatment and 4.3 mm (SI), 1.5 mm (AP), and 0.5 mm (LR) within each treatment, showing substantial intra-fractional motion uncertainties that warrant on-board image guidance [46].

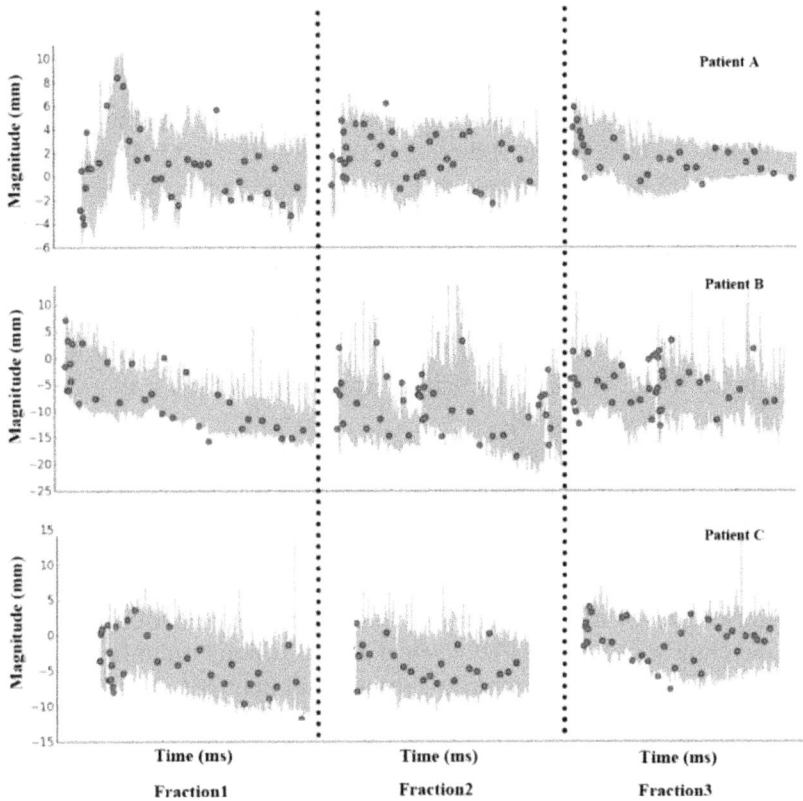

Figure 14.9. Tracked liver tumor motion of different fractions (columns) and patients (rows). The green curves show the tracking results via external markers through an internal–external correlation model. The red dots show the tracking results using intermittent x-ray imaging plus implanted fiducial markers. Considerable intra-fractional motion amplitude variation and tumor baseline drift can be observed. [39] John Wiley & Sons.

In summary, the discrepancy between the intra-fractional motion estimated during the simulation stage and that which actually occurred at each treatment contributes to another source of uncertainty. The intra-fractional motion variations within each treatment, including both free-breathing motion amplitude variation and tumor baseline drift, also add to the uncertainties.

Although the motion-encompassing, free breathing–based treatment technique can address the intra-fractional tumor motion, it inevitably has to treat a large field that damages more normal soft tissues surrounding the tumor. It is also susceptible to many sources of uncertainties of intra-fractional motion, as described above. To reduce the treatment field size as well as the motion uncertainties, there are several strategies employed by today's radiotherapy practices:

(A) *Abdominal compression.* Exerting pressure onto the abdominal area has been reported to effectively reduce the magnitudes of respiration-induced motion. The study of Dolde *et al* found that the mean pancreatic motion along the SI direction can be effectively reduced by 48% with an abdominal corset, as measured on 4D-MRI [43]. Another liver study used fluoroscopic imaging plus fiducial markers to characterize and compare the tumor's respiratory motion between simulation and actual treatment, both with and without abdominal compression [47]. It reported that, by applying abdominal compression, the median excursion has been reduced by 62% in the SI direction and 38% in the AP direction, with the mean excursions <5 mm in all directions. The residual excursions with abdominal compression also reproduced well between the simulation and the actual treatment and within each treatment. With abdominal compression, the uncertainties of intra-fractional respiratory motion can be effectively reduced via simultaneously reduced motion amplitude and enhanced motion reproducibility. However, it should be noted that abdominal compression may introduce additional inter-fractional organ fixation and deformation variations, which are affected by the location of the compression [48].

(B) *Target tracking.* Instead of using a large treatment field to encompass the full range of target motion, an alternative treatment technique moves and synchronizes the treatment field along with the target motion. Such a technique, if successfully implemented, will be immune to intra-fractional target motion variations with the corresponding uncertainties minimized. For conventional L-shaped Linacs, target tracking remains mostly investigational and experimental until today due to various technical and clinical challenges [49]. CyberKnife, a robotic radiotherapy system, is able to track real-time target motion during treatment via a hybrid system of stereotactic x-ray imaging and optical imaging [50]. For the abdominal site, implanted fiducial markers are used as surrogates of target motion. The CyberKnife uses optical markers placed on the patient body surface, representing external motion, to build a correlation model with internal target motion that is tracked with x-ray imaging–localized fiducial markers. The model is being periodically updated with intermittent x-ray imaging re-acquisition. The CyberKnife then uses the continuously tracked optical marker motion as a surrogate for internal target motion to maneuver a compact robotic Linac to follow the target for radiation delivery. However, the internal and external motion correlation models may be limited in accuracy, along with other practical issues like the machine's finite response time (latency), introducing uncertainties distinct from those of the conventional free-breathing treatments [51].

(C) *Respiratory gating.* Besides abdominal compression and target tracking, respiratory gating is another technique to reduce the treatment field size and track the intra-fractional target motion for reduced uncertainty. The 4D-CT pancreatic motion study of Mori *et al* [34] also tested a gating approach (30% duty cycle around the end-expiration phase) and found the range of GTV motion reduced to 2.6 mm (SI), 0.9 mm (AP), and 0.5 mm (LR), a substantial decrease compared with the ungated free-breathing approach. Gating can be performed using cine MV x-ray imaging during radiotherapy delivery or via an external surrogate attached to the skin. Such approaches may be subject to the uncertainties from limited soft-tissue m gating) [52]. The recently introduced MRI-Linac allows on-board real-time cine MRI for gated treatments [53], with superior soft-tissue contrast and zero ionizing radiation hazards. However, the current MRI used for gating can only be 2D, failing to capture the motion along all three dimensions. In addition, for all gated treatments, the treatment time is substantially longer due to a low duty cycle (usually around 30%).

14.5.2 Motion and uncertainties in breath hold–based treatment

In addition to free breathing–based treatment and its multiple variations, breath hold is another major motion management technique for respiration-impacted abdominal sites [54]. For breath-hold cases, the patients are instructed to take a deep inspiration and hold their breath towards a level pre-defined during plan simulation. The radiotherapy beam will only deliver when the breath-hold level is maintained within a narrow window. Compared with free breathing–based treatment, the breath-hold technique allows the treatment field to be focused to a much smaller region enclosing a quasi-static tumor. With a smaller treatment field, the normal tissue can be better spared by breath-hold treatment compared to free-breathing cases. The uncertainties of intra-fractional respiratory-motion variations are effectively eliminated by the breath-hold technique.

However, breath hold–based treatments are subject to a new uncertainty caused by inconsistent breath-hold levels [30, 39]. Such inconsistency can occur between treatment simulation and actual treatments. It can also occur within each treatment, since multiple breath holds are usually needed to complete a treatment fraction due to the limited duration of each breath hold. The inconsistency of breath-hold levels can be caused by many factors, including the lack of correlation between external surrogate position and internal anatomy fixation, patients' breathing maneuvers, and the mechanical limits of breath-hold devices. In detail, due to the lack of fast tomographic imaging–based breath-hold level verification, current clinical protocols have to rely on surrogates to indicate the breath-hold level. The real-time position management (RPM) system (Varian Medical Systems, Palo Alto, CA, USA) places a box with reflective markers on patient body surfaces to indicate breath-hold levels (figure 14.10) [39]. The VisionRT system (Vision RT Limited, London, UK) uses optical imaging–based surface rendering as a surrogate for the breath-hold level [55]. Both systems, however, are prone to errors caused by the lack of correlation between

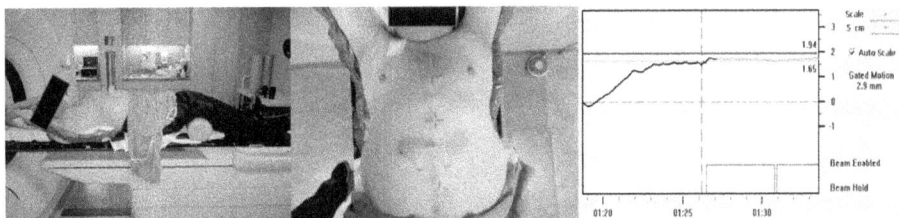

Figure 14.10. Breath-hold setting using the RPM system. The left subfigure shows the overall patient setup. A plastic box was placed on the abdominal region of the patient (middle subfigure). Reflective markers on the box were optically tracked to indicate the breath-hold level. And the treatment beam will only deliver when the breath-hold level reaches and stays within a pre-defined window, where the tracked motion curve turns green (right subfigure). [39] John Wiley & Sons.

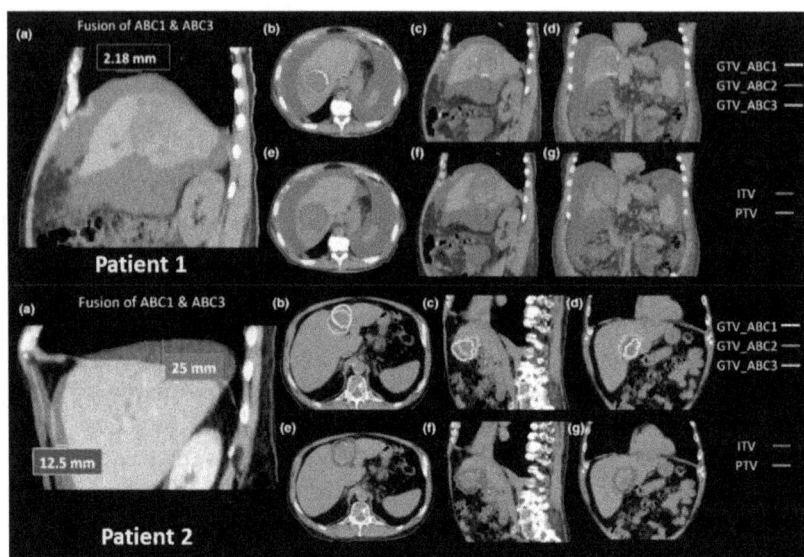

Figure 14.11. Simulation breath-hold CT images of two patient cases with good breath-hold reproducibility (patient 1) and poor breath-hold reproducibility (patient 2). Two simulation CT images under sequential breath holds via ABC were fused together to show their differences (ABC1-CT and ABC3-CT). The GTV contours from the three consecutive simulation CT images (GTV_ABC1, GTV_ABC2, and GTV_ABC3, respectively) were also overlaid on ABC1-CT to show their differences, along with the corresponding ITV and PTV based on a fusion of the three GTV contours. [39] John Wiley & Sons.

external surface position and internal organ fixation. Active breathing coordinator (ABC; Elekta AB, Stockholm, Sweden), another breath-hold system using an involuntary mechanism, gauges the breath-hold level with a spirometer [30]. Its accuracy, however, is affected by patients' breathing maneuver variations (abdominal vs thoracic) and the overshoot effects from high air-flow rate [56]. These uncertainties lead to variations of internal patient anatomy, even under the same 'apparent' breath-hold level as indicated by each breath-hold monitoring system.

A study by the Cleveland Clinic repeated simulation CT images of patients with liver cancer for three consecutive breath holds, which were managed by the ABC system [30]. For some patients, substantial anatomical variations were observed between the repeated CT scans, even though the ABC device reported similar 'apparent' breath-hold levels. For one patient, over 25 mm discrepancy of the liver boundary was observed between CT scans, leading to substantial GTV variations (figure 14.11(b)).

Another study performed by Zeng *et al* used triggered x-ray imaging to track implanted fiducial markers during breath-hold pancreatic treatments and quantified breath-hold level variations for the RPM system [39]. Variations over 1 cm can be observed from some treatment fractions when comparing different breath holds within the same treatment fraction. Motion of the fiducial markers can also be observed within each breath hold due to the instability of the breath hold, which, however, is much smaller in magnitude.

To account for the uncertainties from inconsistent breath-hold levels, an additional margin can be added to the target, although at the cost of increased normal tissue dose. The Cleveland Clinic study [30] uses a more patient-specific method to combine the GTV volumes measured in three repeated breath-hold simulation CT scans (figure 14.11), which, however, is still prone to new variation patterns of the breath-hold level that may occur during the treatment. On-board, real-time tomographic imaging, especially non-ionizing MRI, will be able to directly visualize the anatomy to verify the consistency of breath-hold levels. However, fast, volumetric MRI with high spatial resolution is not available yet and needs to be developed.

14.5.3 Uncertainties from peristaltic motion, gradual organ filling, etc.

In addition to the uncertainties caused by variations from respiratory motion and breath-hold level inconsistencies, there also remain additional sources of uncertainties from intra-fractional motion, including organ peristatic motion, gradual organ filling, gas movement, or gradual organ fixation variation during the treatment [34, 57]. Compared with respiratory motion, these types of motion are non-periodical and heavily patient- and scenario-dependent. For abdominal sites, they are usually mixed with respiratory motion or motion caused by varying breath-hold levels and are difficult to directly estimate or model during plan simulation. Similarly, they are difficult to quantify during the actual treatment. A study of Mostafaei *et al* tried to quantify the peristaltic motion using MRI and 4D-CT and found that the peristaltic motion of ranges was between 0.3 and 1 cm [57]. The results remain preliminary, although, as they could not fully rule out the potential mixture with other types of motion as mentioned above.

In current clinical practice, these types of intra-fractional motion are not separately or explicitly considered for abdominal sites. Practically, they can be partially mitigated through non-radiotherapy interventions such as regulating patients' water/food intake and providing instructions on bladder and bowel preparation prior to each treatment. These types of motion are also usually slowly occurring in nature and can be mitigated through faster radiotherapy delivery.

14.6 Margin recipe

14.6.1 The margin concept

While there are a number of approaches to address these uncertainties from different perspectives, the most straightforward one is using a margin. The ultimate goal of radiotherapy treatment is to deliver the prescribed dose to the CTV under practical constraints, e.g., all the aforementioned geometric uncertainties of the tumor relative to the therapeutic beams. A margin is used to expand the CTV to a larger area of PTV that is intended to be covered in the treatment planning stage so that the CTV will receive sufficient coverage despite its geometry uncertainty.

There are two expansions conceptually corresponding to the uncertainties caused by the 'internal' and 'external' variations of the CTV [58] (see figure 14.12). First, CTV is expanded into ITV with an internal margin (as also mentioned in section 14.5). This expansion takes care of organ motion, which usually occurs for tumors in

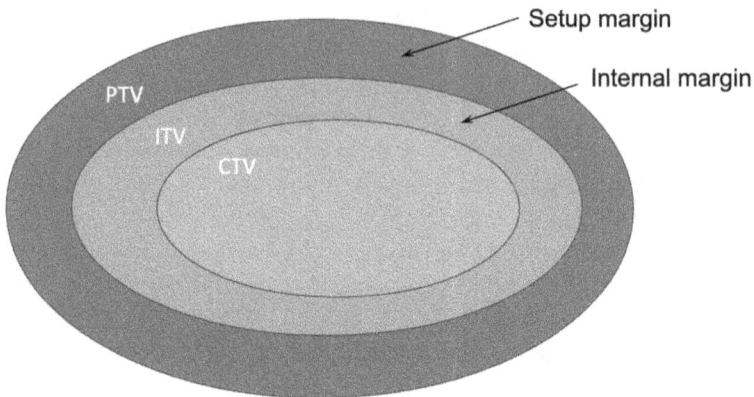

Figure 14.12. Illustration of CTV, ITV, and PTV.

the abdominal region. Here the term motion refers to both intra-fractional motions as well as inter-fractional motions that cause variations in size, shape, and position of the CTV in relation to anatomical reference points, e.g., as the result of the filling of the stomach. From a practical point of view, the uncertainty addressed here is due to physiologic variations and is difficult to control, if not impossible.

The second expansion is to add a setup margin to generate the PTV. This margin accounts for the geometry uncertainty of the CTV due to factors external to the patient, including but not limited to patient positioning under image guidance, beam positioning relative to isocenter, etc. The uncertainty addressed here is largely related to technical factors that can be dealt with by more accurate setup and improved mechanical stability of the machine.

14.6.2 Margin recipe

The next question regarding margin is its size. Intuitively, the larger the uncertainty is, the larger the margin size should be to accommodate it. To quantitatively answer this question, several margin recipes have been published over the years. For instance, Table 2 in [59] gave a comprehensive summary of major published margin recipes.

Before going into detail, let us first revisit the concept of systematic and random uncertainties, because the dosimetric impacts of these uncertainties in a radiotherapy course with a number of fractions are different and, hence, deserve separate treatments in margin calculations. Systematic uncertainty affects all fractions in a treatment course in the same way. Hence, it leads to major impacts in dose distribution coverage to the CTV. Or, conversely, when allowing a certain reduction of the cumulative dose to the target, systematic uncertainty contributes more to the margin size. In contrast, random uncertainty affects different fractions differently. The impact on the cumulative dose distribution is blurring, which generates a small decrease of dose at the edge of the high-dose region. It should hence be weighted less in margin calculations. With this consideration in mind, the margin should be calculated by combining the systematic and random uncertainties, with the systematic uncertainty term being weighted more.

The margin calculation certainly depends on the endpoint: to what extent we would like to preserve the dose coverage of the target. van Herk *et al* used the minimum cumulative CTV dose as the endpoint in their calculations [60]. Specifically, their calculation of the margin is intended to ensure that 90% of the patient population receive a minimum cumulative CTV dose of at least 95% of the prescribed dose. With this goal in mind, they derived the size of margin expanding from the CTV to PTV:

$$m = 2.5\Sigma + 1.64(\sigma - \sigma_p), \tag{14.1}$$

where m is the margin size, Σ is the total standard deviation of systematic error (treatment preparation error, in the paper of van Herk [60]), σ_p is the size of the beam penumbra, and σ is the quadratic combination of the total standard deviation of all random variations (treatment execution error in [60]) and σ_p. Within each of Σ and σ, there may be multiple factors contributing to the standard deviation. The standard deviations should be added in quadrature to compute the total standard deviation. For a typical setup of $\sigma_p = 3.2$ mm, the margin formular can be approximated in a linear form using σ', the standard deviation of all random variations excluding beam penumbra, as

$$m = 2.5\Sigma + 0.7\sigma'. \tag{14.2}$$

The validity of this margin formula has been experimentally tested. It was found that the formula is robust against variations in tumor size and motion, treatment technique, plan conformity, and low tissue density [61].

One final issue requiring clarifications is the combination of internal margin and setup margin, as discussed in the previous subsection and figure 14.12. It seems that the two margins are defined independently, one for internal and one for external errors, and hence, the margin can be linearly added. However, the equation (14.1) clearly indicates that the margin cannot be expressed as the sum of two terms corresponding to the internal and external errors. In fact, the impact of intrafractional organ motion to the CTV dose, due to its randomness, likely manifests as a blur effect. Hence, we should treat the factor of intra-fractional motion as a random error to account for its contribution to the margin size.

14.7 Final remarks and conclusion

One final comment is that the uncertainties discussed in this chapter are mostly restricted to the regime of geometric uncertainty pertaining to image guidance. It is important to keep in mind that there are in fact other types of uncertainties in the treatment. For instance, when delineating a tumor from a treatment planning image, the delineation itself is subject to uncertainty depending on several factors, such as properties of the imaging modality and image quality, as well as human factors such as experience and time to contour, etc. Uncertainties like this are apparently important in the sense that, if not addressed adequately, the success of radiotherapy treatment in terms of tumor control or normal tissue complications would be affected. However, the discussions in this chapter did not put much consideration into these uncertainties and were mostly focused on the uncertainties in the chain of image guidance (figure 14.2). In other words, we assumed the goal here to be delivering the prescribed dose to the already defined CTV following an IGRT workflow.

In conclusion, uncertainty is a critical component that affects the success of radiation therapy. Along the chain of IGRT, every component has uncertainty associated with it. We have reviewed these uncertainties with different causes and characteristics. As a standard and effective approach, margin is used in routine clinical practice to address the uncertainty issue. But, at the same time, care must be given to design an appropriate margin to effectively compensate for uncertainty while achieving the minimal damage to normal tissues.

References

[1] Bijhold J et al 1992 Maximizing setup accuracy using portal images as applied to a conformal boost technique for prostatic cancer Radiother. Oncol. **24** 261–71
[2] Mitine C et al 1991 Is it necessary to repeat quality control procedures for head and neck patients? Radiother. Oncol. **21** 201–10
[3] Bujold A et al 2012 Image-guided radiotherapy: has it influenced patient outcomes? Seminars in Radiation Oncology (Amsterdam: Elsevier)
[4] Craig T et al 2003 The impact of geometric uncertainty on hypofractionated external beam radiation therapy of prostate cancer Int. J. Radiat. Oncol. Biol. Phys. **57** 833–42
[5] Song W Y et al 2006 Evaluation of image-guided radiation therapy (IGRT) technologies and their impact on the outcomes of hypofractionated prostate cancer treatments: a radiobiologic analysis Int. J. Radiat. Oncol. Biol. Phys. **64** 289–300
[6] Zelefsky M J et al 2012 Improved clinical outcomes with high-dose image guided radiotherapy compared with non-IGRT for the treatment of clinically localized prostate cancer Int. J. Radiat. Oncol. Biol. Phys. **84** 125–29
[7] Van der Horst A et al 2013 Interfractional position variation of pancreatic tumors quantified using intratumoral fiducial markers and daily cone beam computed tomography Int. J. Radiat. Oncol. Biol. Phys. **87** 202–8
[8] Zhang T et al 2014 Inter-and intrafractional setup errors and baseline shifts of fiducial markers in patients with liver tumors receiving free-breathing postoperative radiation analyzed by cone-beam computed tomography J. Appl. Clin. Med. Phys. **15** 138–46
[9] Guckenberger M et al 2008 Image-guided radiotherapy for liver cancer using respiratory-correlated computed tomography and cone-beam computed tomography Int. J. Radiat. Oncol. Biol. Phys. **71** 297–304
[10] Schlosser J, Salisbury K and Hristov D 2010 Telerobotic system concept for real-time soft-tissue imaging during radiotherapy beam delivery Med. Phys. **37** 6357–67
[11] Tutkun Sen H et al 2013 A cooperatively controlled robot for ultrasound monitoring of radiation therapy Rep. U. S. **2013** 3071–6
[12] Yu A S et al 2017 Intrafractional tracking accuracy of a transperineal ultrasound image guidance system for prostate radiotherapy Technol. Cancer Res. Treat. **16** 1067–78

[13] Boda-Heggemann J *et al* 2019 Ultrasound-based repositioning and real-time monitoring for abdominal SBRT in DIBH *Phys. Med.* **65** 46–52

[14] Nejad-Davarani S P *et al* 2019 Large field of view distortion assessment in a low-field MR-Linac *Med. Phys.* **46** 2347–55

[15] Tijssen R H *et al* 2019 MRI commissioning of 1.5 T MR-Linac systems–a multi-institutional study *Radiother. Oncol.* **132** 114–20

[16] Arumugam S *et al* 2013 An accuracy assessment of different rigid body image registration methods and robotic couch positional corrections using a novel phantom *Med. Phys.* **40** 031701

[17] Barber J *et al* 2016 Comparison of automatic image registration uncertainty for three IGRT systems using a male pelvis phantom *J. Appl. Clin. Med. Phys.* **17** 283–92

[18] Choi G W *et al* 2019 Assessment of setup uncertainty in hypofractionated liver radiation therapy with a breath-hold technique using automatic image registration–based image guidance *Radiat. Oncol.* **14** 154

[19] Hirose Ta *et al* 2020 Observer uncertainties of soft tissue-based patient positioning in IGRT *J. Appl. Clin. Med. Phys.* **21** 73–81

[20] Ullman K L *et al* 2006 Intra-and inter-radiation therapist reproducibility of daily isocenter verification using prostatic fiducial markers *Radiat. Oncol.* **1** 2

[21] Klein E E *et al* 2009 Task Group 142 report: quality assurance of medical accelerators *Med. Phys.* **36** 4197–212

[22] Gao J and Liu X 2016 Off-isocenter Winston–Lutz test for stereotactic radiosurgery/stereotactic body radiotherapy *Int. J. Med. Phys. Clin. Eng. Radiat. Oncol.* **5** 154–61

[23] Clivio A *et al* 2015 Evaluation of the machine performance check application for TrueBeam Linac *Radiat. Oncol.* **10** 97

[24] Adamson J *et al* 2019 Delivered dose distribution visualized directly with onboard kV-CBCT: proof of principle *Int. J. Radiat. Oncol. Biol. Phys.* **103** 1271–9

[25] Pant K *et al* 2020 Comprehensive radiation and imaging isocenter verification using NIPAM kV-CBCT dosimetry *Med. Phys.* **47** 927–36

[26] Li W *et al* 2009 Accuracy of automatic couch corrections with on-line volumetric imaging *J. Appl. Clin. Med. Phys.* **10** 106–16

[27] Muren L P, Smaaland R and Dahl O 2003 Organ motion, set-up variation and treatment margins in radical radiotherapy of urinary bladder cancer *Radiother. Oncol.* **69** 291–304

[28] Trifiletti D M *et al* 2019 *Stereotactic Radiosurgery and Stereotactic Body Radiation Therapy* (Berlin: Springer)

[29] Abbas H, Chang B and Chen Z J 2014 Motion management in gastrointestinal cancers *J. Gastrointest. Oncol.* **5** 223–35

[30] Lu L *et al* 2020 Dosimetric assessment of patient-specific breath-hold reproducibility on liver motion for SBRT planning *J. Appl. Clin. Med. Phys.* **21** 77–83

[31] Tchelebi L T *et al* 2020 Reducing the toxicity of radiotherapy for pancreatic cancer with magnetic resonance-guided radiotherapy *Toxicol. Sci.* **175** 19–23

[32] Keall P J *et al* 2006 The management of respiratory motion in radiation oncology report of AAPM Task Group 76 a *Med. Phys.* **33** 3874–900

[33] Brandner E D *et al* 2006 Abdominal organ motion measured using 4D CT *Int. J. Radiat. Oncol. Biol. Phys.* **65** 554–60

[34] Mori S *et al* 2009 Four-dimensional measurement of intrafractional respiratory motion of pancreatic tumors using a 256 multi-slice CT scanner *Radiother. Oncol.* **92** 231–7

[35] Watanabe M *et al* 2011 Intrafractional gastric motion and interfractional stomach deformity using CT images *J. Radiat. Res.* **52** 660–5

[36] Zhang Y *et al* 2019 4D liver tumor localization using cone-beam projections and a biomechanical model *Radiother. Oncol.* **133** 183–92

[37] Beddar A S *et al* 2007 Correlation between internal fiducial tumor motion and external marker motion for liver tumors imaged with 4D-CT *Int. J. Radiat. Oncol. Biol. Phys.* **67** 630–8

[38] Park J C *et al* 2012 Liver motion during cone beam computed tomography guided stereotactic body radiation therapy *Med. Phys.* **39** 6431–42

[40] Kothary N *et al* 2009 Safety and efficacy of percutaneous fiducial marker implantation for image-guided radiation therapy *J. Vasc. Interv. Radiol.* **20** 235–9

[39] Zeng C *et al* 2019 Intrafraction tumor motion during deep inspiration breath hold pancreatic cancer treatment *J. Appl. Clin. Med. Phys* **20** 37–43

[41] Beddar A S *et al* 2008 4D-CT imaging with synchronized intravenous contrast injection to improve delineation of liver tumors for treatment planning *Radiother. Oncol.* **87** 445–8

[42] Shimizu S *et al* 1999 Three-dimensional movement of a liver tumor detected by high-speed magnetic resonance imaging *Radiother. Oncol.* **50** 367–70

[43] Dolde K *et al* 2019 4DMRI-based analysis of inter—and intrafractional pancreas motion and deformation with different immobilization devices *Biomed. Phys. Eng. Express* **5** 025012

[44] Shimohigashi Y *et al* 2017 Tumor motion changes in stereotactic body radiotherapy for liver tumors: an evaluation based on four-dimensional cone-beam computed tomography and fiducial markers *Radiat. Oncol.* **12** 61

[45] Liang Z *et al* 2018 Evaluation of the intra-and interfractional tumor motion and variability by fiducial-based real-time tracking in liver stereotactic body radiation therapy *J. Appl. Clin. Med. Phys.* **19** 94–100

[46] Xing L *et al* 2006 Overview of image-guided radiation therapy *Med. Dosim.* **31** 91–112

[47] Wunderink W *et al* 2008 Reduction of respiratory liver tumor motion by abdominal compression in stereotactic body frame, analyzed by tracking fiducial markers implanted in liver *Int. J. Radiat. Oncol. Biol. Phys.* **71** 907–15

[48] Eccles C L *et al* 2011 Interfraction liver shape variability and impact on GTV position during liver stereotactic radiotherapy using abdominal compression *Int. J. Radiat. Oncol. Biol. Phys.* **80** 938–46

[49] Falk M *et al* 2010 Real-time dynamic MLC tracking for inversely optimized arc radiotherapy *Radiother. Oncol.* **94** 218–23

[50] Sayeh S *et al* 2007 Respiratory *motion tracking for robotic radiosurgery Treating Tumors that Move with Respiration* (Berlin: Springer) pp 15–29

[51] Pepin E W *et al* 2011 Correlation and prediction uncertainties in the cyberknife synchrony respiratory tracking system *Med. Phys.* **38** 4036–44

[52] Ge J *et al* 2013 Accuracy and consistency of respiratory gating in abdominal cancer patients *Int. J. Radiat. Oncol. Biol. Phys.* **85** 854–61

[53] Kluter S 2019 Technical design and concept of a 0.35 T MR-Linac *Clin Transl. Radiat. Oncol.* **18** 98–101

[54] Boda-Heggemann J *et al* 2016 Deep inspiration breath hold—based radiation therapy: a clinical review *Int. J. Radiat. Oncol. Biol. Phys.* **94** 478–92

[55] Rice L *et al* 2015 Deep inspiration breath-hold (DIBH) technique applied in right breast radiotherapy to minimize liver radiation *BJR Case Rep.* **1** 20150038

[56] Mittauer K E *et al* 2015 Monitoring ABC-assisted deep inspiration breath hold for left-sided breast radiotherapy with an optical tracking system *Med. Phys.* **42** 134–43

[57] Mostafaei F *et al* 2018 Variations of MRI-assessed peristaltic motions during radiation therapy *PLoS One* **13** e0205917

[58] ICRU 1999 *ICRU Report 62, Prescribing, Recording and Reporting Photon Beam Therapy (Supplement to ICRU 50)* 62 International Commission on Radiation Units and Measurements https://www.icru.org/report/prescribing-recording-and-reporting-photon-beam-therapy-report-62/

[59] Van Herk M 2004 Errors and margins in radiotherapy *Seminars in Radiation Oncology* (Amsterdam: Elsevier)

[60] van Herk M *et al* 2000 The probability of correct target dosage: dose-population histograms for deriving treatment margins in radiotherapy *Int. J. Radiat. Oncol. Biol. Phys.* **47** 1121–35

[61] Ecclestone G, Bissonnette J P and Heath E 2013 Experimental validation of the van Herk margin formula for lung radiation therapy *Med. Phys* **40** 111721

Part IV

Advances in image-guided radiation therapy for abdominal cancer

IOP Publishing

Principles and Practice of Image-Guided Abdominal
Radiation Therapy

Yu Kuang

Chapter 15

Advances in simulation imaging for abdominal radiotherapy

Chia-ho Hua, Hui Wang and Yu Kuang

This chapter describes principles and commercial realization of advanced imaging systems for radiotherapy simulation, including dual-energy and multi-energy computed tomography (CT), magnetic resonance simulators, and digital positron emission tomography/CT. For each imaging modality, key recent advances and functionalities were highlighted to illustrate how they can be applied to radiotherapy simulation for patients with abdominal tumors. Finally, we provide example clinical applications by early adopters of advanced imaging technologies and encourage the utilization of these technologies for improving the treatment design and patient outcomes.

15.1 Introduction

This chapter describes principles of three major advanced imaging systems for radiotherapy simulation: dual-energy/multi-energy computed tomography (CT), magnetic resonance imaging (MRI) simulators, and digital positron emission tomography (PET)/CT. Currently, these new technologies are adopted by only a few radiation oncology departments (figure 15.1). However, each of these technologies offers exciting capabilities not available in conventional simulation imaging systems, i.e., single-energy (single kVp) CT, diagnostic MRI, and conventional analog PET/CT. These new capabilities and tools were recently developed and became commercially available. We introduce to readers those advances relevant to abdominal radiotherapy simulation. At the end of this chapter, selected clinical applications are highlighted to show how these new technologies can be applied to improve treatment planning and radiation delivery.

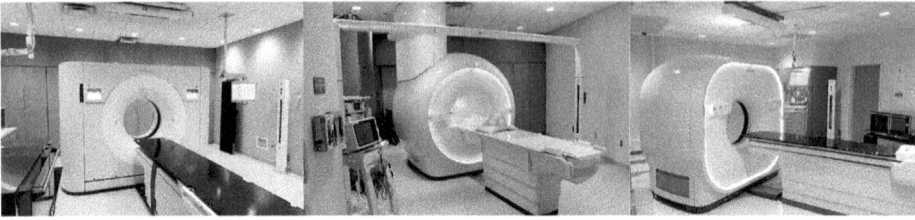

Figure 15.1. Spectral CT simulator (left), MRI simulator (middle), and digital PET/CT simulator (right) at the radiation oncology department of St. Jude Children's Research Hospital in Memphis, Tennessee.

15.2 Principles of advanced imaging systems for radiotherapy simulation

15.2.1 Principle of dual-energy CT and multi-energy CT

CT was invented in 1967 by Sir Godfrey Hounsfield. The concept of using CT to simulate radiation treatments and localize tumor volumes was proposed in 1975 by Chernak *et al.* However, the commercial availability and wide adoption of CT simulators did not occur until early 1990s. Nowadays, CT simulators have become a standard and key equipment for simulation and treatment planning in radiation oncology departments worldwide. CT measures linear attenuation coefficients of tissues, expressed as Hounsfield units (HU), which can be mapped to physical density and electron density as well as relative proton stopping power via calibration curves for radiation dose calculation.

In his 1973 paper (Hounsfield 1973), Hounsfield also described a possibility of acquiring two temporally sequential CT scans with 100 and 140 kV, from which iodine and calcium can be differentiated. This eventually led to the commercialization of the first dual-energy CT in mid 2000s. Current dual-energy CT systems employ one of the following three major designs—dual x-ray sources, rapid switching of x-ray tube potential, and multi-layer detector (McCollough 2015). The dual source technique uses two x-ray tubes approximately 90° apart to acquire high and low kVp images. Examples are Siemens SOMATOM Definition Flash, Force, and Drive (Siemens Healthineers, Erlangen, Germany). An advantage of this design is the relatively higher spectral separation which is crucial for image-based material decomposition. However, adding another pair of a bulky x-ray tube and detector assembly limits the spectral field of view to 26–35 cm, depending on the model. The second design, fast kVp switching, rapidly switches back and forth between the high and low tube potentials as the gantry rotates about the patient. An example is the GE Discovery CT 750HD (GE Healthcare, Milwaukee, Wisconsin, USA). It provides almost simultaneous acquisition of high and low kVp data, which permits material decomposition on the image domain or the projection space. However, it is technically challenging to modulate the tube current for automatic exposure control and apply different spectral filtration as it quickly rotates. The third design, dual-layer detector, maintains the traditional configuration of a single pair of x-ray tube and detector array but replaces the conventional single layer detector with a novel double layer detector. An example is Philips IQon Spectral CT (Philips Healthcare, Cleveland, Ohio, USA). It is a true simultaneous acquisition of high and low projection data, which best handles anatomic sites with internal motion. Another advantage for this type of CT is no need to decide between a dual-energy scan or a conventional scan before acquisition because spectral information is always obtained with each acquisition and can be used to retrospectively reconstruct a variety of spectral images. Finally, some may consider performing sequential CT scans with two tube potentials (dual spiral) or back-to-back axial scans at the same table position (slow kVp switching) is an alternative approach to dual-energy CT. However, severe artifacts and large errors could arise due to image misalignment from patient movement and organ motion. The scan time is much longer. Imaging the contrast flow dynamic is also extremely challenging.

Dual-energy CT produces several spectral images useful for radiotherapy planning (figure 15.2). They include virtual non-contrast images, virtual monoenergetic (monochromatic) images, iodine density images, electron density images, effective atomic number images, and proton stopping-power images. With these new types of images, one can significantly enhance the iodine contrast in tumors and blood vessels or alternatively reduce the injected iodine load for patients with compromised renal functions, reduce the metal artifact with high-energy virtual monoenergetic images, reduce range uncertainty of particle therapy using electron density and effective atomic number images, improve Monte Carlo dose calculation with better tissue assignment, and normal tissue characterization and segmentation. Detailed applications will be described in later sections.

Photon counting CT is the next generation multi-energy CT. Several prototypes already exist, and commercialization is expected within the next few years. Conventional CT and dual-energy CT use energy integrating detectors, while

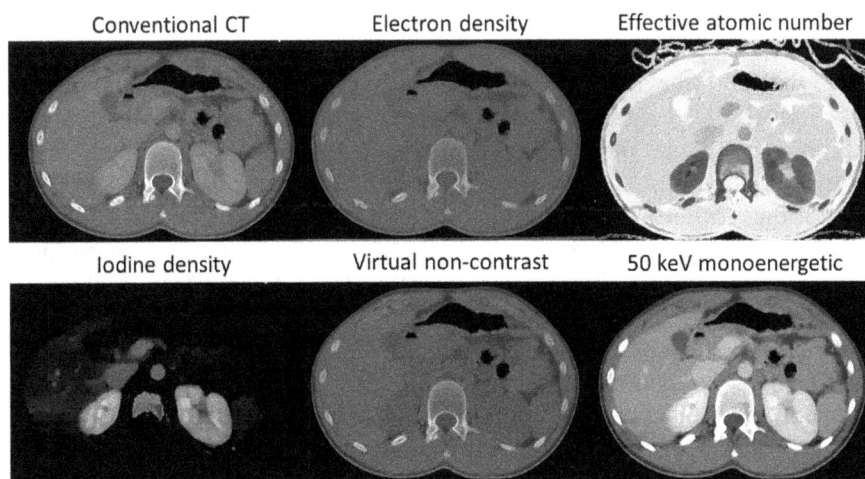

Figure 15.2. Spectral images generated from a single CT scan: conventional HU images, electron density images, effective atomic number images, iodine density map, virtual non-contrast CT, and 50 keV monoenergetic images.

photon counting CT uses energy-resolving photon counting detectors. The entire energy spectrum can be divided into multiple energy bins (windows) from which multi-energy images can be reconstructed. Multi-energy CT can also generate dual-energy imaging when two energy bins are used. Main technical challenges to overcome for photon counting CT before commercialization include cross talk between detector elements and electric pulse pileup due to insufficiently fast detector readout and charge release (Willemink 2018). Multi-energy CT offers many exciting applications, such as significantly reduced beam hardening artifacts, radiation dose reduction, higher spatial resolution, multiple contrast imaging and alternative contrast agents, improved material decomposition, and molecular CT with new types of contrast agents. Because radiation oncology researchers have limited access to photon counting CT so far, clinical applications remain to be explored. Nevertheless, we anticipate clinical impacts on tumor delineation, dose calculation accuracy, and Monte Carlo simulation with improved material decomposition.

15.2.2 Principle of MRI simulators

The value of MRI in delineating soft tissue tumors and critical organs for radio-therapy planning is well recognized. Radiotherapy planning based on MRI alone or in conjunction with CT images has been a longstanding practice for stereotactic radiosurgery. Initial efforts to install a dedicated MRI system in a radiation oncology department for radiotherapy simulation started in late 1990s with low field open systems (0.2T Hitachi MRP-20EX and 0.23T Philips Panorama) (Mizowaki 1996, Mah 2002). Despite the advantages in better patient access, reduced metal artifact, and comfort and favorable safety profiles (e.g., device heating, specific absorption rate) of resistive magnet, the signal-to-noise ratio (SNR) was low and images did not have diagnostic quality. Functional and high gradient imaging also could not be performed. Over the last decade, advances were made by MRI equipment manufacturers, and interest in MRI simulation has increased thanks to the need to better define targets for highly conformal treatments and the advent of combined MRI and accelerator systems. Many academic institutions now have installed dedicated MRI simulators in their radiation oncology departments.

Most of the current MRI simulators are high field (1.5 or 3T), wide bore (70 cm bore size), superconducting, and zero boil-off systems, which are adaptations of diagnostic MRI systems. The unique aspects of these simulators include a flat table top, a laser bridge with movable lasers for patient positioning and marking, flexible radiofrequency coils to accommodate immobilization devices, dedicated pulse sequences to acquire thin slices and high resolution images, and a higher requirement of magnetic field homogeneity to minimize the spatial distortion. Because the coils cannot interfere with the use of immobilization devices and are directly placed on the body to avoid deforming the body shape, the common use of surface coils and the increase in distance from coils to body make the images less uniform and SNR slightly lower compared to their diagnostic counterparts.

Nevertheless, the image quality is deemed adequate for routine tumor and organ delineation in radiotherapy planning (Hua 2018).

There has been a significant effort in recent years by researchers and vendors to make magnetic resonance (MR)–only simulation feasible. Traditionally, MRI and CT studies were performed back to back for radiotherapy simulation, followed by image registration to bring MR images into CT coordinates. MRI data facilitate target and organ delineation, while CT data are needed for accurate dose calculation. Because there is no integrated CT-MRI system, being able to convert MRI into CT-like images for dose calculation would eliminate the need for two imaging studies, the uncertainty introduced by image registration, radiation exposure from CT scans, and prolonged anesthesia/sedation for pediatric patients. MR-only simulation was first implemented in pelvis and brain sites where commercial solutions are available (Tyagi 2017, Bird 2019, Roberge 2020). The abdominal site is more challenging due to respiratory motion, vascular pulsation, intestinal peristalsis, air susceptibility, and radiofrequency inhomogeneity and gradient nonlinearity. Research and development are ongoing. Clinical implementations in the abdominal site are anticipated within the next few years.

15.2.3 Principle of digital PET/CT

The whole-body PET system first appeared in 1977. The hybrid system of PET and CT was introduced in 1998. The latest generation PET/CT systems, which are fully digital, began clinical use in 2018 when Philips introduced Vereos digital PET/CT (Philips Healthcare, Cleveland, Ohio, USA). Other vendors soon followed with their own digital PET/CT systems (Discovery MI, GE Healthcare, Milwaukee, Wisconsin, USA; Biograph Vision, Siemens Healthineers, Knoxville, Tennessee, USA; Cartesion Prime, Canon Medical, Otawara, Japan). Digital PET/CT replaces conventional photomultiplier tubes, which are analog vacuum devices to amplify light signals generated in scintillators, with digital silicon photomultipliers. Because of the large size and the limited timing resolution, analog photomultiplier tubes are a bottleneck for the counting performance of PET data acquisition. Digital silicon photomultipliers are arrays of single-photon avalanche photodiodes (Haemisch 2012). Each cell can detect and convert a single photon into a digital signal, thus the name 'digital photon counting.' It has a higher photon detection efficiency, faster timing, and decreased dead time. Unlike analog photomultiplier tubes in which one tube covers a large area of scintillation detector, digital silicon photomultiplier sensors are 1-to-1 coupled to scintillator elements, which improves spatial resolution. Because of above the mentioned improvements, digital PET/CT provides a significant gain in sensitivity, a reduction in scan time or injected dose, and improved detectability of small lesions when compared to analog PET/CT (Zhang 2020). Although the vast majority of digital PET/CT systems have been installed in diagnostic radiology/nuclear medicine departments, a few are in clinical use for radiotherapy simulation thanks to vendor-provided radiotherapy accessories, respiratory gating and four-dimensional (4D) imaging capability, the external laser positioning and patient marking system, and the rigid patient table compliant with the Task Group No. 66 guidelines issued by American Association of Physicists in Medicine. Some digital PET/CT systems can have a bore size as large as 78 cm. The CT portion of the PET/CT system can be used alone for CT simulation.

15.3 Recent advances in CT simulation technology

15.3.1 Metal artifact reduction

On abdominal CT images, metal artifact often appears as bright and dark streaks in/ near the spinal fusion hardware or implanted electronic devices due to photon starvation. High density objects, such as endovascular coils and the patient's arms in the beam path, could also result in streaking artifacts on reconstructed images. Metal artifact reduction (MAR) can be achieved with software algorithms and/or dual-energy CT. Software algorithms can replace corrupted projection data corresponding to metal pixels with interpolated data from adjacent uncorrupted projections. These types of MAR algorithms are commercially available, including MAR for orthopedic implants (O-MAR, Philips Healthcare, Cleveland, Ohio, USA), iterative MAR (Siemens Healthineers, Erlangen, Germany), Smart MAR (GE Healthcare, Milwaukee, Wisconsin, USA), and single-energy MAR (Canon Medical Systems, Otawara, Japan). All techniques were shown to effectively reduce metal artifacts, including the abdominal site (Jeong 2015, Sofue 2017), except for dental fillings (Chou 2020). Virtual monoenergetic images generated from dual-energy CT simulate how the object would appear when scanned with a single-energy keV x-ray beam. Studies found beam hardening artifacts can be reduced at 70–130 keV, with above 130 keV being helpful for heavier metals (Wellenberg *et al* 2018). Recently, it became feasible to combine projection-based MAR algorithms with the virtual monoenergetic imaging approach of dual-energy CT to provide a greater benefit compared to both single methods. The efficacy of the combined approach has been demonstrated for various anatomic sites, including the abdomen (Lestra 2016, Long 2019).

15.3.2 Iterative and artificial intelligence–based image reconstruction

The concept of iterative image reconstruction was first implemented for single-photon emission CT because of the need to improve image quality with low photon counts. Iterative reconstruction received more attention for transmission CT starting a decade ago thanks to the improved computational capacity and the opportunities to reduce radiation imaging dose and image noise. More recent versions of iterative CT reconstruction are model based, which incorporates accurate models of CT scanner and statistical noise. Commercial solutions include iterative model reconstruction (Philips Healthcare, Cleveland, Ohio, USA), advanced modeled iterative reconstruction (Siemens Heathineers, Erlangen, Germany), Veo (GE Healthcare, Milwaukee, Wisconsin, USA), and forward-projected model-based iterative reconstruction solution (Canon Medical Systems, Otawara, Japan). Compared with traditional filter-back projection reconstruction, model-based methods significantly improve the image quality with low radiation dose. Initial studies showed model-based reconstruction provided better or equivalent performance for auto-segmentation of soft-tissue organs (Miller 2019) compared with filter-back projection reconstruction. However, model-based iterative CT reconstruction has yet to gain wide acceptance in the radiation oncology community because of image appearance, reconstruction speed, and perceived clinical impact. Representing the most advanced CT image reconstruction is the application of artificial intelligence technology. Several vendors have received 510(K) clearance from the U.S. Food and Drug Administration for their deep learning image reconstruction engines, which produce low noise images with improved low contrast detectability from the model trained with a large library of high quality, gold standard images. Examples are GE's TrueFidelity and Canon's AiCE. One major advantage of artificial intelligence–based approaches over iterative reconstruction is the speed. It performs favorably or comparably to iterative reconstruction in terms of noise suppression and structural fidelity but reconstructs much faster for abdomen and chest regions (Shan 2019). Deep learning–based reconstruction has also been shown to produce significantly higher contrast-to-noise ratio and liver lesion detectability than hybrid iterative reconstruction on abdominal CT (Nakamura 2019). Even at submillisievert abdominopelvic CT doses, performance of deep learning–based reconstruction in image quality and lesion detection could be superior to those of filter back projection and iterative reconstruction (Singh 2020).

15.3.3 Iodine contrast enhancement and virtual non-contrast imaging

Simulation CT with intravenous contrast medium to enhance the visualization of tumor and normal anatomy is common for treating cancers in the abdominal site, such as pancreas, kidneys, liver and stomach. On conventional CT, the contrast enhancement is influenced by volume and concentration of the contrast agent, injection rate, time from injection to scanning, and the cardiac output. On dual-energy CT, the iodine contrast can be greatly enhanced on low monoenergetic images (40–50 keV). Because iodine has a K-edge of 33.2 keV, photoelectric attenuation increases when the incident photon energy approaches the K-edge, resulting in increased HU. The gain in iodine contrast enhancement may improve the tumor delineation or allow the iodine contrast dose to be reduced in patients with compromised renal function. A recent study on blood vessel enhancement for radiotherapy planning found that the same contrast-to-noise ratio as 120 kVp images could be achieved with a 2.5-fold lower iodine dose on 50 keV images and 4-fold lower on 40 keV if dual-energy CT is used (Tsang 2017). A quantitative map of iodine concentration and virtual non-contrast (unenhanced) CT images can also be created from dual-energy CT. To avoid incorrectly calculating dose distribution due to the presence of iodine on post-contrast CT, CT simulation for treatment planning typically requires a separate CT scan before contrast administration. This extra scan could be eliminated if the non-contrast CT can be synthesized from the single post-contrast scan. Dual-energy CT makes this possible by subtracting the iodine content from each pixel after material decomposition. Differences in HU were generally small between virtual non-contrast and true non-contrast images, but iodine removal could be insufficient in blood vessels with very high iodine concentrations and excessive in trabecular bones (Borhani 2017, Sauter 2018). Improvement is needed to be applicable for radiotherapy beams traversing the bones.

15.3.4 Direct electron density generation

Mapping HU into electron density via a calibration curve in the treatment planning system is the standard approach to obtain electron density images for dose calculation in photon therapy. Because the calibration curve is sensitive to the choice of tube potential kVp and only one calibration curve for each CT scanner is often allowed in commercial treatment planning systems, radiation oncology departments generally require their patients to be always scanned at 120 kVp with fixed scanner settings, which is restrictive. Dual-energy CT extracts photoelectric-like and Compton scatter–like components of the imaged object from low and high energy projection data, from which relative electron density and effective atomic number images can be reconstructed based on a linear combination of two components. This tool eliminates the need of HU-electron density calibration curve and provides the freedom to optimize imaging protocols. Phantom studies have shown an accuracy of 1% or better for electron density images directly generated from dual-energy CT (Hua 2018, Ohira 2018), although one study found an inferior accuracy for dual-energy CT with the split filter design (Almeida 2017). A new algorithm for single-energy CT (DirectDensity, Siemens Healthineers, Erlangen, Germany) can also reconstruct projection data directly proportional to relative electron density images. Studies showed that the dose differences between the radiotherapy plans designed on the standard 120 kVp CT images and the direct electron density images were less than 1% (van der Heyden 2017). Therefore, directly generated electron density can replace traditional 120 kVp images for dose calculation. Whether the image quality and tissue contrast of electron density are adequate for target and critical organ delineation remains to be determined.

15.3.5 Proton stopping-power imaging

For proton therapy, relative proton stopping-power images are needed for dose calculation. Like relative electron density images for photon therapy, the standard approach is to convert them in treatment planning systems via a calibration curve. Because dual-energy CT can produce electron density and effective atomic number images, proton stopping-power images can be easily calculated based on the Bethe formula. Many studies have experimentally verified the accuracy of stopping-power calculation from dual-energy CT data, showing improved proton range estimation than the conventional approach with single-energy CT (Hünemohr 2014, Möhler 2018, Bär 2018). Another approach to produce the proton stopping-power imaging for treatment planning is proton CT. Proton CT uses high energy protons (typically >200 MeV) passing through the patient as opposed to x-rays in diagnostic CT scanners to reconstruct proton stopping-power tomographic images. Proton CT detectors track positions and directions of individual protons before and after passing through the patient. The reconstructed images directly show how protons interact with tissues, thus reducing the proton range uncertainty from 3.5% with single-energy x-ray CT to ≤1%. Recent phantom experimental results on prototype proton CT performed slightly better than dual-energy CT in terms of relative proton stopping-power accuracy, although both were excellent (Dedes 2019). Low-dose pre-treatment proton CT can be accurately registered to planning proton CT (Cassetta 2019), opening the possibility of image guidance with volumetric proton imaging. Although direct proton stopping-power imaging from both dual-energy CT and proton CT have not yet become routinely used in proton therapy centers, we anticipate clinical implementation to occur within the next 5–10 years.

15.4 Recent advances in MR simulation technology

15.4.1 Imaging in treatment positions

To be able to acquire MRI for radiotherapy patients in treatment positions requires more than simply aligning patient setup marks to room lasers. For consistency, a flat tabletop or overlay is necessary to simulate the flat treatment couch top. Notched style indexing on the tabletop would allow patient immobilization devices to attach. Because the increased distance from the posterior radiofrequency coils underneath the tabletop to the patient's posterior surface, researchers are developing coil arrays that can be embedded into the tabletop or fitted between the patient and the immobilization device to increase the SNR (Tyagi 2020). Instead of directly strapping the anterior surface coils to the patient, which may deform the anatomy, these coils are suspended in air on top of the patient and supported by a rigid coil bridge or arch while maintaining a small distance from the patient (figure 15.3). These anterior coils are also being designed to be applicable for MR-guided radiotherapy, in which the coil elements uniformly and minimally attenuate the therapeutic megavoltage beams (Hoogcarspel 2018). Another advance in patient positioning is the introduction of in-bore optical camera for accelerated patient

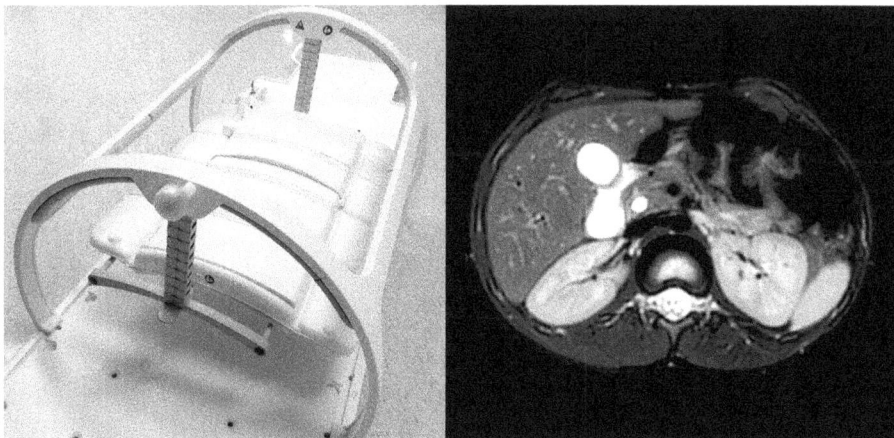

Figure 15.3. Left: photograph of a commercial torso coil supported by the coil bridge, allowing 60 cm coverage. Patients can be scanned in their treatment positions. Right: abdominal MRI acquired using the suspended torso coil and a T2-weighted fat-suppressed MultiVane XD sequence for reducing the motion effect.

setup, motion detection, motion correction, and respiratory motion management. Although this technology was initially developed for diagnostic MRI, impacts are expected on improving the patient setup for repeated MRI (e.g., on-treatment MRI for adaptive replanning) and for establishing the relationship between internal tumor motion and change in external surface in surface-guided radiation therapy.

15.4.2 4D MRI

Respiratory-induced tumor and organ motion is a major challenge to overcome when irradiating abdominal and thoracic tumors with highly conformal radio-therapy techniques. For lung cancers, 4D CT (or respiratory-correlated CT) is helpful in designing internal target margin to compensate for the tumor motion or selecting specific phases of the respiratory cycle for respiratory gated irradiation. For soft-tissue tumors in the abdomen, the ability to visualize their motion patterns with MRI is highly critical for designing effective treatment plans. A decade of research on 4D MRI has resulted in many novel solutions, commercialization, and clinical applications. 4D MRI for radiation oncology can be categorized as respiratory-correlated and time-resolved techniques (Li 2019). The former technique reconstructs data acquired on a free-breathing patient over multiple breathing cycles into eight or ten three-dimensional (3D) image sets, each representing a specific phase of a breathing cycle. It can be prospective gated acquisition or retrospective data binning (Uh *et al* 2019) (figure 15.4). Respiratory-correlated 4D MRI presents average motion over a few minutes of imaging time by assuming the motion periodicity. In contrast, the technique of time-resolved 4D MRI assesses motion over multiple breathing cycles and is more immune to breathing irregularity than respiratory-correlated 4D MRI. Because the spatial resolution is low due to the physical limitation of frame rate, prior high-resolution breath hold MRI is needed to be deformed and mapped to each free breathing acquisition (Li 2017). Alternatively, high-resolution library matching/view sharing (Yip 2017) is used to enhance the image quality. Time-resolved 4D MRI is often performed during treatment on MR-guided linear accelerator systems for motion tracking and treatment adaptation.

15.4.3 Compressed sensing MRI

Sensitivity encoding for fast MRI is a parallel imaging technique to simultaneously acquire multi-slice MRI using multi-array coils. It has been routinely utilized for two decades. The commercial availability of compressed sensing, a new technique to accelerate MRI acquisition, is more recent. It has been implemented for radio-therapy planning and tumor tracking (Dietz 2017, Gao 2018). Compressed sensing acquires and reconstructs images of severely under-sampled k-space data (below the Nyquist rate), thus reducing scan time, with quality almost equivalent to the fully sampled acquisition. Vendors have claimed up to 50% reduction in scanning time compared with those acquired without compressed sensing. Users reported 14%–41% (average 24%) time saving without compromising diagnostic image quality (Sartoretti 2019). Compressed sensing can be combined with parallel imaging to reduce the number of samples and applied to a variety of pulse sequences.

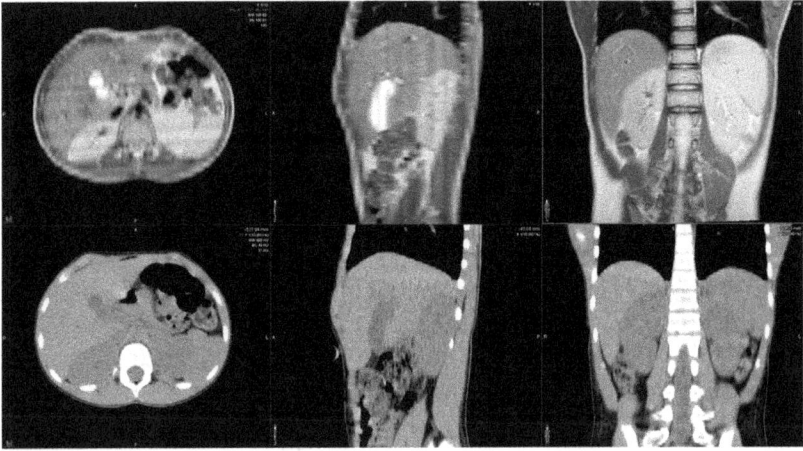

Figure 15.4. Example of prospective gating 4D MRI (top) acquired using a work-in-progress tool before commercial release on a 1.5T MR simulator. The 4D MRI acquisition was triggered by the internal navigator. The 4D CT images (bottom) of the same patient are shown for comparison. The respiratory signal for data binning was obtained from the pneumatic bellows.

Compressed sensing MRI is particularly attractive for abdominal imaging because it allows breath-hold 3D imaging and free-breathing dynamic MRI. Compressed sensing can also be used to produce higher spatial resolution in a reasonable time frame. More detailed descriptions of compressed sensing for abdominal MRI can be found in these publications (Feng 2017, Yoon 2019, Serai 2020).

15.4.4 MR-based synthetic CT

MR-based synthetic CT is commercially available for pelvis and brain sites (MRCAT, Philips Healthcare, Best, The Netherlands; Synthetic CT, Siemens Healthineers, Erlangen, Germany; MriPlanner, Spectronic Medical, Helsingborg, Sweden). The developmental work is ongoing for abdominal MRI. Converting MRI into CT-like images facilitates MR-only simulation without additional CT scanning for treatment planning and dose calculation. The synthetic CT can also be used as a reference dataset for pre-treatment image guidance to be registered with on-board cone beam CT. Research on converting MR to CT images has been conducted for more than a decade, but the last 5 years have seen a renewed interest and commercialization thanks in part to the advent of combined MR-linear accelerator systems. Such systems take advantage of the on-board MRI capability to better localize and track soft tissue tumors during treatment. It is particularly helpful for treating abdominal tumors, such as pancreatic and liver cancers. For MR-guided radiotherapy, performing MR simulation is logical to maximize the benefits of MRI and be consistent from simulation, planning, to treatment. Recently, deep learning has been applied to the generation of synthetic CT from abdominal MRI with great success (Liu 2019, Qian 2020, Fu 2020). Given encouraging results in research studies and the clinical needs, we anticipate MR-based synthetic CT for abdomen to be commercially available soon.

15.4.5 Automatic segmentation

Automatic MRI image segmentation, driven by artificial intelligence, is a recent major advance which impacts both radiation oncology and radiology departments. Although it is not strictly tied to the MR simulation, the efficiency of the workflow from MR simulation to treatment planning is significantly improved. Many publications reported excellent results for segmenting abdominal organs and tumors, including pancreas, liver, spleen, gallbladder, kidneys, and blood vessels (Lenchik 2019, Chen 2019, Cardenas 2019, Liang 2020, CHAOS 2020). The new deep learning–based approaches are considered more time efficient than previous approaches, such as atlas-based segmentation. All major vendors had recently released artificial intelligence segmentation tools, although most are for brain tissue segmentation. There is a trend to apply artificial intelligence techniques to convert a single rapid quantitative MRI scan to multi-contrast MRI from which automatic segmentation is performed. One commercial example is SyntheticMR (Linköping, Sweden) which partners with Siemens Healthineers, Philips Healthcare, and Sectra Medical to generate different contrast images from a 6 min brain scan for subsequent automatic segmentation. GE Healthcare also developed a similar product called MAGiC, which has been applied to brain and prostate MRI.

15.5 Recent advances in PET imaging for guiding radiotherapy

15.5.1 Correcting respiratory motion

Respiratory motion results in blurry PET images, which makes the boundary of lesion hard to determine and affects the accuracy of standard uptake value quantification. Respiratory gating and 4D PET/CT are two available techniques to mitigate the respiratory motion effect on image quality (Aristophanous 2012, Huang 2014). However, the increased acquisition time and the lack of robustness of the external gating device prevent these motion management techniques from being fully utilized in the clinical setting. A new type of data-driven technique is now available without the need of external gating device and can use 100% of the collected counts from the free-breathing patients (Thielemans 2011, Walker 2018, Walker 2020, Buther 2020). An example 20 s segment of the gating trace for a moving phantom shows the visual match between a device-based motion tracking and a data-driven motion tracking, and the blurring of PET images was largely recovered using data-driven or device-based motion management techniques (figure 15.5). An example data-driven technique performs principal component analysis on dynamic sinogram data to extract the respiratory signal during PET acquisition (Walker 2020). Another example deduces the respiratory information by calculating the center of mass of the true coincidence distribution from the list mode data (Buther 2016). Commercial solutions of data-driven deviceless motion correction include MotionFreeze (GE, Milwaukee, Wisconsin, USA) and OncoFreeze AI (Siemens Healthineers, Erlangen, Germany).

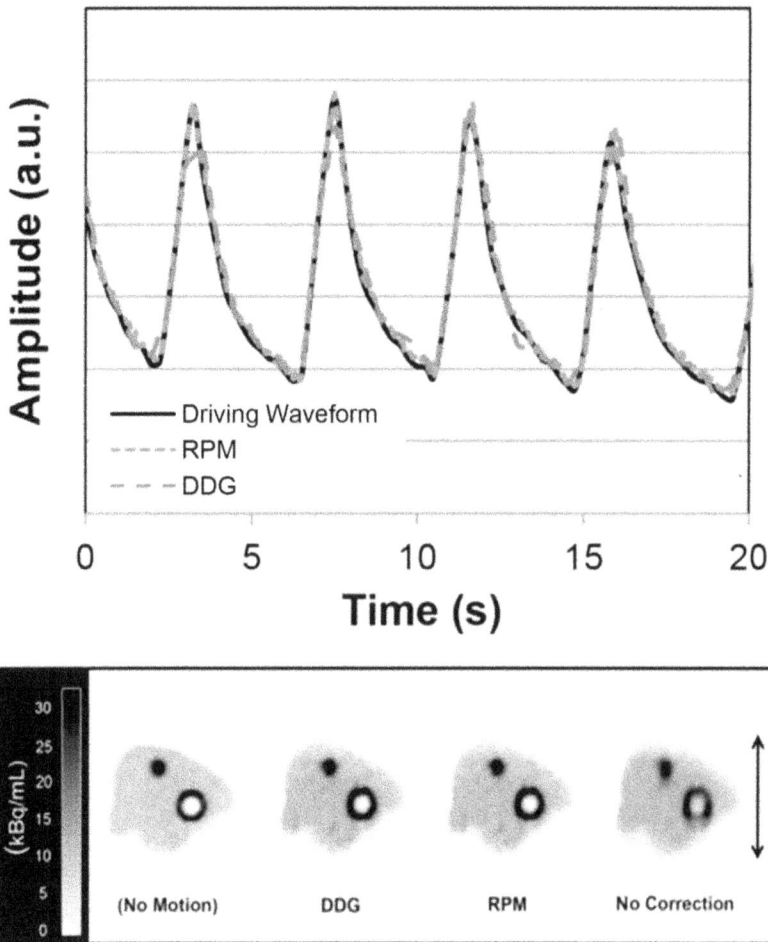

Figure 15.5. An example segment of the gating trace for an Abdo-Man phantom with motion (top) and a representative coronal slice from PET images of the phantom with/without motion management (bottom). DDG: data-driven gating; RPM: Real-Time Position Management™ system. (Reproduced from Walker *et al* 2018.)

15.5.2 On-board PET imaging for biology-guided radiotherapy

Until 2020, PET scans for guiding radiotherapy planning have been exclusively performed on diagnostic PET/CT scanners or PET/CT simulators in radiation oncology departments. A new type of hybrid compact linear accelerator, fan beam CT, and PET on a closed ring structure (X1, RefleXion, Hayward, California, USA) (figure 15.6) became recently available for acquiring planning PET and on-treatment fractional PET on a non-conventional system. The fractional PET consists of a pre-treatment scan and a continuous scan throughout treatment delivery using two 90° PET arc detectors (Fan 2012). The initial design of the hybrid system with the capability of on-board PET imaging and tumor tracking radiotherapy was

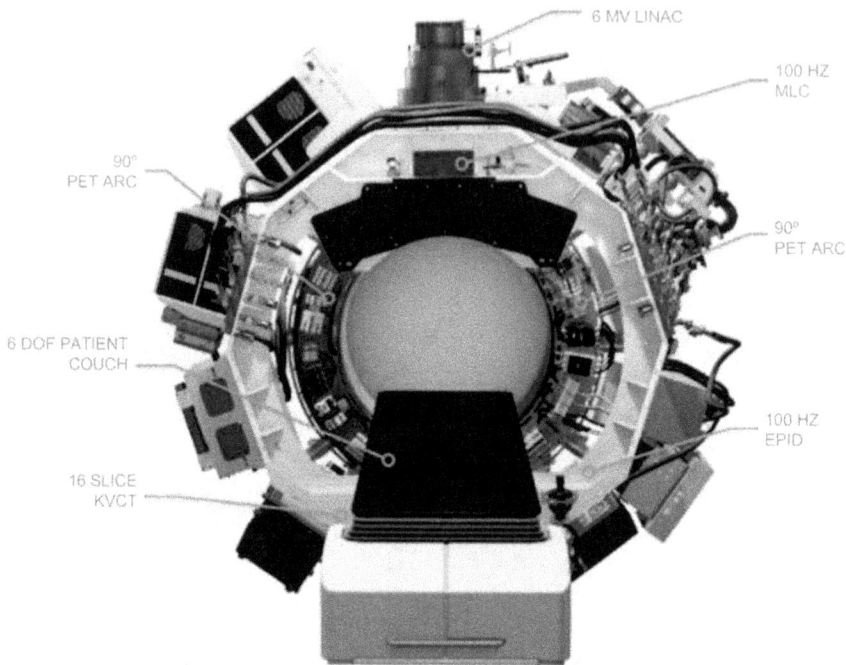

Figure 15.6. The RefleXion™ X1 machine without cover. DOF: degree of freedom; EPID: electronic portal imaging device; KVCT: kilovoltage computed tomography; LINAC: linear accelerator; MLC: multi-leaf collimator; MV: megavolt. (Reproduced from Shirvani *et al* 2021.)

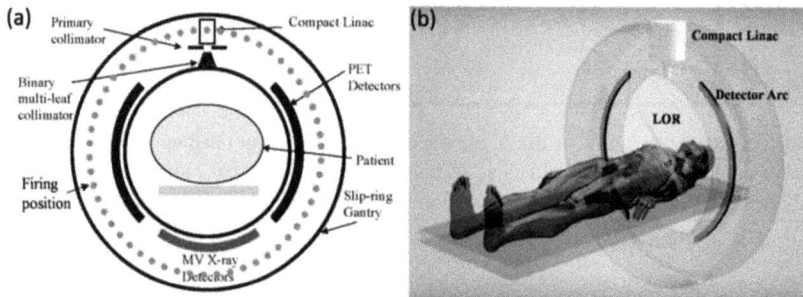

Figure 15.7. The initial design of the hybrid system with the capability of on-board PET imaging and tumor tracking radiotherapy. (a) Cross-sectional diagram of the hybrid system. (b) Snapshot of a line of response (LOR) that directs a beamlet of therapeutic radiation through the emission site inside a tumor. (Reproduced from Fan *et al* 2013.)

illustrated in figure 15.7 (Fan 2012, 2013). The quick pre-treatment scan only acquires limited photon counts and intends to verify the location of tracer accumulation in tumor. Developments are ongoing to utilize the emission photons to track the moving tumors in free-breathing patients (Fan 2013). Applications to abdominal tumors have not yet been reported but are expected.

15.6 Clinical applications of advanced imaging for abdominal radiotherapy

Although some of the advanced technologies have only arrived recently, many institutions have already incorporated them into patient care and shared the initial experience. Dual-energy CT improves conspicuity of the abdominal lesions, such as hypervascular hepatocellular carcinoma, cancer metastases in the liver, pancreatic cancer, lesions in kidneys and adrenal glands. Low energy monoenergetic images enhance the iodine contrast, which improves the visualization and delineation of tumors and adjacent vasculature for treatment planning (Agrawal 2014). Many studies reported that pancreatic tumor contrast and contrast-to-noise ratio significantly improved using virtual monoenergetic images from dual-energy CT, which could lead to more accurate tumor delineation for radiation treatment planning (Di Maso 2018, Noid 2018). Iodine density images of dual-energy CT can be used to identify functional liver and lungs for sparing in stereotactic body radiotherapy (Lapointe 2017, Ohira 2020). Marrow fat fraction derived from dual-energy CT was found to be highly correlated with that from histology samples, thus offering a new tool to allow functional marrow irradiation (Magome 2016). The impact of iodine contrast on dose calculation in proton therapy was mitigated with DECT using two methods (direct proton stopping-power images and virtual non-contrast scans) (Lalonde 2019). It is feasible to use a single post-contrast dual-energy CT scan for accurate dose calculation in photon and proton therapy of pediatric abdominal tumors (Ates 2020).

Advanced MR technologies also provide valuable tools for target delineation and margin design. 4D MRI has been used to assess tumor and organ motion in children and adults who receive radiation therapy (Uh *et al* 2017, Stemkens 2018, Li 2019) as well as the interplay effect for pencil beam proton therapy to abdominal tumors (Boria 2018, Dolde 2019). MR simulation and MR-guided adaptive radiotherapy for liver and pancreatic cancer are increasingly performed (Henke 2018, Witt 2020). Pulse sequences optimized for patients with abdominal cancers receiving MR simulation were reported (Paulson 2015, Hua 2018). Compressed sensing has started to become a routine clinical tool. Institutions have demonstrated on MR simulators that it can be combined with parallel imaging to accelerate the image acquisition with negligible changes in the ability to contour structures for radiotherapy planning. Combining compressed sensing, parallel imaging, and golden-angle radial sampling has made the dynamic contrast-enhanced liver MRI studies a clinical routine in free-breathing patients (Feng 2014). This tool helps assess tumor function for treatment planning and response during and after radiotherapy. MR-derived synthetic CT using a single MR pulse sequence showed promises for MRI-based liver stereotactic body radiotherapy (Bredfeldt 2017, Liu 2019). Multi-organ, auto-contouring methods have also been developed to facilitate online adaptive radiotherapy in the abdomen (Liang 2018).

Recent advances in PET technologies may further enhance the utility of PET imaging for radiotherapy guidance in the clinic with new radiotracers introduced.

Respiratory motion causes blurred PET images, especially notable during the upper abdomen imaging, which makes the PET-based delineation of abdominal treatment targets much more challenging. The data-driven gating (DDG) methods have the potential to provide improved mitigation of respiration motion given that it accounts for the 3D motion of radioactivity within internal organs compared with device-based gating methods with simple tracking of the rise and fall of the chest wall (Rietzel 2005, Hoisak 2004).

The external devices (e.g., optics camera or pressure-sensitive belt) used to track the motion of the chest are suboptimal due to the associated cost, the discomfort for the patient, the added dose on the staff for dealing with the patient, and so on. The removal of a camera-based or pressure-belt–based tracking system would facilitate an efficient workflow for PET used in radiotherapy without needing additional time for setting up the devices. In a recent clinical study, 56 patients with suspected malignancies in the abdomen or thorax who underwent whole-body ^{18}F-FDG PET with continuous-bed-motion acquisition were used to investigate the utility of a commercialized data-driven respiratory gating algorithm. The commercialized data-driven respiratory gating algorithm demonstrated the similar performance with device-based gating (Buther 2020). In another clinical study, Walker *et al* performed the first large-scale clinical evaluation of a commercialized principal component analysis–based DDG algorithm with a total of 144 whole-body ^{18}F-FDG PET/CT examinations. The study demonstrated that the data-driven respiratory gating outperformed device-based gating for clinical ^{18}F-FDG PET/CT (Walker 2020). An example set of PET images of liver metastasis from this study is presented in figure 15.8. The clinical widespread use of data-driven respiratory gating methods in radiation oncology departments worldwide is expected in the next 5 years as some cancer institutions have integrated this novel technique in their radiation therapy practices.

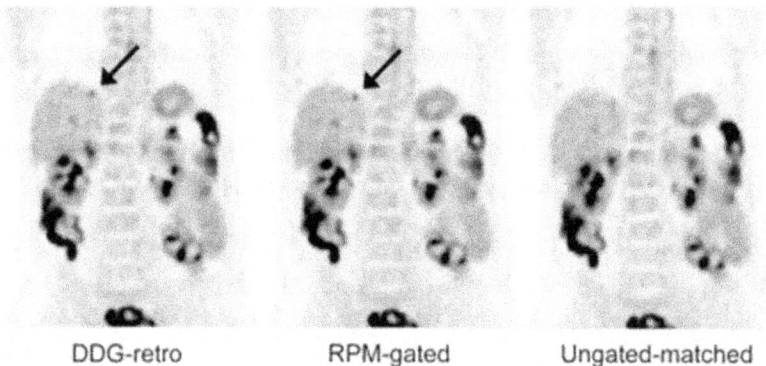

DDG-retro   RPM-gated   Ungated-matched

Figure 15.8. An example set of PET images of liver metastasis with DDG- and RPM-gated respiratory motion mitigation and without motion management. Images are on SUV grayscale of 0–6. DDG: data-driven gating; RPM: Real-Time Position Management™ system. (Adapted from Walker *et al* 2020.)

15.7 Conclusion and outlook

Because of significant advances in technology, computer power, and artificial intelligence capability, we are seeing many exciting developments in the field of medical imaging which impact simulation and delivery of radiotherapy. The general trend is to provide more accurate and informative images and complete the simulation tasks much more efficiently. Automation and more intelligent use of information have received increasing attention in the past decade. It may still take 5–10 years before we see widespread utilization of these highlighted CT and MRI technologies. Many functionalities still require more research developments and clinical testing for the abdominal site. Nevertheless, there are ample opportunities for clinicians and physicists working in research and clinical environments to continuously advance technologies, to adopt technologies that improve the treatment efficiency and patient outcomes, and to promote patient safety while embracing new technologies.

References

Agrawal M D, Pinho D F, Kulkarni N M, Hahn P F, Guimaraes A R and Sahani D V 2014 Oncologic applications of dual-energy CT in the abdomen *Radiographics* **34** 589–612

Almeida I P, Schyns L E J, Öllers M C, van Elmpt W, Parodi K, Landry G, Verhaegen F and Dual-energy C T 2017 Quantitative imaging: a comparison study between twin-beam and dual-source CT scanners *Med. Phys.* **4491** 171–9

Aristophanous M, Berbeco R I, Killoran J H, Yap J T, Sher D J, Allen A M, Larson E and Chen A B 2012 Clinical utility of 4D FDG-PET/CT scans in radiation treatment planning *Int. J. Radiat. Oncol. Biol. Phys.* **82** e99–105

Ates O, Hua C, Zhao L, Shapira N, Yagil Y, Merchant T E and Krasin M J 2020 Feasibility of using post-contrast dual-energy CT for pediatric treatment planning and dose calculation *Br. J. Radiol.* 20200170

Bär E, Lalonde A, Zhang R, Jee K-W, Yang K, Sharp G, Liu B, Royle G, Bouchard H and Lu H-M 2018 Experimental validation of two dual-energy CT methods for proton therapy using heterogeneous tissue samples *Med. Phys.* **45** 45–59

Bird D, Henry A M, Sebag-Montefiore D, Buckley D L, Al-Qaisieh B and Speight R 2019 A systematic review of the clinical implementation of pelvic magnetic resonance imaging-only planning for external beam radiation therapy *Int. J. Radiat. Oncol. Biol. Phys.* **105** 479–92

Borhani A A, Kulzer M, Iranpour N, Ghodadra A, Sparrow M, Furlan A and Tublin M E 2017 Comparison of true unenhanced and virtual unenhanced (VUE) attenuation values in abdominopelvic single-source rapid kilovoltage-switching spectral CT *Abdom. Radiol* **42** 710–7

Boria A, Uh J, Pirlepesov F, Stuckey J C, Axente M, Gargone M A and Hua C 2018 Interplay effect of target motion and pencil-beam scanning in proton therapy for pediatric patients *Int. J. Part. Ther* **5** 1–10

Bredfeldt J S, Liu L, feng M, Cao Y and Balter J M 2017 Synthetic CT for MRI-based liver stereotactic body radiotherapy treatment planning *Phys. Med. Biol.* **62** 2922–34

Büther F, Jones J, Seifert R, Stegger L, Schleyer P and Schäfers M 2020 Clinical evaluation of a data-driven respiratory gating algorithm for whole-body PET with continuous bed motion *J. Nucl. Med.* **61** 1520–7

Büther F, Vehren T, Schafers K P and Schäfers M 2016 Impact of data-driven respiratory gating in clinical PET *Radiology* **281** 229–38

Cardenas C E, Yang J, Anderson B M, Court L E and Brock K B 2019 Advances in auto-segmentation *Semin. Radiat. Oncol.* **29** 185–97

Cassetta R, Piersimoni P, Riboldi M, Giacometti V, Bashkirov V, Baroni G, Ordonez C, Coutrakon G and Schulte R 2019 Accuracy of low-dose proton CT image registration for pretreatment alignment verification in reference to planning proton CT *J. Appl. Clin. Med. Phys* **20** 83–90

CHAOS 2020 CHAOS–Combined (CT-MR) Healthy Abdominal Organ Segmentation https://chaos.grand-challenge.org/

Chen Y, Ruan D, Xiao J, Wang L, Sun B, Saouaf R, Yang W, Li D and Fan Z 2020 Fully automated multiorgan segmentation in abdominal magnetic resonance imaging with deep neural networks *Med. Phys.* **47** 4971–82

Chernak E S, Rodriguez-Antunez A, Jelden G L, Dhaliwal R S and Lavik P S 1975 The use of computed tomography for radiation therapy treatment planning *Radiology* **117** 613–4

Chou R, Chi H-Y, Lin Y-H, Ying L-K, Chao Y-J and Lin C-H 2020 Comparison of quantitative measurements of four manufacturer's metal artifact reduction techniques for CT imaging with a self-made acrylic phantom *Technol. Health Care* **28** 273–87

Dedes G, Dickmann J, Niepel K, Wesp P, Johnson RP, Pankuch M, Bashkirov V, Rit S, Volz L and Schulte RW *et al* 2019 Experimental comparison of proton CT and dual energy x-ray CT for relative stopping power estimation in proton therapy *Phys. Med. Biol.* **64** 165002

Dietz B, Yip E, Yun J, Fallone B G and Wachowicz K 2017 Real-time dynamic MR image reconstruction using compressed sensing and principal component analysis (CS-PCA): demonstration in lung tumor tracking *Med. Phys.* **44** 3978–89

Dolde K, Zhang Y, Chaudhri N, Dávid C, Kachelrieß M, Lomax A J, Naumann P, Saito N, Weber D C and Pfaffenberger A 2019 4DMRI-based investigation on the interplay effect for pencil beam scanning proton therapy of pancreatic cancer patients *Radiat. Oncol.* **14** 30

Fan Q, Nanduri A, Mazin S and Zhu L 2012 Emission guided radiation therapy for lung and prostate cancers: a feasibility study on a digital patient *Med. Phys.* **39** 7140–52

Fan Q, Nanduri A, Yang J, Yamamoto T, Loo B, Graves E, Zhu L and Mazin S 2013 Toward a planning scheme for emission guided radiation therapy (EGRT): FDG based tumor tracking in a metastatic breast cancer patient *Med. Phys.* **40** 081708

Feng L, Benkert T, Block K T, Sodickson D K, Otazo R and Chandarana H 2017 Compressed sensing for body MRI *J. MRI* **45** 966–87

Feng L, Grimm R, Block K T, Chandarana H, Kim S, Xu J, Axel L, Sodickson D K and Otazo R 2014 Golden-angle radial sparse parallel MRI: combination of compressed sensing, parallel imaging, and golden-angle radial sampling for fast and flexible dynamic volumetric MRI *Magn. Reson. Med.* **72** 707–17

Fu J, Singhrao K, Cao M, Yu V, Santhanam A P, Yang Y, Guo M, Raldow A C, Ruan D and Lewis J H 2020 Generation of abdominal synthetic CTs from 0.35T MR images using generative adversarial networks for MR-only liver radiotherapy *Biomed. Phys. Eng. Express* **6** 015033

Gao Y *et al* 2018 Accelerated 3D bSSFP imaging for treatment planning on an MRI-guided radiotherapy system *Med. Phys.* **45** 2595–602

Haemisch Y, Frach T, Degenhardt C and Thon A 2012 Fully digital arrays of silicon photomultipliers (dSiPM)—a scalable alternative to vacuum photomultiplier tubes (PMT) *Phys. Procedia* **37** 1546–60

Henke L *et al* 2018 Phase I trial of stereotactic MR-guided online adaptive radiation therapy (SMART) for the treatment of oligometastatic or unresectable primary malignancies of the abdomen *Radiother. Oncol.* **126** 519–26

van der Heyden B, Öllers M, Ritter A, Verhaegen F and van Elmpt W 2017 Clinical evaluation of a novel CT image reconstruction algorithm for direct dose calculations *Phys. Imag. Radiat. Oncol.* **2** 11–6

Hoisak J D P, Sixel K E, Tirona R, Cheung P C F and Pignol J-P 2004 Correlation of lung tumor motion with external surrogate indicators of respiration *Int. J. Radiat. Oncol. Biol. Phys.* **60** 1298–306

Hoogcarspel S J, Zijlema S E, Tijssen R H N, Kerkmeijer L G W, Jürgenliemk-Schulz I M, Lagendijk J J W and Raaymakers B W 2018 Characterization of the first RF coil dedicated to 1.5 T MR guided radiotherapy *Phys. Med. Biol.* **63** 025014

Hounsfield G N 1973 Computerized transverse axial scanning (tomography). Description of system *Br. J. Radiol.* **46** 1016–22

Hua C, Shapira N, Merchant T E, Klahr P and Yagil Y 2018 Accuracy of electron density, effective atomic number, and iodine concentration determination with a dual-layer dual-energy computed tomography system *Med. Phys.* **45** 2486–97

Hua C, Uh J, Krasin M J, Lucas J T Jr, Tinkle C L, Acharya S, Smith H, Kadbi M and Merchant T 2018 Clinical implementation of magnetic resonance imaging systems for simulation and planning of pediatric radiation therapy *J. Med. Imag. Radiat. Sci.* **49** 153–63

Huang T C, Chou K T, Wang Y C and Zhang G 2014 Motion freeze for respiration motion correction in PET/CT: a preliminary investigation with lung cancer patient data *BioMed. Res. Int.* **2014** 167491

Hünemohr N, Krauss B, Tremmel C, Ackermann B, Jäkel O and Greilich S 2014 Experimental verification of ion stopping power prediction from dual energy CT data in tissue surrogates *Phys. Med. Biol.* **59** 83–96

Jeong S, Kim S H, Hwang E J, Shin C-I, Han J K and Choi B I 2015 Usefulness of a metal artifact reduction algorithm for orthopedic implants in abdominal CT: phantom and clinical study results *AJR Am. J. Roentgenol.* **204** 307–17

Lalonde A, Xie Y, Burgdorf B, O'Reilly S, Ingram W S, Yin L, Zou W, Dong L, Bouchard H and Teo B-K K 2019 Influence of intravenous contrast agent on dose calculation in proton therapy using dual energy CT *Phys. Med. Biol.* **64** 125024

Lapointe A, Bahig H, Blais D, Bouchard H, Filion E, Carrier J-F and Bedwani S 2017 Assessing lung function using contrast-enhanced dual-energy computed tomography for potential applications in radiation therapy *Med. Phys.* **44** 5260–9

Lenchik L *et al* 2019 Automated segmentation of tissues using CT and MRI: a systematic review *Acad. Radiol.* **26** 1695–706

Lestra T, Mulé S, Millet I, Carson-Vu A, Taourel P and Hieffel C 2016 Applications of dual energy computed tomography in abdominal imaging *Diagn. Interv. Imaging* **97** 593–603

Li G, Liu Y and Nie X 2019 Respiratory-correlated (RC) vs. time-resolved (TR) four-dimensional magnetic resonance imaging (4DMRI) for radiotherapy of thoracic and abdominal cancer *Front. Oncol.* **9** 1024

Li G, Wei J, Kadbi M, Moody J, Sun A and Zhang S *et al* 2017 Novel super-resolution approach to time-resolved volumetric 4-dimensional magnetic resonance imaging with high spatio-temporal resolution for multi-breathing cycle motion assessment *Int. J. Radiat. Oncol. Biol. Phys.* **98** 454–62

Liang F, Qian P, Su K-H, Baydoun A, Hedent SV, Kuo J-W, Zhao K, Parikh P, Traughber BJ and Muzic RF Jr 2018 Abdominal, multi-organ, auto-contouring method for online adaptive magnetic resonance guided radiotherapy: an intelligent, multi-level fusion approach *Artif. Intell. Med.* **90** 34–41

Liang Y, Schott D, Zhang Y, Wang Z, Nasief H, Paulson E, Hall W, Knechtges P, Erickson B and Li X A 2020 Auto-segmentation of pancreatic tumor in multi-parametric MRI using deep convolutional neural networks *Radiother. Oncol.* **145** 193–200

Liu Y, Lei Y, Wang T, Kayode O, Tian S, Liu T, Patel P, Curran W J, Ren L and Yang X 2019 MRI-based treatment planning for liver stereotactic body radiotherapy: validation of a deep learning-based synthetic CT generation method *Br. J. Radiol.* **92** 20190067

Long Z, DeLone D R, Kotsenas A L, Lehman V T, Nagelschneider A A, Michalak G J, Fletcher J G, McCollough C H and Yu L 2019 Clinical assessment of metal artifact reduction methods in dual-energy CT examinations of instrumented spines *AJR Am. J. Roentgenol.* **212** 395–401

Magome T *et al* 2016 Evaluation of functional marrow irradiation based on skeletal marrow composition obtained using dual-energy computed tomography *Int. J. Radiat. Oncol. Biol. Phys.* **96** 679–87

Mah D, Steckner M, Palacio E, Mitra R, Richardson T and Hanks G E 2002 Characteristics and quality assurance of a dedicated open 0.23 T MRI for radiation therapy simulation *Med. Phys.* **29** 2541–7

Di Maso L D, Huang J, Bassetti M F, DeWerd L A and Miller J R 2018 Investigating a novel split-filter dual-energy CT technique for improving pancreas tumor visibility for radiation therapy *J. Appl. Clin. Med. Phys.* **19** 676–83

McCollough C H, Leng S, Yu L and Fletcher J G 2015 Dual- and multi-energy CT: principles, technical approaches, and clinical applications *Radiology* **276** 637–53

Miller C *et al* 2019 Impact of CT reconstruction algorithm on auto-segmentation performance *Med. Phys.* **20** 95–103

Mizowaki T, Nagata Y, Okajima K, Murata R, Yamamoto M and Kokubo M *et al* 1996 Development of an MR simulator: experimental verification of geometric distortion and clinical application *Radiology* **199** 855–60

Möhler C, Russ T, Wohlfahrt P, Elter A, Runz A, Richter C and Greilich S 2018 Experimental verification of stopping-power prediction from single- and dual-energy computed tomography in biological tissues *Phys. Med. Biol.* **63** 025001

Nakamura Y, Higaki T, Tatsugami F, Zhou J, Yu Z, Akino N, Ito Y, Iida M and Awai K 2019 Deep learning-based CT image reconstruction: initial evaluation targeting hypovascular hepatic metastases *Radiol. Artif. Intell.* **1** e180011

Noid G, Tai A, Schott D, Mistry N, Liu Y, Gilat-Schmidt T, Robbins J R and Li X A 2018 Technical Note: Enhancing soft tissue contrast and radiation-induced image changes with dual-energy CT for radiation therapy *Med. Phys.* **45** 4238–45

Ohira S, Kanayama N, Toratani M, Ueda Y, Koike Y, Karino T, Shunsuke O, Miyazaki M, Koizumi M and Teshima T 2020 Stereotactic body radiation therapy planning for liver tumors using functional images from dual-energy computed tomography *Radiother. Oncol.* **145** 56–62

Ohira S, Washio H, Yagi M, Karino T, Nakamura K, Ueda Y, Miyazaki M, Koizumi M and Teshima T 2018 Estimation of electron density, effective atomic number and stopping power ratio using dual-layer computed tomography for radiotherapy treatment planning *Phys. Med.* **56** 34–40

Paulson E S, Erickson B, Schultz C and Li X A 2015 Comprehensive MRI simulation methodology using a dedicated MRI scanner in radiation oncology for external beam radiation treatment planning *Med. Phys.* **42** 28–39

Qian P, Xu T, Zheng Q, Yang H, Baydoun A, Zhu J, Traughber B, Muzic R F Jr and Estimating C T 2020 from MR abdominal images using novel generative adversarial networks *J. Grid. Comput.* **18** 211–26

Rietzel E, Chen G T Y, Choi N C and Willet C G 2005 Four-dimensional image-based treatment planning: target volume segmentation and dose calculation in the presence of respiratory motion *Int. J. Radiat. Oncol. Biol. Phys.* **61** 1535–50

Roberge D and Côté J-C 2020 Clinical implementation and evaluation of MR-only radiotherapy planning for brain tumors https://www.magnetomworld.siemens-healthineers.com/clinical-corner/case-studies/clinical-implementation-of-radiotherapy-planning.html

Sartoretti E, Sartoretti T, Binkert C, Najafi A, Schwenk Á, Hinnen M, van Smoorenburg L, Eichenberger B and Sartoretti-Schefer S 2019 Reduction of procedure times in routine clinical practice with Compressed SENSE magnetic resonance imaging technique *PLoS One* **14** e0214887

Sauter A P, Muenzel D, Dangelmaier J, Braren R, Pfeiffer F, Rumment E J, Noël P and Fingerle A A 2018 Dual-layer spectral computed tomography: virtual non-contrast in comparison to true non-contrast images *Eur. J. Radiol.* **104** 108–14

Serai S D, Hu H H, Ahmad R, White S, pednekar A, Anupindi S A and Lee E Y 2020 Newly developed methods for reducing motion artifacts in pediatric abdominal MRI: tips and pearls *AJR Am. J. Roenthenol.* **214** 1042–53

Shan H, Padole A, Homayounieh F, Kruger U, Khera R D, Nitiwarangkul C, Kalra M K and Wang G 2019 Competitive performance of a modularized deep neural network compared to commercial algorithms for low-dose CT image reconstruction *Nat. Mach. Intell.* **1** 269–76

Shirvani S M, Huntzinger C J, Melcher T, Olcott P D, Voronenko Y, Bartlett-Roberto J and Mazin S 2021 Biology-guided radiotherapy: redefining the role of radiotherapy in metastatic cancer *Br. J. Radiol.* **94** 20200873

Singh R *et al* 2020 Image quality and lesion detection on deep learning reconstruction and iterative reconstruction of submillisievert chest and abdominal CT *AJR Am. J. Roentgenol.* **214** 566–73

Sofue K, Yoshikawa T, Ohno Y, Negi N, Inokawa H, Sugihara S and Sugimura K 2017 Improved image quality in abdominal CT in patients who underwent treatment for hepatocellular carcinoma with small metal implants using a raw data-based metal artifact reduction algorithm *Wur. Radiol.* **27** 2978–88

Stemkens B, Paulson E S and Tijssen R H N 2018 Nuts and bolts of 4D-MRI for radiotherapy *Phys. Med. Biol.* **63** 21TR01

Thielemans K, Rathore S, Engbrant F and Razifar P 2011 Device-less gating for PET/CT using PCA *IEEE Nuclear Science Symp. Conf. Record*

Tsang D S, Merchant T E, Merchant S E, Smith H, Yagil Y and Hua C 2017 Estimating potential reduction in contrast dose with mono-energetic images synthesized from double-layer detector spectral CT *Br. J. Radiol.* **90** 20170290

Tyagi N, Fontela S, Zelefsky M, Chong-Ton M, Ostergren K, Shah N, Warner L, Kadbi M, Mechalakos J and Hunt M 2017 Clinical workflow for MR-only simulation and planning in prostate *Radiat. Oncol.* **12** 119

Tyagi N, Zakian K L, Italiaander M, Almujayyaz S, Lis E, Yamada J, Topf J, Hunt M and Deasy J O 2020 Technical Note: A custom-designed flexible MR coil array for spine radiotherapy treatment planning *Med. Phys.* **47** 3143–151

Uh J, Kadbi M and Hua C 2019 Effects of age-related breathing characteristics on the performance of four-dimensional magnetic resonance imaging reconstructed by prospective gating for radiation therapy planning *Phys. Imag. Radiat. Oncol.* **11** 82–7

Uh J, Krasin M J, Li Y, Li X, Tinkle C, Lucas J T Jr., Merchant T E and Hua C 2017 Quantification of pediatric abdominal organ motion with a 4-dimensional magnetic resonance imaging method *Int. J. Radiat. Oncol. Biol. Phys.* **99** 227–37

Walker M D, Bradley K M and McGowan D R 2018 Evaluation of principal component analysis-based data-driven respiratory gating for positron emission tomography *Br. J. Radiol.* **91** 20170793

Walker M D, Morgan A J, Bradley K M and McGowan D R 2020 Data-driven respiratory gating outperforms device-based gating for clinical ^{18}F-FDG PET/CT *J. Nucl. Med.* **61** 1678–83

Wellenberg R H H, Donders J C E, Kloen P, Beenen L F M, Kleipool R P, Maas M and Streekstra G J 2018 Exploring metal artifact reduction using dual-energy CT with pre-metal and post-metal implant cadaver comparison: are implant specific protocols needed? *Skeletal. Radiol.* **47** 839–45

Willemink M J, Persson M, Poumorteza A, Pelc N J and Fleischmann D 2018 Photon-counting CT: technical principles and clinical prospects *Radiology* **289** 293–312

Witt J S, Rosenberg S A and Bassetti M F 2020 MRI-guided adaptive radiotherapy for liver tumours: visualising the future *Lancet Oncol.* **2192** e74–82

Yip E, Yun J, Wachowicz K, Gabos Z, Rathee S and Fallone B G 2017 Sliding window prior data assisted compressed sensing for MRI tracking of lung tumors *Med. Phys.* **44** 84–98

Yoon J, Nickel M D, Peeters J M and Lee J M 2019 Rapid imaging: recent advances in abdominal MRI for reducing acquisition time and its clinical applications *Korean J. Radiol.* **20** 1597–615

Zhang J and Knopp M V 2020 Solid-state digital photon counting PET/CT *Advances in PET* ed J Zhang and M V Knopp (Cham: Springer) pp 53–69

Chapter 16

Advances in treatment planning

Wei Zhao and Lei Xing

This chapter is dedicated to advances and new developments in the treatment planning process for image-guided radiotherapy in the treatment of abdominal cancer. We first introduce data-driven, knowledge-based, and template-based automatic planning. Then, we show how artificial intelligence (AI) is used to predict and calculate radiation dose, with a focus on the usage of deep learning techniques. We also show how AI is used for treatment planning, which is able to yield plan parameters from an automatic decision-making process. Before showing the AI-based techniques, we also provide a short introduction to AI, which aims to present an overview of common concepts and their relationships with each other in the AI field. With the advances in modern imaging modalities, we further discuss the utilization of magnetic resonance imaging, positron emission tomography, and dual-energy CT for treatment planning. Finally, we review the clinical implementation of these advanced treatment planning algorithms.

In the past decade, significant advances in treatment planning for image-guided radiotherapy (IGRT) have been accomplished. In particular, the utilization of advanced imaging modalities and the developments of new algorithms for treatment planning have been extensively investigated to enhance the therapeutic ratio of radiotherapy. These advances have been incorporated into different delivery techniques, including intensity-modulated radiotherapy, volumetric-modulated arc therapy, stereotactic body radiotherapy, [1] and so forth, to improve dose distribution and conformity, reducing the irradiation to nearby organs at risk and therefore minimizing the risk of toxicity. Meanwhile, complicated treatment planning with highly conformal delivery techniques remains challenging to meet the clinical needs, and a manual planning process that involves several trial-and-error iterative optimizations by continuously tuning the dosimetric objectives and weights is usually required for these advanced techniques. The overall quality of the resulting plans is also strongly dependent on the planner experience.

Recent developments in computer science, image processing, and online image modalities are greatly changing the workflow and clinical physics procedures for abdominal radiotherapy using the aforementioned advanced delivery techniques and making it possible to rapidly provide high-quality personalized conformal plans. These changes are mainly focused on the usage of AI and knowledge-based approaches to provide dose-predictive models with simplified procedures and the usage of multi-modality images (such as magnetic resonance image, positron emission tomography, dual-energy CT, and so on) to make planning for better normal tissues sparing with improved soft-tissue contrast and tumor heterogeneity identification. In this chapter, we outline these advances and new developments in the treatment planning process for IGRT in the treatment of abdominal cancer.

16.1 Automatic planning

16.1.1 Data-driven and knowledge-based planning

The concept of knowledge-based treatment planning (KBP) was originally proposed to aid the design of radiation treatment plans in the late 1980s [2, 3]. The original KBP approaches mainly rely on the clinical knowledge and experiences in terms of rules and algorithms, and these sophisticated rules were encoded into complex iterative algorithms to automatically make clinically acceptable [4–7]. These approaches do not rely on predictive models that were built and trained using prior planning data, and thus, they are not data driven in this sense; instead, they can be called 'automatic planning.'

With the accumulation of the carefully designed clinically delivered plans in the past two decades, KBP methods using these high-quality plans in a data-driven fashion have been developed recently and have enjoyed explosive growth over the past years since its commercial implementation release (figure 16.1). Aiming to significantly simplify the procedure of plan optimization and to reduce the time, new developments in the KBP have greatly augmented automatic optimization of the treatment planning. It uses a mathematical model to estimate the dose

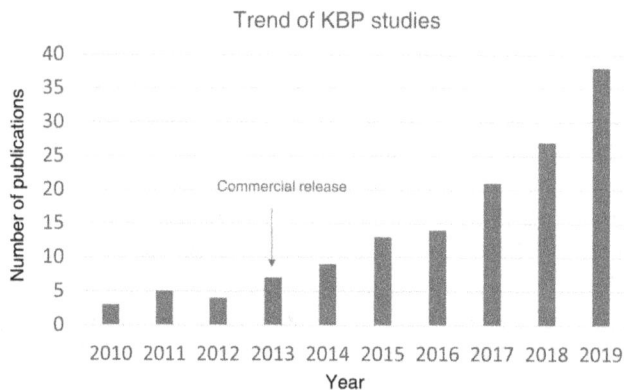

Figure 16.1. Trends of publications related to knowledge-based automatic treatment planning.

distribution and dose-volume histograms (DVHs) for an input patient. The model is built and trained from plenty of previous optimal plans and thus is able to take advantage of prior knowledge from the previously delivered plans. Specifically, the KBP approach uses a database containing information of previously treated patients to generate the DVH objectives for the organs at risk (OARs) for a new patient. The database encompasses the overlap volume histogram (OVH) and DVHs of the previous patients. By using the geometric and dosimetric information extracted from the database, these DVH objectives are able to account for the tradeoffs between the OAR sparing and the target coverage. Meanwhile, the beam characteristics and prescribed dosage of the previous patients and the new patients should be the same. With the KBP-generated achievable DVH objectives, one can predict clinically achievable doses ahead of planning and make the conventional trial-and-error planning optimization process to an automated planning process. Thus, KBP improves the consistency of treatment plan quality and planning efficiency.

While the KBP method was extensively investigated using patients with head-and-neck [8] and prostate [9] cancer, for whom making an optimal plan is often challenging owing to the irregular-shaped target volumes and complex anatomical OARs close to or overlapped to targets, the KBP method was also applied to abdominal radiotherapy such as rectal [10, 11], hepatocellular carcinoma [12], gastrointestinal [13, 14], pancreas [15], and esophageal [16]. For example, in order to predict the rectal dose after hydrogel injection, Yang *et al* evaluated the OVH-predicted dose and used it as the dose objectives for automatic KBP treatment planning. The generated plans were compared with plans that were manually generated using a conventional trial-and-error approach. Quantitative results indicate OVH is a robust predictive indicator to show rectal sparing compared with the volume of hydrogel itself. The KBP approach using the predicted rectum and bladder doses objectives generates acceptable treatment plans with reduced rectal dose compared with conventional plans [10]. The performance of a KBP engine for the optimization of volumetric modulated arc therapy (VMAT) plans was evaluated using patients with hepatocellular cancer [12]. Here, a total of 45 clinically acceptable plans were employed to train a predictive KBP model, which was then tested using 25 plans unseen for the training. The use of KBP optimization results in increases of 0.9% and 0.5% in the pass rate of the clinical objectives compared with the plans optimized using standard methods for two different testing datasets.

The KBP method can also generate DVH constraints to optimize VMAT plans for patients with esophageal cancer. A predicted model was built and trained using a set of 70 patients with esophageal cancer from two different institutions to generate dose-volume constraints for VMAT plans. The model was tested using plan data from another independent institution, and the generated automated plans were compared with the clinically acceptable manual plans. The KBP plans outperform the manual plans in terms of dose-volume parameters [16]. A knowledge-based artificial neural network dose prediction model (ANN-DM) is developed for stereotactic body radiation therapy (SBRT). The model is trained using physician-approved SBRT treatment plans for 43 patients with pancreatic cancer. After training, the model can predict clinically acceptable plans based on plan parameters and a set of geometric parameters. Plans predicted but the ANN-DM provide remarkable improvement by reducing the mean absolute error from >30% to <5%, with respect to the prescribed dose. The dose distributions show excellent consistency with that obtained from the treatment planning system (TPS). In addition, KBP is an efficient radiation therapy clinical trial plan quality control system owing to its plan consistency. In a study using 86 patients with stage IB through IVA cervical cancer from two institutions treated with intensity-modulated radiotherapy (IMRT) according to the protocol that uses a planning target volume (PTV) and two primary OARs [13], the KBP plans outperform manual planning in terms of lower mean normal tissue complication probability for gastrointestinal toxicity. Meanwhile, the mean white blood cell count nadir is also higher by using the KBP plans, which demonstrates the KBP method can achieve lower hematologic toxicity probability.

Since the KBP method can significantly save time to get more patient throughput while providing plans with more consistency, automatic treatment planning using KBP has been translated into clinical practice for different delivery techniques (such as IMRT, VMAT, and SBRT) via several commercial solutions. For example, the Rapidplan module in the Varian Eclipse treatment planning system has successfully provided more superior plans than conventional manual plans for different anatomical sites. The RayStation TPS also incorporates an automatic treatment planning module that can provide a set of Pareto-optimal plans based on user-specified priorities and objectives using multi-criteria optimization [17–19]. By navigating on the Pareto surface, the planner can then choose the plan that balances between clinical tradeoffs.

16.1.2 Template-based planning

The template-based planning (TBP) optimization approach is another automatic treatment planning strategy [20]. Different from the KBP method, which relies on prior patient plan knowledge, the TBP method does not need a database of delivered plans from previously treated patients for model training. Instead, it employs an iterative approach to progressively optimize the plans in a way like an experienced and skillful planner. The auto-planning procedure iteratively optimizes a template of parameters to fulfill the prescribed planning goals. The parameters encompass beam setup and dose prescriptions on the targets and OARs as well as the planning objectives. All these parameters can be customized for the tumor site, treatment protocol, or delivery technique. To achieve better normal tissue sparing and target dose conformity, a set of automatically generated dummy contours and new objectives are added into the planning goals list to manage the steep dose gradient and to force the dose uniformity outside the target.

The TBP approach has been successfully applied to the treatment planning of different delivery techniques for several abdominal treatment sites and has been implemented in commercial solutions. The Auto-planning module in Pinnacle TPS uses the TBP approach, and it has been recently evaluated and tested for VMAT treatments of esophageal cancer [21] and for SBRT treatments of liver cancer [22], showing promising results compared with the manual plans. Gallio *et al* have compared the manual module (MP) and Auto-planning module (AP) of the Pinnacle TPS using 10 SBRT liver patients [22]. While the quality of the plans created using AP was comparable to the manual plans, there are statistically significant differences between the MP and AP for spinal cord doses and planning time. The time saved by the AP method reduces the manpower and enables a larger patient throughput. Moreover, the independence of the planner of the AP module is desirable for plan quality to standardize.

16.2 Artificial intelligence–based automatic planning

16.2.1 Introduction to artificial intelligence

Artificial intelligence (AI), which can also be called machine intelligence, usually denotes intelligence demonstrated by machines [23]. Generally speaking, AI describes machines that mimic cognitive functions (such as learning) of the human mind. AI itself is an old concept and was first proposed back in the 1950s. However, recent advances in computing capability allow the development of high-level feature learning models to represent complex relationships within observational data. Figure 16.2 shows some basic definitions utilized in the AI field and their relationships with each other. The powerful feature extraction and representation learning using the deep neural network have greatly augmented the applications of AI in

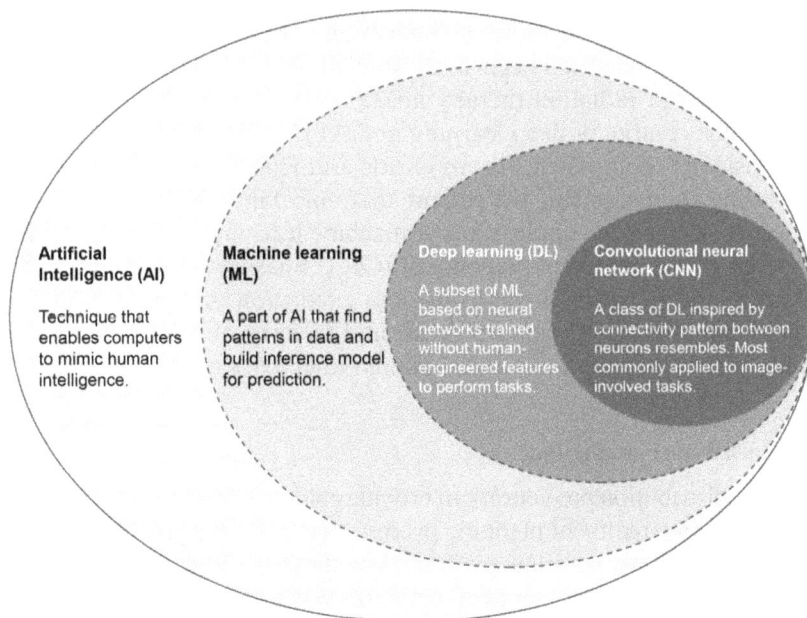

Figure 16.2. Overview of common concepts and their relationships with each in the AI field: artificial intelligence, machine learning, deep learning, convolutional neural network.

many different fields, including computer vision, natural language processing, and biomedicine. Of note, IGRT, which encompasses imaging and diagnosis, modeling, treatment planning, and image-guided patient setup and beam delivery as well as follow up, could particularly benefit from the AI approaches or the powerful representation learning. Furthermore, many steps of the complex workflow of routine IGRT could be simplified via AI modeling (e.g., segmentation, planning, and beam delivery). Several standard routine practices, for example, the delineation of OARs and different target volumes, are labor intensive and dependent on the experience of the clinicians. The quality of conventional manual treatment planning is also dependent on the experience of the planners.

Although recent advances in automatic segmentation using atlas-based methods have mitigated the workload involved in image segmentation, they still require manual updates and reviews before clinical application [24]. Machine learning is the general description of computer algorithms that improve automatically through a data-driven or experienced fashion. As a subset of AI, it can learn from data via human-engineered features and apply what has been learned to make informed decisions. These features are usually intensity, size, and shape based, such as texture features, intensity histogram features, and wavelet features. All features are needed to extract manually to train a predicted model, which may be time consuming. To address this problem, deep learning uses a deep neural network to progressively extract high-level features and perform representation learning. Throughout, the feature extraction procedure can be completed automatically, and no human-engineered feature is required, which is desirable for the high throughput of clinical practice. Since images are heavily involved in the workflow of IGRT, a specific class of deep neural networks, convolutional neural network (CNN), which is suitable for image analysis tasks, has been extensively studied for many steps in IGRT. Indeed, CNN was applied to radiation therapy image segmentation and analysis since the early days of the resurge of deep learning and AI [25, 26].

AI is envisioned to augment human efforts and reasoning logics by comprehensively analyzing and processing the patient anatomical information. In the following section, we will discuss the application of machine learning or deep learning with a focus on CNN in the treatment planning of IGRT. Specifically, we introduce how to use a deep neural network to perform dose prediction and dose calculation, to significantly simplify the planning process, and to improve the planning efficiency and quality consistency with minimum human intervention.

16.2.2 AI-based dose prediction

Accurate dose distribution prediction can provide guidance to improve plan quality and efficiency during the treatment planning process, and it has the potential to facilitate a fully automated treatment planning workflow without the need for inverse optimization. Dose prediction using deep neural networks was first proposed by Mardani *et al* [27], which represents a major advance toward automatic treatment planning. Different from KBP methods, which are usually based on geometry/anatomical features (such as OAR volumes, target volumes, volume overlapping, distance-to-target histogram, out-of-field volume, etc), machine learning–based methods are able to exploit the most descriptive features (e.g., dosimetric features) to build a predictive model for treatment planning, and these features have the potential to uniquely describe the resultant plan. Ma *et al* used a deep learning model to predict DVH for VMAT planning from features with dosimetric capability [28, 29]. In addition to a database of prior clinical plans, the input dosimetric features of the model also include the planning target volume (PTV)-only plans, which correspond to the best scenario of tumor targeting and worst scenario of OARs sparing. Hence, the DVHs of the PTV plan can not only provide dosimetric information but also implicitly encode the anatomical/geometric information. The DVH of the clinical treatment plan, which is the output of the predictive model, shows superior performance to the KBP-based RapidPlan (figure 16.3), indicating a machine learning–based dosimetric feature-driven method has the potential to improve planning and quality control in clinical practice.

While the DVH is primarily employed to statistically summarize the isodose distribution for a plan, it only provides first-order dosimetry information and cannot provide spatial dose distribution. To provide dose conformity and gradient measurements and other spatial information of dosimetric endpoints, the deep neural network, especially the CNN, is extensively exploited to predict dose distributions in planning. While most of the CNN-based spatial dose distribution prediction approaches are targeting the challenging sites (such as head and neck), these

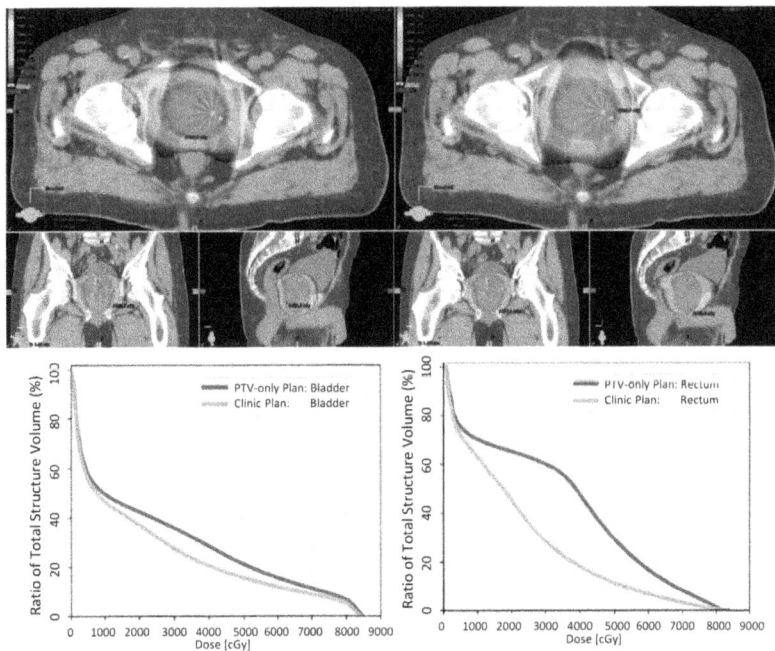

Figure 16.3. Comparison of dose distributions and DVHs of bladder and rectum between clinical plan (left) and PTV-only plan for a case from the training cohort. Reprinted with permission from [29].

approaches should also be applied to the abdominal tumors straightforward. Mardani and Xing developed a learning-empowered approach to predict the achievable 3D dose distribution using multi-layer convolutional auto-encoders (CAEs) [27, 30]. This is the first study that used a deep learning approach to predict dose distribution for VMAT/IMRT treatment planning. The deep CAEs automatically extract the representation features that can be used to derive a contour-to-dose relation map. Campbell *et al* used an ANN-DM to predict dose distribution for pancreatic SBRT treatment [15]. The ANN-DM was trained by physician-approved treatment plans with plan parameters and voxel-wise geometric parameters. The results show ANN-DMs could predict desirable dose distribution.

U-net-type models, which are based on a popular fully convolutional network structure U-net, were extensively studied to predict both 2D and 3D dose distribution [31–39] and image analysis [26, 40]. Nguyen *et al* used a modified U-net for dose prediction from delineated PTV and OAR contour images of patients, and they found the average of the absolute differences between the predicted dose and the prescribed dose is smaller than 5% for all the contours [31]. Kearney *et al* developed a U-net structure model, DoseNet, to predict 3D dose distribution for SBRT patients [32]. This model utilized 3D convolutional downsampling with corresponding 3D transposed convolutional upsampling to preserve image resolution. A comparison study with clinically approved delivered dose distribution in terms of clinically relevant metrics shows high-quality 3D dose distribution can be reconstructed, and the proposed model is superior to the classical U-net. By using the isodose feature-preserving voxelization (IFPV) to simplify the representation of the dose distribution, Ma *et al* implement a deep convolutional neural network to predict dose distribution for VMAT treatment planning. The network was trained using contour-to-dose relationship from the prior patient database and was able to predict dose distribution in the IFPV domain with input contours [33]. Even with a significantly reduced number of dose representation points, this IFPV-based method has comparable results with respect to the voxel-based method in terms of DVH and dose difference maps. The method can greatly improve the efficiency of dose prediction and facilitate the routine treatment planning workflow. Dose prediction is also performed using other popular networks such as ResNet [41, 42] and GAN [43, 44].

16.2.3 AI-based dose calculation

For a given patient with anatomy information and a linear accelerator (Linac) multi-leaf collimator (MLC), a shape or segment with different beam delivery techniques, dose calculation using the Linac parameters and underlying anatomy is an essential part of the treatment planning process for the specific patient. This process is accomplished using a dose calculation engine embedded in routine 3D TPS. The dose calculation engine is usually based on different types of algorithms, including pencil beam convolution [45], collapsed cone convolution [46], and Monte Carlo (MC) simulation [47]. Due to the tissue heterogeneity, MC-based methods are able to provide more accurate dose calculation compared with the convolution-based methods. However, dose calculation using MC simulation is typically time consuming, and fast dose calculation is desirable for clinical practice.

Different from dose prediction that only relies on the correlation of the target volumes and OARs to the prescribed dose and does not incorporate the actual Linac physical parameters (such as MLC and jaw positions and MLC/gantry angles for each segment), dose calculation takes into account the physical parameters. Based on its powerful representation learning ability, the deep neural network is able to correlate these Linac physical parameters to the dose distribution. Recently, Dong and Xing pioneered the use of the deep neural network for photons [48]. A similar approach was extended to proton dose calculation [49]. Kontaxis *et al* developed a deep learning framework, DeepDose, to derive the dose distribution by using the input MLC shape with underlying patient anatomy [50]. Specifically, a 3D U-net model was employed to map five different 3D volumes, including segment binary mask, distance from source, central beamline distance, radiological depth, and volume density, to a 3D dose distribution of the specific segment. Each of these input 3D volumes models a physical quantity or feature implicitly used by dose engines in TPS. The results show a trained model can infer the dose distribution for a segment in 0.6 s and 1 min for all segments for a patient treatment. The accuracy and inference time of the dose calculation framework are very desirable for the workflow of online adaptive radiotherapy.

Instead of directly obtaining a high-quality dose plan, Dong and Xing's approach can be employed for general-purpose dose calculation by mapping a dose distribution calculated using a low-cost algorithm [48]. Specifically, dose profiles calculated using the anisotropic analytical algorithm (AAA) in 5 mm resolution and Acuros XB algorithm (AXB) in 1.25 mm resolution were employed to train a three-layer-deep convolutional neural network. The super-resolution-type CNN can then map the low-resolution dose images to high-resolution dose images. A comparison study with respect to the AXB dose map in 1.25 mm high resolution suggests the proposed model can be used as dose calculation acceleration engine across various treatment planning platforms. It has been proved that the dose calculated by the linear Boltzmann transport equation-based AXB is able to theoretically converge to that simulated by the gold-standard MC algorithm. More importantly, the difference between a high-accuracy AXB dose and low-accuracy AAA dose is mainly attributed to the inhomogeneous regions. The difference can be learned and correlated to the anatomical information shown in the CT images from high-level features extracted by a deep neural network. To this end, Xing *et al* developed a dense U-net model to map the analytical AAA dose to high-definition AXB dose [51]. The model can synergistically learn the AAA and AXB dose difference together with the CT images both locally and globally. The trained model was trained using the AAA dose together with the CT image, and the AXB doses as the input and output, respectively. The gamma passing rate (1 mm/1%) of the dose distribution calculated using the trained model is 97.6% ± 2.4% (mean ± standard deviation), compared with 87.8% ± 9.0% of the original AAA doses. This scheme can greatly boost radiotherapy dose calculation, and the inference procedure can be potentially accomplished within 1 s.

16.2.4 AI-based treatment planning

The AI-based dose prediction has provided guidance for the treatment planning process and is able to facilitate a fully automatic plan generation [42]. On the other hand, AI-based dose calculation allows dose computation with both high efficiency and high accuracy and thus expedites the plan optimization and plan checks. In addition to these advances, direct prediction of plan parameters could also benefit from AI approaches. In this case, AI approaches have played the role of the reasoning process in treatment planning. Different from human planners that vary from each other when solving a specific dosimetric-tradeoff problem in the manual treatment planning process, the decision-making of AI-based treatment planning can be performed without human intervention; thus, it is able to power the implementation of automatic treatment planning with fully automated workflow.

The decision-making process in manual treatment planning is basically a series of actions that regard to dose trade-offs, which is particularly suitable for several AI algorithms, such as Monte Carlo tree search (MCTS) and reinforcement learning (RL). MCTS is a heuristic search algorithm, and it performs analysis of the most promising move and rapidly takes an action in a complex sequential decision process. Recently, Dong and Xing proposed using the MCTS algorithm to optimize treatment trajectory strategy, which is able to determine the beam delivery trajectory that can provide a truly optimal treatment plan with the consideration of the synergism between the trajectory and other beam parameters [52]. They discretized the entire angular space into 2701 nodes in a $5°$ angular interval in both gantry and couch rotations. The algorithm is designed to select the node that can yield a higher average objective value and fewer visits. For each iteration of the MCTS optimization, the algorithm performs the following procedures: (1) select node from the root level to the leaf level, (2) grow the leaf node in a random fashion, (3) select a newly grown node according to the trajectory optimization policy, and (4) the objective function value is backpropagated to the root node. The trajectory is optimized using the alternating direction method of multipliers with beamlet delivery. Evaluation studies using chest well and braincases show the MCTS method offers much better OARs sparing while maintaining PTV coverage compared with the coplanar plans.

In the trial-and-error treatment planning process, the planner adjusts plan parameters (such as DVH constraints' locations and weights) to achieve an acceptable plan by using the TPS. This process can be automatically accomplished using the trial-and-error search and the delayed reward in RL. Shen *et al* used an RL-based virtual treatment planner network (VTPN) to model the human planner behaviors for the TPS [53]. The VTPN is able to use the intermediate plan DVHs together with weights and threshold doses in the objective function to decide the action to optimize the plan. The results on the patient treated with IMRT show the VTPN can yield satisfactory plans compared with human planners.

16.3 Multimodality imaging-based planning

16.3.1 Dual-energy CT-based planning

Classical CT imaging falls short in accurate quantification of tissue material composition because the pixel value or Hounsfield units in a CT image represents the effective linear attenuation coefficient, which is an averaged contribution of all materials or chemical elements within the pixel. To alleviate the problem, dual-energy CT (DECT) 2–4, which acquires two sets of CT data with different energy spectra, has been realized as a result of advancements in hardware and post-processing capabilities. DECT, which is also known as a major implementation of 'spectral imaging,' takes advantage of the energy dependence of the linear attenuation coefficients of the tissue to yield material-specific images, such as blood, iodine, or water maps. Different implementations have been realized for DECT imaging, including dual-source DECT implemented by Siemens Healthineers (figure 16.4(a)); fast kV switching DECT scanner adopted by GE Healthcare, which acquires projections of low- and high-energy spectra alternatively (figure 16.4(b)); and dual-layer DECT approach adopted by the Philips Healthcare scanner, which acquires DECT projection data using a layered detector, with the low- and high-energy data collected by the front- and back-detector layer, respectively (figure 16.4(c)).

The introduction of the DECT scanner has greatly augmented the power of cancer diagnosis and radiation treatment. In recent years, DECT imaging has been extensively used in radiation oncology departments, and new strategies exploiting DECT imaging are being integrated into the entire chain of radiotherapy. These strategies are focused on improving dose calculation accuracy, enabling better tumor delineation and metal artifacts reduction. In this section, we will focus on the recent advance in the use of DECT for improving dose calculation and treatment planning for radiotherapy.

DECT is able to provide accurate effective atomic number Z_{eff} and electron density ρ_e. In proton therapy, the stopping power ratio (SPR) image is required by the pencil beam algorithm for proton range calculation. While using conventional single-energy

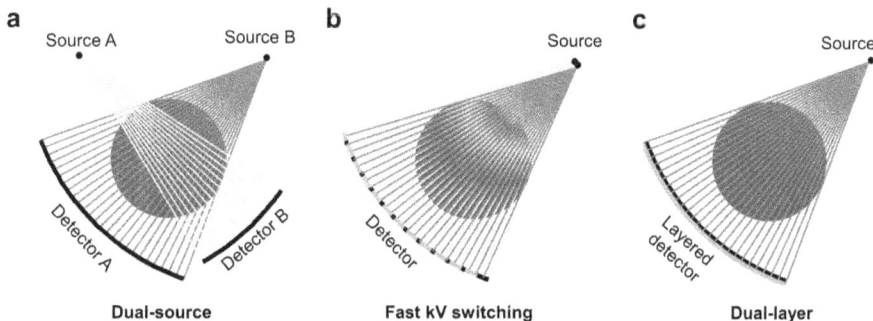

Figure 16.4. A schematic diagram of different DECT imaging strategies. (a) Dual-source DECT; (b) fast kV switching DECT; and (c) dual-layer DECT.

CT (SECT) results in an uncertainty in the SPR, which has to be taken into account in the treatment margin, DECT imaging can be applied to improve the accuracy of SPR and proton range calculation. By fitting the relationship between Z_{eff} of human tissues and the logarithm of mean excitation energy using linear functions, Yang *et al* investigated a patient-specific SPR calculation for pencil beam dose calculation [54]. They found the root-mean-square error of SPR calculated using the DECT images is within 1% from standard human tissue value. Their DECT method is compared with the standard clinical practice, the stoichiometric calibration method, which is based on the treatment planning CT scan. The DECT method shows superior SPR results compared with the standard clinical practice. Extensive experimental validation of the DECT method using phantom scans also demonstrated superior tissue substitutes SPR accuracy to SECT-based methods [55, 56].

For dose calculation in external beam photon radiotherapy, Tsukihara *et al* proposed a DECT-subtracted CT number to ρ_e conversion approach for treatment planning [57]. Two radiotherapy plans are employed to quantify the reliability of dose calculation for both the DECT-subtracted conversion method and the SECT conversion method. The reliability of dose calculation of the DECT conversion method outperforms that of the SECT-based method, especially for dentition and bony structures. The dose distributions and DVHs of the calibration measurements match with each other quite well for the DECT scenario, and it reduces dose uncertainties from 11% to 1% for specific cases. It thereby has the potential to provide an accurate and reliable inhomogeneity correction in treatment planning and can be implemented into the current TPS system without modification or interruption of routine workflow. Since DECT can provide virtual non-contrast (VNC) enhanced CT images, another application for DECT in radiotherapy is to provide more accurate dose calculation for treatment planning using a VNC image, which can exclude the influence of iodine-enhanced structures from a contrast-enhanced CT scan [58].

16.3.2 Magnetic resonance imaging–based treatment planning

Compared with the standard practice that uses CT imaging for treatment planning, magnetic resonance imaging (MRI) allows improved OAR segmentation and target delineation. Due to its non-ionizing radiation nature, MRI is of particular benefit for pediatric patients. The use of MRI in radiotherapy treatment planning is therefore rapidly expanding. In this section, we introduce treatment planning based on the multi-contrast mechanisms and superior soft-tissue contrast of MRI imaging. Challenges including estimation of electron density and geometric inconsistency are also considered.

Compared with CT imaging, MRI can identify substantially larger tumor volumes for liver metastases treatment planning. Depending on different contrast mechanisms, the target volume can be increased by 180%, 178%, and 246% for T1w-MRI, contrast-enhanced T1w-MRI, and T2w-MRI, respectively [59]. The superior soft-tissue contrast of MRI also decreases OARs and target volumes delineation uncertainties [60], and it can also account for variations between breathing cycles by using multiple and extended acquisitions based on its dose-free nature. In addition, functional MRI provides tumor heterogeneity and thus enables functionally weighted planning and improved treatment [61, 62].

MRI is also suitable for monitoring intrafractional motion. Tumors in many sites undergo both interfraction and intrafraction motion. While the interfraction motion can be mitigated by using either daily con-beam CT or orthogonal kV fiducial images, intrafraction motion is much more challenging. Breath holding can be one of the approaches, but there can be residual displacement between tumor positions during beam delivery. Ultrasound has the potential for real-time RT guidance, but its accuracy can suffer from poor spatial resolution [63, 64]. MRI is able to be used for real-time intrafraction motion management. For abdominal cancers, the RT treatment is very likely to be performed in the presence of respiratory motion, and accurate modeling of organ motion is required. Prior effects have incorporated time-resolved 2D and 4D MRI into treatment planning to address organ motion, either together with CT or as the sole imaging modality. A typical application of MRI-based planning is the gated treatment approach that is designed to deliver a radiation dose when the tumor is in a pre-defined position [65]. In this scenario, real-time monitoring of the target position is performed using MRI and is compared with planning. Different from x-ray–based imaging methods that need surrogates, MRI enables a more effective solution to directly visualize the target in a non-invasive way. For example, in-room MRI integrated systems were used to derive an internal surrogate for gate purposes [66]. Although MRI-Linac helps gate upper abdominal tumor motion and therefore reduces the margins in radiotherapy, it must be noted that the 'freezing' tumor motion in one phase of the breathing cycle for planning and treating is performed at the expense of reduced treatment efficiency and prolonged workflow.

Instead of using a gating window in RT, the target volume is usually expanded to encompass the full track of tumor motion during planning, resulting in a volume called 'internal target volume' (ITV). The radiation beam is then delivered based on the assumption that the breathing cycle and tumor motion are consistent and reproducible with respect to the planning throughout treatment. Since breathing cycle variations may occur and ITV obtained from a single planning cycle may not be sufficient, based on improved margins definition, cine-MRI has been proposed to reduce uncertainties associated with cycle-to-cycle breathing variations [67].

Despite its superior soft-tissue contrast and multi-contrast mechanisms for target volume definition and motion modeling in planning, there are also limitations to the use of MRI in planning. The most substantial limitation is that MRI cannot provide the electron density map needed for dose calculation. To address this problem, CT images synthesized from MRI images are extensively studied using atlas-based or machine learning solutions [68], especially the recently boomed deep learning approaches [69, 70]. In the deep learning approaches, an end-to-end neural network (such as GAN and U-net-type) is usually trained using paired MRI and co-registered

CT images. The network takes the MRI images as input and output CT images for electron density calculation. While most of the studies are targeting relatively homogeneous sites such as the brain and pelvis, synthetic-CT generation in sites suffering from respiratory motion are less investigated [71]. The reasons are mainly attributed to the complex and homogeneous anatomical structures and respiratory motion that make the accurate registration and electron density calculation extremely challenging.

16.3.3 Positron emission tomography–based planning

Positron emission tomography (PET) is an imaging modality that uses radio-pharmaceuticals to visualize and measure metabolic processes within the patient. In contrast to other widely used image modalities such as CT and MRI, which provide morphologic information, PET provides physiological activities from the three-dimensional distribution of activity concentration of the radiopharmaceutical. Based on its highly sensitive features, PET is widely used in cancer staging, restaging, and the monitoring of treatment response. The high metabolic activity of a tumor is associated with worse prognosis, and the metabolic activity changes before and after treatment can reflect tumor response. Different from routine practice using CT imaging in which the target volumes are delineated based on x-ray attenuation coefficient difference, in PET imaging, a metabolic tumor volume (MTV) can be delineated according to the metabolism, which is quantified using standardized uptake value (SUV). Compared with CT- or MRI-based anatomical measurement, tumor volume measurement using PET shows several advantages. First, MTV-based measurement is much easier and faster owing to the superior contrast compared to anatomic imaging. Second, MTV is more sensitive to reflect the treatment response than anatomic imaging.

While current clinical studies on the integration of PET in radiotherapy treatment planning are mainly focused on the brain, head-and-neck, lung, and genitourinary tumors, integrating PET in the target volume definition for abdominal tumor entities such as pancreatic cancer and liver cancer is much more challenging owing to the respiratory motion, which can cause target displacement as large as 20 mm. To address the respiratory motion, 4D-PET has been employed for pancreatic [72] and liver [73] cancer RT planning. Riou et al assessed the target volume definition using non-gated PET and 4D-PET for liver SBRT planning [73]. For the 4D-PET scenario, a gated planning target volume (PTVg) was delineated by adding a 3 mm setup margin to a biological internal target volume. For the non-gated PET scenario, a manual planning target volume (PTV) was generated from a semi-automatic biological target volume with a 5 mm radial and a 10 mm craniocaudal margin that accounted for tumor motion and setup uncertainty. It was found that the 4D-PET–enabled PTVg is statistically significantly smaller than the PTV. Compared with non-gated PET, 4D-PET is able to provide better target definition with respiratory motion and improve treatment planning for liver metastases. For pancreatic cancer radiotherapy treatment planning, Kishi et al also showed that the averaged ITV generated using non-gated PET was twofold larger than those generated using 4D-PET, and the latter can reduce normal tissue irradiation and improve therapeutic ratio [72].

Of note, PET-guided radiotherapy, also called biology-guided radiotherapy (BgRT), has been developed and released recently [74, 75]. BgRT uses emissions generated from injected radiopharmaceuticals to guide the RT beam during treatment. Different from conventional RT PTV, which encompasses the clinical target volume (CTV), an internal margin that accounts for tumor motion, and a setup margin that accounts for setup uncertainty, a unique biological guidance margin (BgM) is proposed to encompass target tracking and registration uncertainties, and the PTV for BgRT only includes the CTV and the BgM. While conventional RT PTV is fixed relative to the anatomy background, the BgRT PTV moves within a tracking region, and the radiation beam fires according to the moving PTV. Hence, BgRT can significantly reduce the treatment target volume. Based on that unique feature, BgRT has the potential to overcome the current motion management and toxicity challenges for patients with metastatic abdominal cancer. It also enables the treatment of multiple tumors in parallel during a single session.

16.4 Integrating advances in treatment planning system

Treatment planning plays an essential role in the success of modern radiation therapy, and timely translation of technical developments is critical for patients to benefit from the technical advances in a timely fashion. Along this line, vendors are actively integrating the new developments into the treatment planning system to enable rapid, efficient, and high-performance radiotherapy. Fully automated plans for different delivery techniques have been successfully implemented for clinical practice using novel vendor-specific commercial solutions. For example, the iPlan RT treatment planning system from the Brainlab offers precise radiation therapy dose calculations using Monte Carlo simulation. Elekta's Monaco system incorporates the biological effect into the cost functions of a multi-criteria constrained optimization problem to enable functional planning. It also features a robust Monte Carlo dose calculation module and a leaf sequence optimizer for treatment planning for IMRT, VMAT, and SBRT. RaySearch released a machine learning–based treatment planning system that is able to greatly expedite the clinical workflow and get more patient throughput. The system creates the treatment plan for a new patient using a model trained with a curated set of high-quality, previously delivered plans. Varian's Eclipse also enables the simplification of contouring tasks, field setup, and rapid dose calculation. It also features machine learning–based planning for intelligent treatment planning of external beam and proton therapy, enabling clinicians to leverage knowledge from prior treatment plans to rapidly create high-performance personalized plans for new patients using a knowledge-based algorithm. For MRI-based treatment planning, both Viewray-MRIdian and Elekta-Unity MRI-Linac have been adopted clinically for IGRT.

16.5 Summary

The last few decades have witnessed tremendous progress in radiation therapy treatment planning and delivery. In this chapter, we have provided a comprehensive overview of modern techniques related to radiation therapy treatment planning. Leveraging from the emerging AI and continuous surging of high-performance computing technologies, fully automated image analysis and treatment planning, as well as integrated clinical decision-making, are becoming a clinical reality. The development and applications of advanced AI methods for radiation therapy afford unique opportunities to greatly enhance the quality, safety, and efficiency of radiation treatment planning and dose delivery. Ultimately, these developments will greatly facilitate personalized radiotherapy decision-making and substantially advance patient care.

Acknowledgments

The authors would like to thank the many researchers who were involved in the related work at our institute.

References

[1] Timmerman R and Xing L 2009 *Image Guided and Adaptive Radiation Therapy* (Baltimore, MD: Lippincott Williams & Wilkins)

[2] Kalet I J and Paluszynski W 1990 Knowledge-based computer systems for radiotherapy planning *Am. J. Clin. Oncol.* **13** 344–51

[3] Shwe M, Tu S and Fagan L 1989 Validating the knowledge base of a therapy planning system *Methods Inf. Med.* **28** 36–50

[4] Yang Y and Xing L 2004 Clinical knowledge-based inverse treatment planning *Phys. Med. Biol.* **49** 5101

[5] Xing L and Chen G T 1996 Iterative methods for inverse treatment planning *Phys. Med. Biol.* **41** 2107

[6] Xing L *et al* 1999 Estimation theory and model parameter selection for therapeutic treatment plan optimization *Med. Phys.* **26** 2348–58

[7] Yang Y and Xing L 2004 Inverse treatment planning with adaptively evolving voxel-dependent penalty scheme: adaptive voxel-dependent penalty scheme for inverse planning *Med. Phys.* **31** 2839–44

[8] Wu B *et al* 2011 Data-driven approach to generating achievable dose–volume histogram objectives in intensity-modulated radiotherapy planning *Int. J. Radiat. Oncol. Biol. Phys.* **79** 1241–7

[9] Moore K L *et al* 2011 Experience-based quality control of clinical intensity-modulated radiotherapy planning *Int. J. Radiat. Oncol. Biol. Phys.* **81** 545–51

[10] Yang Y *et al* 2013 An overlap-volume-histogram based method for rectal dose prediction and automated treatment planning in the external beam prostate radiotherapy following hydrogel injection. *Med. Phys.* **40** 011709

[11] Fan J *et al* 2017 Iterative dataset optimization in automated planning: implementation for breast and rectal cancer radiotherapy *Med. Phys.* **44** 2515–31

[12] Fogliata A *et al* 2014 Assessment of a model based optimization engine for volumetric modulated arc therapy for patients with advanced hepatocellular cancer *Radiat. Oncol.* **9** 236

[13] Li N *et al* 2017 Highly efficient training, refinement, and validation of a knowledge-based planning quality-control system for radiation therapy clinical trials *Int. J. Radiat. Oncol. Biol. Phys.* **97** 164–72

[14] Sharfo A W M *et al* 2018 Automated VMAT planning for postoperative adjuvant treatment of advanced gastric cancer *Radiat. Oncol.* **13** 74

[15] Campbell W G *et al* 2017 Neural network dose models for knowledge-based planning in pancreatic SBRT *Med. Phys.* **44** 6148–58

[16] Fogliata A *et al* 2015 A broad scope knowledge based model for optimization of VMAT in esophageal cancer: validation and assessment of plan quality among different treatment centers *Radiat. Oncol.* **10** 220

[17] Monz M *et al* 2008 Pareto navigation—algorithmic foundation of interactive multi-criteria IMRT planning *Phys. Med. Biol.* **53** 985

[18] Lahanas M, Schreibmann E and Baltas D 2003 Multiobjective inverse planning for intensity modulated radiotherapy with constraint-free gradient-based optimization algorithms *Phys. Med. Biol.* **48** 2843

[19] Breedveld S *et al* 2012 iCycle: integrated, multicriterial beam angle, and profile optimization for generation of coplanar and noncoplanar IMRT plans *Med. Phys.* **39** 951–63

[20] Cilla S *et al* 2020 Template-based automation of treatment planning in advanced radiotherapy: a comprehensive dosimetric and clinical evaluation *Sci. Rep.* **10** 1–13

[21] Li X *et al* 2017 Dosimetric benefits of automation in the treatment of lower thoracic esophageal cancer: is manual planning still an alternative option? *Med. Dosim.* **42** 289–95

[22] Gallio E *et al* 2018 Evaluation of a commercial automatic treatment planning system for liver stereotactic body radiation therapy treatments *Physica Med.* **46** 153–59

[23] Xing L, Giger M L and Min J K 2020 *Artificial Intelligence in Medicine: Technical Basis and Clinical Applications* (St. Louis, MO: Elsevier S&T Books)

[24] Eldesoky A R *et al* 2016 Internal and external validation of an ESTRO delineation guideline–dependent automated segmentation tool for loco-regional radiation therapy of early breast cancer *Radiother. Oncol.* **121** 424–30

[25] Ibragimov B and Xing L 2017 Segmentation of organs-at-risks in head and neck CT images using convolutional neural networks *Med. Phys.* **44** 547–57

[26] Seo H *et al* 2019 Machine learning techniques for biomedical image segmentation: an overview *Med. Phys.* **47** e148–167

[27] Mardani M, Dong P and Xing L 2016 Deep-learning based prediction of achievable dose for personalizing inverse treatment planning *Int. J. Radiat. Oncol. Biol. Phys.* **96** E419–20

[28] Yang Y, Xing L and Ma M 2020 Dosimetric features-driven machine learning model for DVHs/dose prediction *US Patent* US20200171325A1

[29] Ma M *et al* 2019 Dosimetric features-driven machine learning model for DVH prediction in VMAT treatment planning *Med. Phys.* **46** 857–67

[30] Mardani Korani M, Dong P and Xing L 2016 MO-G-201-03: deep-learning based prediction of achievable dose for personalizing inverse treatment planning *Med. Phys.* **43** 3724–4

[31] Nguyen D *et al* 2017 Dose prediction with U-net: a feasibility study for predicting dose distributions from contours using deep learning on prostate IMRT patients arXiv:1709.09233

[32] Kearney V *et al* 2018 DoseNet: a volumetric dose prediction algorithm using 3D fully-convolutional neural networks *Phys. Med. Biol.* **63** 235022

[33] Ma M *et al* 2019 Dose distribution prediction in isodose feature-preserving voxelization domain using deep convolutional neural network *Med. Phys.* **46** 2978–87

[34] Liu Z *et al* 2019 A deep learning method for prediction of three-dimensional dose distribution of helical tomotherapy *Med. Phys.* **46** 1972–83

[35] Barragán-Montero A M *et al* 2019 Three-dimensional dose prediction for lung IMRT patients with deep neural networks: robust learning from heterogeneous beam configurations *Med. Phys.* **46** 3679–91

[36] Nguyen D *et al* 2019 A feasibility study for predicting optimal radiation therapy dose distributions of prostate cancer patients from patient anatomy using deep learning *Sci. Rep.* **9** 1–10

[37] Ma M *et al* 2019 Incorporating dosimetric features into the prediction of 3D VMAT dose distributions using deep convolutional neural network *Phys. Med. Biol.* **64** 125017

[38] Fan J *et al* 2020 Data-driven dose calculation algorithm based on deep learning arXiv:2006.15485

[39] Lee M S *et al* 2019 Deep-dose: a voxel dose estimation method using deep convolutional neural network for personalized internal dosimetry *Sci. Rep.* **9** 1–9

[40] Seo H *et al* 2019 Modified U-Net (mU-Net) with incorporation of object-dependent high level features for improved liver and liver-tumor segmentation in CT images *IEEE Trans. Med. Imaging* **39** 1316–25

[41] Chen X *et al* 2019 A feasibility study on an automated method to generate patient-specific dose distributions for radiotherapy using deep learning *Med. Phys.* **46** 56–64

[42] Fan J *et al* 2019 Automatic treatment planning based on three-dimensional dose distribution predicted from deep learning technique *Med. Phys.* **46** 370–81

[43] Babier A *et al* 2020 Knowledge-based automated planning with three-dimensional generative adversarial networks *Med. Phys.* **47** 297–306

[44] Murakami Y *et al* 2020 Fully automated dose prediction using generative adversarial networks in prostate cancer patients *PLoS One* **15** e0232697

[45] Mohan R, Chui C and Lidofsky L 1986 Differential pencil beam dose computation model for photons *Med. Phys.* **13** 64–73

[46] Ahnesjö A 1989 Collapsed cone convolution of radiant energy for photon dose calculation in heterogeneous media *Med. Phys.* **16** 577–92

[47] Rogers D 2006 Fifty years of Monte Carlo simulations for medical physics *Phys. Med. Biol.* **51** R287

[48] Dong P and Xing L 2020 Deep DoseNet: a deep neural network for accurate dosimetric transformation between different spatial resolutions and/or different dose calculation algorithms for precision radiation therapy *Phys. Med. Biol.* **65** 035010

[49] Nomura Y *et al* 2020 Fast spot-scanning proton dose calculation method with uncertainty quantification using a three-dimensional convolutional neural network *Phys. Med. Biol.* **65** 215007

[50] Kontaxis C *et al* 2020 DeepDose: towards a fast dose calculation engine for radiation therapy using deep learning *Phys. Med. Biol.* **65** 075013

[51] Xing Y *et al* 2020 Boosting radiotherapy dose calculation accuracy with deep learning arXiv:2005.03065

[52] Dong P, Liu H and Xing L 2018 Monte Carlo tree search-based non-coplanar trajectory design for station parameter optimized radiation therapy (SPORT) *Phys. Med. Biol.* **63** 135014

[53] Shen C *et al* 2020 Operating a treatment planning system using a deep-reinforcement learning-based virtual treatment planner for prostate cancer intensity-modulated radiation therapy treatment planning *Med. Phys.*

[54] Yang M *et al* 2010 Theoretical variance analysis of single-and dual-energy computed tomography methods for calculating proton stopping power ratios of biological tissues *Phys. Med. Biol.* **55** 1343

[55] Bourque A E, Carrier J-F and Bouchard H 2014 A stoichiometric calibration method for dual energy computed tomography *Phys. Med. Biol.* **59** 2059

[56] Hünemohr N *et al* 2013 Experimental verification of ion stopping power prediction from dual energy CT data in tissue surrogates *Phys. Med. Biol.* **59** 83

[57] Tsukihara M *et al* 2015 Initial implementation of the conversion from the energy-subtracted CT number to electron density in tissue inhomogeneity corrections: an anthropomorphic phantom study of radiotherapy treatment planning *Med. Phys.* **42** 1378–88

[58] Yamada S *et al* 2014 Radiotherapy treatment planning with contrast-enhanced computed tomography: feasibility of dual-energy virtual unenhanced imaging for improved dose calculations *Radiat. Oncol.* **9** 1–10

[59] Pech M *et al* 2008 Radiotherapy of liver metastases *Strahlenther. Onkol.* **184** 256–61

[60] Schmidt M A and Payne G S 2015 Radiotherapy planning using MRI *Phys. Med. Biol.* **60** R323

[61] Bainbridge H E *et al* 2017 Treating locally advanced lung cancer with a 1.5 T MR-Linac – effects of the magnetic field and irradiation geometry on conventionally fractionated and isotoxic dose-escalated radiotherapy *Radiother. Oncol.* **125** 280–5

[62] Menten M J, Wetscherek A and Fast M F 2017 MRI-guided lung SBRT: present and future developments *Physica Med.* **44** 139–49

[63] Fontanarosa D *et al* 2015 Review of ultrasound image guidance in external beam radio-therapy: I. Treatment planning and inter-fraction motion management *Phys. Med. Biol.* **60** R77

[64] O'Shea T *et al* 2016 Review of ultrasound image guidance in external beam radiotherapy part II: intra-fraction motion management and novel applications *Phys. Med. Biol.* **61** R90

[65] Keall P J *et al* 2006 The management of respiratory motion in radiation oncology report of AAPM Task Group 76 *Med. Phys.* **33** 3874–900

[66] Mutic S and Dempsey J F 2014 *The ViewRay system: magnetic resonance–guided and controlled radiotherapy Semin. Radiat. Oncol.* **24** 196–9

[67] Cai J, Read P W and Sheng K 2008 The effect of respiratory motion variability and tumor size on the accuracy of average intensity projection from four-dimensional computed tomography: an investigation based on dynamic MRI *Med. Phys.* **35** 4974–81

[68] Edmund J M and Nyholm T 2017 A review of substitute CT generation for MRI-only radiation therapy *Radiat. Oncol.* **12** 28

[69] Liu F *et al* 2019 MR-based treatment planning in radiation therapy using a deep learning approach *J. Appl. Clin. Med. Phys.* **20** 105–14

[70] Kazemifar S *et al* 2019 MRI-only brain radiotherapy: assessing the dosimetric accuracy of synthetic CT images generated using a deep learning approach *Radiother. Oncol.* **136** 56–63

[71] Jonsson J H *et al* 2010 Treatment planning using MRI data: an analysis of the dose calculation accuracy for different treatment regions *Radiat. Oncol.* **5** 1–8

[72] Kishi T *et al* 2016 Comparative evaluation of respiratory-gated and ungated FDG-PET for target volume definition in radiotherapy treatment planning for pancreatic cancer *Radiother. Oncol.* **120** 217–21

[73] Riou O *et al* 2014 Integrating respiratory-gated PET-based target volume delineation in liver SBRT planning, a pilot study *Radiat. Oncol.* **9** 127

[74] Fan Q *et al* 2013 Toward a planning scheme for emission guided radiation therapy (EGRT): FDG based tumor tracking in a metastatic breast cancer patient *Med. Phys.* **40** 081708

[75] Partouche J *et al* 2019 Evaluation of a prototype treatment planning system (TPS) designed for biology-guided radiotherapy for SBRT of oligmetastases *Int. J. Radiat. Oncol. Biol. Phys.* **104** 1196–97

IOP Publishing

Principles and Practice of Image-Guided Abdominal Radiation Therapy

Yu Kuang

Chapter 17

Advances in verification and delivery techniques

Ting Chen and Hesheng Wang

Treatment verification ensures geometric accuracy of delivered radiation dose. This chapter is focused on several clinically valuable techniques for interfraction patient setup and intrafraction motion management during radiation delivery. We first present the widely implemented ionizing radiation image guidance systems and then elaborate the emerging technique of magnetic resonance (MR)–guided radiotherapy (MRgRT) for patient localization and positioning. We also describe nonionizing radiation motion management solutions such as surface guided radiation therapy (SGRT) and beacon-based electromagnetic tracking system. Commercialized clinical applications of each system are introduced. Quality assurance (QA) procedures and clinical workflow of these systems are discussed in fine detail.

Advances of radiotherapy techniques such as intensity modulated radiation therapy (IMRT), volumetric modulated arc therapy (VMAT), and stereotactic body radiation therapy (SBRT) enable the delivery of highly conformal dose around a tumor with steep dose gradients. These techniques deliver higher doses to the targets for better tumor control while sparing adjacent tissue minimizing normal tissue toxicities. It becomes crucial to set up the patient on a treatment couch as accurately as possible to that in the treatment plan.

Geometric missing is one of the prominent causes for radiotherapy incidents, leading to substantial underdose to a target and/or significant overdose to healthy tissue. The effect is amplified by the tighter conformities, sharper dose gradients, and higher fractionated doses of advanced treatment techniques. The missing may arise from daily setup variation prior to radiation. Patient movement and internal motion such as breathing during treatment can result in significant discrepancies between planned and delivered dose distribution. To minimize interfraction and intrafraction localization uncertainty, pre-treatment position verification and during-treatment motion monitoring have become indispensable components of treatment delivery workflow.

doi:10.1088/978-0-7503-2468-7ch17

Treatment verification is a process to ensure that geometric accuracy of delivered dose is within a tolerance. It is predominantly achieved by comparing localization information at the delivery against those in the treatment plan. Image-guided RT employs various imaging techniques to provide interfraction and/or intrafraction localization and motion information. Those could be two-dimensional (2D) or three-dimensional (3D) anatomy from in-room kilovoltage (kV), megavoltage (MV), cone beam computed tomography (CBCT), and magnetic resonance imaging (MRI); patient surface from body surface imaging; and tumor position from radiofrequency (RF) tracking of implanted transponders. Advances in these imaging techniques of recent years have significantly improved the accuracy and precision of radiation delivery.

17.1 X-ray–based in-room imaging

The past two decades have seen the beginning and dramatic increase of radiation treatment guidance using in-room imaging [1]. Multiple in-room imaging techniques [2–9] have been proposed, developed, and quickly integrated into the clinical workflow of radiation therapy for pre-treatment patient setup and intrafraction motion management. Among all these new techniques, portal imaging (MV), on board kV imaging, MV computed tomography (MVCT), and CBCT, after proper commissioning and calibration, share the common advantage of taking the isocenter of the treatment unit (linear accelerator [Linac]) as the origin in the image domain. Patient images acquired before and during the treatment via these in room imaging systems can be used to locate the target volumes relative to the treatment unit. Under the assumption that the isocenter in the treatment plan conforms to the isocenter of the treatment unit, images acquired using in-room imaging systems are geometrically registerable to treatment planning images and/or beam's eye views (BEVs) digitally reconstructed from the planning images. The registration results quantify the interfraction residual setup error, infrafraction patient motion, and internal anatomy changes. Treatment couch shifts needed for accurate delivery of the treatment plan can be derived from online registration performed at the treatment console to achieve the desired dose distribution.

A typical clinical workflow of an imaging-based patient setup verification was suggested by American Association of Physicists in Medicine (AAPM) Task Group (TG) Report No. 104 [1]. The recommended general clinical workflow (or process) of image guidance in the treatment room using radiographic imaging includes steps shown in figure 17.1, although the actual workflow varies among different institutions, users, imaging systems, and applications for optimal outcome. There are two main correction strategies: online correction (figure 17.1(a)) and offline correction (figure 17.1(b)). With either strategy selected, radiographic imaging could be achieved using any type of in-room imaging systems and any type of imaging methods (kV, MV, CBCT, MVCT, etc). This image-guided patient setup and verification workflow includes three stages of in room imaging: (1) imaging after initial setup (in-room imaging I); (2) imaging after correction (in-room imaging II); and (3) imaging during and/or after treatment (in-room imaging III). Depending on individual application, imaging at stage II and stage III may not be applied to improve overall treatment efficiency and reduce imaging radiation dose to the patient.

(a)

(b)

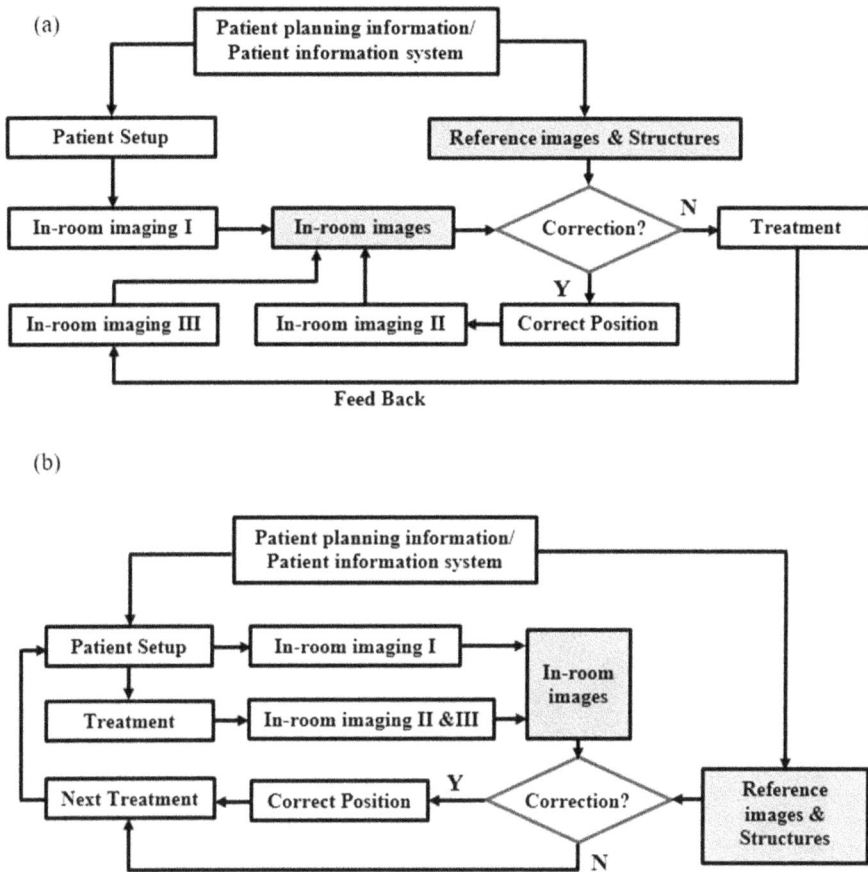

Figure 17.1. General scheme for in-room image-guidance with (a) online and (b) offline corrections. (Based on figure I-C-1 in AAPM TG report 104 [1].)

In the following we will discuss the implementation of each one of these in room imaging system in the treatment setup and verification process for abdominal cancer patients.

17.1.1 MV portal imaging

Portal imaging [10] on a Linac radiation treatment unit made use of the treatment beam. For portal imaging, the x-ray source is the target in the Linac and the imaging x-ray is the MV treatment beam. Thus, it is also known as MV imaging in the clinical environment. MV is a transmission imaging protocol. In current clinical practice, the patient lies on the treatment couch as the MV x-ray penetrates the body before hitting the detector panel known as electronic portal imaging device (EPID) (shown with other OBI components in figure 17.2).

Traditionally, portal images were acquired with film. The advantages of using EPID instead of film for portal imaging is that the images are immediately available

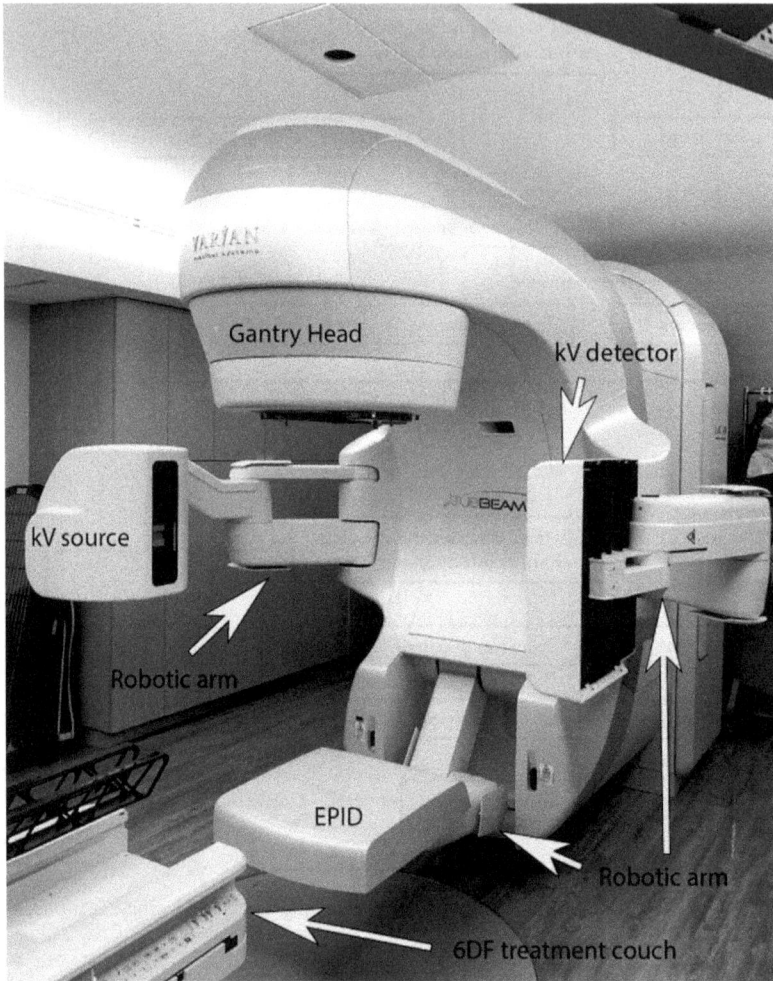

Figure 17.2. OBI components on a Varian TrueBeam Linac.

without the need for developing film (which is costly and time consuming) and that they are digitally ready for the following image processing and enhancement as well as allowing easy access over a computer network to facilitate clinical and research analysis [11]. The motion and positioning of the modern EPID is controlled by robotic arm with high accuracy. The relative position of the center of the EPID against the machine isocenter is calibrated and fine-tuned as part of the imaging system acceptance and commissioning [12] and can be determined precisely during imaging. Therefore, the matching between the MV images acquired via EPID and the 2D planning-image–based digitally reconstructed radiograph (DRR) has intrinsically aligned both images into the BEVs sharing the same coordinate system with identical origin and scale. The spatial difference between the anatomy in the MV image and the DRR can be used to calculate the setup error.

Portal images are acquired using the radiotherapy treatment beam so they exactly show the irradiated area from the BEV. For this reason, portal imaging is useful for treatment setup verification. However, because of MV radiation, portal imaging has inherently lower image quality and contrast compared with kV x-ray images. When used for patient setup verification, portal image is matched with a reference DRR image derived from planning computed tomography (CT) in order to verify that the patient is positioned consistent to the treatment plan. To compensate for the relatively poor image contrast and quality of MV, the matching is usually performed between bony structures or on radiopaque markers implanted prior to radiation therapy [13].

One of the most useful applications of MV imaging is the patient setup verification of tangential fields in whole breast radiation therapy. For tangential fields, MV captures both the ribs (good indication of chest wall-lung boundary) and air flash to warrant a whole breast irradiation. Currently, MV is not commonly used for abdominal imaging if there are other imaging techniques available. However, for facilities in which other imaging options are limited, MV portal imaging can be used to guide 3D conformal treatments. MV is also a good choice for L-spine treatment, as it provides adequate information for the alignment of bony structures, which in this case is the target volume of the treatment (figure 17.3).

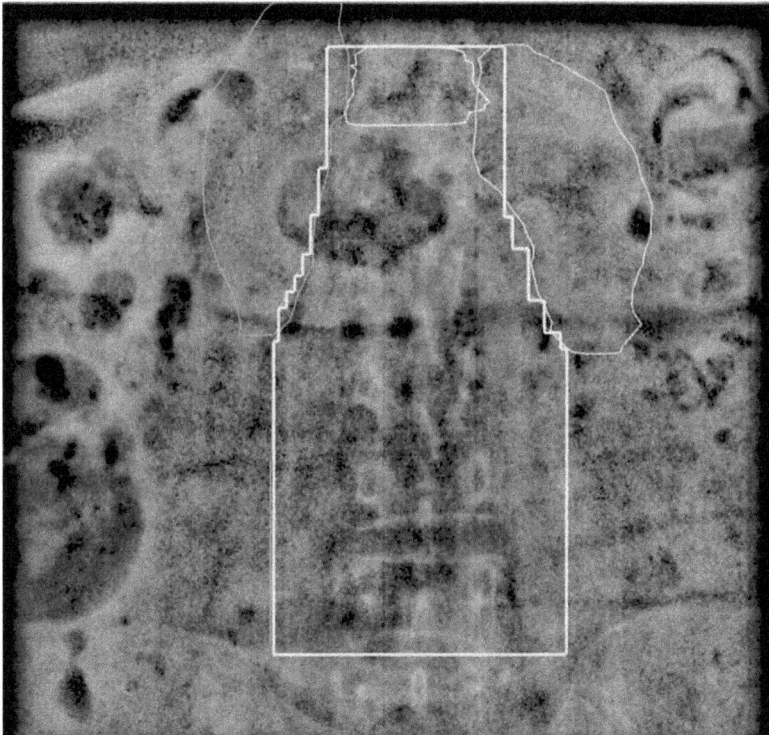

Figure 17.3. L-spine MV image with planned beam aperture (yellow) and critical organ (orange and cyan contour of kidneys) overlaid.

The clinical workflow for MV-based patient setup verification starts during the treatment planning when DRR is generated for each field based on beam geometry and the treatment planning CT. On the treatment/setup day, the patient is initially set up based on triangulation/biangulation markers on the skin surface using stationary lasers to reproduce the patient position during the CT simulation. In the next step, additional translational couch shifts, which are usually provided by the treatment planner, are applied according to the treatment plan. After the shifts, the patient position relative to the machine isocenter should be close to the definition in the treatment plan. However, residual setup error may still exist due to the interfractional patient anatomy change and setup uncertainties. The portal imaging is utilized at this point by moding up the treatment field and adding imaging at treatment console. At the treatment/imaging console, the DRR of the field is loaded as the reference image. In the treatment room, the EPID moves to the detecting position, which usually is located >50 cm away from the treatment machine isocenter to avoid collision with the patient's body or treatment couch top. For the purpose of imaging, usually 2–5 machine units (MU) of treatment beam (beam energy usually <6 MV) will be delivered. The MV beam penetrates the patient's body and interacts with body tissue mostly in the form of Compton scattering. As the beam attenuates in the patient's body, the beam fluence varies spatially according to the anatomy variation in the specific beam path. When the MV beam hits the EPID, the beam fluence will be converted into a 2D array of digital signals, which is the MV portal image. The MV image reflects the patient's internal anatomy the beam transmits. Online image matching can be conducted by therapists to determine the shifts needed to minimize the different between the acquired MV portal image and the corresponding DRR of the loaded field. The image guidance concludes by accepting and applying the shifts via moving the patient or the underlying treatment couch.

MV imaging can be paired with kV for orthogonal imaging-based patient setup in order to avoid gantry rotation of the Linac between orthogonal imaging. Application of kV–MV orthogonal imaging can reduce the risk of gantry collision and enable orthogonal imaging for patients with clearance risks during gantry rotation.

To improve the accuracy of MV-based image guidance, at some clinical practices, fiducial markers are pre-implanted close to the site of the tumor. During image guidance, the high density of the implant will result in high contrast against surrounding soft tissues in the MV images. Under the assumption of minimal migration of the marker away from the tumor during the treatment course, the location of the marker can be used as the indicator of target when matching the MV portal image to the DRR. The use of fiducial markers helps to compensate for the inherent low contrast between target and surrounding soft tissues in MV images.

17.1.2 kV x-ray imaging

kV imaging has been widely used in diagnostic radiology for many years. The clinically implementable solution for the integration of kV imager with modern Linac, on the other hand, was established relatively late [8]. Early commercialized kV imaging integrated radiation treatment systems include Synergy® accelerator, first marketed by Elekta Inc. (Stockholm, Sweden), On-Board Imager® (OBI) kV imaging system marketed by Varian Medical Systems (Palo Alto, CA) starting in 2003, and the ARTISTE™ solution proposed by Siemens (Concord, CA) in 2007. In modern IGRT systems, kV source is a separated kV x-ray tube mounted 90° apart from the treatment gantry on the Linac, and the kV detector is mounted 180° from the kV source, opposing the kV source across the treatment unit isocenter. An example of the OBI system on a Varian TrueBeam system is illustrated in figure 17.2.

Most clinics use a pair of orthogonal kV images either independently for treatment guidance of targets in close approximation of bony structures or as a preliminary imaging step before CBCT. Similar to MV imaging, kV imaging uses planning CT-derived DRR for reference. However, because the kV x-ray source is independent to the treatment beam path, the reference DRR for kV imaging needs to be either explicitly generated for a verification field (independent to treatment fields) in the treatment plan or generated online at the treatment and imaging console. The commonly used imaging angles for kV orthogonal are usually AP (kV source at 0°, treatment system gantry at 90° in the standard isocentric coordinate system) and lateral (kV source at 90° or 270° depending on the imaging site for easier gantry clearance). Between orthogonal imaging, the treatment gantry (and the integrated kV imaging system) needs to rotate around the isocenter to the corresponding imaging position. The positions of the kV source and the kV detector panel are carefully calibrated so that the kV imaging system shares the same rotational isocenter with the treatment unit. The geometry offset between the imaging isocenter and the treatment isocenter is of the amplitude of submillimeter. Unlike portal imaging, which directly determines if the anatomy matches within the beam aperture, kV imaging usually consists the acquisition of a pair of orthogonal images. Orthogonal imaging can localize the target in the 3D spatial domain by retrieving reversely via 3D to 2D projection.

During kV orthogonal imaging, the motion of the kV source and kV detector panel is enabled from the treatment and imaging console for precise positioning. The kV source is usually set at 100 cm away from the isocenter; the position of the kV detector panel varies but is usually set to be at least 50 cm away from the isocenter for clearance during rotation. With given information on the location of kV source and imager, the kV images are automatically scaled to the image plane intersecting the isocenter so that the couch shifts calculated based on image matching are 1:1 to the physical world.

The modern integrated kV imaging system usually has multiple imaging modes, each of which has its unique combination of imaging parameters, e.g. kilovoltage peak (kVp) energy and milliampere-second (mAs). Different kV imaging modes are designed to better image targets located at different part of the patient's body and for patients with different body sizes.

When the kV images are acquired, automated translational image registration and 2D matching to the reference views are performed, followed by online evaluation and manual adjustment to the matching result. To help with viewing and evaluation, tools such as superposition with DRRs, alpha blending, contour overlays, moving spy-glass window, and split windows are usually integrated as part of the image reviewing system.

Similar to MV portal imaging, pre-implanted radiopaque markers were widely used with kV imaging to help the localization of the target. For pre-treatment patient setup verification, the matching between the kV images and the reference view is performed either manually or automatically using software, focusing on the alignment of the markers. The radiopaque markers can also be used with kV imaging for intrafraction target motion tracking in gated treatment. In both clinical scenarios, multiple pre-implanted radiopaque markers are contoured during treatment planning. The 2D projections of marker contours are reconstructed as additional layers in reference DRRs. During radiation treatment, kV images are acquired with relatively high frequency counting in frames per second, and the radiopaque markers with high contrast against surrounding soft tissues in kV images are automatically detected and matched to marker contours overlaid in DRR to determine if the target, which may not be directly visible, is still within close proximity of its position in the treatment plan.

17.1.3 Cone-beam computed tomography

kV imaging systems integrated to treatment unit are also intended, most likely, for the functionality of CBCT. In the CBCT acquisition mode, the kV x-ray source and imager panel rotate around the patient, taking and acquiring projection images at a constant speed along a full or partial arc trajectory of gantry rotation. The projection images will be sent to the CBCT reconstructor workstation to generate the 3D volumetric CBCT of the patient on the couch top. Compared with portal imaging and orthogonal imaging, CBCT can provide real 3D target localization information and with a superior contrast between different types of tissues.

To achieve better image quality and to reduce patient imaging dose, CBCT systems usually have multiple preset acquisition modes corresponding to different parts of the human body. Each acquisition mode has its own set of imaging parameters including energy (kVp), output (mAs), field of view (FOV), length of trajectory (full or partial), and filter (half-fan bowtie, or full-fan bowtie), based on the body size and anatomy component at the targeted human body part. For example, CBCT acquisition mode 'head' has a small FOV, which requires the use of full-fan bowtie filter during imaging, whereas acquisition mode 'pelvic' images has a higher energy x-ray, larger output, and larger FOV, which usually requires the use of full trajectory and half-fan bowtie filter.

As shown in figure 17.4, there are two different bowtie filters mounted in front of the kV source x-ray tube on the filter deck: the half-fan bowtie and the full-fan bowtie. Bowtie filters can be switched between different CBCT modes and are not used for orthogonal kV imaging. The main purpose of the usage of bowtie filter is to improve quality of CBCT projections by reducing x-ray scatter. With the bowtie filter, the CBCT will have better image quality, and the patient will receive a lower skin dose.

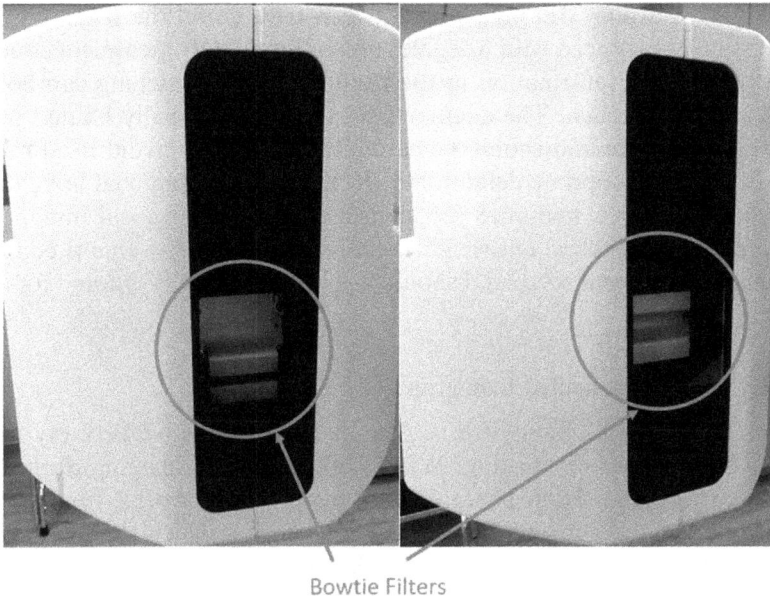

Bowtie Filters

Figure 17.4. Half (left) and full (right) bowtie filters.

CBCT acquisition usually needs to be explicitly requested beforehand by adding a CBCT setup verification field in the treatment plan. Unlike portal imaging and kV orthogonal imaging, the entire 3D planning CT volume as well as planning structure set will be passed to the imaging console as the reference for patient setup. Therapists set up the patient first based on permanent markers on the skin to roughly reproduce the patient position at simulation. Before CBCT, orthogonal imaging, which is usually composed of either a pair of kV or kV–MV, can be performed to further align the boney structures in the setup image and the DRR reference via 2D to 2D matching. Translational couch shifts will be derived based on the matching result and performed. This preliminary imaging step can effectively reduce the residual setup error so that the location of the CBCT target volume during imaging is close to its position in the treatment plan. It is known that automated image matching is more accurate when the difference between images is relatively small. Instead of 2D matching, 3D matching is implemented for CBCT to planning CT matching. In currently commercialized systems, rigid 3D matching can be performed either automatically or manually. For automatic 3D matching, it is critical to carefully define the region of interest (ROI) around the target to match, otherwise, because of the non-rigid deformation of the patient's anatomy, the matching may focus on other part of the patient's body and misalign the target.

In addition to the 3D translational couch shifts (namely vertical, longitudinal, and lateral), angular displacement including the pitch, roll, and yaw needed to align the target anatomy in the acquired image to the reference are also part of the information provided by 3D matching between CBCT and the planning CT. For treatment systems equipped with 6 degrees of freedom (DOF) treatment couch, all 6-DOF transformation information in the result of image matching can be used to adjust and shift the couch. The angular adjustments are usually limited to be less than a preset tolerance and should be carefully reviewed to avoid local minima in image matching. Appropriate definition of ROI and the orthogonal imaging before CBCT can also help to minimize the chance of trapping in local minima during image matching. Due to patient weight, relatively large pitch value is common for abdominal cancer patient setup and should not be intentionally ignored for accurate treatment delivery.

17.1.4 Megavoltage computed tomography

TomoTherapy (Accuray, Sunnyvale, CA) makes use of the 6 MV x-ray treatment source for the purpose of imaging. When being used for imaging during patient setup verification, the TomoTherapy treatment source works similarly to the imaging source in the conventional helical CT, during which the imaging source continuously rotates around the patient while the treatment table is processing at a constant speed. The imaging modality and the 3D image volumes acquired via this modality are usually called MVCT.

MVCT has a robust performance against metal artifacts. This advantage is particularly important when imaging abdominal cancer patients with prostheses implanted. On the other hand, MVCT has a low soft tissue contrast compared with conventional kV CT. Although during TomoTherapy imaging, the dose rate of MV x-ray source output is reduced, the imaging dose of MVCT is still significantly higher than both conventional CT and CBCT.

The MVCT guided patient setup verification process follows the TG 104 guideline. TomoTherapy has a dual laser positioning system: a stationary green laser system that represents the virtual imaging isocenter, and a movable red laser system that will be used for patient repositioning based on image registration between MVCT and the treatment planning CT. In TomoTherapy clinical workflow, the patient will be initially positioned on the treatment table by matching the tattoo or fiducial markers on the skin with the intersections of green lasers. Then the therapists will acquire MVCT and align the target volumes using the imaging interface at the treatment/imaging console. When the image registration is completed and the result is accepted, the treatment couch shift needed for the treatment delivery will be indicated by the red lasers, which will move to the location for therapists to align the surface tattoo/fiducials to their intersections.

MVCT as an imaging modality delivers additional radiation dose to the entire volume imaged. For long treatment courses, it is an optional practice to reduce one fraction from the prescribed treatment to account for the MVCT dose.

17.2 Magnetic resonance imaging

In comparison with CT, MR images have superior contrasts for soft tissue, enabling direct visualization of target volume and anatomy before and during treatment delivery. Combining MR imaging and radiation delivery, MRgRT provides the potentials to monitor intrafractional motion and interfractional anatomical changes without marker implantation or additional radiation to the patient. Accurate patient setup in radiotherapy for abdominal targets is often limited by the low CT contrasts of tumors such as pancreas cancer. Furthermore, the treatment delivery is challenged by the significant motion and deformation of the tumors and abdominal organs. By real-timely showing the targets and organs, MRgRT could become a 'game changer' for radiotherapy in the abdomen [14].

17.2.1 MR-Linac systems

Currently there are four types of MR-guided delivery systems with varying magnetic strengths [14, 15]. The ViewRay MRIdian (ViewRay Inc., Oakwood, USA) integrated 0.35 T magnet with three cobalt-60 heads in its first version and replaces the cobalt sources with a 6 MV flattening filter–free Linac [16]. The Elekta Unity system (Elekta AB, Stockholm, Sweden) combines a 6 MV Linac with 1.5 T magnet for better image quality resulting from the higher field strength. The Canadian Aurora-RT system places a 6 MV Linac in the central opening of a 0.5 T magnet. The Australian magnetic resonance imaging (MRI)–Linac system is being developed by using a 1.0 T magnet with a 4 and 6 MV portable Linac system. These MRgRT systems have demonstrated improved visualization of patient anatomy in RT even with a low field strength of 0.35 T [17]. With these systems, MRgRT has been increasingly gaining acceptance and clinic use [15, 18].

17.2.2 MR-guided treatment verification and delivery

One of the prominent driving forces for MR guidance is the precise tumor localization of MR images, especially for abdominal tumors that typically appear with low contrast in CT images. On-board MR imaging in an MR-Linac enables accurate patient positioning prior to radiation by aligning the target volume and/or soft tissue structures with those in the plan. In comparison with CBCT and MVCT, 0.35 T MR images from a ViewRay MRIdian system were superior on 71% and equivalent on 14% of the structures for visualization of the thorax, head and neck, abdomen, and pelvis [17]. Contrast-enhanced MR images on a ViewRay system can be confidently used for delineating hepatic tumors for MR-guided SBRT [19]. MR simulation is being increasingly adopted in clinics to establish an MR-only radiotherapy workflow in which the treatment is planned on MR images [20]. Online registration of on-board MR images with simulation MR will allow further improvement of patient setup accuracy and reduction of the margin from CTV to PTV. However, MRgRT cannot be used on the patients with metal implants. Additionally, MR guidance could be inferior in assessing high-density structures such as the ribs and vertebral bodies.

MRgRT has an inherent ability to address the intrafraction motion by non-invasively, continuously imaging the target and structures with high soft-tissue contrasts during irradiation. The liver motion in radiotherapy can be monitored with fast cine-MR imaging [21]. Compared with a surrogate-based tracking, the MR guidance showed a higher gain in motion tracking accuracy. Motion tracking using 0.35 T cine-MRI from a ViewRay system was effective in radiotherapy for various sites and potentially optimized gating parameters [22]. Gating on MR motion tracking can be an effective method to manage intrafraction motion [23, 24]. The ViewRay MRIdian system provides a tool to continuously acquire a sagittal plane at four frames per second and, subsequently, real-time deform and segment the ROI for motion tracking and gating [24]. The performance of gated delivery depends on the speed of image acquisition and the time resolution of motion tracking. Scanning one or multiple 2D planes with 2D cine-MR imaging is most commonly used for real-time motion tracking. Right now, it is still practically difficult to acquire, reconstruct, and postprocess 3D MR images at an adequate speed for tracking fast motion. With the rapid development of advanced MR sequences, image acquisition, and postprocessing, MRgRT will greatly mitigate the challenge of intrafraction motion in radiotherapy for sites with substantial motion such as the abdomen.

Beyond daily imaging guidance to minimize intrafraction uncertainty, MRgRT has enabled the holy grail of RT delivery—adaptive re-planning (ART), in which a treatment plan is adjusted online to account for interfraction anatomical changes, tumor and organ deformation, and even tumor physiology and tissue function changes [14, 25]. An online ART workflow using an MRIdian system [26] consisted of (1) acquiring an MR scan of a patient after positioning on the treatment couch; (2) reviewing and manually contouring the target and critical structures on the MR images; (3) evaluating the necessity of re-planning and re-planning if needed; (4) performing QA check for the new plan; and (5) delivering treatment for a fraction. The re-planning compensated for patient anatomy of the day. It could be implemented by simple repositioning of the couch or shifting the pre-treatment plan accordingly. A full plan reoptimization may be performed when significant, non-rigid anatomical changes present [27]. The process from re-contouring to online plan QA took a median time of 26 min [26].

The Elekta Unity MR-Linac provides two workflows for adaption planning in its Monaco treatment planning system (figure 17.5) [28]. One workflow adapts to patient online position, which shifts the isocenter of a pre-treatment plan based on the rigid registration of planning CT with online MR images. Using the pre-treatment CT and contours, the adaptive plan is recalculated and/or reoptimized to achieve optimal dose coverage for current fraction. The other workflow adapts to the patient anatomy. The contours on the initial plan propagate to the daily MR images after deformable registration with simulation CT. Electron densities of the structures on the planning CT is assigned to the corresponding structures on the daily MR Re-planning on the daily MRI is subsequently performed based on the pre-treatment planning objectives. The adaption technique can meet clinical domestic criteria within a planning time between 17 and 485 seconds [28].

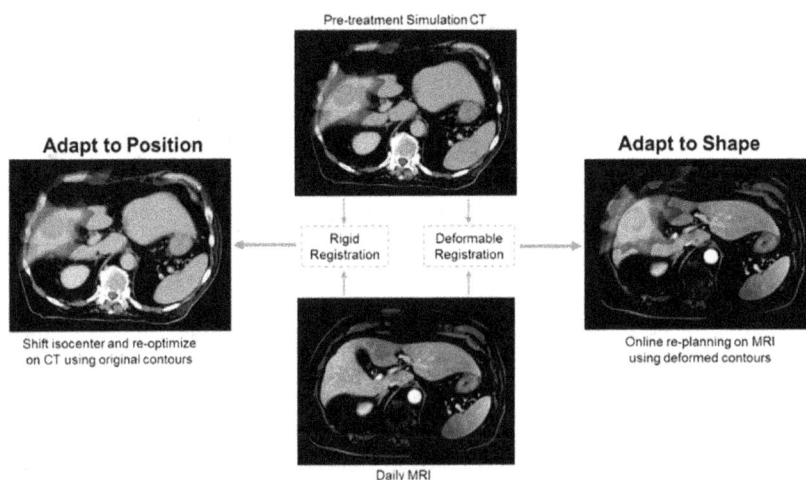

Figure 17.5. Elekta unity MR-Linac online adaptive planning.

17.2.3 MR imaging system QA

QA for a conventional MR scanner and Linac machine in the guidelines of respective TG reports should be performed for an MR-Linac. Additionally, the integration of an MR scanner and a Linac has to be thoroughly tested for safe, proper, and accurate treatment delivery. The smooth and correct communication between subsystems including MR imaging, radiation delivery, treatment planning, record and verify (R&V), and online QA system is essential for commissioning of an MR-Linac. The modification of MR hardware for a hybrid system impacts MR image quality, which should be routinely assessed for different treatment sites, imaging, and radiation settings. In particular, radiotherapy places a higher requirement for geometric accuracy of MR images in a large FOV. The static and dynamic localization accuracies of MR tracking should be characterized in commissioning and verified in periodic QA. Only MR-safe or compatible equipment can be used.

There are presently no QA guidelines specific for the MR system in an MR-Linac machine. The reports on clinical implementation of MR simulation suggested requirements and tests for radiotherapy treatment planning [29, 30], which could be referenced in developing an MR-Linac QA process. A comprehensive MRI commissioning protocol was presented in a four-institutional study using the Elekta Unity system [31]. The protocol consisted of three parts: MR scanner functionality and connectivity check; quality control tests for MR subsystems (static magnetic field, imaging gradients, RF fields) and their interactions with the Linac; and QA for MR image quality. The results can be used as initial bench mark data for the systems which has been installed in many institutions. Recently, a daily end-to-end QA workflow was reported for MR-Linac adaptive RT [32]. The 10-min workflow tested each step in an online ART including daily MR image acquisition and registration, online re-planning, independent dose calculation, physics plan check, and dose delivery verification.

17.3 Real-time position management

To account for a patient's respiratory motion during simulation as well as during the actual treatment, multiple systems have been proposed and developed to provide clinical solutions for respiratory gating. The Real-time Position Management (RPM) system [33] clinically implemented by Varian is one of these systems and has been successfully implemented into various clinical environments.

17.3.1 RPM system

Previous studies have shown there is correlation between the respiratory motion and the movement of surrogates placed on a patient's body surface [34]. The RPM system was developed to track internal respiratory motion utilizing this relationship. The RPM system includes an infrared tracking camera and a reflective marker. The infrared tracking camera is a video camera equipped with an array of LEDs that emit infrared light in the direction in which the camera is pointing. The reflective marker is a lightweight plastic box with multiple (2 or 6) infrared reflective dots on one side. During CT simulation, the tracking camera is mounted on the end of the imaging couch so it moves with the couch during the simulation. The reflective marker is placed on the patient within the view of the infrared camera, usually between umbilicus and xiphoid depending on the treatment site. The relative position of the camera and reflective marker is independent of the couch movement during CT simulation, as both move with the CT couch (and the patient). The infrared LED array is on during simulation, emitting light toward the reflective marker. As the patient breathes, the reflective dots move along with the marker, which moves vertically with the patient surface during the respiratory cycle. Infrared light will be reflected by the moving dots and then captured by the tracking camera. The motion of the patient surface during breathing will be converted into the motion of the marker and eventually the motion of the reflective dots in the captured video signal. The RPM system automatically detects, converts, and measures the range of the motion of the reflective dots and displays them as a waveform representing the patient's respiratory pattern. This way a 2D or 3D (when 6 reflective dots are used) motion signal can be generated for the patient's respiration.

When the RPM system is integrated with the CT simulator, the RPM captured motion signal can be synchronized with the acquisition of each individual CT images based on the time stamp. Therefore, it is possible to determine corresponding respiratory status of the reconstructed CT using the RPM signal. The information about the respiratory status can be useful during treatment planning as it enables the dosimetrists/physicists to plan with CT images acquired at specific respiratory phase or amplitude. The respiratory information of the planning CT can be used to reproduce the patient's position and breathe pattern in the treatment room with the help of the treatment room RPM system.

For RPM to reproduce the patient respiratory signal in the treatment room, the reflective marker needs to be placed on the patient surface at the exact location during simulation. Instead of the mountable tracking camera, a stationary tracking camera is permanently installed on the ceiling in the treatment room, pointing towards the direction of the treatment machine (Linac) isocenter, as a component of the RPM system. The camera is stationary because the treatment couch does not move during in-room imaging and treatment, so the relative position between the camera and the marker is fixed when the patient has been set up on the couch. During gated imaging or treatment, the tracking camera will capture the patient surface motion signal via the reflection of the reflective marker similarly as during the simulation. An amplitude or respiratory phase range for gating should have been pre-determined during treatment planning based on the institutional clinical and physics practice. The radiotherapy is gated as the radiation beam is enabled by the RPM system only when the patient breathing signal is within the beam-on range.

17.3.2 Clinical applications of RPM in abdomen

To provide image guidance for abdominal radiotherapy, RPM has been widely used in the reconstruction of four-dimensional CT (4DCT) [35–37] and to manage the beam delivery during gated radiotherapy as well as deep inspirational breath holding (DIBH) [38] treatment.

The 4DCT consists of multiple CT images of the same target acquired over time. When imaging over respiratory cycle(s) as the patient breathes normally, the 4DCT can be used to record the internal anatomy at each respiratory phase. The internal organ motion over the respiratory cycle can be tracked if 4DCT is played in sequential order. The theory and specific acquisition and reconstruction methodology of 4DCT is beyond the scope of this chapter. However, it is of our interest to discuss the role the RPM system played in 4DCT reconstruction.

The reconstruction of 4DCT relies on accurately regrouping CT images into corresponding respiratory phases based on the time they were acquired. RPM system is one of the solutions that provide the phase information to the reconstruction process. The RPM system communicates with the CT simulator so that the RPM-generated amplitude-temporal motion curve is recorded using the timer of the CT. All the CT images have the time stamp of their own acquisition time so the surface motion at the acquisition time of each CT image can be identified using the RPM motion curve. There are different methods to determine respiratory phases given the motion curve. Currently, the RPM system determines the respiratory phase of any given point on the motion curve by picking the peak (local maxima) points on the curve. The picking of these cusp points is performed by the RPM system automatically and can be manually corrected later if wrong points are picked or points are missed, as shown in figure 17.6. The peak points represent max inhalation and can be assigned as 0% phase or the beginning of a new respiratory cycle and, simultaneously, 100% phase or the end of the last respiratory cycle. The phase of other points on the motion curve can be determined using interpolation between peak points (from 0% to 100% phase). When phase is determined for points on the amplitude–temporal motion curve, it is also determined for the time when the point on the motion curve was detected. As explained before, the motion curve shares the same timer with CT acquisition. Therefore, the phase information for

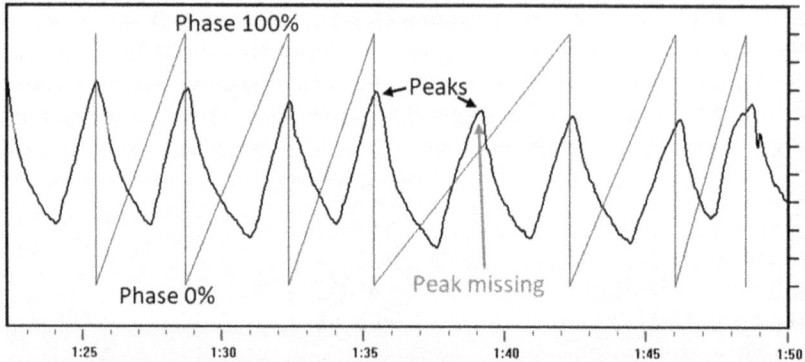

Figure 17.6. Determine phase of points on the motion curve. Peak points are automatically selected by the RPM system. There are missing points which need to be manually corrected to generate correct phase information for points on the motion curve. Once the phase is determined for a point on the motion curve, it is also determined for the time (*x*-axis) of the point on the motion curve.

each individual CT images can be determined through the RPM system using the acquisition time.

If the treatment plan is based on only certain respiratory phase(s) of the 4DCT, then during treatment the radiation will be enabled for delivery only when the patient's breath is in the same phase(s). This is called the gated radiotherapy. Depending on the signal for beam on, gated radiotherapy can be divided into phase-gated and amplitude-gated radiotherapy, respectively.

The RPM system in the treatment room serves the purpose to trigger the beam-on signal when the patient's breath pattern and phase/amplitude matches the planning CT and block beam delivery when the patient's breath is out of phase/amplitude during gated treatment. To enable the radiation beam at the right phase/amplitude, during treatment planning, the planner needs to enable 'gating' for each beam and define the range of gating signal. The RPM system is integrated with the treatment and imaging console and has an interface where the therapists can edit the beam-on phase/amplitude at the console. During gated treatment and imaging, the RPM will capture the patient motion and the phase/amplitude of motion curve will be determined in real time to enable or disable radiation. For phase-gated radio-therapy, a training stage is needed for the system to collect prior information about the patient respiratory cycle in order to calculate the respiratory phase before the completion of one respiratory cycle.

Another way of gated radiotherapy is to ask the patient to hold their breath after taking a max inhalation during CT simulation and treatment delivery. This approach is called the DIBH. The major benefits of using DIBH for abdominal cancer treatment is to reduce the GTV to PTV margin compared against the use of internal target volume (ITV). Although there are no periodic respiratory motion signals to catch, RPM is still a critical component for DIBH as it can be used to record the amplitude of max inhalation. For DIBH simulation, the patient will be asked to take a deep breath and hold it under coaching. The amplitude of

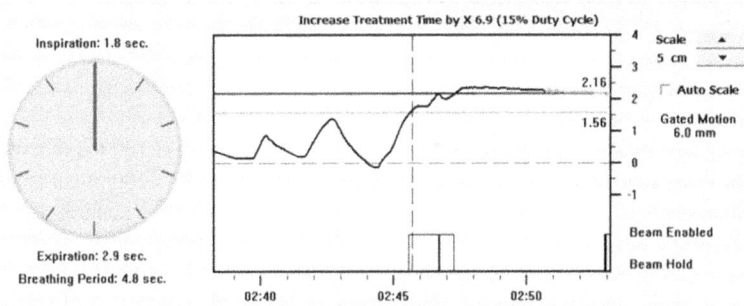

Figure 17.7. Cropped screenshots of RPM interface for DIBH treatment. As shown in the figure, the radiation beam is enabled only when the amplitude of the capture motion of the reflective marker (in black) is within the DIBH range (1.56–2.16 cm, indicated by the orange and blue line, respectively). The motion of the marker has high correlation to the respiratory motion.

Table 17.1. AAPM TG142 monthly QA recommendations on respiratory gating system.

Respiratory gating	
Beam output constancy	2%
Phase, amplitude beam control	Functional
In-room respiratory monitoring system	Functional
Gating interlock	Functional

the max inhalation is recorded and will be used to determine a range in which the radiation beam is enabled. For treatment planning, the dose distribution in the plan will be calculated using the CT acquired with DIBH. The patient will be asked to hold their breath at max inhalation in the treatment room to reproduce the anatomy in the planning CT, and the breath motion will be tracked by RPM to enable beam on only when the breath pattern is in the range of DIBH, as shown in figure 17.7.

17.3.3 QA of RPM

As a respiratory gating system, RPM requires periodic QA to warranty its functionality and accuracy for clinical application. Deteriorating or loss of the gating functionality and/or imperfect integration with the beam delivery system can cause severe damage to normal tissues and mistreatment to the target volume. In AAPM TG142 [39], recommendations on monthly (table 17.1) and annual QA checklists (table 17.2) for respiratory gating system have been given. QA procedures for RPM systems in the clinical use may adopt these recommendations. In additional, end-to-end QA making use of 4D motion phantom can be used for QA and commissioning of clinical procedures that require higher accuracy such as SBRT.

Table 17.2. AAPM TG142 annual QA recommendations on respiratory gating system.

Respiratory gating	
Beam energy constancy	2%
Temporal accuracy of phase/amplitude gate on	100 ms of expected
Calibration of surrogate for respiratory phase/amplitude	100 ms of expected
Interlock testing	Functional

17.4 Optical surface imaging

Optical surface imaging systems continuously render a patient's body surface by touchless optical scanning. Surface guided radiation therapy (SGRT) compares the online surface with the patient's body surface derived from simulation images, enabling accurate patient positioning and motion monitoring without additional radiation to the patient. SGRT has been increasingly finding applications in RT for the breast, head and neck, abdomen, prostate, and extremities as well as intracranial SRS and SBRT [40, 41]. The respiratory surrogate created from surface imaging correlated well with spirometry measurement [42]. Used as respiratory surrogates, surface imaging allows treatment of liver and gastric cancer in breath-holds.

17.4.1 Optical surface imaging systems

Although several open-source imaging solutions have been reported for clinical or research uses [43], most clinical SGRT uses commercially available optical surface imaging systems including AlignRT (VisionRT, London, UK), Catalyst (C-Rad, Upsalla, Sweden), and Identify (Humediq, Grunwald, Germany). These systems have been developed independently from the vendors manufacturing treatment delivery systems but are integrated with current gantry-based delivery systems to support treatment guidance.

A clinical AlignRT system typically mounts three camera units on the ceiling of treatment room to visualize the patient's body in a maximum view (figure 17.8). A camera unit consists of a projector and high-definition camera together placed on a pod. The projector emits red light containing a pseudo-random speckle pattern projected on patient skin, and the camera real-time sensors the textural patterns to generate body surface images. The surfaces are fed into the software station to evaluate position differences based on rigid registrations with a reference surface image. The reference surface can be a body contour extracted from the treatment planning images or a captured surface image of the patient at a confirmed position on treatment couch. The registration performed on a user-defined ROI generates the three axis shifts and rotations for a 6-DOF couch which is displayed in real time to therapists in the treatment room (figure 17.8). Thereby, the system allows setup a patient as closely as possible to the planned position. It also alerts therapists if the patient position is out of a preset tolerance during monitoring.

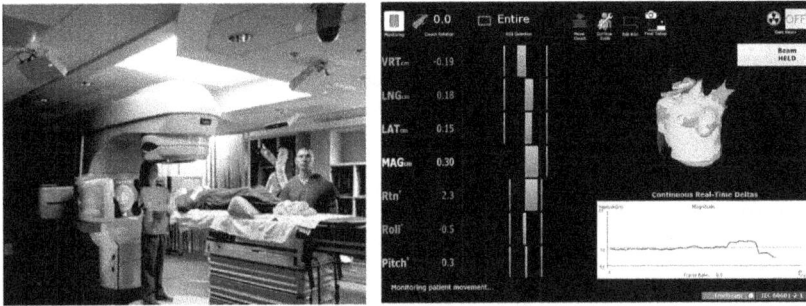

Figure 17.8. Left: AlignRT in treatment room. Right: AlignRT user interface shows 6-DOF displacements, reference surface and live detected surface and dynamic motion. Reprinted from [40], Copyright (1991), with permission from Elsevier.

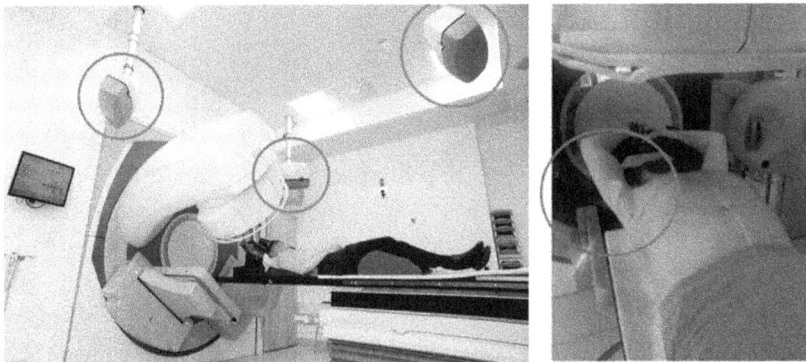

Figure 17.9. Left: Catalyst system in the treatment room. Right: Position deviations projected on the patient's surface. Reprinted from [44], Copyright (2002), with permission from SAGE.

The Catalyst system employs three ceiling-mounted units of a light projection and an optical surface scanning device (figure 17.9). The projector emits light in a wavelength of 405 nm and the camera captures the projection to form a map of the patient's topography. The three units are arranged at an angle of approximately 120° to each other in the treatment room to enable continuous scanning of the patient during gantry rotation. The Catalyst software online performs finite element model-based deformable registration between acquired surfaces and the reference surface to provide local shifts of the patient body. Six DOF shifts of the couch are derived from the calculated displacement of the isocenter. The Catalyst presents not only couch shifts but also patient posture differences from the reference treatment position. The surface differences are colormap projected on the patient body with red and green light to visually guide therapists to correct patient posture and position (figure 17.9).

The Identify is another surface imaging system that was recently acquired by Varian Medical Systems and integrated with the Varian Halcyon and Edge LINAC machines. The Identify system reconstructs body surfaces using four ceiling-mounted camera pods, including one central, two lateral, and one time of flight, each of which has a projector and high-definition camera. Additionally, the system features an RF identification reader for accessary verification. The 6-DOF couch shifts for patient positioning are calculated from rigid registration between the captured and reference surfaces and real-time displayed as error bars. The shifts simultaneously show as live color-overlap video to provide augmented reality assistance for adjusting the patient.

All three systems can provide positioning accuracy below 1.0 mm and 1.0° [45]. Meanwhile, they all can operate in a gating mode in both simulation and treatment room for respiratory motion management. AlignRT obtains breathing signal by tracking the motion of a patient's abdomen surface in simulation. Surface monitoring the motion in treatment, AlignRT can trigger beam on/off based on pre-defined phases on the respiratory signal. The Catalyst system uses a laser to measure the rise and fall of a patient's surface as a surrogate for the respiration cycle for 4D-CT and breath-hold scan. It has been integrated with Elekta Synergy Linac machine for gating delivery [46].

17.4.2 SGRT verification and delivery

All the surface imaging systems functionally provide 6-DOF shifts of the treatment couch and visual guidance for patient positioning. Furthermore, the systems track a patient's surface motion in real time and can operate in gating mode to hold beam accordingly. The systems generally play a similar role in radiotherapy workflow including aiding in initial patient setup and monitoring patient motion in the duration of radiation delivery.

Initial setup of a patient typically starts from aligning patient skin markers (biangulation or triangulation) tattooed at simulation with treatment-room lasers that intersect at the machine isocenter. The skin markers where radio-opaque markers (BBs) are attached for imaging present as bright landmarks in simulation CT. A plan is designed on the simulation CT, and the treatment isocenter is set with known couch shifts from the landmark intersection. Moving the couch with the planed shifts supposedly aligns the treatment isocenter with machine isocenter.

Subsequently, the surface imaging system is used to verify matching of the patient's posture and position with those in the simulation CT. The therapists first select whole body or a large body surface for gross alignment and then a smaller ROI relating to the target volume for fine tuning. The ROI needs to be selected with enough size and curvature for reliable surface registration. Surface guidance does not provide direct localization of the patient's anatomy and target volume. MV/kV orthogonal imaging is typically followed to set up the patient based on alignment of projection images with DRRs created from the planning CT. Additionally, volumetric CBCT that visualizes soft tissue may be acquired for final alignment of target volume and patient anatomy.

Patient setup from radiographic and/or volumetric CBCT verification is usually considered the gold standard. Afterward, the surface imaging system acquires the patient surface on an ROI as a reference to monitor patient movement or respiratory

Index immobilization on treatment couch

Position patient using the skin tattoos and planned couch shifts

Define surface ROI for positioning using surface imaging

Adjust patient posture and position based on surface alignment

Perform MV/kV and/or CBCT-guided couch shifts

Capture a surface image as reference for motion tracking

Deliver radiation with monitoring or gating by surface imaging

Beam hold if tracking motion out of tolerance is detected

Redo imaging-guided patient positioning

Re-capture a surface image as reference

Resume radiation delivery with surface imaging motion tracking

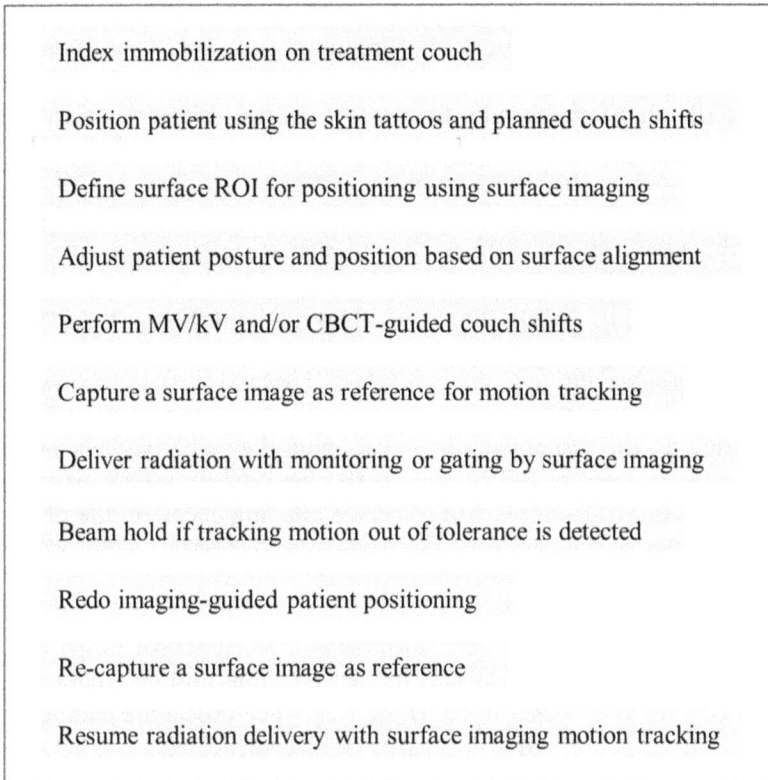

Figure 17.10. SGRT treatment delivery workflow.

motion in the duration of radiation delivery. The system pauses radiation beam automatically if in gating mode or manually by the therapist when a motion out of tolerance is detected. If the motion is continuously out of tolerance or the patient position becomes unreliable, re-adjustment and verification of the patient position need to be done with radiographic or CBCT imaging. A new reference surface should be subsequently acquired and the SGRT delivery can resume for the remaining treatment fraction. A possible workflow of SGRT is shown in figure 17.10.

17.4.3 Surface imaging system QA

AAPM TG-147 provides guidelines for QA for nonradiographic localization and positioning systems including the optical surface imaging technique [47]. According to TG-147, the following evaluations are needed for the commissioning of a surface imaging system.

- *Verify the integration of the system in a radiotherapy environment*: This evaluation ensures the communication with the treatment planning system,

R&V system, and the Linac, such as faithful data transfer, correct patient selection with R&V, interlock functionalities, etc. End-to-end tests on anthropomorphic phantoms such as the IMT MAX-HD head phantom or MAX-EI phantom should be performed in different orientations for targets at various locations. Functionalities of the system should be assessed, such as the image quality and position accuracy of the optical surfaces created under different camera exposure rates, lighting conditions, and the phantom in various clinical scenarios with obstruction for the cameras. Thresholds to create body surfaces from simulation images need be established by comparing with the actual surfaces generated from the camera system.

- *Assess reproducibility and temporal drift*: Thermal drift of the system can be evaluated by monitoring motion of a fixed phantom immediately from powering up the camera system to reaching the stability. This assessment is important to prevent SGRT use in the period for system warm-up. After stability is established, repeating surface imaging check on the phantom at a periodic interval such as every 5 min characterizes system's reproducibility for spatial localization.

- *Evaluate the accuracy of static localization*: Accuracy of target localization should be assessed under various patient positions and obstruction scenarios. The accuracy is expected to be below 1 cm. The procedure can be an end-to-end test to compare SGRT-measured displacements with applied couch shifts on an anatomical phantom. Their coincidences over a full clinical range need be established with a shift increment of 1 cm in each axis.

- *Determine the dynamic tracking capabilities:* A dynamic radiation delivery phantom such as QUASAR respiratory motion phantom can be used for this test. An end-to-end dynamic localization and delivery SGRT test evaluates the temporal and spatial accuracies of the guidance system for motion tracking. Meanwhile, gating delivery with a motion threshold typically used in clinic should be tested. The AAPM TG-142 report set the tolerance for output accuracy of gate mode.

Routine periodic QA including daily, monthly, and annual tests use a part of or similar procedures in the commissioning to ensure safety and accuracy of the system in clinical use. The methods, frequencies, and tolerances are also recommended in the TG-147 report. Typically, the system vendor provides phantoms and procedures for routine QA that should be performed in addition to any tests specified in the individual institution.

17.5 Electromagnetic tracking

Electromagnetic tracking is a market-based method that wirelessly localizes the markers and monitors their motion by sensing RF signal emitted from the marker transponders. These small transponders are surgically implanted in a volume of interest within the body, by which the position and motion of the volume can be continuously tracked. Such direct localization of the target volume can reduce interfraction and intrafraction positioning uncertainties, allowing a smaller margin from CTV to PTV for better sparing of normal tissue.

17.5.1 Electromagnetic tracking systems

The Calypso 4D system (Calypso Medical Technologies, Inc. Seattle, WA) is an electromagnetic tracking system that has been US Food and Drug Administration approved for use in prostate radiotherapy and is undergoing clinical trials for the liver [44] and other sites. The system consists of five main components: beacon transponders, electromagnetic array, a mobile console, an optical system, and a tracking station (figure 17.11).

A beacon transponder with a diameter of approximately 2 mm and a length of 8.0–8.7 mm contains a passive alternating current (AC) electromagnetic resonance circuit encapsulated in glass. Three beacons are typically implanted in target volume for localization in simulation and treatment. In the treatment room, the electromagnetic array mounted in a flat panel is placed above the patient. The array consists of four RF signaling coils and 32 receiving coils: the RF coils emit RF waves to excite the transponders, excited transponders send out RF signals in their unique resonant frequencies, and the receiving coils sensor the resonant signals. The array panel connects with a mobile console that communicates with the array to generate and detect the signals for transponders localization.

The panel also has nine optical targets embedded in its top that enable the in-room optical system to measure the array's position relative to the machine isocenter. The optical system is composed of three ceiling-mounted infrared cameras connected to a tracking station in the control room. The tracking station receives data from the array mobile console and the optical system, online interprets the beacon positions relative to the machine isocenter. Thereby, actual treatment

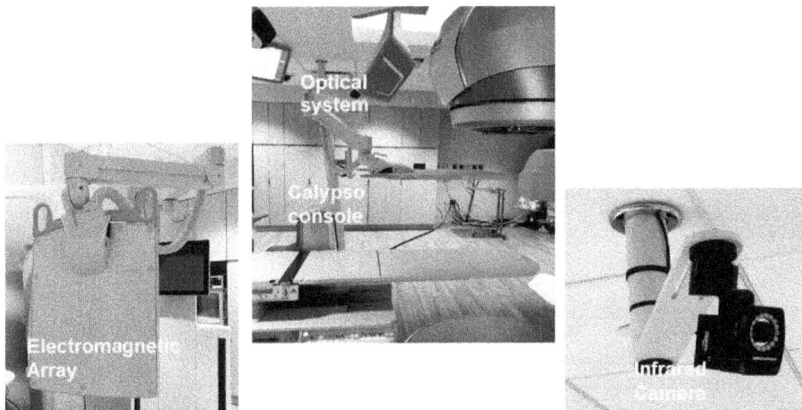

Figure 17.11. Calypso 4D localization system.

isocenter is continuously determined and compared with the planning value for a 6-DOF shift. The shift displays in real time on the tracking console for localization and motion guidance.

17.5.2 RF tracking-guided verification and delivery

Electromagnetic tracking starts with surgically implanting the transponders in the tissue of interest a few days before simulation for position stabilization. The simulation CT need be acquired in a high resolution to ensure that a transponder can be accurately contoured in the images to determine its radiographic center. The coordinates of the transponders and treatment isocenter identified in treatment planning are imported and used as the reference in RF tracking systems for treatment delivery. It is recommended that the treatment isocenter is placed at or near the centroid of the transponders.

The tracking system can be used in the mode of localization and/or tracking with/without gating in a similar workflow as the SGRT (figure 17.12). Localizing the transponders and treatment isocenter, the system dynamically displays target displacement and guides patient setup prior to irradiation. MV/kV imaging and/or CBCT-guided patient positioning is typically followed and considered as the gold standard because RF tracking is limited within the implanted volume and the

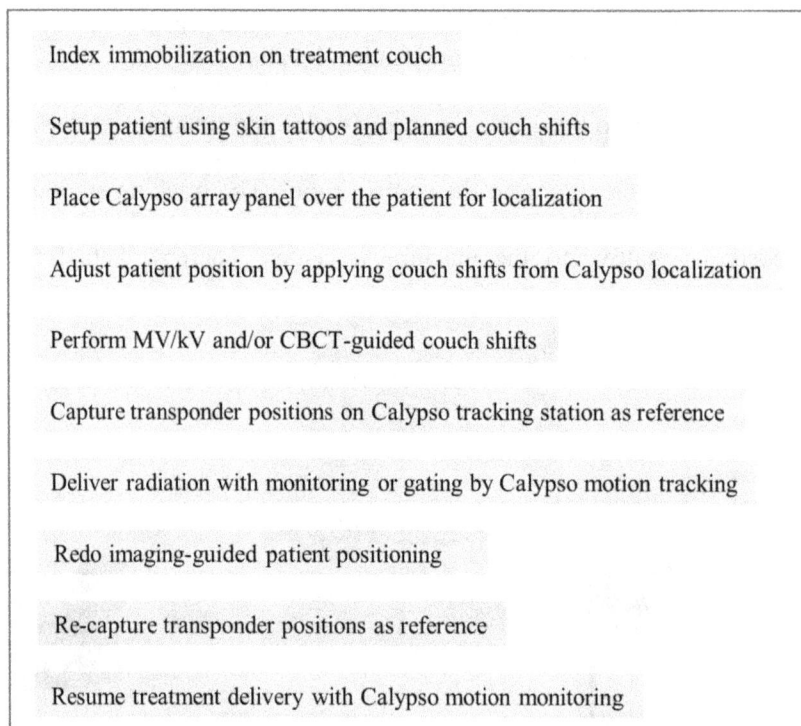

Index immobilization on treatment couch

Setup patient using skin tattoos and planned couch shifts

Place Calypso array panel over the patient for localization

Adjust patient position by applying couch shifts from Calypso localization

Perform MV/kV and/or CBCT-guided couch shifts

Capture transponder positions on Calypso tracking station as reference

Deliver radiation with monitoring or gating by Calypso motion tracking

Redo imaging-guided patient positioning

Re-capture transponder positions as reference

Resume treatment delivery with Calypso motion monitoring

Figure 17.12. A Calypso guided radiation delivery workflow.

transponders may migrate from fraction to fraction. Immediately afterwards, the transponder positions are captured as the reference for subsequent motion monitoring during treatment. The target motion is displaced with reference to a motion tolerance in real time on the tracking station. When a motion exceeds the limit, radiation beam is held if gating mode is enabled. If not gating, the beam can be manually paused and then resumed when the motion is back to the limit. However, if the motion is out of the tolerances for a substantial time, imaging-guided repositioning and recapturing reference for RF tracking should be performed before resuming radiation delivery.

A tracking limit describes an acceptable range of displacement from the reference position for each axis, which need be pre-defined based on clinical practice in the individual institution. One of the factors is whether the tumor motion is incorporated in expanding CTV to PTV for treatment planning. For radiotherapy to the abdomen, the impact of patient respiration on a target is important in deciding the motion tolerance. RF tracking can follow respiratory cycle with a frequency of 10 Hz. It is suggested for monitoring treatment in free breathing to capture RF tracking reference at the exhalation when extreme motion presents. With a limit based on the full range of target motion anticipated in respiration cycles, this reference maximizes the time for beam-on while catching abnormal patient movement.

Integrating with a RF tracking system, a dynamic MLC tracking technique is being evaluated in a multi-institutional and industrial collaboration [48]. The technique dynamically repositions treatment beams to follow target position provided in real time by a tracking system. This delivery had a higher time efficiency than gating while compensating intrafraction motion at the same time. In addition to translational shift, the pitch, roll, and yaw from a RF system may enable MLC tracking for a more complex motion of a target [49].

17.5.3 RF tracking system QA

The TG-147 report described commissioning and QA for electromagnetic tracking systems. The commissioning verifies the full integration with a Linac system, measures system accuracies for static localization and dynamic motion tracking, and understands its performances and limitations in various clinical scenarios. Most of the tests are performed with end-to-end procedures on vendor-provided phantoms such as the QA phantom for the Calypso system [50]. The procedures involve simulation CT of the Calypso QA phantom which has three beacon transponders implanted. Treatments are planned on the CT for treatment isocenter placed at the centroid of contoured beacons and at locations with offsets from the centroid. Calypso guided deliveries of the plans allow verification of the localization accuracy in which the actual phantom position should be confirmed by onboard MV/kV imaging [51]. The accuracy for localization displacement can be evaluated on the phantom by applying known 6-DOF couch shifts.

Periodic calibration is essential for proper operation and constant accuracy of a RF tracking system. It includes camera calibration that ensures each camera has clear visualization of the array in full FOV surrounding the machine isocenter, and system calibration that aligns the optical camera system with machine isocenter. For the Calypso system, camera calibration checks its correct and uniform detection of optical targets on a L-frame fixture while moving the fixture throughout the camera's field of view [50]. System calibration is performed by accurately localizing an isocenter fixture that has implanted beacons and is aligned at the treatment isocenter. The vendor recommends that system calibration is performed as a part of monthly QA and camera calibration is conducted when it deems necessary such as system calibration failure.

References

[1] Yin F-F, Wong J and Balter J *et al* 2009 *The Role of In-Room kV X-Ray Imaging for Patient Setup and Target Localization: Report of Task Group 104 of the Therapy Imaging Committee American Association of Physicists in Medicine* 104 American Association of Physicists in Medicine

[2] Cho Y and Munro P 2002 Kilovision: thermal modeling of a kilovoltage x-ray source integrated into a medical linear accelerator *Med. Phys.* **29** 2101–8

[3] Ostapiak O Z, O'Brien P F and Faddegon B A 1998 Megavoltage imaging with low Z targets: implementation and characterization of an investigational system *Med. Phys.* **25** 1910–8

[4] Biggs P J, Goitein M and Russell M D 1985 A diagnostic x ray field verification device for a 10 MV linear accelerator *Int. J. Radiat. Oncol. Biol. Phys.* **11** 635–43

[5] Uematsu M, Fukui T and Shioda A *et al* 1996 A dual computed tomography linear accelerator unit for stereotactic radiation therapy: a new approach without cranially fixated stereotactic frames *Int. J. Radiat. Oncol. Biol. Phys.* **35** 587–92

[6] Faddegon B A, Wu V, Pouliot J, Gangadharan B and Bani-Hashemi A 2008 Low dose megavoltage cone beam computed tomography with an unflattened 4 MV beam from a carbon target *Med. Phys.* **35** 5777–86

[7] Mackie T R, Holmes T and Swerdloff S *et al* 1993 Tomotherapy: a new concept for the delivery of dynamic conformal radiotherapy *Med. Phys.* **20** 1709–19

[8] Jaffray D A, Drake D G, Moreau M, Martinez A A and Wong J W 1999 A radiographic and tomographic imaging system integrated into a medical linear accelerator for localization of bone and soft-tissue targets *Int. J. Radiat. Oncol. Biol. Phys.* **45** 773–89

[9] Jaffray D A and Siewerdsen J H 2000 Cone-beam computed tomography with a flat-panel imager: initial performance characterization *Med. Phys.* **27** 1311–23

[10] Langmack K A 2001 Portal imaging *Br. J. Radiol.* **74** 789–804

[11] Leszczynski K W and Shalev S 1989 Digital contrast enhancement for online portal imaging *Med. Biol. Eng. Comput.* **27** 507–12

[12] Herman M G, Balter J M and Jaffray D A *et al* 2001 Clinical use of electronic portal imaging: report of AAPM Radiation Therapy Committee Task Group 58 *Med. Phys.* **28** 712–37

[13] Vigneault E, Pouliot J, Laverdiere J, Roy J and Dorion M 1997 Electronic portal imaging device detection of radioopaque markers for the evaluation of prostate position during megavoltage irradiation: a clinical study *Int. J. Radiat. Oncol. Biol. Phys.* **37** 205–12

[14] Pathmanathan A U, van As N J and Kerkmeijer L G W *et al* 2018 Magnetic resonance imaging-guided adaptive radiation therapy: a 'game changer' for prostate treatment? *Int. J. Radiat. Oncol. Biol. Phys.* **100** 361–73

[15] Das I J, McGee K P, Tyagi N and Wang H 2019 Role and future of MRI in radiation oncology *Br. J. Radiol.* **92** 20180505

[16] Kluter S 2019 Technical design and concept of a 0.35 T MR-Linac *Clin. Transl. Radiat. Oncol.* **18** 98–101

[17] Noel C E, Parikh P J and Spencer C R *et al* 2015 Comparison of onboard low-field magnetic resonance imaging versus onboard computed tomography for anatomy visualization in radiotherapy *Acta Oncol.* **54** 1474–82

[18] Corradini S, Alongi F and Andratschke N *et al* 2019 MR-guidance in clinical reality: current treatment challenges and future perspectives *Radiat. Oncol.* **14** 92

[19] Wojcieszynski A P, Rosenberg S A and Brower J V *et al* 2016 Gadoxetate for direct tumor therapy and tracking with real-time MRI-guided stereotactic body radiation therapy of the liver *Radiother. Oncol.* **118** 416–18

[20] Schmidt M A and Payne G S 2015 Radiotherapy planning using MRI *Phys. Med. Biol.* **60** R323–61

[21] Paganelli C, Seregni M and Fattori G *et al* 2015 Magnetic resonance imaging-guided versus surrogate-based motion tracking in liver radiation therapy: a prospective comparative study *Int. J. Radiat. Oncol. Biol. Phys.* **91** 840–8

[22] Mazur T R, Fischer-Valuck B W, Wang Y, Yang D, Mutic S and Li H H 2016 SIFT-based dense pixel tracking on 0.35 T cine-MR images acquired during image-guided radiation therapy with application to gating optimization *Med. Phys.* **43** 279

[23] Fischer-Valuck B W, Henke L and Green O *et al* 2017 Two-and-a-half-year clinical experience with the world's first magnetic resonance image guided radiation therapy system *Adv. Radiat. Oncol.* **2** 485–93

[24] Henke L, Kashani R and Robinson C *et al* 2018 Phase I trial of stereotactic MR-guided online adaptive radiation therapy (SMART) for the treatment of oligometastatic or unresectable primary malignancies of the abdomen *Radiother. Oncol.* **126** 519–26

[25] Kupelian P and Sonke J J 2014 Magnetic resonance-guided adaptive radiotherapy: a solution to the future *Semin. Radiat. Oncol.* **24** 227–32

[26] Acharya S, Fischer-Valuck B W and Kashani R *et al* 2016 Online magnetic resonance image guided adaptive radiation therapy: first clinical applications *Int. J. Radiat. Oncol. Biol. Phys.* **94** 394–403

[27] Bohoudi O, Bruynzeel A M E and Senan S *et al* 2017 Fast and robust online adaptive planning in stereotactic MR-guided adaptive radiation therapy (SMART) for pancreatic cancer *Radiother. Oncol.* **125** 439–44

[28] Winkel D, Bol G H and Kroon P S *et al* 2019 Adaptive radiotherapy: the Elekta Unity MR-linac concept *Clin. Transl. Radiat. Oncol.* **18** 54–9

[29] Kapanen M, Collan J, Beule A, Seppala T, Saarilahti K and Tenhunen M 2013 Commissioning of MRI-only based treatment planning procedure for external beam radiotherapy of prostate *Magn. Reson. Med.* **70** 127–35

[30] Paulson E S, Erickson B, Schultz C and Allen Li X 2015 Comprehensive MRI simulation methodology using a dedicated MRI scanner in radiation oncology for external beam radiation treatment planning *Med. Phys.* **42** 28–39

[31] Tijssen R H N, Philippens M E P and Paulson E S *et al* 2019 MRI commissioning of 1.5T MR-linac systems – a multi-institutional study *Radiother. Oncol.* **132** 114–20

[32] Chen X, Ahunbay E, Paulson E S, Chen G and Li X A 2020 A daily end-to-end quality assurance workflow for MR-guided online adaptive radiation therapy on MR-Linac *J Appl. Clin. Med. Phys.* **21** 205–12

[33] Kubo H D, Len P M, Minohara S and Mostafavi H 2000 Breathing-synchronized radiotherapy program at the University of California Davis Cancer Center *Med. Phys.* **27** 346–53

[34] Hoisak J D, Sixel K E, Tirona R, Cheung P C and Pignol J P 2004 Correlation of lung tumor motion with external surrogate indicators of respiration *Int. J. Radiat. Oncol. Biol. Phys.* **60** 1298–306

[35] Low D A, Nystrom M and Kalinin E *et al* 2003 A method for the reconstruction of four-dimensional synchronized CT scans acquired during free breathing *Med. Phys.* **30** 1254–63

[36] Pan T, Lee T Y, Rietzel E and Chen G T 2004 4D-CT imaging of a volume influenced by respiratory motion on multi-slice CT *Med. Phys.* **31** 333–40

[37] Rietzel E, Pan T and Chen G T 2005 Four-dimensional computed tomography: image formation and clinical protocol *Med. Phys.* **32** 874–89

[38] Korreman S S, Pedersen A N, Nottrup T J, Specht L and Nystrom H 2005 Breathing adapted radiotherapy for breast cancer: comparison of free breathing gating with the breath-hold technique *Radiother. Oncol.* **76** 311–8

[39] Klein E E, Hanley J and Bayouth J *et al* 2009 Task Group 142 report: quality assurance of medical accelerators *Med. Phys.* **36** 4197–212

[40] Hoisak J D P and Pawlicki T 2018 The role of optical surface imaging systems in radiation therapy *Semin. Radiat. Oncol.* **28** 185–93

[41] Krengli M, Gaiano S and Mones E *et al* 2009 Reproducibility of patient setup by surface image registration system in conformal radiotherapy of prostate cancer *Radiat. Oncol.* **4** 9

[42] Hughes S, McClelland J and Tarte S *et al* 2009 Assessment of two novel ventilatory surrogates for use in the delivery of gated/tracked radiotherapy for non-small cell lung cancer *Radiother. Oncol.* **91** 336–41

[43] Gilles M, Fayad H and Miglierini P *et al* 2016 Patient positioning in radiotherapy based on surface imaging using time of flight cameras *Med. Phys.* **43** 4833

[44] Carl G, Reitz D and Schonecker S *et al* 2018 Optical surface scanning for patient positioning in radiation therapy: a prospective analysis of 1902 fractions *Technol. Cancer Res. Treat.* **17** 1533033818806002

[45] Stanley D N, McConnell K A, Kirby N, Gutierrez A N, Papanikolaou N and Rasmussen K 2017 Comparison of initial patient setup accuracy between surface imaging and three point localization: a retrospective analysis *J. Appl. Clin. Med. Phys.* **18** 58–61

[46] Freislederer P, Reiner M and Hoischen W *et al* 2015 Characteristics of gated treatment using an optical surface imaging and gating system on an Elekta linac *Radiat. Oncol.* **10** 68

[47] Willoughby T, Lehmann J and Bencomo J A *et al* 2012 Quality assurance for nonradiographic radiotherapy localization and positioning systems: report of Task Group 147 *Med. Phys.* **39** 1728–47

[48] Cho B, Poulsen P R, Sloutsky A, Sawant A and Keall P J 2009 First demonstration of combined kV/MV image-guided real-time dynamic multileaf-collimator target tracking *Int. J. Radiat. Oncol. Biol. Phys.* **74** 859–67

[49] Sawant A, Smith R L and Venkat R B *et al* 2009 Toward submillimeter accuracy in the management of intrafraction motion: the integration of real-time internal position monitoring and multileaf collimator target tracking *Int. J. Radiat. Oncol. Biol. Phys.* **74** 575–82

[50] Muralidhar K R, Komanduri K, Rout B K and Ramesh K K 2013 Commissioning and quality assurance of Calypso four-dimensional target localization system in linear accelerator facility *J. Med. Phys.* **38** 143–7

[51] Hamilton D G, McKenzie D P and Perkins A E 2017 Comparison between electromagnetic transponders and radiographic imaging for prostate localization: a pelvic phantom study with rotations and translations *J. Appl. Clin. Med. Phys.* **18** 43–53

IOP Publishing

Principles and Practice of Image-Guided Abdominal
Radiation Therapy

Yu Kuang

Chapter 18

Treatment response assessment and response-guided adaptive treatment

Xiao Wang, Chi Ma and Yin Zhang

This chapter is dedicated to treatment response assessment and response-guided adaptive treatment, focused on imaging-based methods. We first outline treatment response assessment techniques in abdominal cancers aiming to identify anatomical or functional changes, including tumor shrinkage, disease progression, and normal tissue changes. Then, we present the strategies that have been applied in response-guided adaptive treatment. This includes offline adaptive strategies accommodating the systematic changes of identified by in-room imaging or other biomarkers using radiation therapy during the treatment course, as well as online adaptive strategies while the patient is still on the treatment couch.

18.1 Introduction

Treatment response assessment for abdominal cancer during treatment or after the completion of therapy has been widely adopted in radiation therapy (RT). Response-guided adaptive RT is a process to identify anatomical or functional changes, including tumor shrinkage, disease progression, or normal tissue changes, and to correspondingly modify the treatment plan guided by treatment response.

18.2 Treatment response assessment techniques in abdominal cancers

Medical imaging as a tool to assess treatment response in abdominal cancer has been established for solid tumors, according to RECIST (Response Evaluation Criteria in Solid Tumors) [1]. Imaging-based evaluation should always be prioritized over clinical examination, except for the situation where the tumor being followed through the treatment course cannot be seen on imaging but is assessable by clinical examination. In this section, we will cover treatment response assessment techniques in abdominal cancers.

Diffusion weighted (DWI) and dynamic contrast-enhanced (DCE) magnetic resonance imaging (MRI) have shown potential as early response biomarkers for renal cell carcinoma. In a recent preliminary study of 12 patients, Reynolds *et al* [2] explored the utility of DWI and DCE MRI in primary renal cell carcinoma after stereotactic ablative body radiotherapy (SABR). They found that the washout contrast enhancement curve behavior at 70 days after SABR significantly correlated with the percentage change in tumor volume shown on computed tomography (CT) at 12 months and onwards.

Using medical imaging techniques to assess treatment response in pancreatic radiotherapy treatment has seen great development in recent years. Choi *et al* [3] investigated the correlation of fluorodeoxyglucose (FDG)–positron emission tomography (PET) measurements of maximum standardized uptake value and kinetic parameters with the clinical outcomes of 20 patients. The study showed that FDG-PET scans may aid in monitoring clinical outcome such as the feasibility of complete surgical resection as early as one month after neoadjuvant treatment. Future studies using larger patient populations are needed.

A single-institution study [4] utilized apparent diffusion coefficient (ADC) to assess treatment response in pancreatic ductal adenocarcinoma (PDAC) after neoadjuvant chemoradiation. With MRI and pathological data collected from 25 patients with resectable and borderline resectable PDAC, changes between pre- and post-treatment ADC radiomic parameters (e.g., mean ADC) were found to correlate with pathological treatment response. In an exploratory study by the same research group, Chen *et al* [5] investigated radiation-induced changes in quantitative radiomics features on CT images of tumors during the delivery of chemoradiation therapy for pancreatic cancer. Analysis of the daily diagnostic-quality CTs showed significant changes in most of the 20 patients in various histogram-based radiomics metrics, including mean CT number, peak position, volume, standard deviation, entropy, skewness, and kurtosis.

Medical imaging plays a crucial role in treatment response assessment of hepatic malignancies. Objective assessment of anatomical changes of liver tumor has been widely adopted in clinical practice. Criteria such as the World Health Organization criteria, RECIST 1.0, and RECIST 1.1 are frequently used. However, anatomical changes defined in these criteria may not occur until late in the treatment course or may not occur at all despite effective treatment. Recent development of functional imaging may open the door to early determination of treatment effectiveness.

For example, in a study by Eccles *et al* [6] involving 11 patients who underwent six-fraction conformal liver RT, DWI MRI was performed at the time of treatment planning, during RT (weeks one and two), and one month following RT. This small cohort of patients consisted of four with hepatocellular carcinoma, five with liver metastases, and two with cholangiocarcinoma. The researchers observed changes in tumor ADC even during the short course of RT. They also found that larger increases in ADC were correlated with higher total radiation doses and increased likelihood of response. A recent study by Schmeel *et al* [7] focused on DWI MRI imaging for patients following selective internal RT with yttrium-90 microspheres (SIRT). Forty-four patients with liver-predominant metastatic colorectal cancer underwent DWI MRI imaging 19 ± 16 days before and 36 ± 10 days after SIRT treatment. Intratumoral minimal ADCs were recorded for 132 liver tumors. It was found that the functional imaging response (an increase in minimal ADC by 22%) is an independent predictor for overall survival.

Radiomics techniques have also been adapted to facilitate treatment assessment in liver cancer. In a retrospective feasibility study, Cozzi *et al* [8] looked at the CT data from 138 consecutive patients with hepatocellular carcinoma. These patients were treated with volumetric modulated arc therapy (VMAT). Treatment planning non–contrast-enhanced CT images from 106 out of 138 patients with complete information about treatment outcome were analyzed. Six first-level histogram-based and 29 texture-based high-level features were used for the analysis. The researchers were able to identify a single first-level feature, compacity, to fit a model predicting local control and survival.

Treatment response using medical imaging methods for radiotherapy of gastric cancer is still at an early stage. Hou *et al* [9] recently published a study regarding the use of radiomic analysis on pulsed low dose rate radiotherapy. Forty-three patients included in this study underwent contrast-enhanced CT 3–5 days prior to RT. A total of 1117 image features were extracted from arterial phase CT images. Six radiomic features demonstrated significant difference between responders and non-responders. Both predictive models used in the study show potential in predicting treatment response with acceptable accuracy.

18.3 Offline adaptive

Adaptive radiotherapy (ART) has been suggested as a possible solution to anatomical changes or functional changes during the course of radiotherapy [10–12]. ART allows modification of the treatment plan to account for changes in target and normal tissues (size, shape, function, etc) with the goal to maximize dose to target and minimize normal tissue exposure. ART can be applied at three timescales: offline between fractions; online immediately prior to a fraction; and inline, or real-time, during a fraction [13]. Due to limited software tools, computational power, and other logistical reasons, offline ART was introduced before online ART [11].

In this section, we will cover the response-guided offline ART in abdominal cancers. Offline ART aims to accommodate the systematic changes to the target or normal tissues identified by in-room imaging or other biomarkers during the course of treatment. Offline ART typically monitors morphological and biological/functional changes from serial imaging scans to assess anatomical change or treatment response. Offline plan modification takes place following the same workflow as the original plan creation including repeated patient simulation, volume delineation, plan optimization, and plan quality assurance (QA). It is often triggered when changes are seen on daily in-room imaging, such as cone beam CT (CBCT), megavoltage CT, CT-on-rail, or on-board MRI, or when changes are measured using other functional imaging techniques. These changes trigger the creation of a new plan to achieve the intended prescription and improve dosimetry for the remaining fractions. Offline response-guided adaptive treatment has been applied to lung cancer [14, 15], brain cancer [16], head-and-neck cancer [17, 18], prostate cancer [19, 20], and cervical cancer [21, 22]. A few clinical studies have been performed in abdominal cancers as well, and some characteristic examples are summarized below.

In a phase 2 clinical trial that included 90 patients with intrahepatic cancers, Feng *et al* demonstrated that a new strategy of biomarker-based individualized adaptive stereotactic body radiotherapy (SBRT) can be used to achieve high rates of control and safety [23]. In their previous study [24], it was found that the subclinical decline in a patient's liver function after radiotherapy can be estimated by assessing indocyanine green (ICG) extraction, and hence it was used as a direct measurement of dynamic liver function in this trial. A baseline measurement of ICG retention rate at 15 min (ICGR15) was calculated prior to initiation of SBRT. Patients then received three of five planned SBRT treatments. Measurement of ICGR15 was repeated four weeks after the third fraction of SBRT treatment. The dose would be adjusted for the final two treatments depending on the ICGR15 results. If the ICGR15 exceeded the upper limit at four weeks to allow for treatment, the ICGR15 value would be re-measured one month later. If the repeated ICGR15 value was deemed to decrease sufficiently, the patients would have the opportunity to receive remaining radiation treatment. The initial treatment plan was generated for five fractions to a maximum predicted rate of radiation-induced liver disease of 15% based on a prior mathematical model, or a maximum total dose of 60 Gy. In this study, the treatment was adapted for 52 out of 116 tumors, and 26 tumors were treated with only three fractions due to the increased ICGR15 at the mid-treatment assessment. Treatment was adapted based on the change in ICGR15 value after the first three fractions for 26 tumors, and the SBRT dose would be lower for the last two fractions. The other 64 tumors received full dose in five planned fractions. The median delivered prescription dose was 49 Gy (range 23–60 Gy). The treatment was well tolerated with no radiation-induced liver disease and a lower complication rate.

The use of functional MRI in ART is still in its infancy. In the new clinical trial (NCT02460835) by the same group at the University of Michigan, radiotherapy is adapted based on the spatial distribution of liver function as determined with portal venous perfusion DCE MRI, so that high-functioning portions of the liver are preferentially spared during ART.

Single photon emission computer tomography (SPECT) imaging is also used for offline ART for liver cancer. Fourteen patients with liver cancer treated with RT received dynamic ^{99m}Tc-labeled iminodiacetic acid (^{99m}Tc-IDA) SPECT scan before, during, and one month after radiotherapy to assess regional hepatic function and to predict post-radiotherapy regional liver function reserve [25]. The researchers validated the hepatic extraction fraction (HEF) derived from SPECT imaging by using measurement of ICG clearance. Reduction in the HEF was observed one

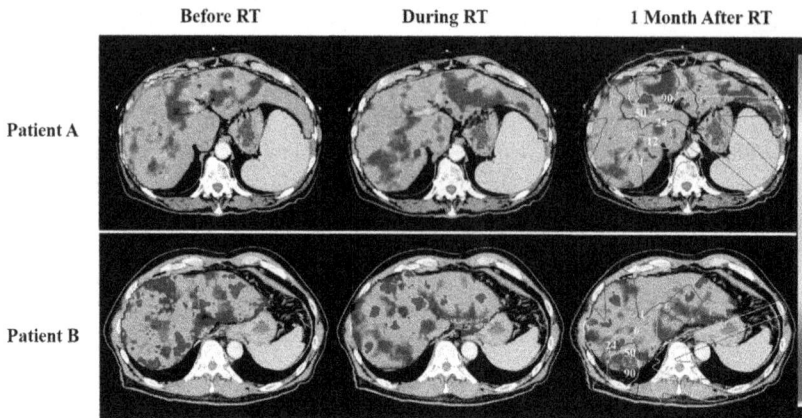

Figure 18.1. HEF before, during, and one month after RT for two patients with intrahepatic cancers. HEF derived from dynamic ^{99m}Tc-IDA SPECT scans. A reduction in the HEF one month after RT can be observed in the volume receiving dose greater than 50 Gy. Reprinted from [25], Copyright (2013), with permission from Elsevier.

month after RT in the volume receiving dose greater than 50 Gy, as shown in figure 18.1. The findings in this study could be potentially used to adapt the treatment plan midway to spare the functional parts of the liver. Bowen *et al* introduced a novel radiotherapy planning paradigm of differential hepatic avoidance RT (DHART) whereby regions of functional liver as defined by ^{99m}Tc-sulphur colloid (^{99m}Tc-SC) SPECT images were differentially spared through dose painting techniques in proton Pencil Beam Scanning (PBS) and photon VMAT RT [26]. This study demonstrated that DHART is technically feasible.

For locally advanced pancreatic cancer, PET imaging showed potential for focal boosting strategies in radiation treatment. [^{18}F]-fluoromisonidazole ([^{18}F]-FMISO) and [^{18}F]-flortanidazole ([^{18}F]HX4) hypoxia PET imaging was performed in pancreatic cancer, and both tracers showed potential as imaging tools for treatment response evaluation, but no clinical studies have been performed for ART application. In 17 patients with pancreatic cancer, Wilson *et al* showed that the pretreatment PET could predict area of residual disease for boosting after chemoradiation [27]. Another study demonstrated that in 25 of 28 patients with borderline resectable pancreatic cancer, a simultaneous integrated boost on the gross tumor volume (GTV) as defined on FDG-PET/CT could be completed with acceptable toxicity with a dose escalation protocol [28]. This was observed to achieve negative margin in 95% of cases undergoing subsequent pancreatectomy. In a study performed by Magallon-Baro *et al*, a motion model was built to extract the movement directions of the gastrointestinal organs using in-room CT scans [29]. This model could be potentially used for future adaptive ART.

Different adaptive strategies have shown to be possible in clinical settings for bladder cancer radiotherapy. In a multicenter study [30], an average anatomy model, using the first five or six daily CBCTs to generate an ITV expanded by 1 cm, was used and reduced the treatment volume on average by 40% with similar planning target volume (PTV) coverage. In an interesting study [31], Cha *et al* used radiomics information from the pre- and post-treatment CT of patients who had undergone neoadjuvant chemotherapy for bladder cancer to assist in assessment of treatment response. They developed radiomics-based predictive models to distinguish between bladder cancers that have fully responded to chemotherapy and those that have not. This methodology can be theoretically used in patients receiving RT to evaluate treatment response.

Once the decision to re-plan the treatment has been made, tools to enable this re-planning must be available. While the standard treatment planning system can be employed for offline ART, online ART requires a highly integrated, specialized system due to the compressed timeline. For example, re-planning tools integrated into the treatment delivery unit may also reduce the QA burden, as they can reduce the need to send and receive data between different systems. Finally, and importantly, QA must be integrated throughout the ART process. ART can be a highly complex process, and robust and efficient QA is therefore critical to ensure accurate, consistent, and safe delivery of ART.

18.4 Online adaptive

Online ART is the process of adapting the RT treatment plan while the patient is still on the treatment couch [32]. It is a collaborative process that requires real-time assessment of patient anatomy, recalculation of the treatment plan based on the current anatomy, possible re-plan due to anatomy change, and real-time re-evaluation of the resultant plan. Online adaptation of the plan normally occurs when target coverage or predetermined organ-at-risk constraint is not met, though thresholds for re-planning are often patient-specific and site-specific.

Treatment management for pancreatic cancer faces a great challenge from significant movement of target volume, including interfractional motion related to the anatomical variability of the surrounding organs at risk and intrafractional motion affected by breathing cycle phases and physiological movements throughout the delivery of radiotherapy [33–35]. Li *et al* [36] used in-room CT as respiration-synchronized diagnostic-quality imaging to guide online repositioning to achieve ART, combined with respiratory-gated technique, and proved it can effectively correct for both inter- and intrafraction variations.

Online ART provides the opportunity to adapt the plan based on the variation of the target volume observed during the course of treatment, thereby reducing the irradiated volume. In bladder cancer, significant variation of bladder shape and volume is expected and commonly observed throughout the course of treatment due to internal organ motion and the extent of bladder filling [37–39]. As a result, large PTV margins have been used, leading to unnecessary dose spillage to normal tissue. With the capability to visualize soft tissue and reduce setup errors [40], CBCT has been widely used as the imaging modality for online ART in bladder cancer treatment. To account for large interfraction and interpatient variations in bladder RT, three adaptive strategies, namely 'plan of the day' (POD) [41–51], 'patient-specific PTV' [30, 52], and daily reoptimization (ReOpt) [53], have been proposed and investigated on the PTV margins, coverage, and normal tissue sparing. Kong *et al* [54] compared the dosimetric differences between these three approaches and found ReOpt to perform best at reducing the irradiated volume as a consequence of its frequent adaptation based on the daily bladder geometry. Burridge *et al* [41] and Tuomikoski *et al* [44] both used the POD approach and have shown that CBCT-assisted online ART can significantly reduce the PTV margins in some directions and the volume of small bowel receiving high doses. The outcome of a multicenter trial [Trans-Tasman Radiation Oncology Group (TROG) 10.01 BOLART] [48] on muscle-invasive bladder cancer used the POD approach and showed feasibility of online ART across multiple radiation oncology departments using different imaging, delivery, and recording technology. Vestergaard *et al* [55] showed the potential of MRI-guided ART in the ReOpt approach, in addition to conventional CBCT, in bladder cancer treatment to considerably spare normal tissue.

Online ART has shown an ability to adapt to small changes in patient anatomy to account for mobile organs at risk and thereby to deliver high-dose radiation adjacent to critical structures. Magnetic resonance (MR)–guided radiotherapy (MRgRT) has been adopted in abdominal cancer due to the superior soft tissue contrast to precisely identify tumor volumes and the subsequent reduction of PTV margin. New hybrid systems that integrate MRI scanners with radiation delivery units (either cobalt source or Linac) [56–59] provide various significant advantages for radiation treatment delivery, especially for the treatment of abdominal cancer. The two commercial MRI-guided radiation treatment systems that are available for clinical use and approved by the US Food and Drug Administration are Elekta Unity system (Elekta, Stockholm, Sweden) and the ViewRay MRIdian system (ViewRay, Oakwood Village, OH, USA). Unity uses a 1.5T MRI scanner with a 7 MV flattening filter free (FFF) Linac, while MRIdian joins a 0.35T MRI scanner with three Co-60 γ-ray sources or a 6 MV FFF Linac. The ViewRay/MR-Linac systems offer the possibilities of real-time MRI imaging before and during treatment without surrogates and of adapting the RT treatment plan online with the patient remaining on the treatment couch, and they successfully address most of the sources of variability in treatment management.

Online MR-guided ART (MRgART) has been shown to improve organ-at-risk sparing and target coverage in patients with pancreatic cancer treated with SBRT [60]. Mittauer *et al* [61] has published case reports using MRgART in a patient with liver SBRT, a patient with kidney SBRT, and a patient with gastric lymphoma treatment. Henke *et al* [62] showed the potential advantage of MRgART in treating oligometastatic disease of the non-liver abdomen and central thorax with SBRT. Rosenberg *et al* [63] published multi-institutional retrospective data on the use of MRgART to treat liver tumors with SBRT, exhibiting the efficacy of online adaptive treatment for liver-based primary malignancies and metastatic disease. Luterstein *et al* [64] reported a case of a 69-year-old patient with locally advanced pancreatic cancer (LAPC) treated with

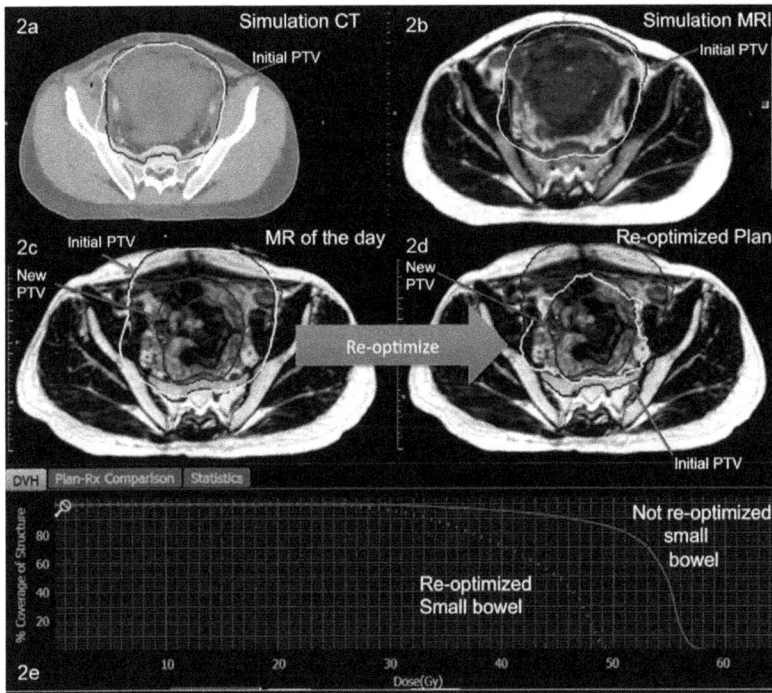

Figure 18.2. CT simulation (a), simulation MRI scan (b), and dose verification on daily MRI scan before (c) and after (d) reoptimization (95% isodose line, yellow; initial PTV receiving 50 Gy, red [color wash]; new PTV receiving 50 Gy, blue [color wash]; small bowel, green). (e) Dose-volume histograms of reoptimized and not reoptimized small bowel. Reprinted from [65], Copyright (2016), with permission from Elsevier.

SBRT using MRgART combined with gated delivery. Acharya *et al* [65] demonstrated the feasibility of online MRgART in five patients with abdominopelvic malignancies, including three neoadjuvant colorectal malignancies, one unresectable gastric malignancy, and one unresectable pheochromocytoma (figure 18.2). All above published treatments were well tolerated by patients with excellent local control. In order to achieve fast and standardized online MRgART workflow in LAPC, two different optimization approaches have been proposed [66, 67].

The ability to adapt radiation treatment plans online to the anatomy of the day has opened the way for previously difficult dose escalation in abdominal cancer treatment. A phase I trial assessed stereotactic MR-guided online ART in abdominal cancer [68]. In this trial, the cohort consisted of 20 patients with oligometastatic or unresectable cancer, ten of whom suffered from primary or secondary liver lesions, five from pancreatic cancer (three recurrences and two primary), and five from abdominal secondary nodal lesions. Treatment plans were adapted online daily as needed, based on anatomy of the day, to preserve organ-at-risk constraints, escalate PTV dose, or both. Plan adaptation increased PTV coverage in 64/97 fractions. Dose escalation beyond the originally prescribed dose was achieved only in three patients with liver lesions. Grade 3 or worse treatment-related toxicities were not observed in the 15-month median follow-up period. The two patients with primary LAPC were both alive with no progression after 14 months of follow-up.

An ongoing prospective phase IA/IB trial (NCT04020276) studies the side effects and best dose for MRgART in liver SBRT. The trial proposed a strategy to escalate dose for liver metastases by theoretically delivering the highest degree of dose to the tumor, as constrained by the total of all limiting organs at risk across each individual treatment. Retrospective data from 44 patients with inoperable pancreatic cancer showed that adopting MRgART to escalate dose to tumor resulted in improved overall survival [69]. Although this retrospective study is only a hypothesis-generating study, an ongoing phase 2 trial (NCT03621644) to test this approach prospectively in patients with inoperable and borderline operable pancreatic cancer is currently recruiting.

18.5 Summary

In summary, we have outlined the techniques used in assessing treatment response of abdominal cancer, focused on imaging-based methods. Various imaging techniques have played important roles in assessing treatment response. We also summarized the strategies that have been applied in response-guided adaptive treatment, both offline and online. The current developments in auto-segmentation and rapid plan re-optimization significantly reduce the time for offline and online ART and make more efficient plan adjustment possible. The development in radiomics-based treatment response assessment is encouraging, and radiomics potentially provides a fundamental methodology for future personalized treatment in the era of precision medicine. The availability of MR-Linac or MR Co-60 systems has substantial impact on response-guided adaptive therapy not only in abdominal cancer but also other cancer types. A few clinical trials on MRgART are ongoing to demonstrate the efficacy and clinical benefits.

References

[1] Eisenhauer E A et al 2009 New response evaluation criteria in solid tumours: revised RECIST guideline (version 1.1) Eur. J. Cancer 45 228–47
[2] Reynolds H M, Parameswaran B K, Finnegan M E, Roettger D, Lau E, Kron T, Shaw M, Chander S and Siva S 2018 Diffusion weighted and dynamic contrast enhanced MRI as an imaging biomarker for stereotactic ablative body radiotherapy (SABR) of primary renal cell carcinoma PLoS One 13 e0202387
[3] Choi M, Heilbrun L K, Venkatramanamoorthy R, Lawhorn-Crews J M, Zalupski M M and Shields A F 2010 Using 18F-fluorodeoxyglucose positron emission tomography to monitor clinical outcomes in patients treated with neoadjuvant chemo-radiotherapy for locally advanced pancreatic cancer Am. J. Clin. Oncol. 33 257–61
[4] Dalah E, Erickson B, Oshima K, Schott D, Hall W A, Paulson E, Tai A, Knechtges P and Li X A 2018 Correlation of ADC with pathological treatment response for radiation therapy of pancreatic cancer Transl. Oncol. 11 391–8

[5] Chen X *et al* 2017 Assessment of treatment response during chemoradiation therapy for pancreatic cancer based on quantitative radiomic analysis of daily CTs: an exploratory study *PLoS One* **12** e0178961

[6] Eccles C L, Haider E A, Haider M A, Fung S, Lockwood G and Dawson L A 2009 Change in diffusion weighted MRI during liver cancer radiotherapy: preliminary observations *Acta Oncol.* **48** 1034–43

[7] Schmeel F C, Simon B, Sabet A, Luetkens J A, Traber F, Schmeel L C, Ezziddin S, Schild H H and Hadizadeh D R 2017 Diffusion-weighted magnetic resonance imaging predicts survival in patients with liver-predominant metastatic colorectal cancer shortly after selective internal radiation therapy *Eur. Radiol.* **27** 966–75

[8] Cozzi L *et al* 2017 Radiomics based analysis to predict local control and survival in hepatocellular carcinoma patients treated with volumetric modulated arc therapy *BMC Cancer* **17** 829

[9] Hou Z, Yang Y, Li S, Yan J, Ren W, Liu J, Wang K, Liu B and Wan S 2018 Radiomic analysis using contrast-enhanced CT: predict treatment response to pulsed low dose rate radiotherapy in gastric carcinoma with abdominal cavity metastasis *Quant. Imag. Med. Surg.* **8** 410–20

[10] Yan D, Vicini F, Wong J and Martinez A 1997 Adaptive radiation therapy *Phys. Med. Biol.* **42** 123–32

[11] Sonke J J, Aznar M and Rasch C 2019 Adaptive radiotherapy for anatomical changes *Semin. Radiat. Oncol.* **29** 245–57

[12] Matuszak M M, Kashani R, Green M, Owen D, Jolly S and Mierzwa M 2019 Functional adaptation in radiation therapy *Semin. Radiat. Oncol.* **29** 236–44

[13] Green O L, Henke L E and Hugo G D 2019 Practical clinical workflows for online and offline adaptive radiation therapy *Semin. Radiat. Oncol.* **29** 219–27

[14] Kong F M *et al* 2017 Effect of midtreatment PET/CT-adapted radiation therapy with concurrent chemotherapy in patients with locally advanced non-small-cell lung cancer: a phase 2 clinical trial *JAMA Oncol.* **3** 1358–65

[15] Moller D S, Holt M I, Alber M, Tvilum M, Khalil A A, Knap M M and Hoffmann L 2016 Adaptive radiotherapy for advanced lung cancer ensures target coverage and decreases lung dose *Radiother. Oncol.* **121** 32–8

[16] Darázs B, Ruskó L, Vegvary Z, Ferenczi L, Varga Z, Fodor E, Kis D, Barzo P and Hideghety K 2017 Adaptive radiation therapy for high grade brain tumors: impact on the dose distribution and disease outcome *Int. J. Radiat. Oncol. Biol. Phys.* **99** E79

[17] Teng F, Aryal M, Lee J, Lee C, Shen X, Hawkins P G, Mierzwa M, Eisbruch A and Cao Y 2018 Adaptive boost target definition in high-risk head and neck cancer based on multi-imaging risk biomarkers *Int. J. Radiat. Oncol. Biol. Phys.* **102** 969–77

[18] Lee N *et al* 2016 Strategy of using intratreatment hypoxia imaging to selectively and safely guide radiation dose de-escalation concurrent with chemotherapy for locoregionally advanced human papillomavirus–related oropharyngeal carcinoma *Int. J. Radiat. Oncol. Biol. Phys.* **96** 9–17

[19] Park S S, Yan D, McGrath S, Dilworth J T, Liang J, Ye H, Krauss D J, Martinez A A and Kestin L L 2012 Adaptive image-guided radiotherapy (IGRT) eliminates the risk of biochemical failure caused by the bias of rectal distension in prostate cancer treatment planning: clinical evidence *Int. J. Radiat. Oncol. Biol. Phys.* **83** 947–52

[20] McPartlin A J *et al* 2016 MRI-guided prostate adaptive radiotherapy—a systematic review *Radiother. Oncol.* **119** 371–80

[21] Tan L T Mbbs Mrcp Frcr Md *et al* 2019 Image-guided adaptive radiotherapy in cervical cancer *Semin. Radiat. Oncol.* **29** 284–98

[22] Tanderup K, Georg D, Potter R, Kirisits C, Grau C and Lindegaard J C 2010 Adaptive management of cervical cancer radiotherapy *Semin. Radiat. Oncol.* **20** 121–9

[23] Feng M 2018 Individualized adaptive stereotactic body radiotherapy for liver tumors in patients at high risk for liver damage: a phase 2 clinical trial *JAMA Oncol.* **4** 40–7

[24] Stenmark M H, Cao Y, Wang H, Jackson A, Ben-Josef E, Ten Haken R K, Lawrence T S and Feng M 2014 Estimating functional liver reserve following hepatic irradiation: adaptive normal tissue response models *Radiother. Oncol.* **111** 418–23

[25] Wang H, Feng M, Frey K A, Ten Haken R K, Lawrence T S and Cao Y 2013 Predictive models for regional hepatic function based on 99mTc-IDA SPECT and local radiation dose for physiologic adaptive radiation therapy *Int. J. Radiat. Oncol. Biol. Phys.* **86** 1000–6

[26] Bowen S R, Saini J, Chapman T R, Miyaoka R S, Kinahan P E, Sandison G A, Wong T, Vesselle H J, Nyflot M J and Apisarnthanarax S 2015 Differential hepatic avoidance radiation therapy: proof of concept in hepatocellular carcinoma patients *Radiother. Oncol.* **115** 203–10

[27] Wilson J M, Mukherjee S, Chu K-Y, Brunner T B, Partridge M and Hawkins M 2014 Challenges in using 18 F-fluorodeoxyglucose-PET-CT to define a biological radiotherapy boost in locally advanced pancreatic cancer *Radiat. Oncol.* **9** 146

[28] Huang X, Knoble J L, Zeng M, Aguila F N, Patel T, Chambers L W, Hu H and Liu H 2016 Neoadjuvant gemcitabine chemotherapy followed by concurrent IMRT simultaneous boost achieves high r0 resection in borderline resectable pancreatic cancer patients *PLoS One* **11** e0166606

[29] Magallon-Baro A, Loi M, Milder M T W, Granton P V, Zolnay A G, Nuyttens J J and Hoogeman M S 2019 Modeling daily changes in organ-at-risk anatomy in a cohort of pancreatic cancer patients *Radiother. Oncol.* **134** 127–34

[30] Pos F J, Hulshof M, Lebesque J, Lotz H, van Tienhoven G, Moonen L and Remeijer P 2006 Adaptive radiotherapy for invasive bladder cancer: a feasibility study *Int. J. Radiat. Oncol. Biol. Phys.* **64** 862–8

[31] Cha K H, Hadjiiski L, Chan H-P, Weizer A Z, Alva A, Cohan R H, Caoili E M, Paramagul C and Samala R K 2017 Bladder cancer treatment response assessment in CT using radiomics with deep-learning *Sci. Rep.* **7** 8738–8

[32] Wu Q J, Li T, Wu Q and Yin F F 2011 Adaptive radiation therapy: technical components and clinical applications *Cancer J.* **17** 182–9

[33] Heerkens H D, van Vulpen M, van den Berg C A, Tijssen R H, Crijns S P, Molenaar I Q, van Santvoort H C, Reerink O and Meijer G J 2014 MRI-based tumor motion characterization and gating schemes for radiation therapy of pancreatic cancer *Radiother. Oncol.* **111** 252–7

[34] Knybel L, Cvek J, Otahal B, Jonszta T, Molenda L, Czerny D, Skacelikova E, Rybar M, Dvorak P and Feltl D 2014 The analysis of respiration-induced pancreatic tumor motion based on reference measurement *Radiat. Oncol.* **9** 192

[35] Karava K, Ehrbar S, Riesterer O, Roesch J, Glatz S, Klock S, Guckenberger M and Tanadini-Lang S 2017 Potential dosimetric benefits of adaptive tumor tracking over the

internal target concept for stereotactic body radiation therapy of pancreatic cancer *Radiat. Oncol.* **12** 175

[36] Li X A, Liu F, Tai A, Ahunbay E, Chen G, Kelly T, Lawton C and Erickson B 2011 Development of an online adaptive solution to account for inter- and intra-fractional variations *Radiother. Oncol.* **100** 370–4

[37] Turner S L, Swindell R, Bowl N, Marrs J, Brookes B, Read G and Cowan R A 1997 Bladder movement during radiation therapy for bladder cancer: implications for treatment planning *Int. J. Radiat. Oncol. Biol. Phys.* **39** 355–60

[38] Pos F J, Koedooder K, Hulshof M C, van Tienhoven G and Gonzalez Gonzalez D 2003 Influence of bladder and rectal on spatial variability of a bladder tumor during radical radiotherapy *Int. J. Radiat. Oncol. Biol. Phys.* **55** 835–41

[39] Fokdal L, Honore H, Hoyer M, Meldgaard P, Fode K and von der Maase H 2004 Impact of changes in bladder and rectal filling on organ motion and dose distribution of the bladder in radiotherapy for urinary bladder cancer *Int. J. Radiat. Oncol. Biol. Phys.* **59** 436–44

[40] Dawson L A and Jaffray D A 2007 Advances in image-guided radiation therapy *J. Clin. Oncol.* **25** 938–46

[41] Burridge N, Amer A, Marchant T, Sykes J, Stratford J, Henry A, McBain C, Price P and Moore C 2006 Online adaptive radiotherapy of the bladder: small bowel irradiated-reduction *Int. J. Radiat. Oncol. Biol. Phys.* **66** 892–7

[42] Lalondrelle S, Huddart R, Warren-Oseni K, Hansen V N, McNair H, Thomas K, Dearnaley D, Horwich A and Khoo V 2011 Adaptive-predictive organ localization using cone-beam computed tomography for improved accuracy in external beam radiotherapy for bladder cancer *Int. J. Radiat. Oncol. Biol. Phys.* **79** 705–12

[43] Murthy V, Master Z, Adurkar P, Mallick I, Mahantshetty U, Bakshi G, Tongaonkar H and Shrivastava S 2011 'Plan of the day' adaptive radiotherapy for bladder cancer using helical tomotherapy *Radiother. Oncol.* **99** 55–60

[44] Tuomikoski L, Collan J, Keyrilainen J, Visapaa H, Saarilahti K and Tenhunen M 2011 Adaptive radiotherapy in muscle invasive urinary bladder cancer – an effective method to reduce the irradiated bowel volume *Radiother. Oncol.* **99** 61–6

[45] Kuyumcian A, Pham D, Thomas J M, Law A, Willis D, Kron T and Foroudi F 2012 Adaptive radiotherapy for muscle-invasive bladder cancer: optimisation of plan sizes *J. Med. Imag. Radiat. Oncol.* **56** 661–7

[46] Meijer G J, van der Toorn P P, Bal M, Schuring D, Weterings J and de Wildt M 2012 High precision bladder cancer irradiation by integrating a library planning procedure of 6 prospectively generated SIB IMRT plans with image guidance using lipiodol markers *Radiother. Oncol.* **105** 174–9

[47] Foroudi F *et al* 2011 Online adaptive radiotherapy for muscle-invasive bladder cancer: results of a pilot study *Int. J. Radiat. Oncol. Biol. Phys.* **81** 765–71

[48] Foroudi F *et al* 2014 The outcome of a multi-centre feasibility study of online adaptive radiotherapy for muscle-invasive bladder cancer TROG 10.01 BOLART *Radiother. Oncol.* **111** 316–20

[49] Vestergaard A, Kallehauge J F, Petersen J B, Hoyer M, Sondergaard J and Muren L P 2014 An adaptive radiotherapy planning strategy for bladder cancer using deformation vector fields *Radiother. Oncol.* **112** 371–5

[50] Lutkenhaus L J, Visser J, de Jong R, Hulshof M C and Bel A 2015 Evaluation of delivered dose for a clinical daily adaptive plan selection strategy for bladder cancer radiotherapy *Radiother. Oncol.* **116** 51–6

[51] Tuomikoski L, Valli A, Tenhunen M, Muren L and Vestergaard A 2015 A comparison between two clinically applied plan library strategies in adaptive radiotherapy of bladder cancer *Radiother. Oncol.* **117** 448–52

[52] Tolan S, Kong V, Rosewall T, Craig T, Bristow R, Milosevic M, Gospodarowicz M and Chung P 2011 Patient-specific PTV margins in radiotherapy for bladder cancer—a feasibility study using cone beam CT *Radiother. Oncol.* **99** 131–6

[53] Vestergaard A, Muren L P, Sondergaard J, Elstrom U V, Hoyer M and Petersen J B 2013 Adaptive plan selection vs. re-optimisation in radiotherapy for bladder cancer: a dose accumulation comparison *Radiother. Oncol.* **109** 457–62

[54] Kong V C, Taylor A, Chung P, Craig T and Rosewall T 2019 Comparison of 3 image-guided adaptive strategies for bladder locoregional radiotherapy *Med. Dosim.* **44** 111–6

[55] Vestergaard A *et al* 2016 The potential of MRI-guided online adaptive re-optimisation in radiotherapy of urinary bladder cancer *Radiother. Oncol.* **118** 154–9

[56] Lagendijk J J, Raaymakers B W and van Vulpen M 2014 The magnetic resonance imaging-linac system *Semin. Radiat. Oncol.* **24** 207–9

[57] Mutic S and Dempsey J F 2014 The ViewRay system: magnetic resonance-guided and controlled radiotherapy *Semin. Radiat. Oncol.* **24** 196–9

[58] Raaymakers B W *et al* 2009 Integrating a 1.5 T MRI scanner with a 6 MV accelerator: proof of concept *Phys. Med. Biol.* **54** N229–37

[59] Yun J, Wachowicz K, Mackenzie M, Rathee S, Robinson D and Fallone B G 2013 First demonstration of intrafractional tumor-tracked irradiation using 2D phantom MR images on a prototype linac-MR *Med. Phys.* **40** 051718

[60] El-Bared N, Portelance L, Spieler B O, Kwon D, Padgett K R, Brown K M and Mellon E A 2019 Dosimetric benefits and practical pitfalls of daily online adaptive MRI-guided stereotactic radiation therapy for pancreatic cancer *Pract. Radiat. Oncol.* **9** e46–54

[61] Mittauer K *et al* 2018 A new era of image guidance with magnetic resonance-guided radiation therapy for abdominal and thoracic malignancies *Cureus* **10** e2422

[62] Henke L *et al* 2016 Simulated online adaptive magnetic resonance-guided stereotactic body radiation therapy for the treatment of oligometastatic disease of the abdomen and central thorax: characterization of potential advantages *Int. J. Radiat. Oncol. Biol. Phys.* **96** 1078–86

[63] Rosenberg S A *et al* 2019 A multi-institutional experience of mr-guided liver stereotactic body radiation therapy *Adv. Radiat. Oncol.* **4** 142–9

[64] Luterstein E, Cao M, Lamb J, Raldow A C, Low D A, Steinberg M L and Lee P 2018 Stereotactic MRI-guided adaptive radiation therapy (SMART) for locally advanced pancreatic cancer: a promising approach *Cureus* **10** e2324

[65] Acharya S *et al* 2016 Online magnetic resonance image guided adaptive radiation therapy: first clinical applications *Int. J. Radiat. Oncol. Biol. Phys.* **94** 394–403

[66] Bohoudi O, Bruynzeel A M E, Senan S, Cuijpers J P, Slotman B J, Lagerwaard F J and Palacios M A 2017 Fast and robust online adaptive planning in stereotactic MR-guided adaptive radiation therapy (SMART) for pancreatic cancer *Radiother. Oncol.* **125** 439–44

[67] Olberg S, Green O, Cai B, Yang D, Rodriguez V, Zhang H, Kim J S, Parikh P J, Mutic S and Park J C 2018 Optimization of treatment planning workflow and tumor coverage during

daily adaptive magnetic resonance image guided radiation therapy (MR-IGRT) of pancreatic cancer *Radiat. Oncol.* **13** 51

[68] Henke L *et al* 2018 Phase I trial of stereotactic MR-guided online adaptive radiation therapy (SMART) for the treatment of oligometastatic or unresectable primary malignancies of the abdomen *Radiother. Oncol.* **126** 519–26

[69] Rudra S *et al* 2019 Using adaptive magnetic resonance image-guided radiation therapy for treatment of inoperable pancreatic cancer *Cancer Med* **8** 2123–32

IOP Publishing

Principles and Practice of Image-Guided Abdominal
Radiation Therapy

Yu Kuang

Chapter 19

Adaptive radiation therapy for abdominal cancer

Hui Wang, Xiadong Li, Shenglin Ma and Yu Kuang

This chapter is dedicated to current advancements in adaptive radiation therapy for abdominal cancer. We first introduce the rationale of adaptive radiation therapy to address anatomical and functional variations in patients with abdominal cancer during the course of radiation treatment, especially the concept of biologically adapted radiation therapy. We also describe the three typical workflows of adaptive radiation therapy implementation, including offline, online and inline or real-time adaption methods. We particularly showcase several examples of dosimetric benefits from inline adaption implementation. Then, we present the tools for adaptive radiation therapy of abdominal cancer and the main advances in each tool. We especially include the new imaging methods for adaptive radiation therapy, such as ^{99m}Tc-sulphur colloid (SC) single photon emission computed tomography (SPECT)/ computed tomography (CT) imaging for differential hepatic avoidance radiation therapy and the hybrid SPECT/spectral-CT/cone-beam CT (CBCT) imaging method for on-board function adaptive radiation therapy.

19.1 Introduction

Adaptive radiation therapy (ART) is generally a process to address anatomical and functional variations in patients during the course of radiation treatment. Currently, ART can be more specifically defined as radiation therapy (RT) in which the treatment plan is modified as needed to improve clinical outcomes through accounting for a variety of uncertainties from many different sources (e.g., changes in tumor size, function and responses) [1]. ART was firstly proposed by Yan *et al* over two decades ago as a process to initially adapt the daily setup changes using megavoltage portal imaging and CT imaging [2]. Along with the evolution of the underlying technologies as discussed in the previous chapters, including imaging

doi:10.1088/978-0-7503-2468-7ch19

simulation, tools for treatment planning, image registration and methods for treatment response assessment, ART continues to evolve with a variety of clinical trials designed and performed to test the fundamental concepts of ART [3–10].

There are many potential anatomical, physiological and/or functional changes (i.e., uncertainties) during a course of RT for patients with abdominal cancers. As is well known, intrafraction respiration motion and interfraction physiological organ variations cause inherent positional uncertainty of abdominal structures. Singh *et al* characterized small bowel organ motion between fractions of pancreatic RT. The results show that the volume of irradiated small bowel excluding duodenum changed significantly between fractions during the process of pancreatic RT, which suggests dose escalation to the pancreas might be possible to potentially improve the treatment outcome because no single segment of nonduodenal small bowel is likely to receive the full prescription dose [11]. Liu *et al* observed substantial interfractional anatomic variations, particularly the organ deformation, in pancreatic cancer radiotherapy. The results show that the average maximum overlap ratios of all daily CTs for pancreatic head, duodenum and stomach were 80.2%, 61.7% and 72.2%, respectively, for the patients studied [12]. In this study, the authors concluded that the dosimetric advantages of using adaptive replanning accounting for these anatomic variations might enable safe dose escalation in radiotherapy for pancreatic cancer.

Tumors inherently appear to be highly heterogeneous, partly originating from a variety of gene mutations in patients with cancer. The microenvironment of a malignant tumor is characterized by significant heterogeneity in oxygen concentration, nutrient distribution, cellular density, metabolism and so on [13, 14]. The tumor microenvironment also undergoes considerable changes over the treatment course caused by radiation. The antitumor effects of RT rely heavily on cytotoxicity from damage to genomic DNA by radiation irradiation. For RT using x-rays, γ-rays and proton beams, damage to genomic DNA is primarily attributed to the free radicals produced by radiation in the cells and becomes permanent with the help of oxygen. Tumor hypoxia has been investigated to be strongly associated with radioresistance, recurrence of malignant tumors and poor prognosis of patients with cancer after radiation treatment [15, 16].

It is known that tumor hypoxia subvolumes may present an increased level of radiation resistance by 2.5–3 times [14, 17]. Thus, radiation dose escalations targeted to the radioresistant subvolumes may help increase tumor local control rates of RT. Today's radiation treatment machines have the flexibility and the accuracy to deliver the locally escalated doses to these subvolumes and spare organs at risk (OAR) and healthy surrounding tissue. This technique can be referred to as biologically adapted RT. For clinical implementation of biologically adapted RT, several technical components or steps are necessitated, including radiobiology, molecular/functional imaging, selection of relevant imaging parameters, defining of an appropriate prescription function relating imaging signals to dose prescriptions, dose painting treatment planning and delivery. These steps are illustrated in figure 19.1.

To realize biologically adapted RT and potentially improve the outcome of RT, the relevant radiobiological effects strongly related to the efficiency of radiation

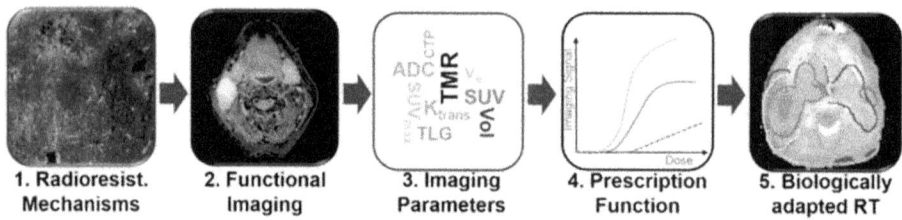

| 1. Radioresist. Mechanisms | 2. Functional Imaging | 3. Imaging Parameters | 4. Prescription Function | 5. Biologically adapted RT |

Figure 19.1. Schematic illustration of steps to clinically perform biologically adapted RT. (1) Biological radioresistance mechanisms (immunohistochemical staining of a human tumor section; green: hypoxia; blue: proliferation; red: vessels). (2) Functional imaging (^{18}F-fluoromisonidazole [^{18}F-MISO] positron emission tomography [PET]/magnetic resonance [MR]). (3) Selection of relevant imaging parameters (SUV: standardized uptake value). (4) Definition of prescription functions. (5) Planning and delivery of biologically adapted RT plans (hypoxia dose painting in photon radiotherapy). Reproduced from Thorwarth *et al* [14]. Copyright (2018), with permission from Elsevier.

treatment and the success of RT should be targeted through dose painting RT. Radiobiology researchers have found several potential effects that can be targeted for improved outcome, and most of these effects can be imaged noninvasively by modern imaging techniques, such as PET, SPECT, magnetic resonance imaging (MRI) and so forth.

Tumor control probability (TCP), which describes the efficacy of radiation treatment, can be simply modeled using the Poisson model as [18]

$$-\ln(TCP) = n\rho\exp(-\alpha D), \tag{19.1}$$

where ρ, n, α and D denote the number of cells per voxel, the number of voxels, the radiation sensitivity and the applied dose, respectively. In conventional RT, the parameters of ρ, n and α are assumed to be constant over the whole tumor volume, and the applied dose D therefore should be homogeneous throughout the tumor volume consequently to maintain the similar cell kill over the whole tumor. By contrast, in biologically adapted RT, both the cell density ρ and the radiation sensitivity α that are usually derived from molecular/functional imaging are heterogeneous in a tumor volume. It means dose prescriptions should be also heterogeneous accordingly to obtain a constant level of TCP over the tumor.

Multiple medical imaging techniques have been performed to visualize tumor biological factors that can be used for biologically dose adaption purposes, such as tumor hypoxia, cellular density, metabolism, and so on. With dedicated radio-tracers, such as [^{18}F]-fluoromisonidazole (^{18}F-MISO), [^{18}F]-HX4 or [^{18}F]-fluoroazomycin arabinoside (FAZA) [19–22], PET has been used for tumor hypoxia imaging. The radiotracer of ^{123}I-iodoazomycin arabinoside (IAZA) has been used for the measurement of hypoxia in human tumors by noninvasive SPECT imaging [23]. Dynamic contrast enhanced (DCE) MRI has also been used to assess tumor hypoxia. Based on the hypoxia imaging signals, radiation sensitivity can be derived. Thus, hypoxia imaging can be used for biologically adapted RT. [^{18}F]-Fluorodesoxyglucose (FDG) PET and diffusion weighted (DW) MRI can be used for tumor cellular density and metabolism imaging. The prognostic feature of FDG PET and DW-MRI has been demonstrated from different tumor entities [24, 25]. Therefore, the cellular heterogeneity of a tumor volume can be addressed by biologically adapted RT through relating the imaging signals to the local cell densities. The first clinical trials using biological dose adaption, particularly biological imaging-based radiation dose painting, show promising results with better outcome rates and comparable toxicity after biologically adapted RT [14].

ART uses imaging information to adapt treatment plans to anatomic and/or functional/biological changes of tumor to potentially improve radiation treatment outcome. Based on the timescale applied, ART currently is categorized into three types or modes: offline between treatment fractions, usually on a timescale of hours to days; online, with the patient in the treatment position immediately before a treatment fraction on a timescale of minutes; and inline or real time in a treatment fraction on a timescale of seconds to minutes. ART is an emerging and evolving radiation treatment paradigm, and its use has been investigated in a variety of patients with cancer. In this chapter, we focus on ART applied to manage the abdominal cancers, mainly upper abdomen cancers, such as pancreas cancer, liver cancer, etc.

19.2 ART workflow

Because ART is a highly advanced RT paradigm that adapts treatment plans to anatomy and/or physiology changes in patients with cancer during the course of radiation treatment, more technical components are required than those needed for conventional RT. In other words, ART is essentially a process streamlining a variety of techniques or tools with a diverse and still laborious effort involved, and thereby highly efficient workflows are required because of the high complexity of ART implementation. The key technical components for ART are imaging, assessment, replanning and quality assurance (QA). Different imaging techniques were used to accurately measure the changes in patients. Assessment is a decision-making process that determines whether to adapt the plan or not. After the decision of adaption is made, the replanning process can be done using a variety of treatment planning tools. Finally, QA has to be performed to ensure accurate and safe delivery of ART. These processes are discussed in details in the next section of this chapter. Example workflows for the three types of ART are illustrated in figure 19.2. The choice of a particular ART type is always scenario dependent, with advantages and disadvantages for each of them that have to be accounted for in practical implementations.

19.2.1 Offline ART

Offline ART is generally performed between treatment fractions on a relatively longer timescale as compared with online and real-time ART processes. As such, there is the flexibility of using conventional treatment tools without highly specialized and/or integrated tools with a small amendment to the conventional tools, thus minimizing the extra costs. For example, a conventional CT or MRI simulator or diagnostic scanner can be used in the treatment replanning process, and

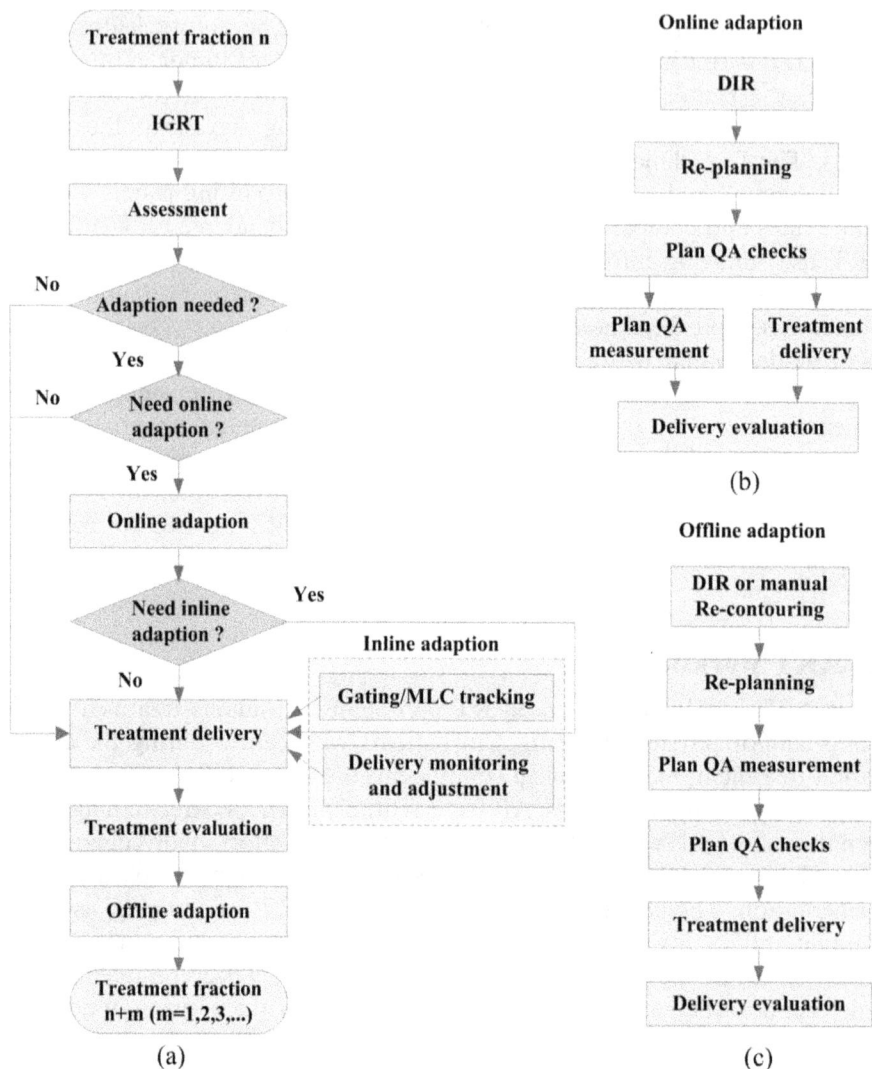

Figure 19.2. Illustration of exemplary workflows of ART. (a) A typical workflow of the ART process; (b) an example of online ART process; (c) an example of offline ART process (IGRT: image-guided radiation therapy; DIR: deformable image registration; QA: quality assurance; MLC: multileaf collimator.) Adapted from Lim-Reinders *et al* [26]. Copyright (2017), with permission from Elsevier.

in-room imaging (e.g., cone-beam CT or on-board MRI) can also be used if sufficient image quality can be achieved. More importantly, offline functional/molecular imaging such as PET or SPECT can be performed in the fraction intervals outside the treatment room because there are limited in-room or on-board functional imaging machines available for clinical usage currently in the treatment room. In the offline implementation, the standard QA tools, such as the phantom-based measurement, can be employed to ensure accurate dose delivery.

Offline ART can effectively address the relatively slow changes in tumors or the surrounding tissue, such as tumor responses to radiation, patient weight loss, etc. However, offline ART encounters difficulties in catching the relatively frequent changes between fractions and even fast changes during a fraction, such as respiration motion, internal organ rearrangement (especially in abdomen) and baseline shift of tumors/tissue. If offline ART does not have the capability to respond to such changes quickly enough, the unexpected alignment errors could be induced rather than corrected.

Once the decision to use offline ART is made, the implementation of offline ART is similar to conventional RT, although usually at a compressed time frame, mainly including deformable image registration (DIR) or manual re-contouring, replanning, plan QA measurement, plan QA checks, treatment delivery and delivery evaluation.

19.2.2 Online ART

Online ART is conducted with the patient on the treatment couch on a relatively tight timeline as compared with offline ART. In-room/on-board images are acquired first, and then DIR is performed to assess the need for ART. If an adaption is needed, the replanning and plan QA processes should be performed efficiently with the patient remaining on the treatment couch. The plan QA measurement should be conducted during the treatment delivery while the patient remains in the treatment position. Online ART requires an efficient workflow that seamlessly streamlines the multiple technical components because of potential patient movement and the consideration of patient tolerance of prolonged on-table time. As such, specialized and well-integrated tools are necessary for this process.

As compared with offline ART, online ART has the ability to immediately adapt the treatment plan to the observed changes based on in-room imaging just before treatment delivery. It helps address the frequent interfraction changes (day-to-day) in patients, such as deformation and shifting of abdominal organs from day to day, baseline variation of tumors/tissue, etc. One challenge of online ART is that the patient-specific QA measurement (e.g., phantom measurement) cannot be conducted with the patient on the treatment couch as the conventional QA process. As such, a modified QA check procedure is required. Another challenge to online ART is the limited ability to address fast intrafraction changes in patients after the plan has been adapted, such as liver and pancreas movement, along with the respiration motion and stomach emptying within minutes, although some of these motions can be managed by the beam gating technique or patient surveillance.

19.2.3 Inline or real-time ART

Inline or real-time ART adapts the plan in real time during radiation treatment delivery to address the intrafractional variations in patients through using real-time or near–real-time imaging of patients or surrogates. Therefore, inline ART is potentially the most accurate form of ART and could see wide application in clinical practices in the near future with the world embracing the information age, in which fast decisions are made based on real-time data, such as driverless vehicle technology, etc [27]. The benefits of the adoption of inline ART include reducing the radiation toxicity or treatment-related side effects and therefore the human and economic costs, improving tumor local control, increasing patient throughput and so on. The challenges of inline ART lie in substantially more automation than offline or online ART and a higher plan QA burden. The current clinical state-of-the-art inline ART adapts the plans to the geometric changes of targets only, whereas the functional adaptions are usually performed through using online or more likely offline ART.

The reason to use inline ART is that tumor motion caused by respiratory, digestive, circulatory and muscular systems can happen in subseconds to minutes, which challenges any assumptions from pretreatment measurements. An example of three-dimensional patient motion trace of a tumor measured during liver stereotactic ablative radiation therapy (SABR) beam delivery is shown in figure 19.3 [28]. Through integrating the motion measurement in figure 19.3 into dose reconstruction, the dose delivered with inline adaption and without adaption are compared with the planned dose in figure 19.4. As shown in the figure 19.4, the high dose missed part of the clinical target volume (CTV) and moved into healthy liver without

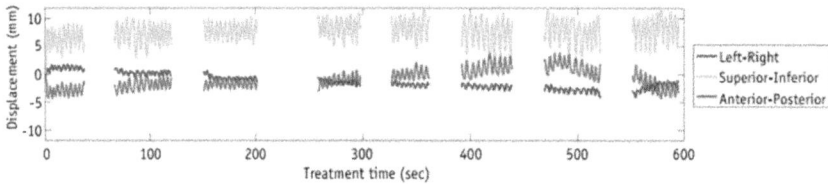

Figure 19.3. An example of three-dimensional patient motion trace of a liver tumor during SABR beam delivery measured with the system of COSMIK (combined optical and sparse monoscopic imaging with kilovoltage x-rays) on a conventional linear accelerator. Figure adapted from Keall *et al* [28]. Copyright (2018), with permission from Elsevier.

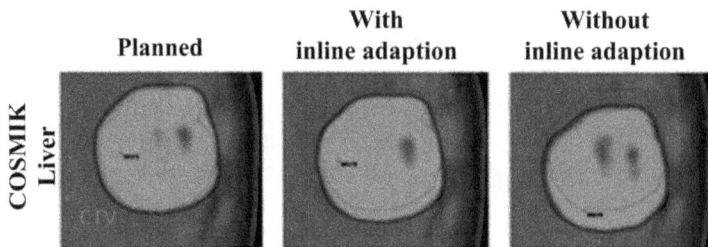

Figure 19.4. An example of the dosimetric impact of three-dimensional motion of the tumor during liver SBRT. The planned patient dose (left), dose with inline ART (center) and the dose without inline ART (right) demonstrate the dose distribution advantage of inline ART, in which the motion was accounted for through (simulated) MLC tracking for the COSMIK (combined optical and sparse monoscopic imaging with kilovoltage x-rays) guidance system in a conventional linear accelerator. Color wash spans dose interval of 95%–108% of the prescribed dose. Figure adapted from Keall *et al* [28]. Copyright (2018), with permission from Elsevier.

accounting for the motion measured. The CTV D95/D99 was reduced by 11%/15% without inline adaption and by 2.6%/4.2% using (simulated) multileaf collimator (MLC) tracking with inline adaption compared with the planned dose.

To account for the motion of tumors, the target volume is typically increased by the internal target volume (ITV) method. As a result, the mean prescription dose to the target is usually constrained to reduce the probability of radiation-induced toxicity. Through eliminating motion of tumors (i.e., motion management) using gating or tumor tracking methods, dose escalation is allowed to be prescribed to the target to potentially increase tumor TCP. Gargett *et al* studied the clinical impact of removing respiratory motion during liver SABR [8]. The study concluded that substantial increased TCP could be achieved using dose escalation in the majority of patients studied with the management of respiratory motion involved (i.e., planning target volume [PTV] reduction). Figure 19.5 shows an example of comparison of dose distributions for a lesion in a liver prescribed with ITV-based plan, motion-managed plan and dose-escalated motion-managed plan, in which a dose escalation of 20.7 Gy_{10} was achieved using the dose-escalated motion-managed plan without compromising OAR tolerances compared with the ITV-based plan. Meanwhile, the mean dose to the liver for the dose-escalated motion-managed plan remained similar to that of the ITV-based plan, whereas the mean liver dose for the motion-managed plan was 1.8 Gy lower than that of the ITV-based plan.

19.2.4 ART implementation

The current techniques of ART implementation vary from center to center world-wide because of the differences in available resources and the lack of international

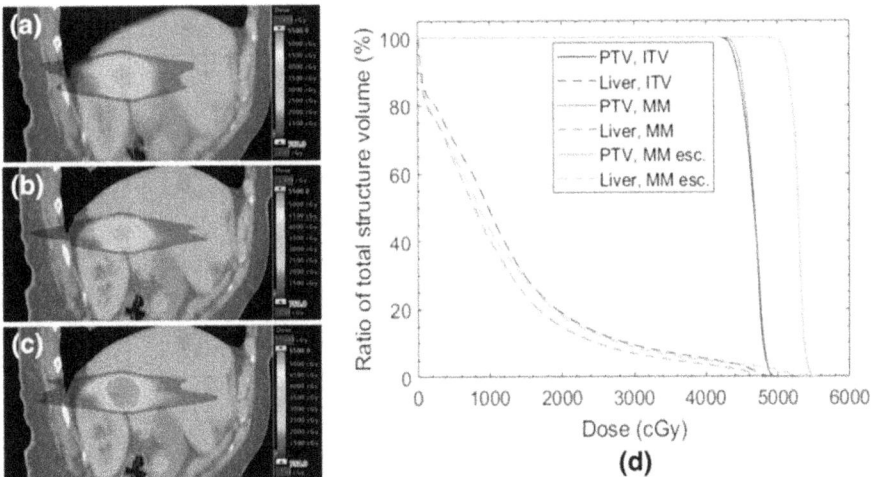

Figure 19.5. Comparison of dose distributions for a lesion in liver prescribed with ITV-based plan (a), motion-managed plan (b) and dose-escalated motion-managed plan (c). Dose-volume histogram (DVH) demonstrated PTV coverage (solid lines) and liver dose (broken lines) for the cased shown in (a)–(c). ITV—ITV-based, MM —motion managed, MM esc—dose escalated motion management. Reproduced from Gargett *et al* [8].

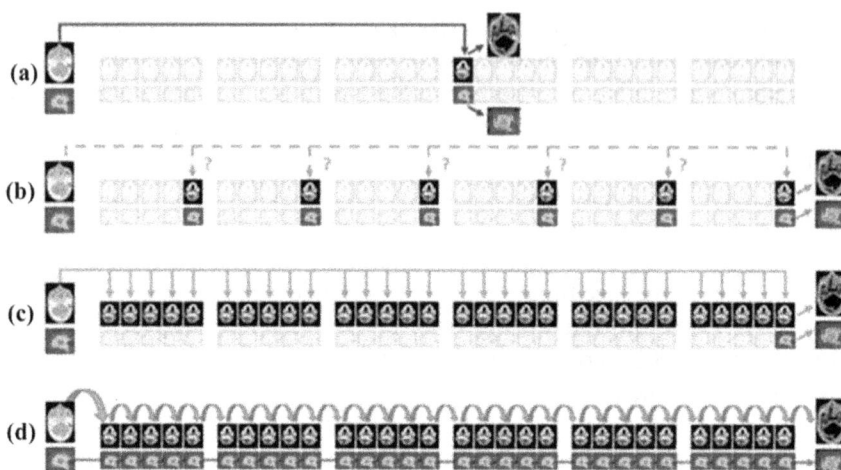

Figure 19.6. Illustration of ART implementation frames. (a) Fixed interval ART; (b) 'triggered' ART; (c) sequential ART; (d) cascade ART. Reproduced from Heukelom *et al* [29]. Copyright (2019), with permission from Elsevier.

guidelines on various aspects of ART. The four possible types of ART implementation are illustrated in figure 19.6 with different adaption frequencies during the course of radiation treatment [29].

Figure 19.6(a) illustrates a 'fixed interval' ART implementation, in which the pretreatment imaging data obtained at a single (often midtreatment) time point are registered with the initial plan data to perform a single adaption if needed. This method is usually implemented using midtreatment anatomy/function imaging and thus is relatively computationally efficient. A 'triggered' adaption is illustrated in figure 19.6(b). In this implementation frame, typically weekly imaging data are acquired for dose adaption purpose with the qualitative or quantitative trigger thresholds preset. A 'sequential' adaption is illustrated in figure 19.6(c), wherein the repeated Image$_{Fraction}$ and Image$_{planning}$ assessment are performed for dose adaption purpose based on high-frequency (e.g., day-to-day) pretreatment imaging. In a cascade adaption implementation illustrated in figure 19.6(d), daily deformation in anatomy/function and set-up error are incorporated subsequently to all fractions. This approach is computationally expensive and needs substantial automation processes; therefore, wide implementation of such a data-rich method might need vendor support.

19.3 ART processes and tools

Because ART is an advanced RT technique, ART shares the basic technical components of conventional radiotherapy, such as imaging modalities, treatment planning process and treatment delivery. For ART particularly as mentioned above, the key technologies include imaging, assessment, replanning and QA [1].

19.3.1 Imaging

Imaging is required to detect anatomy and/or physiology changes in patients, which trigger dose adaption during the treatment course. The imaging objects can be internal anatomy/function targets, embedded fiducial markers or external surrogates. For inline or online ART for patients with abdominal cancer, imaging is commonly performed in the treatment room with cone-beam CT (CBCT), MRI, in-room CT/four-dimensional CD (4DCT), combined kilovoltage-optical imaging or other methods. For offline ART, imaging can be performed with a CT or MRI simulator or diagnostic CT, MRI, PET or SPECT scanner usually located outside the treatment room.

Electron density information provided by CT imaging can be used for dose calculation for initial and adaptive treatment planning. For patients with abdominal cancers, which are commonly involved with respiration motion, in-room 4DCT helps evaluate changes of tumor/OARs during treatment. The work by Li *et al* demonstrated the effect of respiration gating in treatment planning using daily 4DCT [3]. The gated planning significantly reduced the organ positioning errors (e.g., kidney translation and deformation) during the radiation delivery as compared with nongated planning. Although in-room CT or 'CT-on-rails' systems can provide diagnostic-quality daily CT images for online ART, there are potential alignment errors due to the couch shifting between imaging and treatment systems. Furthermore, the extra size and infrastructure requirements of the treatment room might hamper this technique in widely spreading clinically.

Standard CBCT on most modern medical linear accelerators is the most widespread on-board imaging system for online IGRT/ART. CBCT has good image contrast for anatomy of patients with or without using image contrast agents under reduced motion conditions, especially for structures with relatively large density differences, such as bone tissue, tissue lung, tissue air, etc. Bony anatomy–based rigid image registration is routinely performed clinically for position correction. For online ART using DIR, especially in abdominal tissue, CBCT often suffers from low soft-tissue contrast, scatter artifacts, cone beam artifacts, etc [30]. Consequently, CBCT might be insufficient to accurately determine electron density for dose calculation for online ART without appropriate image correction. To apply DIR to the daily images, related corrections for CBCT image artifacts (e.g., scatter artifacts) may be required before the DIR process [31]. With technical advancement, these difficulties can be overcome, which renders CBCT a powerful tool for online ART based on its widespread availability. For example, there have been new methods proposed to accurately derive electron density from dual- or multiple-energy CBCT scan for dose calculation [32, 33].

For internal changes in the abdomen that occur on a relatively rapid timescale, such as the constant peristaltic motion of the bowel, online ART is the common choice. This poses a request for high-quality soft-tissue imaging to identify changes of the target and surrounding OARs. MRI provides superior soft-tissue imaging quality in the abdomen even with low field strengths (e.g., 0.35 T) as compared to CBCT, though CBCT has been improved considerably [34]. As such, MRI enables robust and accurate DIR for dose adaption. MRI has been increasingly integrated with Linac as MRI-Linac for MRI-guided RT along with numerous relevant concepts emerging [35, 36]. For a commercially available low-field MRI-Linac, a shorter MRI scan time (less than 60 s) than that of CBCT scan mitigates motion blurring of imaging for accurate contouring. Figure 19.7 demonstrates MRI-based delineation occurred between consecutive days of treatment for the case of the upper abdomen [37]. It is observed that the pancreatic tumor and adjacent duodenum can be well identified with sufficient image contrast.

MRI does not provide electron density information as CT does, so there is a need to assign an electron density value for each voxel of the images through using density assignment methods, such as a bulk density override of the patient's anatomy and atlas-based fusion of a 'pseudo-CT' scan to the MR image [38, 39]. For MRI-Linac systems, the major challenge is to handle the cross-talk between the magnetic field generated by the MRI hardware components and the radiofrequency produced by the Linac. As great improvement has been made to address the challenges for MRI-guided RT, MR-Linac systems have gained increasingly interest for potentially widespread clinical implementation in the near future.

PET imaging has established its role in diagnosis, (re-)staging and response assessment of several abdominal tumors in patients, and its role for biological imaging–guided target volume adaption and ART strategies in these tumors has gained increasing interest in the last decades. Noninvasive functional PET imaging displays the biological changes of tumors during the course of treatment, which helps clinicians adapt the irradiated target volume in an ART protocol. In addition to the most commonly used tracer of [^{18}F]-FDG, the hypoxia tracers as mentioned above could be used in abdominal tumors for target delineation [40]. For target volume segmentation/delineation using the PET/CT images, the visual judgment and automated contouring methods based on standardized uptake value (SUV) thresholds are commonly used. These methods have a limitation of variability due to interobserver bias, variable tumor-to-background activity, heterogeneous tracer distribution within the tumor and variations in tumor size. The gradient-based methods, however, might be more reliable than other methods through using the

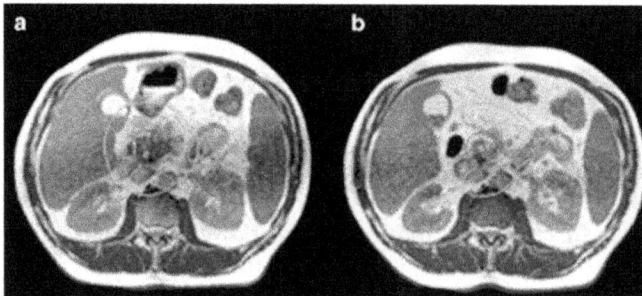

Figure 19.7. An example of MRI-based delineation occurred between consecutive days of treatment of a pancreatic cancer in a patient. (Red: pancreatic cancer; orange: duodenum; green: 3 cm around the gross tumor volume [GTV].) Reproduced from Boldrini *et al* [37].

maximum spatial gradient to identify boundaries between normal tissue and tumors because of its robustness to the variations induced by different reconstruction algorithms and image device.

Patients with abdominal tumors (e.g., liver tumor, pancreas tumor) show highly heterogeneous tracer uptake within tumors, which represents the heterogeneous tumor microenvironment and indicates a need for individualized radiation treatment strategies. During the course of treatment, the tumor microenvironment also undergoes considerable changes, which necessitates dose adaption to improve the tumor local control rate. As a biological imaging modality, PET can be used to identify tumor subvolumes with different radiobiology features for optimized treatment planning. As mentioned above, online ART has been investigated for radiation treatment of pancreatic cancer using daily CT or even MRI to correct for interfraction changes during the course of treatment. Instead, offline ART currently implemented with the fixed interval (often midtreatment) or even 'triggered' schemes has been performed using PET imaging to adapt the treatment plan in the treatment of pancreatic cancer.

There are several studies investigating the use of PET imaging for optimized target volume delineation and radiation dose escalation to improve the tumor local control rate for patients with pancreatic cancer [41–45]. A previous study found that the GTV needed to be adapted in 5 of 14 patients with locally advanced pancreatic cancer based on acquired [18F]FDG-PET imaging, and the resultant GTV was considerably larger than the planning GTV based on CT only (104.5 cm^3 vs. 92.5 cm^3) [44]. From the phase II clinical trial investigating the effect of using [18F]FDG-PET/CT-based GTV for patients with locally advanced pancreatic cancer treated with chemoradiotherapy, the study found that the GTV size defined by [18F]FDG-PET/CT can potentially predict the outcome of patients [45]. In contrast, Li et al found the considerably smaller mean GTV size (49.3 cm^3) using an [18F]FDG-PET/CT-based target delineation method as compared with the mean GTV size (64.1 cm^3) delineated by a CT-based method [41]. The difference in the GTV sizes in these studies could be attributed to the differences in scale setting for the PET images.

One major challenge for the implementation of [18F]FDG-PET-based target delineation in pancreatic cancer is that a pancreatic tumor undergoes significant movement variability due to respiration motion. Kishi et al found that the average 4D or respiration-gated [18F]FDG-PET avid tumor volume was two times smaller than that defined using non–respiration-gated [18F]FDG-PET in 14 patients, which suggests that 4D [18F]FDG-PET has the potential to improve the accuracy of radiation delivery compared with conventional [18F]FDG-PET for patients with pancreas cancer [42].

The potential of PET imaging used for local radiation dose escalation strategies for patients with locally advanced pancreatic cancer has been investigated. A study of 17 patients by Wilson et al shows that the regions with residual metabolic activity [i.e., biological target volume (BTV)] after chemoradiotherapy correlated strongly with the maximum SUV before treatment, indicating the dose boosting for these regions [46]. Huang et al shows that a dose escalation protocol using concurrent intensity-modulated radiation therapy (IMRT) simultaneous integrated boost on the GTV defined by [18F]FDG-PET/CT could be completed with acceptable toxicity and negative margins with R0 resection for 25 of 28 patients undergoing subsequently pancreatectomy [47].

Radiation therapy, including stereotactic body radiation therapy (SBRT), has gained growing interest in less invasive treatment for patients with liver cancer. There were studies that investigated the use of [18F]FDG-PET/CT for target delineation and treatment planning in liver radiotherapy, but only a few studies investigated the role of PET in treatment adaption in patients with liver cancer during external beam radiotherapy (EBRT) [48, 49]. Bundschuh *et al* found that the PET-based GTV were 13.8% larger than MRI-based GTV in 14 patients with 16 metastases scheduled for SBRT [48]. They also found a considerable difference in patients with a previous local treatment, suggesting the usefulness of [18F]FDG-PET/CT for differentiation between vital tumor tissue and scar tissue in the patients. The respiration-gated PET provides superior imaging co-registration with CT or MR images as compared with nongated PET. Using respiration-gated PET with an appropriate evaluation of internal target volume, Riou *et al* found the statistically significant decrease in PTV in 8 patients with 14 liver lesions scheduled for SBRT through using the adapted margins [50]. The investigation of PET in radiation treatment planning and ART of liver tumors is ongoing research, and the role of PET in management of liver cancer requires further studies.

SPECT/CT plays an increasing role in diagnosis, staging, treatment and follow-up of abdominal cancers. An example of ^{111}In-somatostatin SPECT/CT for assessment of neuroendocrine neoplasms is shown in figure 19.8, in which the SPECT image helped define the expansion of the primary lesion in the pancreas of a patient [51]. Wang *et al* performed ^{99m}Tc-labeled iminodiacetic acid SPECT and indocyanine green clearance tests before, during and after RT for intrahepatic tumors [52]. They found that the adaptive model including dose and middle-RT regional functional imaging based on SPECT was the optimal model to potentially predict regional hepatic function reserve after RT, indicating that the changes detected by functional liver imaging during RT could guide personalized midcourse adaption to maximize tumor control and minimize the risk of liver injury by radiation irradiations. This work opened a new opportunity for midway plan adaption through the RT course to be performed in patients with intrahepatic cancers.

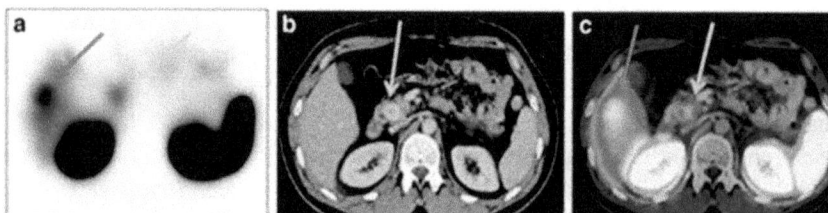

Figure 19.8. ^{111}In-somatostatin SPECT/CT of abdominal lesions in a patient: (a) SPECT image shows two lesions in the upper abdomen; (b) CT shows a hypervascular primary lesion in the head of the pancreas; (c) fused SPECT/CT localized an isodense metastasis in the right lobe of the liver. Reproduced from Israel *et al* [51].

Using ^{99m}Tc-sulphur colloid (SC) SPECT/CT imaging to define the functional liver regions, Bowen *et al* evaluated the feasibility of a novel planning concept that differentially redistributes RT dose away from defined functional liver regions, referred to as differential hepatic avoidance RT (DHART), for proton pencil beam scanning (PBS) therapy and photon-beam volumetrically modulated arc therapy (VMAT) [53]. The workflow to perform DHART is illustrated in figure 19.9. The discrete functional liver volumes (FLVs) were defined by the thresholds of percentage of maximum tracer uptake ratio (such as the thresholds of 43%, 70% and 90% in this illustration). The DHART planning was generated/optimized from the conventional planning through altering the beam angle configuration for proton PBS therapy (or arc configuration for photon-beam VMAT plan therapy) to avoid delivery through FLVs. An example of a plan comparison between conventional radiotherapy and DHART is shown in figure 19.10.

The offline PET/SPECT-guided dose painting IMRT technique to selectively boost the radiation dose to the BTV has increasingly been demonstrated to promisingly improve tumor local control rate and reduce radiation toxicity in normal tissue. The online/on-board functional ART has also gained great interest because of its potential to improve the accuracy of radiation treatment delivery by detecting the functional changes within the tumor in a more timely and convenient manner. Roper *et al* has demonstrated the potential for on-board SPECT imaging used in the treatment room in simulation studies [54–56]. In one simulation, a camera with a parallel-hole, low-energy, high-resolution collimation was modeled to rotate around the patient on the treatment couch to acquire on-board SPECT images of the patient. In another simulation, a 49-pinhole SPECT camera system was designed with a robot maneuvering the multipinhole SPECT system around the patient in treatment position for RT. However, there are challenges in translating these designs into practical implementation. For the parallel-hole collimator-based SPECT camera, the imaging localization errors for deep targets are relatively high, and it is difficult to integrate the SPECT scanner into Linacs with on-board kilovoltage/megavoltage imaging devices mounted. For the robotic on-board

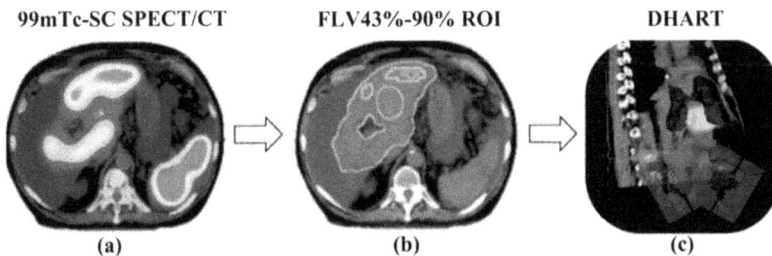

99mTc-SC SPECT/CT **FLV43%-90% ROI** **DHART**

(a) (b) (c)

Figure 19.9. Demonstration of workflow for differential hepatic avoidance RT (DHART): (a) the ^{99m}Tc-sulphur colloid (SC) SPECT/CT uptake in the tumor-subtracted liver; (b) discrete functional liver volumes (FLVs) defined by the thresholds of percentage of maximum uptake ratio (FLV$_{43\%}$—forest green contour, FLV$_{70\%}$—light green contour, FLV$_{90\%}$—light blue contour, ROI—region of interest); (c) the FLV regions were used to generate the DHART plans. Reproduced from Bowen *et al* [53]. Copyright (2015), with permission from Elsevier.

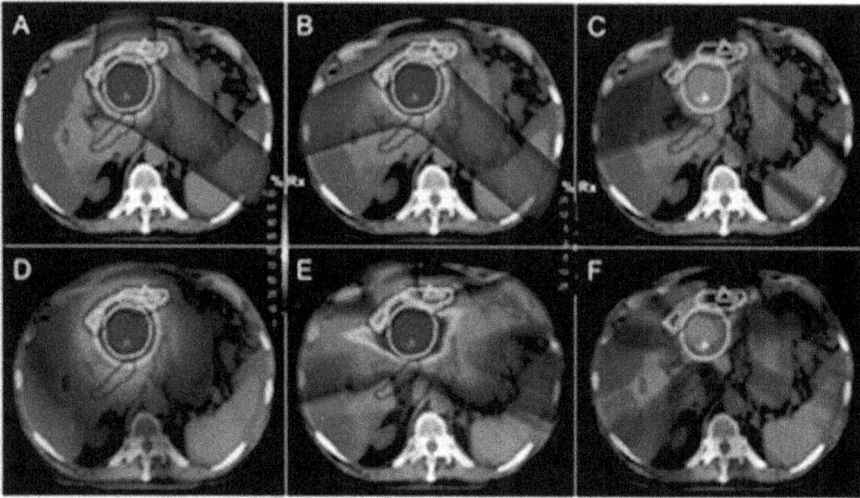

Figure 19.10. An example of a plan comparison between conventional radiotherapy and DHART. (A–C) Proton pencil beam scanning dose distributions in an axial plane generated for conventional plan (A), functional liver avoidance plan (B) and the resultant dose difference distribution (C). (D–F) Photon VMAT dose distributions in the same axial plane for conventional plan (D), functional liver avoidance plan (E) and the resultant dose difference distribution (F). Contours are shown for PTV (white), $FLV_{43\%}$ (forest green), $FLV_{70\%}$ (light green) and $FLV_{90\%}$ (cyan). Reproduced from Bowen *et al* [53]. Copyright (2015), with permission from Elsevier.

multipinhole-based SPECT camera used in the treatment room with improved target localization accuracy and flexibility of scan trajectories, alignment uncertainties from the co-registration of SPECT images and CBCT images and the trajectory error of the robots might be concerns.

To potentially address the challenges mentioned above for on-board SPECT imaging, Wang *et al* proposed a trimodal SPECT/spectral-CT/CBCT on-board imager (OBI) for online functional ART using a novel photon counting detector panel design in a Monte Carlo simulation study [57–59]. Using a single trimodal imager integrated into Linacs, SPECT images intrinsically registered with CBCT and spectral CT images, which makes the alignment of on-board triple images with the treatment coordinates much easier and thus minimizes the alignment uncertainties. Furthermore, the relatively limited deep target localization accuracy of SPECT can be addressed by on-board contrast-enhanced spectral-CT of the OBI with a high relatively uniform localization accuracy. In addition, the distribution of biomarker-tagged contrast agents within tumors can be used to identify BTV and therefore facilitates target localization and verification for re-contouring and replanning in ART processes. Along with CBCT providing anatomy context and spectral CT providing radiotherapy-related parameter decomposition, this trimodal imager can work complementarily and could provide one-stop on-board image guidance for online ART with the flexibility of choosing one, two or all of the three modalities for personalized medicine.

The treatment scheme/workflow based on the trimodal OBI proposed by Wang *et al* [57] is illustrated in figure 19.11, in which multiple RT components, including

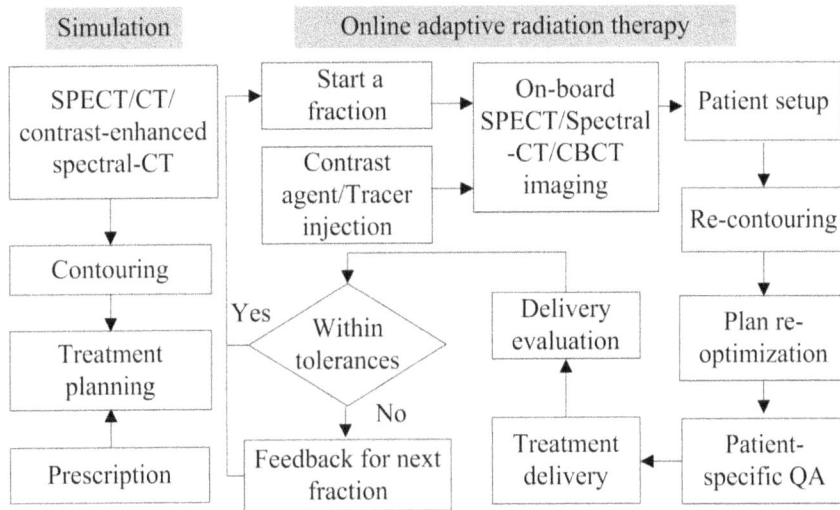

Figure 19.11. Online treatment scheme based on a trimodal SPECT/spectral-CT/CBCT on-board imager system. Reproduced from Wang *et al* [57]. Copyright (2020), with permission from John Wiley & Sons.

pretreatment triple imaging, assessment, re-contouring, plan re-optimization and patient-specific QA, among others, were streamlined in an efficient way. The translation of this design to a clinical study is envisioned with rapid advances of a variety of technical components needed and seamlessly streamlining all these elements in the highly complex workflow.

19.3.2 Assessment

After detecting a variation of anatomy and/or physiology in the patient with cancer treated with RT, a decision of whether to adapt the plan or not should be made through the assessment process. The ART assessment process can be a straightforward manual review of pretreatment imaging based on a decision protocol with predefined criteria, recalculation of the dose distribution using the new imaging or complex deformation dose estimation of the cumulative dose [60–63]. The tools used for the assessment include image review, delineation of targets and OARs, recalculation of dose and evaluation of dose distribution and rigid/deformable imaging registration, which can be manual, semi-automated, or fully automated as needed.

For offline adaption, the need to adapt can only be assessed during the course of treatment. Typically, identification of anatomic change triggers the reassessment of performance of the initial plan on current anatomy. An anatomic change can be identified by a surrogate, such as fiducial marker motion, weight loss and misfit of thermoplastic mask, or by systemic changes (e.g., shrinking of gross tumor) visualized on daily imaging. In some cases, the midcourse functional imaging (e.g., PET, SPECT) can be acquired to identify the functional changes for the assessment process. The obtained daily CBCT, a fixed or preplanned interval CT scan (e.g., weekly obtained CT scan during proton beam therapy) or a requested resimulation scan indicated by significant changes can be used to reassess the performance of the initial plan. If the performance of the initial plan violates the predefined criteria, offline adaption is required. Depending on the degree of the violation of the rules, the initial plan may be halted until the offline adaptive plan is generated, or the current treatment may continue while the offline adaptive plan is prepared and applied to the next fraction. This is usually subjective because the judgment is based on multiple factors, including dosimetric change, the treatment intent and the treatment fractions remaining to be delivered.

Online adaption otherwise should be chosen before the treatment course because of the relatively tight time constraint as the patient remains in the treatment position, during which the online adaption plan should be generated with sufficient robustness. It is also necessary to allocate time and resources before the initiation of a treatment course; for example, the physician and physicist should be present at each treatment delivery. Therefore, online adaption is suitable for the situations in which the need to adapt is known or predictable before the start of the treatment course. In these scenarios, the physician can predict that daily unplanned OAR constraint or dose coverage violations will occur unless online adaption is applied. For patients with abdominal cancer treated with ablative radiation doses, such as SBRT, the motion of the target and the adjacent OARs usually occur within a high-dose gradient, which directly links to the considerable change of dose distribution. Therefore, online adaption should be preplanned and prepared before the start of the treatment course.

19.3.3 Replanning

The replanning process for ART is based on the imaging information that indicates a requirement of dose adaption. This imaging information can be acquired in different ways and time points: (1) imaging during the treatment delivery; (2) just prior to on-table replanning; (3) using the on-board images that originally triggers a dose replan; (4) CT or MRI resimulation; (5) additional diagnostic imaging. For inline/real-time ART, the first option is applied. For online ART, the second option is utilized. For offline ART, several of these methods may be used to obtain the imaging information needed for adaption, depending on anatomical or functional ART performed, available resources, and so on. Replanning for ART can vary slightly depending on the types of adaption selected, that is, offline, online or real-time ART.

19.3.3.1 Replanning for offline ART

The replanning process for offline ART is similar to that of the standard treatment planning process, which requires much finessing of a treatment plan to be completed. Because the patient is not waiting on-table during the replanning process, it is possible to transfer a regenerated image in diagnostic quality or the image of the day to a separate treatment planning system without using specialized tools. It has the advantage of being able to perform standard QA on any adapted plans to reduce the possibility of adaption failure as compared with online replanning ART. This allows dosimetrist to create the best possible plan by selecting the appropriate optimization parameters within a needed time. The image of the initial plan is fused with the new image through rapid or deformable imaging registration, and the contours are propagated from the original image to the new image. If functional adaption is employed, the functional subvolume (i.e., BTV) defined by the offline functional image is also contoured and registered with the original image of the initial plan for the dose escalation process.

19.3.3.2 Replanning for online ART

For online ART, the image of the initial plan is registered with the daily image either rigidly or deformably. In the current online low-field MRI-guided ART, the contours and underlying relative electron density are transferred from the original images in the same way with additional manual or automatic corrections if needed for sufficient accuracy. For online adaptive replanning, a lean workflow is desirable because the patient is waiting on the treatment couch. The physicist can re-optimize the daily plan through simply using the original inverse optimization algorithm, and the only major changes are the contours and relative electron densities. Reducing the optimization steps help further minimize computational time for online replanning. A two-step technique referred to as segment aperture morphing (SAM)-segment weight optimization (SWO) that has been developed has simplified the online optimization step [64]. In the SAM step, the segment aperture is adapted to the varied target contours of the daily image as projected at the isocenter for a specific IMRT beam. In the next SWO step, the segmented weights are optimized using the new adapted segments, which avoids any shifts in the couch. A similar method to SAM is the virtual couch shift (VCS), in which the patient is rotated and translated to find an optimal projected aperture to which the MLC leaves adapt [65]. SAM-SWO is superior in adaption because of the ability to account for both patient deformations and target shifting compared with couch shifting, for which only the rigid-body transformation is compensated. However, these segment-shifting methods generally do not account for the relative motion between targets and OARs.

19.3.3.3 Replanning for real-time ART

Replanning for real-time ART is a challenging task because of the highly efficient and specialized tools needed. Currently, clinical state-of-the-art real-time ART only accounts for the changes in target geometry, which is basically a first-order correction that navigates the beam to point at the moving target [27]. The further improved strategies may be used for higher-order corrections, such as correction of the delivered dose to the patient. For real-time dose adaption, the dose delivered to the moving anatomy has to be estimated on the fly during the treatment delivery. As such, fast motion–including dose-reconstruction algorithms for real-time adaption are necessary. So far, fast dose-reconstruction algorithms have been developed through either using the precalculated doses or reducing the complexity of dose calculation. Fast *et al* presented a real-time dose-reconstruction solution based on precalculated dose influence data, and it can calculate the actually delivered dose in less than 10 ms at a rate of 25 Hz [66]. Before treatment delivery, all treatment beams are divided into beamlets, and the dose distribution of each possible beamlet in the patient's anatomy is precalculated. During treatment delivery, through assigning the precalculated doses at an increment calculation of each dose to different tissue elements that depends on the current target position and beam shape, the real-time motion-including dose reconstruction could be achieved.

Alternatively, the fast motion–including dose reconstruction may be realized through reducing the complexity of dose calculation. Ravkilde *et al* developed a simplified pencil beam algorithm implemented in the DoseTracker software that assumes homogeneous tissue density (e.g., water), flat patient surfaces, flat dose profiles and the same depth dose curve for all fields [67, 68]. The accuracy of the algorithm for reconstruction of dose and motion-induced errors in tracking and nontracking beam deliveries was quantified, in which the dose error was shown to agree well with phantom measurement. The mean computation time of 295 ms was used for each calculation of dose and dose error [68]. This fast motion–including dose reconstruction approach enables temporal and spatial pinpointing of errors in the absorbed dose, which facilitates the real-time replanning process. Using the DoseTracker software, the dose delivered to moving liver tumors during SBRT was tested in a previous study [69]. Figure 19.12 shows reconstructed doses using DoseTracker compared with doses calculated using the Eclipse software for the fraction with largest motion in a patient with liver cancer treated with SBRT. This study shows promising results for liver SBRT, which suggested that the real-time dose reconstruction could be used for more elaborate dose analysis and dose adaption because of the ability to immediately evaluate the treatment quality.

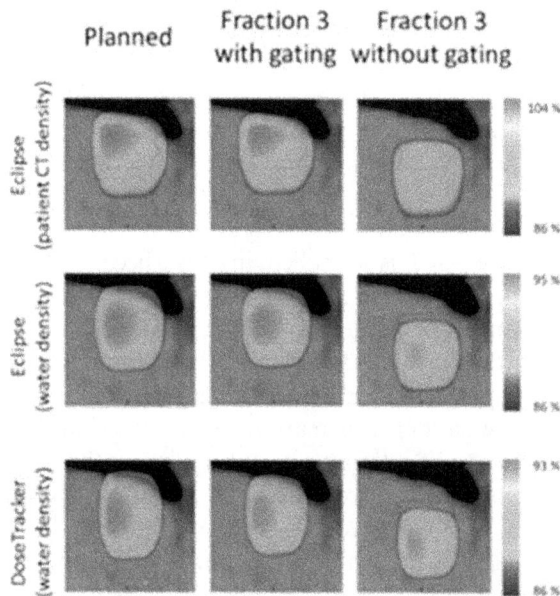

Figure 19.12. Demonstration of real-time dose reconstruction for moving liver tumors using the DoseTracker software. Reproduced from Skouboe *et al* [69]. Copyright (2018), with permission from Elsevier.

For all three types of ART, one of the ongoing challenges is the calculation of dose accumulation to estimate the actual total delivered dose, for which both anatomic and dosimetric variations should be taken into account. It is particularly challenging for online and real-time ART because online and real-time ART require more frequent generation of new plans during the treatment course compared with offline ART. It is difficult or currently impossible to robustly identify and track corresponding point volumes of tissue between treatment fractions or even in a treatment fraction because of the motion of tumors and OARs, especially for the malleable abdomen organs. Most OARs are not sufficiently tracked to allow reliable dose accumulation using current technology, although it is relatively straightforward to track immobile tissue. As such, a conservative solution is to use an isotoxicity and 'parameter adding' method [70].

In the parameter adding approach, rather than trying to estimate the location of the maximum point dose regions to an OAR between treatment deliveries, the maximum point dose to an OAR is summed over all treatment deliveries. If the contour for the GTV is not changed during the course of treatment, dose accumulation for the GTV can be estimated in some situations [71]. Using the conservative isotoxicity method, the previously delivered dose is not considered in the current treatment, and each new plan is evaluated all over again as if it were to be delivered for all the past and present fractions. The violation of OAR constraints is not allowed in this method. Therefore, this method requires the physician to be cautious of the minimum acceptable PTV coverage, as the proximity of OARs may reduce PTV coverage considerably to maintain OAR constraints.

19.4 Patient-specific QA

The patient-specific QA for an offline adaptive plan is performed using the same steps as a standard new plan but usually on a compressed timescale, in which detailed plan review and a test of the dose to be delivered through a phantom measurement are conducted. In contrast, the patient-specific QA for an online or real-time adaptive plan should be different, as pretreatment measurement-based QA is not feasible with the patient on the treatment position. However, the error caught in the patient-specific measurement-based QA process does not contribute to the majority of errors in an RT process [72]. As such, an independent dose calculation approach or a secondary dose calculation platform may be used instead of the measurement-based dose test [5, 73]. Actually, the accuracy of predicted dose delivery is not the most important safety concern for online ART if a comprehensive commission of the treatment planning system and treatment devices is carefully conducted. Instead, the fundamental concerns for a safe and robust online ART replan are fidelity and reliability of the pretreatment imaging, accurate description of relative electron density and high quality re-contouring to reflect the shifting and deformation of the target and/or OARs. Therefore, a well-prepared and experienced team as well as automation checks for potential errors in contouring and plan parameters may be the most important factors to fulfill a comprehensive and robust patient-specific QA for online ART.

Because the patient-specific QA crosses all three types of ART, the verification of the good or expected performance of the machine components is another important aspect. For online and offline ART, this can be done through analyzing the machine delivery log files after treatment or after the offline patient-specific phantom-based delivery. In this regard, establishment of a correlation between the independent dosimetric QA for online ART and the phantom-based delivery for the first set of patients could help build confidence in the QA process. For real-time ART, the machine log analysis should be done in real time, ideally combined with the method of real-time feedback, such as via portal dosimetry and exit dose analysis.

Online patient-specific QA measurements are possible through using both the Linac trajectory log files and the Linac's electron portal imaging device (EPID). Using the Linac's log files, the adaptive plan can be reconstructed as delivered, and the dose is calculated on the patient's daily image. However, analysis of log files does not provide end-to-end verification of the actual dose delivered to the patient, although it can provide information on the Linac system performance. A group from the Medical College of Wisconsin has developed a QA solution allowing a fast online patient-specific QA performed within 1 min. This QA package checks transfer of plan data to the verification system and confirms post-treatment delivery parameters during the treatment [74].

Using a real-time dose-monitoring device is another potential solution to online patient-specific QA. A group from Netherlands Cancer Institute in Amsterdam has led the work of *in vivo* EPID dosimetry using the megavoltage flat panel detector to directly measure the dose delivered the patient [75]. This system has helped avoid or reduce many treatment errors arising from patient anatomic variations or technical error for years. A commercial system of this type has been developed using both log files and EPID imaging to compare the measured fluence with a reference. The higher radiation load on the imaging panels for online or real-time QA purpose during the treatment delivery may require more frequent QA servicing.

There are challenges for patient-specific QA for online, especially for real-time ART. There is a considerably higher workload for physics staff to evaluate daily or even more frequent QA results. Meanwhile, the higher-frequency usage of QA devices for checking adaptive plans would cause wear-and-tear issues and more frequent servicing as mentioned above. It is difficult to verify the dose delivery of the generated adaptive plans online or real time because the patient is waiting on-table in the treatment position, which precludes QA phantom-based measurement during the treatment delivery. Although the solutions discussed above have partially addressed some of these challenges in a variety of degrees of success, improvements and/or innovations on this ongoing research topic are certainly expected.

References

[1] Green O L, Henke L E and Hugo G D 2019 Practical clinical workflows for online and offline adaptive radiation therapy *Semin. Radiat. Oncol.* **29** 219–27

[2] Yan D, Vicini F, Wong J and Martinez A 1997 Adaptive radiation therapy *Phys. Med. Biol.* **42** 123–32

[3] Li X A, Liu F and Tai A *et al* 2011 Development of an online adaptive solution to account for inter- and intra-fractional variations *Radiother. Oncol.* **100** 370–4

[4] Stemkens B, Tijssen R H and de Senneville B D *et al* 2015 Optimizing 4-dimensional magnetic resonance imaging data sampling for respiratory motion analysis of pancreatic tumors *Int. J. Radiat. Oncol. Biol. Phys.* **91** 571–8

[5] Lamb J, Cao M and Kishan A *et al* 2017 Online adaptive radiation therapy: implementation of a new process of care *Cureus.* **9** e1618

[6] Feng M, Suresh K and Schipper M J *et al* 2018 Individualized adaptive stereotactic body radiotherapy for liver tumors in patients at high risk for liver damage: a phase 2 clinical trial *JAMA Oncol.* **4** 40–7

[7] Long D E, Tann M and Huang K C *et al* 2018 Functional liver image guided hepatic therapy (FLIGHT) with hepatobiliary iminodiacetic acid (HIDA) scans *Pract. Radiat. Oncol.* **8** 429–36

[8] Gargett M, Haddad C, Kneebone A, Booth J T and Hardcastle N 2019 Clinical impact of removing respiratory motion during liver SABR *Radiat. Oncol.* **14** 93

[9] Chun S J, Jeon S H and Chie E K 2018 A case report of salvage radiotherapy for a patient with recurrent gastric cancer and multiple comorbidities using real-time MRI-guided adaptive treatment system *Cureus* **10** e2471

[10] Choi R and Yu J 2019 Radiation therapy for renal cell carcinoma *Kidney Cancer* **3** 1–6

[11] Singh A K, Tierney R M and Low D A *et al* 2006 A prospective study of differences in duodenum compared to remaining small bowel motion between radiation treatments: implications for radiation dose escalation in carcinoma of the pancreas *Radiat. Oncol.* **1** 33

[12] Liu F, Erickson B, Peng C and Li X A 2012 Characterization and management of interfractional anatomic changes for pancreatic cancer radiotherapy *Int. J. Radiat. Oncol. Biol. Phys.* **83** e423–9

[13] Yoshimura M, Itasaka S, Harada H and Hiraoka M 2013 Microenvironment and radiation therapy *BioMed. Res. Int.* **2013** 685308

[14] Thorwarth D 2018 Biologically adapted radiation therapy *Z. Med. Phys.* **28** 177–83

[15] Brown J M and Wilson W R 2004 Exploiting tumour hypoxia in cancer treatment *Nat. Rev. Cancer* **4** 437–47

[16] Aebersold D M, Burri P and Beer K T *et al* 2001 Expression of hypoxia-inducible factor-1alpha: a novel predictive and prognostic parameter in the radiotherapy of oropharyngeal cancer *Cancer Res.* **61** 2911–6

[17] Gray L H, Conger A D, Ebert M, Hornsey S and Scott O C 1953 The concentration of oxygen dissolved in tissues at the time of irradiation as a factor in radiotherapy *Br. J. Radiol.* **26** 638–48

[18] Webb S and Nahum A E 1993 A model for calculating tumour control probability in radiotherapy including the effects of inhomogeneous distributions of dose and clonogenic cell density *Phys. Med. Biol.* **38** 653–66

[19] Thorwarth D, Eschmann S M, Paulsen F and Alber M 2007 Hypoxia dose painting by numbers: a planning study *Int. J. Radiat. Oncol. Biol. Phys.* **68** 291–300

[20] Zips D, Zophel K and Abolmaali N *et al* 2012 Exploratory prospective trial of hypoxia-specific PET imaging during radiochemotherapy in patients with locally advanced head-and-neck cancer *Radiother. Oncol.* **105** 21–8

[21] Savi A, Incerti E and Fallanca F *et al* 2017 First evaluation of PET-based human biodistribution and dosimetry of (18)F-FAZA, a tracer for imaging tumor hypoxia *J. Nucl. Med.* **58** 1224–9

[22] Peeters S G, Zegers C M and Lieuwes N G *et al* 2015 A comparative study of the hypoxia PET tracers [(18)F]HX4, [(18)F]FAZA, and [(18)F]FMISO in a preclinical tumor model *Int. J. Radiat. Oncol. Biol. Phys.* **91** 351–9

[23] Urtasun R C, Parliament M B and McEwan A J *et al* 1996 Measurement of hypoxia in human tumours by non-invasive spect imaging of iodoazomycin arabinoside *Br. J. Cancer Suppl.* **27** S209–12

[24] Min M, Lin P and Lee M T *et al* 2015 Prognostic role of metabolic parameters of (18)F-FDG PET-CT scan performed during radiation therapy in locally advanced head and neck squamous cell carcinoma *Eur. J. Nucl. Med. Mol. Imaging* **42** 1984–94

[25] Lambrecht M, Van Calster B and Vandecaveye V *et al* 2014 Integrating pretreatment diffusion weighted MRI into a multivariable prognostic model for head and neck squamous cell carcinoma *Radiother. Oncol.* **110** 429–34

[26] Lim-Reinders S, Keller B M, Al-Ward S, Sahgal A and Kim A 2017 Online adaptive radiation therapy *Int. J. Radiat. Oncol. Biol. Phys.* **99** 994–1003

[27] Keall P, Poulsen P and Booth J T 2019 See, think, and act: real-time adaptive radiotherapy *Semin. Radiat. Oncol.* **29** 228–35

[28] Keall P J, Nguyen D T and O'Brien R *et al* 2018 Review of real-time 3-dimensional image guided radiation therapy on standard-equipped cancer radiation therapy systems: are we at the tipping point for the era of real-time radiation therapy? *Int. J. Radiat. Oncol. Biol. Phys.* **102** 922–31

[29] Heukelom J and Fuller C D 2019 Head and neck cancer adaptive radiation therapy (ART): conceptual considerations for the informed clinician *Semin. Radiat. Oncol.* **29** 258–73

[30] Schulze R, Heil U and Gross D *et al* 2011 Artefacts in CBCT: a review *Dentomaxillofac. Radiol.* **40** 265–73

[31] Niu T, Al-Basheer A and Zhu L 2012 Quantitative cone-beam CT imaging in radiation therapy using planning CT as a prior: first patient studies *Med. Phys.* **39** 1991–2000

[32] Simard M, Lapointe A, Lalonde A, Bahig H and Bouchard H 2019 The potential of photon-counting CT for quantitative contrast-enhanced imaging in radiotherapy *Phys. Med. Biol.* **64** 115020

[33] Lapointe A, Lalonde A, Bahig H, Carrier J F, Bedwani S and Bouchard H 2018 Robust quantitative contrast-enhanced dual-energy CT for radiotherapy applications *Med. Phys.* **45** 3086–96

[34] Noel C E, Parikh P J and Spencer C R *et al* 2015 Comparison of onboard low-field magnetic resonance imaging versus onboard computed tomography for anatomy visualization in radiotherapy *Acta Oncol.* **54** 1474–82

[35] Jelen U, Dong B and Begg J *et al* 2020 Dosimetric optimization and commissioning of a high field inline MRI-Linac *Front. Oncol.* **10** 136

[36] Mutic S and Dempsey J F 2014 The ViewRay system: magnetic resonance-guided and controlled radiotherapy *Semin. Radiat. Oncol.* **24** 196–9

[37] Boldrini L, Cusumano D, Cellini F, Azario L, Mattiucci G C and Valentini V 2019 Online adaptive magnetic resonance guided radiotherapy for pancreatic cancer: state of the art, pearls and pitfalls *Radiat. Oncol.* **14** 71

[38] Uh J, Merchant T E, Li Y, Li X and Hua C 2014 MRI-based treatment planning with pseudo CT generated through atlas registration *Med. Phys.* **41** 051711

[39] Karotki A, Mah K, Meijer G and Meltsner M 2011 Comparison of bulk electron density and voxel-based electron density treatment planning *J. Appl. Clin. Med. Phys.* **12** 3522

[40] Bulens P, Thomas M, Deroose C M and Haustermans K 2018 PET imaging in adaptive radiotherapy of gastrointestinal tumors *Q. J. Nucl. Med. Mol. Imaging.* **62** 385–403

[41] Li X X, Liu N B and Zhu L *et al* 2015 Consequences of additional use of contrast-enhanced (18)F-FDG PET/CT in target volume delineation and dose distribution for pancreatic cancer *Br. J. Radiol.* **88** 20140590

[42] Kishi T, Matsuo Y and Nakamura A *et al* 2016 Comparative evaluation of respiratory-gated and ungated FDG-PET for target volume definition in radiotherapy treatment planning for pancreatic cancer *Radiother. Oncol.* **120** 217–21

[43] Pretz J L, Blake M A and Killoran J H *et al* 2018 Pilot study on the impact of F18-labeled thymidine PET/CT on gross tumor volume identification and definition for pancreatic cancer *Pract. Radiat. Oncol.* **8** 179–84

[44] Topkan E, Yavuz A A, Aydin M, Onal C, Yapar F and Yavuz M N 2008 Comparison of CT and PET-CT based planning of radiation therapy in locally advanced pancreatic carcinoma *J. Exp. Clin. Cancer Res.* **27** 41

[45] Parlak C, Topkan E, Onal C, Reyhan M and Selek U 2012 Prognostic value of gross tumor volume delineated by FDG-PET-CT based radiotherapy treatment planning in patients with locally advanced pancreatic cancer treated with chemoradiotherapy *Radiat. Oncol.* **7** 37

[46] Wilson J M, Mukherjee S, Chu K Y, Brunner T B, Partridge M and Hawkins M 2014 Challenges in using (18)F-fluorodeoxyglucose-PET-CT to define a biological radiotherapy boost volume in locally advanced pancreatic cancer *Radiat. Oncol.* **9** 146

[47] Huang X, Knoble J L and Zeng M *et al* 2016 Neoadjuvant gemcitabine chemotherapy followed by concurrent IMRT simultaneous boost achieves high R0 resection in borderline resectable pancreatic cancer patients *PLoS One.* **11** e0166606

[48] Bundschuh R A, Andratschke N and Dinges J *et al* 2012 Respiratory gated [18F]FDG PET/CT for target volume delineation in stereotactic radiation treatment of liver metastases *Strahlenther. Onkol.* **188** 592–8

[49] Van De Voorde L, Vanneste B and Houben R *et al* 2015 Image-guided stereotactic ablative radiotherapy for the liver: a safe and effective treatment *Eur. J. Surg. Oncol.* **41** 249–56

[50] Riou O, Serrano B and Azria D *et al* 2014 Integrating respiratory-gated PET-based target volume delineation in liver SBRT planning, a pilot study *Radiat. Oncol.* **9** 127

[51] Israel O, Pellet O and Biassoni L *et al* 2019 Two decades of SPECT/CT—the coming of age of a technology: an updated review of literature evidence *Eur. J. Nucl. Med. Mol. Imaging* **46** 1990–2012

[52] Wang H, Feng M, Frey K A, Ten Haken R K, Lawrence T S and Cao Y 2013 Predictive models for regional hepatic function based on 99mTc-IDA SPECT and local radiation dose for physiologic adaptive radiation therapy *Int. J. Radiat. Oncol. Biol. Phys.* **86** 1000 6

[53] Bowen S R, Saini J and Chapman T R *et al* 2015 Differential hepatic avoidance radiation therapy: proof of concept in hepatocellular carcinoma patients *Radiother. Oncol.* **115** 203–10

[54] Roper J, Bowsher J and Yin F F 2009 On-board SPECT for localizing functional targets: a simulation study *Med. Phys.* **36** 1727–35

[55] Roper J R, Bowsher J E, Wilson J M, Turkington T G and Yin F F 2012 Target localization using scanner-acquired SPECT data *J. Appl. Clin. Med. Phys.* **13** 3724

[56] Bowsher J, Yan S, Roper J, Giles W and Yin F F 2014 Onboard functional and molecular imaging: a design investigation for robotic multipinhole SPECT *Med. Phys.* **41** 010701

[57] Wang H, Nie K, Chang J and Kuang Y 2020 A Monte Carlo study to investigate the feasibility of an on-board SPECT/spectral-CT/CBCT imager for medical linear accelerator *Med. Phys.* **47** 5112–22

[58] Wang H and Kuang Y 2017 A novel linac gantry-based onboard imager for simultaneous CBCT, spectral CT and SPECT nnline imaging *Int. J. Radiat. Oncol. Biol. Phys.* **99** S225

[59] Wang H and Kuang Y 2017 The Feasibility of simultaneous CBCT, spectral CT and SPECT online imaging using a novel linac gantry-based on-board imager *Med. Phys.* **44** 3184

[60] Kwint M, Conijn S and Schaake E *et al* 2014 Intra thoracic anatomical changes in lung cancer patients during the course of radiotherapy *Radiother. Oncol.* **113** 392–7

[61] Acharya S, Fischer-Valuck B W and Kashani R *et al* 2016 Online magnetic resonance image guided adaptive radiation therapy: first clinical applications *Int. J. Radiat. Oncol. Biol. Phys.* **94** 394–403

[62] Yan D, Jaffray D A and Wong J W 1999 A model to accumulate fractionated dose in a deforming organ *Int. J. Radiat. Oncol. Biol. Phys.* **44** 665–75

[63] Jaffray D A, Lindsay P E, Brock K K, Deasy J O and Tome W A 2010 Accurate accumulation of dose for improved understanding of radiation effects in normal tissue *Int. J. Radiat. Oncol. Biol. Phys.* **76** S135–9

[64] Ahunbay E E, Peng C and Chen G P *et al* 2008 An on-line replanning scheme for interfractional variations *Med. Phys.* **35** 3607–15

[65] Bol G H, Lagendijk J J and Raaymakers B W 2013 Virtual couch shift (VCS): accounting for patient translation and rotation by online IMRT re-optimization *Phys. Med. Biol.* **58** 2989–3000

[66] Fast M F, Kamerling C P and Ziegenhein P *et al* 2016 Assessment of MLC tracking performance during hypofractionated prostate radiotherapy using real-time dose reconstruction *Phys. Med. Biol.* **61** 1546–62

[67] Ravkilde T, Skouboe S, Hansen R, Worm E and Poulsen P R 2018 First online real-time evaluation of motion-induced 4D dose errors during radiotherapy delivery *Med. Phys.* **45** 3893–903

[68] Ravkilde T, Keall P J, Grau C, Hoyer M and Poulsen P R 2014 Fast motion-including dose error reconstruction for VMAT with and without MLC tracking *Phys. Med. Biol.* **59** 7279–96

[69] Skouboe S, Ravkilde T and Muurholm C G *et al* 2018 Real-time dose reconstruction for moving tumours in stereotactic liver radiotherapy *Radiother. Oncol.* **127** S214–5

[70] Teo B K, Bonner Millar L P, Ding X and Lin L L 2015 Assessment of cumulative external beam and intracavitary brachytherapy organ doses in gynecologic cancers using deformable dose summation *Radiother. Oncol.* **115** 195–202

[71] Henke L, Kashani R and Robinson C *et al* 2018 Phase I trial of stereotactic MR-guided online adaptive radiation therapy (SMART) for the treatment of oligometastatic or unresectable primary malignancies of the abdomen *Radiother. Oncol.* **126** 519–26

[72] Ford E C, Terezakis S, Souranis A, Harris K, Gay H and Mutic S 2012 Quality control quantification (QCQ): a tool to measure the value of quality control checks in radiation oncology *Int. J. Radiat. Oncol. Biol. Phys.* **84** e263–9

[73] Cai B, Green O L, Kashani R, Rodriguez V L, Mutic S and Yang D 2018 A practical implementation of physics quality assurance for photon adaptive radiotherapy *Z. Med. Phys.* **28** 211–23

[74] Chen G P, Ahunbay E and Li X A 2016 Technical note: development and performance of a software tool for quality assurance of online replanning with a conventional Linac or MR-Linac *Med. Phys.* **43** 1713

[75] Mans A, Wendling M and McDermott L N *et al* 2010 Catching errors with *in vivo* EPID dosimetry *Med. Phys.* **37** 2638–44

IOP Publishing

Principles and Practice of Image-Guided Abdominal Radiation Therapy

Yu Kuang

Chapter 20

Cone-beam computed tomography–guided adaptive radiation therapy for abdominal cancer

Tonghe Wang, Yang Lei, Walter J Curran, Tian Liu and Xiaofeng Yang

This chapter introduced the role of cone-beam computed tomography (CBCT) in adaptive radiation therapy for abdomen and its challenges in implementation. Recent studies in improvement of quantification accuracy and automatic segmentation on CBCT have been reviewed by highlighting the proposed method, study designs and reported performance. A summary and outlook on this topic are provided in the last section.

20.1 Introduction

CBCT has been widely used in image-guided radiation therapy to improve treatment outcomes. CBCT images are acquired at the time of treatment delivery to provide the anatomic information in the treatment position. Currently, CBCT is primarily used to determine the error of patient setup and inter-fraction motion by comparing the displacement of anatomic landmarks from the treatment planning computed tomography (CT) images [1]. Couch shift and motion management can then be made based on the difference between CBCT and planning CT.

Recently, more demanding applications of CBCT have been proposed with the increasing interest in adaptive radiation therapy. For example, daily dose estimation before treatment is promising by recalculating plan dose on CBCTs since CBCT—acquired and reconstructed with the same basic physical principle of x-ray attenuation as CT—can potentially be related to electron density as well as dose calculation. Meanwhile, the anatomic information provided by daily CBCT images demonstrates the patient geometry of each fraction, where targets and organs can be re-contoured by manual delineation, automatic segmentation or contour propagation from CT by deformable registration. The updated dose and contours can then be used to calculate dose volume histogram (DVH) and compared with DVH of planning, which is critical in evaluating the dosimetric impact of the anatomical variations [2, 3]. The recently implemented four-dimensional (4D) CBCT allows the imaging of the patient at different timed breathing phases, which can be used to quantify the dosimetric uncertainty resulting from the intra-fraction motion [4].

However, these potential uses of CBCT have been limited due, in part, to its degraded image quality when compared to planning CTs. One major source of degradation is the streaking and cupping artifacts caused by scatter contamination [5, 6]. These artifacts lead to large CT number errors, which complicates the calibration process of CBCT Hounsfield unit (HU) to electron density, an essential step in dose calculation [6]. The degraded image contrast and the suppression of bone CT number can also cause large errors in segmentation using conventional intensity-based segmentation methods as well as in deformable registration for contour propagation from planning CT to CBCT [7]. Thus, the severe distortion caused by scatter is considered one of the fundamental limitations of CBCT and prevents CBCT from quantitative usage in radiation therapy.

Many correction methods for CBCT shading artifacts have been proposed in the literature, and, in general, these methods are in two major categories. The first category includes hardware-based pre-processing methods, including air-gap [8], bowtie filter [9] and anti-scatter grid methods [10]. These add-on devices successfully mitigate severe shading artifacts by preventing part of the scattered photons from reaching the detector, while to maintain the signal-to-noise ratio, the correction efficacy is limited by the resulted increase in imaging dose since primary photons are attenuated by these devices as well as scattered photons. The second category includes post-processing techniques, which remove image artifacts by estimating the scatter in the projection domain or the image domain. Typical methods of this type include analytical modeling [11, 12], Monte Carlo simulation [13, 14], measurement-based methods [15] and modulation methods [16]. For example, high-quality planning CT images can be used as prior knowledge to enhance the CBCT images of the same patient in either image domain [6, 17–19] or projection domain [20, 21]. Other methods mitigate the shading artifacts by estimating the low-frequency shading field from the images, which is achieved by sophisticated image segmentation methods [22–24] or ring-correction methods [25]. These methods improve the scatter correction performance, while their implementations require combined considerations of computational complexity, imaging dose, scan time, practicality and efficacy.

A combination of the above methods has been implemented in current commercial CBCT imaging systems. For example, on the Varian TrueBeam On-Board Imager CBCT system, a 10:1 anti-scatter grid is mounted on the flat-panel detector, and a model-based adaptive-scatter-kernel-superposition method is implemented in the image reconstruction process. However, it is still common to observe residual artifacts in clinical CBCT images, which limits a further utility in delineating soft tissue organs. These methods cannot restore the HU value in CBCT images; thus, the pixel values in CBCT images are not calibrated identically to planning CT images in the treatment planning system for dose calculation.

Recently, the applications of machine learning and deep learning have permeated the fields of radiation oncology. Image segmentation and translation techniques have been adapted from computer vision field for specific clinical tasks. In this chapter, a brief review on the current image processing studies for CBCT-guided adaptive radiation therapy for abdominal cancer is presented. As mentioned previously, the CBCT plays a potentially important role in adaptive therapy in providing accurate HU numbers for dose calculation and contour information of the target and organs at risk. The studies can then be categorized by their focuses into image quantitative accuracy improvement and automatic segmentation, which are presented as follows.

20.2 CBCT image quality improvement

In addition to the general correction on the shading artifacts caused by scatter photons, a lot of effort has been made to correct the respiratory motion artifact, which is more severe in thorax and abdomen than in other sites. By using the recently introduced 4D CBCT, Zhang *et al* proposed a motion model which consists of deformable image registration that maps each CBCT to a reference image in the respiration-correlated CBCT image set [26]. The model is patient specific and derived from the 4D CT during the CT simulation. A principal component analysis was applied to reduce the registration error. The deformed images were then combined into a single CBCT image set with improved image quality. They implemented the proposed method on two patients in upper abdomen and showed that the blurry artifacts were visibly reduced, and the tumor boundaries became sharper and more discernible. Kincaid Jr *et al* [27] reported that by the method of Zhang *et al*, the localization error was reduced from 0.43 cm of uncorrected CBCT to 0.21 cm on average among gastroesophageal junction cancer and pancreatic patients.

X-ray scatter can be estimated using Monte-Carlo (MC) simulations. However, this approach requires the information of the patient's entire body, which may not be available in abdomen cases. The limited field-of-view (FOV) of CBCT scanners may truncate the bi-lateral part. To address this issue, Bertram *et al* proposed a volumetric extension method to approximate the image object outside the FOV. Their results showed that even a rough volume extension with homogeneous water can be sufficient for MC-based scatter correction [28]. Without volumetric extension, the scatter tends to be over-estimated due to the missing absorption by the image truncation.

The application of artificial intelligence has been investigated into generating synthetic CT images from CBCT using the latest image translation techniques to restore the HU numbers as well as to improve the image quality. Most of the current studies are focused on brain and pelvis CBCT [29–38]. Studies showed that learning-based methods feature better image quality than conventional CBCT correction methods using the same datasets [30, 31, 33, 39, 40]. Adrian *et al* found their U-Net-based method outperformed two conventional methods—deformable registration method and analytical image-based correction method—with the lowest mean absolute error (MAE) of synthetic CT, the lowest spatial non-uniformity and the most accurate bone geometry [39]. Harms *et al* observed a lower noise of their synthetic CT and a more similar appearance as a real CT when compared with a conventional image-based correction method [33]. Conventional correction methods are designed to enhance a specific aspect of image quality, while the learning-based methods, which aim to generate synthetic CT from CBCT, would change every aspect of image quality to be close to CT, such as noise level that is usually not considered in conventional correction methods. Compared with brain and pelvis, generating synthetic CT from abdomen CBCT is more challenging due to the variation introduced by respiratory and peristalsis. Liu *et al* recently developed a deep learning-based method to generate synthetic CT from CBCT for pancreatic adaptive radiation therapy [41]. An exemplary case is shown in figure 20.1.

In this study by Liu *et al* [41], the CBCT and planning CT were paired for each training patient using deformable registration. A cycle generative adversarial network (CycleGAN) architecture was adopted. CycleGAN is a variant of generative adversarial network (GAN), which is composed of a generative network and a discriminative network that are trained simultaneously. The generative network would generate synthetic images, and the discriminative network would classify an input image as real or synthetic. The training goal of GAN is to let the generative network produce synthetic images as realistic as possible to fool the discriminator while allowing the discriminative network to distinguish the synthetic images from real images. In this way, blurry synthetic images can be easily detected and rejected since they look considerably fake. This conflict goal explains the name of 'adversarial.' Both networks are trained better and better when they compete against one another until equilibrium is reached. In the prediction stage, the trained generative is applied on new incoming images.

Based on the basic architecture of GAN, many variants have been designed and investigated. As shown in figure 20.2, the CycleGAN includes two generators—CBCT-CT generator and CT-CBCT generator—and two discriminators—real CT–synthetic CT discriminator and real CBCT–synthetic CBCT discriminator. In the first cycle, the input CBCT is fed into the CBCT-CT generator to synthesize CT, and then the synthetic CT is fed into the CT-CBCT generator to generate cycle CBCT, which is supposed to be the same as the input CBCT. The cycle CBCT is compared

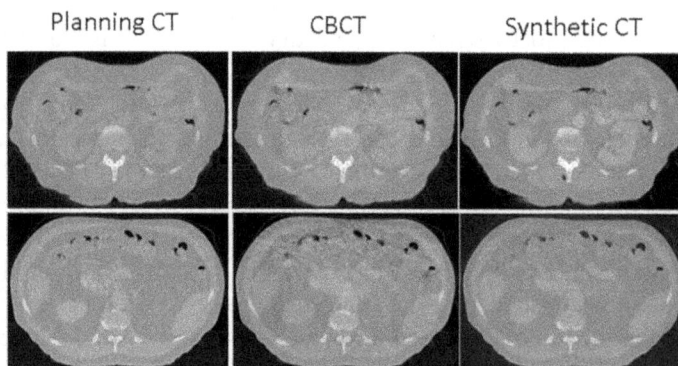

Figure 20.1. Synthetic CT images from CBCT images with planning CT images compared side by side.

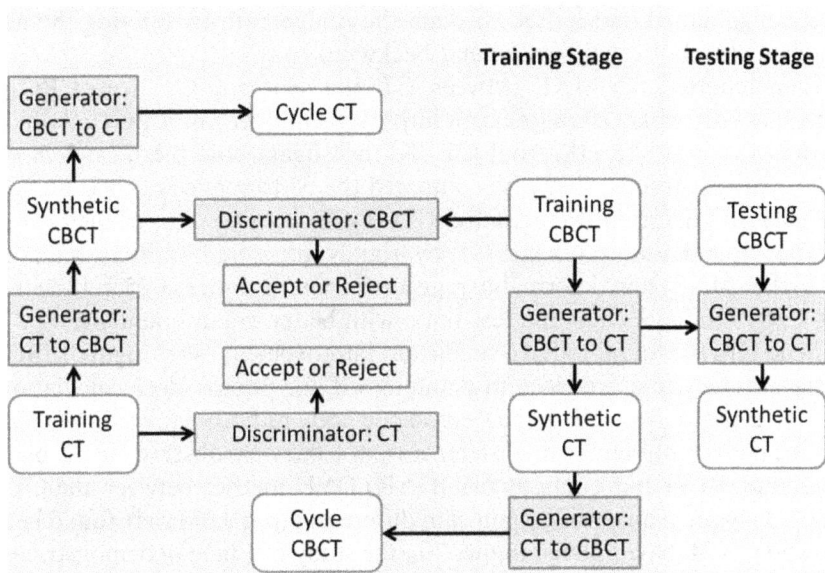

Figure 20.2. Schematic flowchart of CycleGAN.

to the original input CBCT to generate CBCT cycle consistency loss. Meanwhile, the real CT–synthetic CT discriminator distinguishes between the real CT and the synthetic CT to generate CT adversarial loss. To encourage one-to-one mapping between CT and CBCT, a second cycle transformation from CT to CBCT is performed. The second cycle is the same as the first cycle, except the roles of CBCT and CT are swapped, i.e. real CT is fed into the same CT-CBCT generator to synthesize CBCT, and then the synthetic CBCT is fed into the same CBCT-CT generator to generate cycle CT. The cycle CT is compared to the real CT to generate CT cycle consistency loss. The real CBCT–synthetic CBCT discriminator distinguishes between the CBCT and the synthetic CBCT to generate CBCT adversarial loss. Unlike GAN, the CycleGAN couples an inverse mapping network by introducing a cycle consistency loss which enhances the network performance, especially when paired training CT/CBCT images are absent. As a result, CycleGAN can tolerate a certain level of misalignment in the paired training dataset. This property of CycleGAN is attractive to inter-modality synthesis since misalignment in the training datasets is sometimes inevitable due to the unavailability of exact matching image pairs. In many studies, training images are still paired by registration to preserve quantitative pixel values, remove large geometric mismatch to allow network to focus on mapping details and accelerate training [33].

To deal with the potential mismatch of the training datasets, attention gate was used in the generator network to learn the structure variations. With the attention gate, the feature maps extracted from the low-resolution scale were used in gating to distinguish irrelevant responses in long skip connections rather than being used directly. The attention gate also allows the most salient features from compression path to be highlighted and passed through the bridge path by filtering the neuron activations during both the forward and backward pass.

Liu *et al* reported the MAE between CT and synthetic CT from CBCT was 56.89 ± 13.84 HU, which is a significant improvement over raw CBCT of which the MAE was 81.06 ± 15.86 HU from CT. To investigate how the mismatch would affect the synthesis results, they also compared the performance of their method by using rigidly and deformably registered CBCT-CT training data in their pancreas study. They found that synthetic CT by rigidly registered training data had a slightly higher MAE than deformably registered training data (58.45 ± 13.88 HU vs 56.89 ± 13.84 HU, $p > 0.05$) and less noise with better organ boundaries.

Synthetic CTs are found to have significant improvement over original CBCTs in dosimetry accuracy and are close to planning CT for photon dose calculation. As reported by Liu *et al*, with the same stereotactic body radiotherapy (SBRT) plan for each testing patient, no significant differences ($p > 0.05$) were observed in the planning target volume (PTV) and organ at risk (OAR) DVH metrics between the CT- and synthetic CT–based plans, while significant differences ($p < 0.05$) were found between the CT- and CBCT-based plans. Figure 3 in the study of Liu *et al* demonstrates that large local dose calculation error happened at locations with severe artifacts, and synthetic CT successfully mitigated the artifacts and, therefore, the dosimetry error [41]. It should be noted that the dose calculated on the CBCT and synthetic CT images had the same air pocket status as the planning CT to exclude the effect from the different body size and air pocket status. Thus, the dose discrepancy was completed caused by the HU error of CBCT and synthetic CT. It is true that the soft tissue organ variation can be significantly different between CT and CBCT/synthetic CT, which makes the PTV and OAR DVH comparison questionable when directly transferring contours from CT to CBCT/synthetic CT. However, other than bone and air, the HU differences among the other organs are relatively small such that the dosimetric impact resulting from the different locations of those organs would be minimal. Moreover, the SBRT treatment plans use the volumetric-modulated arc therapy (VMAT) technique that employs multiple entry points that will ultimately mitigate the impact of different organ size and locations. The dose comparison in the study by Liu *et al* only served the purpose to show that the synthetic CT generated by their proposed method is capable to provide relatively accurate HU numbers that show little dosimetric difference compared to the planning CT. Future work can focus on the impact of patient size and motion on the PTV and OAR DVHs with the use of the true-size CBCT/synthetic CT and the online registration that was performed prior to each treatment with the updated contours showing the real organ locations.

An alternative approach has also been proposed for daily dose calculation, which is registering the planning CT to the CBCT by deformable registration, and using the registered planning CT for dose calculation. Liu *et al* compared the results of this approach with their results from synthetic CT. They found that although the deformable registration-based method can offer better HU accuracy, it cannot provide accurate daily anatomy of the patient because of the registration error, especially around the regions with large variations such as gas and soft tissue organs. The synthetic CTs, which are generated directly from CBCT, are in the same geometry as the CBCT, thus they can introduce fewer errors in re-contouring of the next step. The unmatched geometry on registered planning CT may lead to under- and overdose at the level of prescription dose in target and OAR due to the re-contouring error, which is much larger than the dose calculation error that is less than 1Gy as reported by Liu *et al.*

The study of Liu *et al* is based on photon therapy using VMAT. The dose calculation on photon plans is quite forgiving to image inaccuracy [38, 42]. For the widely studied VMAT, the contribution from errors in images to dose tends to cancel out in an arc. However, the small dosimetric improvement may still be worthwhile in cases such as abdomen SBRT, where a large amount of dose is to be delivered into a small volume. In such cases, the dose calculation accuracy could be sensitive to the errors on synthetic CT around the target volume [42]. The recent adoption of non-coplanar beams may also be challenging to synthetic CT since the beam path length can be sensitive to the prediction error of patient surface due to the beam obliquity, which is worth further investigation. The introduction of proton therapy provides an alternative way to treat abdominal cancer. Unlike photon beams, proton beams deposit dose with a very high dose gradient at the distal end of the beam [43–45]. The proton treatment plan thus has highly conformal dose distribution to the target by proton beams coming from several angles. The local HU inaccuracy along the beam path on the planning CT would lead to a shift of the highly conformal high-dose area, which may cause the tumor to be substantially underdosed or the organs-at-risk to be overdosed [46]. Future study can investigate the CBCT in adaptive radiation proton therapy with a thorough evaluation of the beam range difference on the CBCTs.

The synthetic CT from CBCT has also been proposed for image registration with CT for potential contour propagation. Fu *et al* generated the synthetic CT images from CBCT images as Liu *et al* did in the previous example and used a fast Demons deformable image registration to register the planning CT with the synthetic CT [47]. The deformed synthetic CT images were then compared with the fixed CT images after image registration. A similar registration was also performed between original CBCT and planning CT. It was shown that the MAE and correlation were improved by synthetic CT, and visually, the registration error was also less on synthetic CT–planning CT registration.

20.3 CBCT image segmentation

Image segmentation has been widely studied in the radiation therapy field to facilitate current treatment planning workflow. In adaptive radiation therapy, automatic image segmentation on CBCT aims to provide prompt and accurate contours for target and OARs. The accuracy of contours is essential for the DVH evaluation and treatment replanning. The speed of the segmentation algorithm would determine its clinical practicality due to the high patient volume in current clinics. Compared with other sites, organs in the abdomen contain high anatomical heterogeneities, large variance of contrast and size and inter-patient variability, which is challenging for automatic segmentation algorithms.

Li *et al* developed a probabilistic atlas to generate liver contours from daily CBCT [48]. The atlas was built from 50 high-contrast planning CT images with manual contour from physicians. The incoming CBCT images were then deformably registered with the atlas; thus the liver contours on the atlas can be propagated to the CBCTs. The Dice similarity coefficient was reported to be 0.83. Based on this work, an improved version was then proposed by Li *et al* later [49]. Unlike the previous study that used the entire volume of the target organs, the improved one only used the information along the contours from the atlas images to guide segmentation. The atlas, built from 139 CT images with manual contours, included the coordinates of contour points as well as the image features close to the contour. The CT images in the atlas were first registered with each other. Between the atlas case and a new case, matching voxels were selected within a narrow shell along the atlas contours based on which deformable registration was then performed. They evaluated the proposed method on liver CBCT and demonstrated that the Jaccard similarity metric was 0.901 ± 0.073, significantly better than that by the previous method (0.453 ± 0.031).

As mentioned previously, the recent advancement in learning-based synthetic image techniques allows the synthesis of realistic and high-quality CT images from CBCT images. CBCT images with high quantification accuracy not only enable daily dose calculation but also expedite the re-contouring task. CT-based abdominal multi-organ segmentation has been reported through the atlas-based method [50, 51] as well as the learning-based method [52, 53]. When the synthetic CTs are close to real CTs, the segmentation methods developed for abdomen CT are supposed to have good performance on the synthetic CTs as well. Segmentation on abdomen CT has been studied for decades. Inspired by the success of deep learning in image segmentation [54–62], Liu *et al* proposed a three-dimensional (3D) deep attention U-Net–based structure to automatically segment multiple OARs for pancreatic SBRT [63]. An exemplary result is shown in figure 20.3. The U-Net architecture, as shown in figure 20.4, has an encoding and a decoding part. The encoding part has three convolutional layers with the first layer followed by a max pooling layer. The decoding part has three up-sampling blocks, each of which has a deconvolution layer followed by a convolutional layer. The two parts are connected through long skip connections that concatenate the feature maps. Therefore, both high-frequency information (such as textural information) and low-frequency information

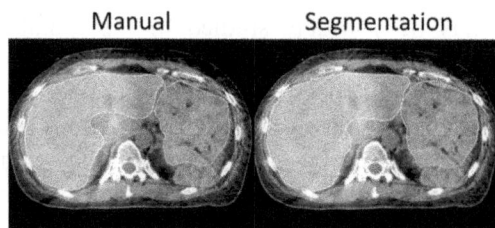

Figure 20.3. CBCT images with liver and stomach maps of manual contours and segmentations resulting from learning-based methods.

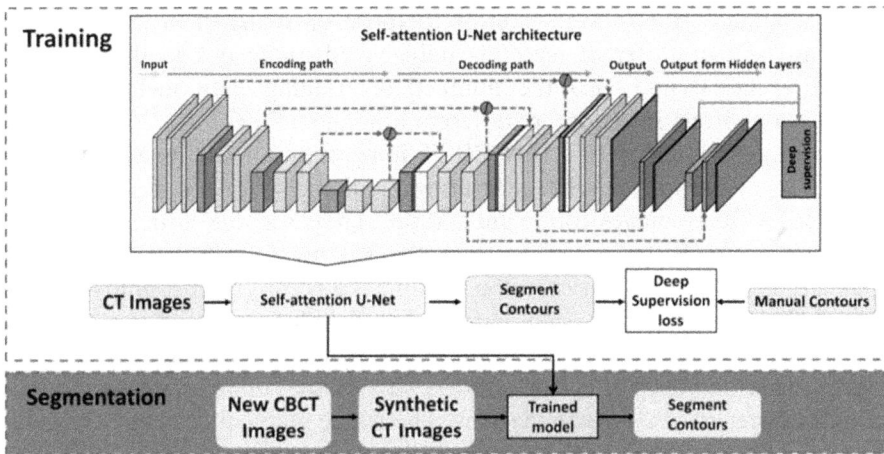

Figure 20.4. Schematic flowchart of self-attention U-Net.

(such as structural information) would be included. Attention gate is also used such that the feature maps extracted from the low-resolution scale were used in gating to distinguish noisy responses in long skip connections rather than being used directly.

To address the over-fitting caused by the limited training data, a deep supervision strategy was used with a compound loss function. Deep supervision was applied to supervise the final prediction stage and each of the decoding stages such that the residual information became semantically meaningful for both the early and final stages in the architecture, which can reduce the convergence time and improve the segmentation performance. In the compound loss function, weighted cross-entropy and Dice similarity coefficient were combined.

In the evaluation, datasets from 30 patients, each with images and manual contours, were trained and tested using a five-fold cross-validation. A total of organs were selected as segmentation targets. After training was finished, the automatic generation of multi-organ contours could be done in less than 1 s. The results showed that the average Dice similarity coefficients were larger than 0.85 for large bowel, small bowel, left kidney, right kidney, liver and stomach, while they were less than 0.8 for duodenum and spinal cord. Duodenum is challenging due to its irregular shape and low contrast of boundaries against surrounding organs. Spinal cord, on the other hand, is in a very small size on axial slices, and thus a little discrepancy in segmentation contour can introduce a large decrease in Dice similarity coefficient. Comparison between their method and other state-of-the-art methods demonstrated a comparable accuracy in most organs with higher accuracy in bowel and duodenum. The segmentation accuracy of bowel and duodenum is important since they are the OARs most concerned in pancreatic radiotherapy. The authors commented that the CT images in comparing studies were acquired during the radiological process for disease diagnosis and thus feature finer voxel size and higher exposure dose than the CT scans used for the radiotherapy treatment planning purpose. This might partly explain why for some organs such as liver and stomach, their results were inferior to those of Wang *et al* [52]. The other reasons can be that different methods may be advantageous in different organ segmentation tasks, and also the number of training data included and the uniformity of contrast injection are different among studies.

Instead of generating high-quality CBCT images, an alternative way for CBCT automatic segmentation can be generating synthetic magnetic resonance (MR) images from CBCT images for segmentation. The rationale is that MR images contain superior soft tissue to contrast the CT or CBCT, which can better distinguish the organ boundaries in the abdomen. Recently, Lei *et al* [64] proposed a method that combines the deep learning-based image synthesis method, which generates synthetic MR images from on-board setup CBCT images to aid CBCT segmentation with a deep attention strategy that focuses on learning discriminative features for differentiating organ margins. The whole segmentation method consists of three major steps. First, synthetic MR images were generated from CBCT images by a CycleGAN. Second, a deep attention network was trained based on synthetic MR images and their corresponding manual contours. Third, the contours for a new patient were obtained by feeding the patient's CBCT images into the trained synthetic MR image estimation and segmentation model. Fu *et al* commented that synthetic MR images are inferior to CBCT for bony structure segmentation [65]. In their study, they propose using both the original CBCT and the CBCT-generated synthetic magnetic resonance imaging (MRI) as inputs for multi-organ segmentation. The purpose is to combine the superior bony structure contrast of CBCTs and the superior soft tissue contrast of synthetic MR images for improved segmentation performance. They used the same CycleGAN to generate synthetic MR images from CBCTs, but both the synthetic MR images and the original CBCT images were processed by two U-Net–like networks, respectively, and combined via a deep fusion network. The proposed networks are called dual pyramid networks. The CBCT–U-Net and synthetic-MRI–U-Net could independently learn relevant features from CBCT and synthetic MR images, respectively, for multi-organ segmentation. The features from corresponding pyramid levels were then concatenated using additive attention gates in the final deep fusion net for further processing. Except for different feature sizes and the attention gates involved, the deep fusion net was the same as the decoding path of the two U-Net. The deep fusion process was designed to combine the advantages of CBCT-specific features in bony structure segmentation and synthetic MRI–specific features in soft tissue segmentation. Finally, a deep fusion net was introduced to combine the features learned from corresponding levels of CBCT–U-Net and synthetic-MRI–U-Net via attention gates. These attention gates were used since they could retrieve the most relevant features of organ boundaries. Since the synthetic MRI is directly dependent on the CBCT, the information provided by the synthetic MRI may not be as complementary as that of the original MRI. Despite being less complementary as compared to the original MRI, the synthetic MRI could introduce certain additional structural information that is helpful to the segmentation task. Although these methods have only been evaluated in pelvis CBCT multi-organ segmentation, the results are encouraging such that they can also be promising for abdomen CBCT.

20.4 Summary and outlook

Adaptive radiation therapy has been actively studied in recent years due to its potential improvement in treatment outcomes. For the abdomen, studies showed the dosimetric benefit of adaptive re-planning. For example, Li *et al* reported that for pancreatic cancer SBRT, treatment re-optimization was able to improve the coverage of PTV and reduce dose to duodenum. Specifically, the hot spot in PTV decreased from 4.5% to 0.5% after daily adaptive re-planning, and the volume of duodenum receiving prescription dose decreased from 0.9% to 0.3%. The air in the stomach was found to be a key factor that can substantially enlarge the discrepancy from the planned dose [66]. Similarly, inter-fraction deformation is also found to perturb the delivered dose distribution for liver cases. Velec *et al* reported that 70% of liver SBRT patients in their study have more than 5% dose discrepancy from planned dose with CBCT only for 3D daily repositioning [67]. These examples demonstrated the dosimetric necessity of adaptive radiation therapy for abdomen cases, especially when the SBRT has been widely used in clinic.

The achievable benefits of adaptive radiation therapy highly depend on its clinical implementations. As presented previously, adaptive radiation therapy based on CBCT is challenging since the inaccurate quantification and unsatisfactory image quality on current CBCT pose problems in the dose calculation task and re-contouring task. These undesired features of CBCT are a combined effect by the suboptimal acquisition scheme of flat panel detector affected by scatter photons and slow scan speed that increases motion artifacts and large respiratory motion inherently in the abdomen. To address these issues, hardware such as the anti-scatter grid has been applied to alleviate the scatter photons reaching the detector. Motion management on patients has also been proposed to mitigate the motion artifacts by using the recent commercially accessible 4D CBCT or gated CBCT. These hardware-based solutions have been also extended to other sites such as lung and pelvis.

The studies introduced in this chapter are presented in the context of abdominal cancer radiation therapy. They provide additional solutions from software to facilitate the adaptive radiotherapy workflow. Improvement in the quantitative accuracy of CBCT voxel values allow the CBCT to be viable for dose calculation as well as make the next step of contouring easier. Automatic segmentation methods further enable an expedited DVH evaluation and treatment re-planning. Recent years have witnessed the trend of deep learning being increasingly used in the application of medical imaging. The latest networks and techniques have been borrowed from computer vision fields and adapted to specific clinical tasks for radiology and radiation oncology. Learning-based image synthesis and segmentation is an emerging and active field. Although a few pieces of literature show the success of deep learning-based image synthesis in various applications, there are still some common open questions that need to be answered in future studies.

The reviewed studies show the advantages of learning-based methods over conventional methods in clinical applications. In implementation, depending on the hardware, training a model usually takes several hours to days for learning-based methods. However, once the model is trained, it can be applied to new patients to generate synthetic CT or contours within a few seconds or minutes. Although the advantages of learning-based methods have been demonstrated, it should be noted that their performance can be unpredictable when the input images are very different from their training datasets. In most of the current studies, unusual cases are excluded. However, these unusual cases can happen from time to time in clinic and should be handled with caution. For example, it is not uncommon to see a patient with fiducial markers or stent in an abdominal scan. These methods create severe artifacts on CBCT; thus, it can be of clinical interest to see the related effect of its inclusion in training or testing dataset, which has not been studied yet. Similar unusual cases can also be seen in other forms such as obese patients that present much higher noise level on image than average and patients with anatomical abnormality. Moreover, the small to intermediate number of patients in training/testing included in current feasibility studies is far from enough in evaluating clinical utility and potential impact. The absence of diverse demographics may reduce the robustness in the performance of the model. It is also of great clinical interest to evaluate the performance in multi-scanner, multi-vendor and multi-center situations to validate the generality.

Before being deployed into clinical workflow, there are still a few challenges to be addressed. To account for the potential unpredictable synthetic images that can result from noncompliance with imaging protocols as training data or unexpected anatomic structures, an additional quality assurance (QA) step would be essential in clinical practice. The QA procedure would aim to check the consistency on the performance of the model routinely or after upgrade by re-training the network with more patient datasets as well as to check the quality of patient-specific cases. For re-contouring, although the human inspection would still be necessary after the automatic segmentation, the total time required for inspection and modification is supposed to be much less than the time spent on manual contouring, provided that the model is high in accuracy and robustness.

In addition to the high-quality CBCT images, rapid dose calculation is also essential to online dose calculation. Besides the fast development of graphics processing unit–based dose calculation engine [68], learning-based automatic dose distribution generation has been an active field [69, 70]. However, compared with the research on CT that is widely investigated, related studies on CBCT are still sparse. As discussed before, CBCT has additional challenges in that CBCT has smaller FOV compared to the CT images due to its limited size in detector and the longer acquisition time that causes larger motion artifacts. Considering coplanar plan is usually used for abdomen SBRT, the limited FOV would pose a problem on dose calculation accuracy due to the lack of body information.

References

[1] Barney B M, Lee R J, Handrahan D, Welsh K T, Cook J T and Sause W T 2011 Image-guided radiotherapy (IGRT) for prostate cancer comparing kV imaging of fiducial markers with cone beam computed tomography (CBCT) *Int. J. Radiat. Oncol. Biol. Phys.* **80** 301–5
[2] Zhu L, Xie Y, Wang J and Xing L 2009 Scatter correction for cone-beam CT in radiation therapy *Med. Phys.* **36** 2258–68

[3] de la Zerda A, Armbruster B and Xing Lei 2007 Formulating adaptive radiation therapy (ART) treatment planning into a closed-loop control framework *Phys. Med. Biol.* **52** 4137

[4] Qin A *et al* 2018 A clinical 3D/4D CBCT-based treatment dose monitoring system *J. Appl. Clin. Med. Phys.* **19** 166–76

[5] Grimmer R and Kachelriess M 2011 Empirical binary tomography calibration (EBTC) for the precorrection of beam hardening and scatter for flat panel CT *Med. Phys.* **38** 2233–40

[6] Marchant T E, Moore C J, Rowbottom C G, MacKay R I and Williams P C 2008 Shading correction algorithm for improvement of cone-beam CT images in radiotherapy *Phys. Med. Biol.* **53** 5719

[7] Hou J, Guerrero M, Chen W and D'Souza W D 2011 Deformable planning CT to cone-beam CT image registration in head-and-neck cancer *Med. Phys.* **38** 2088–94

[8] Siewerdsen J H and Jaffray D A 2000 Optimization of x-ray imaging geometry (with specific application to flat-panel cone-beam computed tomography) *Med. Phys.* **27** 1903–14

[9] Mail N, Moseley D J, Siewerdsen J H and Jaffray D A 2008 The influence of bowtie filtration on cone-beam CT image quality *Med. Phys.* **36** 22–32

[10] Siewerdsen J H, Moseley D J, Bakhtiar B, Richard S and Jaffray D A 2004 The influence of antiscatter grids on soft-tissue detectability in cone-beam computed tomography with flat-panel detectors *Med. Phys.* **31** 3506–20

[11] Boone J M and Seibert J A 1988 An analytical model of the scattered radiation distribution in diagnostic radiology *Med. Phys.* **15** 721–5

[12] Yang X, Wu S, Sechopoulos I and Fei B 2012 Cupping artifact correction and automated classification for high-resolution dedicated breast CT images *Med. Phys.* **39** 6397–406

[13] Colijn A P and Beekman F J 2004 Accelerated simulation of cone beam X-ray scatter projections *IEEE Trans. Med. Imaging* **23** 584–90

[14] Kyriakou Y, Riedel T and Kalender W A 2006 Combining deterministic and Monte Carlo calculations for fast estimation of scatter intensities in CT *Phys. Med. Biol.* **51** 4567

[15] Ning R, Tang X and Conover D 2004 X-ray scatter correction algorithm for cone beam CT imaging *Med. Phys.* **31** 1195–202

[16] Zhu L, Bennett N R and Fahrig R 2006 Scatter correction method for X-ray CT using primary modulation: theory and preliminary results *IEEE Trans. Med. Imaging* **25** 1573–87

[17] Brunner S, Nett B E, Tolakanahalli R and Chen G-H 2011 Prior image constrained scatter correction in cone-beam computed tomography image-guided radiation therapy *Phys. Med. Biol.* **56** 1015

[18] Yu L, Bruesewitz M R, Thomas K B, Fletcher J G, Kofler J M and McCollough C H 2011 Optimal tube potential for radiation dose reduction in pediatric CT: Principles, clinical implementations, and pitfalls *RadioGraphics.* **31** 835–48

[19] Shi L, Tsui T, Wei J and Zhu L 2017 Fast shading correction for cone beam CT in radiation therapy via sparse sampling on planning CT *Med. Phys.* **44** 1796–808

[20] Niu T, Sun M, Star-Lack J, Gao H, Fan Q and Zhu L 2010 Shading correction for on-board cone-beam CT in radiation therapy using planning MDCT images *Med. Phys.* **37** 5395–406

[21] Niu T, Al-Basheer A and Zhu L 2012 Quantitative cone-beam CT imaging in radiation therapy using planning CT as a prior: first patient studies *Med. Phys.* **39** 1991–2000

[22] Wu P *et al* 2015 Iterative CT shading correction with no prior information *Phys. Med. Biol.* **60** 8437

[23] Zhao W, Vernekohl D, Zhu J, Wang L and Xing L 2016 A model-based scatter artifacts correction for cone beam CT *Med. Phys.* **43** 1736–53

[24] Wang T and Zhu L 2017 Image-domain non-uniformity correction for cone-beam CT *2017 IEEE 14th Int. Symp. on Biomedical Imaging (ISBI 2017)*

[25] Fan Q *et al* 2015 Image-domain shading correction for cone-beam CT without prior patient information *J. Appl. Clin. Med. Phys.* **16** 65–75

[26] Zhang Q, Hu Y-C, Liu F, Goodman K, Rosenzweig K E and Mageras G S 2010 Correction of motion artifacts in cone-beam CT using a patient-specific respiratory motion model *Med. Phys.* **37** 2901–9

[27] Kincaid R E Jr *et al* 2018 Evaluation of respiratory motion-corrected cone-beam CT at end expiration in abdominal radiotherapy sites: a prospective study *Acta Oncol.* **57** 1017–24

[28] Bertram M, Sattel T, Hohmann S and Wiegert J 2008 Monte-Carlo scatter correction for cone-beam computed tomography with limited scan field-of-view *Proc. SPIE* **6913** 69131Y

[29] Hansen D C *et al* 2018 ScatterNet: a convolutional neural network for cone-beam CT intensity correction *Med. Phys.* **45** 4916–26

[30] Kida S *et al* 2018 Cone beam computed tomography image quality improvement using a deep convolutional neural network *Cureus* **10** e2548

[31] Xie S, Yang C, Zhang Z and Li H 2018 Scatter artifacts removal using learning-based method for CBCT in IGRT system *IEEE Access* **6** 78031–7

[32] Chen L, Liang X, Shen C, Jiang S and Wang J 2019 Synthetic CT generation from CBCT images via deep learning *Med. Phys.* **47** 1115–25

[33] Harms J *et al* 2019 Paired cycle-GAN-based image correction for quantitative cone-beam computed tomography *Med. Phys.* **46** 3998–4009

[34] Kurz C *et al* 2019 CBCT correction using a cycle-consistent generative adversarial network and unpaired training to enable photon and proton dose calculation *Phys. Med. Biol.* **64** 225004

[35] Landry G *et al* 2019 Comparing Unet training with three different datasets to correct CBCT images for prostate radiotherapy dose calculations *Phys. Med. Biol.* **64** 035011

[36] Kida S *et al* 2019 Visual enhancement of cone-beam CT by use of CycleGAN *Med. Phys.* **47** 998–1010

[37] Lei Y *et al* 2018 Learning-based CBCT correction using alternating random forest based on auto-context model *Med. Phys.* **46** 601–18

[38] Wang T *et al* 2019 Dosimetric study on learning-based cone-beam CT correction in adaptive radiation therapy *Med. Dosim.* **44** e71–9

[39] Thummerer A *et al* 2020 Comparison of CBCT based synthetic CT methods suitable for proton dose calculations in adaptive proton therapy *Phys. Med. Biol.* **65** 095002

[40] Nomura Y, Xu Q, Shirato H, Shimizu S and Xing L 2019 Projection-domain scatter correction for cone beam computed tomography using a residual convolutional neural network *Med. Phys.* **46** 3142–55

[41] Liu Y *et al* 2020 CBCT-based synthetic CT generation using deep-attention cycleGAN for pancreatic adaptive radiotherapy *Med. Phys.* **47** 2472–83

[42] Wang T *et al* 2019 MRI-based treatment planning for brain stereotactic radiosurgery: dosimetric validation of a learning-based pseudo-CT generation method *Med. Dosim.* **44** 199–204

[43] Liu Y *et al* 2019 Evaluation of a deep learning-based pelvic synthetic CT generation technique for MRI-based prostate proton treatment planning *Phys. Med. Biol.* **64** 205022

[44] Liu Y *et al* 2019 MRI-based treatment planning for proton radiotherapy: dosimetric validation of a deep learning-based liver synthetic CT generation method *Phys. Med. Biol.* **64** 145015

[45] Shafai-Erfani G *et al* 2019 MRI-based proton treatment planning for base of skull tumors *Int. J. Part. Ther.* **6** 12–25

[46] Li B *et al* 2017 Comprehensive analysis of proton range uncertainties related to stopping-power-ratio estimation using dual-energy CT imaging *Phys. Med. Biol.* **62** 7056

[47] Fu Y *et al* 2020 Cone-beam computed tomography (CBCT) and CT image registration aided by CBCT-based synthetic CT *Proc. SPIE* **11313** 113132U

[48] Li D W, Wang H J, Chen D and Yin Y 2012 Automated liver segmentation for cone beam CT dataset by probabilistic atlas construction *Appl. Mech. Mater.* **195–196** 583–8

[49] Li D *et al* 2016 Augmenting atlas-based liver segmentation for radiotherapy treatment planning by incorporating image features proximal to the atlas contours *Phys. Med. Biol.* **62** 272

[50] Karasawa K *et al* 2017 Multi-atlas pancreas segmentation: atlas selection based on vessel structure *Med. Image Anal.* **39** 18–28

[51] Wolz R, Chu C, Misawa K, Fujiwara M, Mori K and Rueckert D 2013 Automated abdominal multi-organ segmentation with subject-specific atlas generation *IEEE Trans. Med. Imaging* **32** 1723–30

[52] Wang Y, Zhou Y, Shen W, Park S, Fishman E K and Yuille A L 2019 Abdominal multi-organ segmentation with organ-attention networks and statistical fusion *Med. Image Anal.* **55** 88–102

[53] Okada T, Linguraru M G, Hori M, Summers R M, Tomiyama N and Sato Y 2015 Abdominal multi-organ segmentation from CT images using conditional shape-location and unsupervised intensity priors *Med. Image Anal.* **26** 1–18

[54] Dong X *et al* 2019 Synthetic MRI-aided multi-organ segmentation on male pelvic CT using cycle consistent deep attention network *Radiother. Oncol.* **141** 192–9

[55] Dong X *et al* 2019 Automatic multiorgan segmentation in thorax CT images using U-net-GAN *Med. Phys.* **46** 2157–68

[56] He X *et al* 2020 Automatic segmentation and quantification of epicardial adipose tissue from coronary computed tomography angiography *Phys. Med. Biol.* **65** 095012

[57] Jun Guo B *et al* 2020 Automated left ventricular myocardium segmentation using 3D deeply supervised attention U-net for coronary computed tomography angiography; CT myocardium segmentation *Med. Phys.* **47** 1775–85

[58] Lei Y *et al* 2019 CT prostate segmentation based on synthetic MRI-aided deep attention fully convolution network *Med. Phys.* **47** 530–40

[59] Lei Y *et al* 2019 Ultrasound prostate segmentation based on multidirectional deeply supervised V-Net *Med. Phys.* **46** 3194–206

[60] Wang B *et al* 2019 Deeply supervised 3D fully convolutional networks with group dilated convolution for automatic MRI prostate segmentation *Med. Phys.* **46** 1707–18

[61] Wang T *et al* 2019 A lea et al.rning-based automatic segmentation and quantification method on left ventricle in gated myocardial perfusion SPECT imaging: a feasibility study *J. Nucl. Cardiol.* **27** 976–87

[62] Wang T *et al* 2019 Learning-based automatic segmentation of arteriovenous malformations on contrast CT images in brain stereotactic radiosurgery *Med. Phys.* **46** 3133–41

[63] Liu Y *et al* 2020 CT-based pancreatic multi-organ segmentation by a 3D deep attention U-net network *Proc. SPIE* **11318** 1131813

[64] Lei Y *et al* 2020 Male pelvic multi-organ segmentation aided by CBCT-based synthetic MRI *Phys. Med. Biol.* **65** 035013

[65] Fu Y *et al* 2020 Pelvic multi-organ segmentation on cone-beam CT for prostate adaptive radiotherapy *Med. Phys.* **47** 3415–22

[66] Li Y *et al* 2015 Dosimetric benefit of adaptive re-planning in pancreatic cancer stereotactic body radiotherapy *Med. Dosim.* **40** 318–24

[67] Velec M, Moseley J L, Craig T, Dawson L A and Brock K K 2012 Accumulated dose in liver stereotactic body radiotherapy: positioning, breathing, and deformation effects *Int. J. Radiat. Oncol. Biol. Phys.* **83** 1132–40

[68] Tian Z, Li Y, Hassan-Rezaeian N, Jiang S B and Jia X 2017 Moving GPU-OpenCL-based Monte Carlo dose calculation toward clinical use: automatic beam commissioning and source sampling for treatment plan dose calculation *J. Appl. Clin. Med. Phys.* **18** 69–84

[69] Chen X, Men K, Li Y, Yi J and Dai J 2018 A feasibility study on an automated method to generate patient-specific dose distributions for radiotherapy using deep learning *Med. Phys.* **46** 56–64

[70] Liu Z *et al* 2019 A deep learning method for prediction of three-dimensional dose distribution of helical tomotherapy *Med. Phys.* **46** 1972–83

IOP Publishing

Principles and Practice of Image-Guided Abdominal
Radiation Therapy

Yu Kuang

Chapter 21

Magnetic resonance–guided online adaptation for abdominal tumors: technical aspects

Ergun E Ahunbay, Xinfeng Chen, Eric Paulson and X Allen Li

This chapter summarizes the application of magnetic resonance (MR)–guided adaptive radiation therapy (MRgART) for abdominal tumors. We start with a brief overview of the MR-Linac technology and the major clinical concepts in its clinical implementation. This is followed by explaining the main benefits of MRgART specifically for abdominal radiation therapy, namely (1) the real-time imaging capability with MR while the radiation is being delivered, (2) improved soft tissue contrast to facilitate accurate target and organ delineation for online replanning, and (3) potential improvements for better target identification or response assessment with functional MR imaging performed online. These properties of MR-Linac–based treatments can allow target doses to be escalated safely which could potentially be beneficial for tumors of the abdomen with otherwise poor prognosis. Several issues related to the implementation of MRgART are also covered in the rest of the chapter.

21.1 Introduction

Magnetic resonance (MR)–guided adaptive radiation therapy (MRgART) is often delivered with a hybrid MR-Linac machine that integrates magnetic resonance imaging (MRI) and Linac. MR-Linac technology provides high-quality MRI of the patient in the treatment position before, during, and after the delivery of daily radiation. Compared with conventional radiation therapy (RT) with on-board computed tomography (CT)/cone-beam CT-based Linac, MRgART offers distinct advantages (see table 1 in reference [1]) particularly (1) higher soft tissue contrast for better target and/or delineation of organs at risks (OARs), (2) real-time imaging during RT delivery showing the anatomy with no extra imaging dose, and (3) quantitative and/or functional MRI on a daily basis with no additional cost. These advantages are highly desirable for radiation treatment of a variety of tumor sites including the abdomen [2–4]. In this chapter, we will present an overview and technical implementation of MRgART and applications of MRgART for abdominal tumors.

21.2 Overview of MRgART

21.2.1 MR-Linac systems

There are two commercial MR-Linac systems, ViewRay MRIdian [3] and Elekta Unity [4], currently available. Several other systems are currently under development [5, 6]. The main difference between the two commercial systems is their magnetic field strength, 0.35T for MRIdian versus 1.5T for Unity. High field strength allows for higher signal-to-noise ratio, which provides a wider range of imaging possibilities, and allows for shorter scans for the same image quality. However, the difference in the contrast-to-noise ratio between the two systems is not substantial [7]. The higher field strength also results in higher susceptibility-induced distortion for certain types of MR sequences and has higher specific absorption rate, which could be a hindrance during real-time monitoring with MR. Another problem only affecting high-field MR-Linac is the electron return effect (ERE) [8], which can cause an unwanted dosimetric effect near air regions when anatomy changes during RT delivery. This chapter primarily describes MRgART for abdominal tumors using the high-field Unity system. The general methodology presented should be applicable to any MRgART system.

21.2.2 High-field MR-Linac

Elekta's Unity system (Elekta AB, Stockholm, Sweden) integrates a Philips 1.5T magnet with a Linac on a rotating gantry single energy 7 MV flattening filter free beam. It uses the Monaco treatment planning system (Elekta) with a Monte Carlo dose calculation engine for the offline and online plan generation. The online adaptive plan generation happens in two different workflows: adapt to position (ATP) and adapt to shape (ATS). The ATP workflow is mainly an equivalence of conventional image-guided radiation therapy (IGRT), where the original plan dose is preserved with the daily isocenter shifts applied [9]. Unlike the conventional IGRT where the patient (couch) is moved based on the daily image registration, the ATP moves the plan (multileaf collimator (MLC)/segment) to account for patient positioning difference from the simulation position. As the treatment beam is flattening filter free, shifting the segments may not preserve the original dose; thus, a simplified plan optimization is necessary for ATP. Since the ATP workflow does not utilize the full functionalities of Unity, it will not be discussed further. The remaining chapter will focus on the ATS workflow, in which the full adaptive planning workflow based on daily imaging is employed.

21.2.3 MRgART workflow

The general process of MRgART (i.e., ATS) include the following: (1) setting up the patient in the treatment position, (2) acquiring daily MRIs and transferring the images to the treatment planning system (TPS), (3) delineating the target and OARs based on the daily MRI, (4) creating an electron density (ED) map (e.g., synthetic CT) from the MRI set for dose calculation, (5) generating an adaptive plan optimized based on the daily contours created from the daily MRI, (6) evaluating and verifying the new plan, and (7) delivering the adaptive plan to the patient. Prior to the MRgART, a set of reference images (e.g., CT, MRI, or positron emission tomography) is often acquired during RT simulation and is used to create a reference plan which will be used in the subsequent MRgART process.

The online ATS workflow uses a reference plan, which needs to be created beforehand offline. Multiple plans may be generated and saved as a library of plans and one of them picked for online ATS planning. When a reference plan is selected for online replanning, several pieces of information are imported from the reference plan to the daily plan. These include intensity-modulated radiation therapy (IMRT) constraints and an ED map for dose calculation. The IMRT constraints define the optimization goals for the inverse planning, also known as objective function. At the time this chapter is written, the ED map is created by bulk density overriding, where selected structures are assigned with their representative ED value, and all voxels of that structure are overridden with the same ED value.

To speed up the online adaptation process, we implemented a workflow to allow simultaneous contouring at different workstations by multiple operators [10, 11], e.g., radiation oncologist, medical physicist and radiation therapist, utilizing an external software tool (MIM, Beachwood, OH, USA) (figure 21.1). Typical times for this workflow for our patients with abdominal tumors are listed in figure 21.7. The main benefit of our workflow is to reduce the contour delineation times by dividing the contouring process among multiple users who can work simultaneously.

The adaptive plan created needs to be verified for its accuracy before its delivery. As the patient is lying on the table, the online plan QA cannot be done with measurement. In our clinic, we use a software tool developed in house, ArtQA [12], to verify the adaptive plan. Using ArtQA, we perform a secondary calculation based

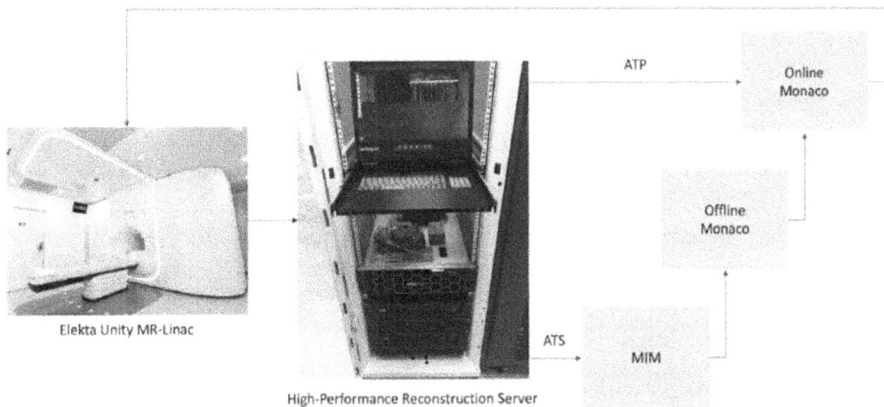

Figure 21.1. System architecture employed to reconstruct four-dimensional (4D) MRI data and adjusted workflow at the Medical College of Wisconsin for ATS treatment of abdominal MRgART. Reprinted from [10], Copyright (2020), with permission from Elsevier.

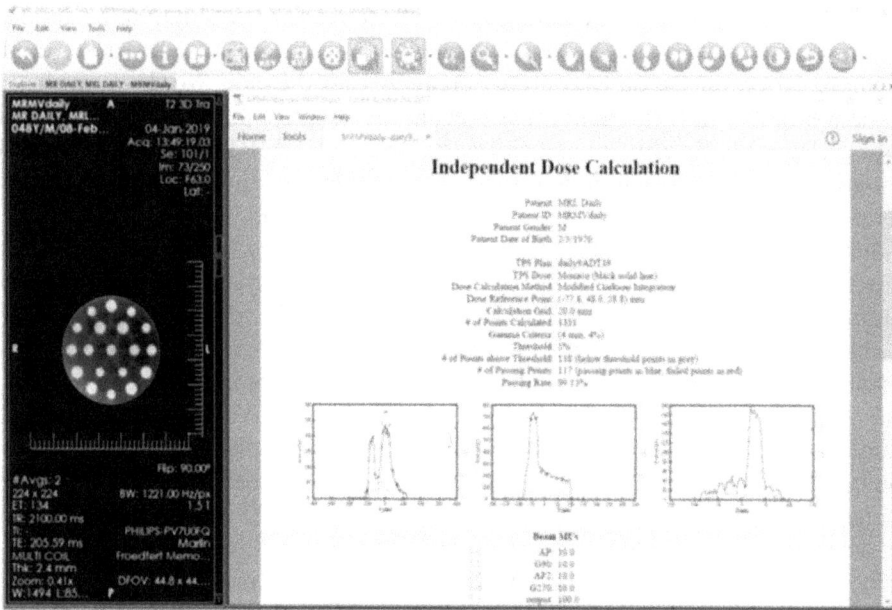

Figure 21.2. ArtQA dose verification software. From [13], Figure 3, John Wiley & Sons.

on a modified Clarkson algorithm, on a grid of 5×5×5 points via gamma analysis (figure 21.2), and check all required conditions that the adaptive plan must meet, as well as the consistency of plan parameters between those in the planning system (Monaco, Elekta) and those transferred into delivery system (Mosaiq). This software QA process is considered to be acceptable, as the original reference plan generated form the reference images is fully verified with measurement using ArcCheck (Sun Nuclear), after the dry run but before the first fraction of the patient plan.

The Unity 1.5T MR-Linac system has been operational in our clinic at the Medical College of Wisconsin since January 2019. As of July 2020, 80 patients had been treated; 55 of them had abdominal tumors, of which 32 had RT delivered with the ATS workflow. All of the abdominal ATS treatments were stereotactic body radiotherapy (SBRT) with five or fewer fractions. The disease sites included 10 liver metastases, 16 pancreas, 5 abdominal lymph nodes, and 1 retroperitoneum. Due to the long plan preparation times of ATS, verification imaging is performed before the start of the beam delivery, and an ATP plan is performed in case the target has moved during the plan preparation. Images for the ATS planning of abdominal treatments are either motion-averaged or mid-position MRI from 4D MRI acquisition.

21.3 Clinical implementation of MRgART

21.3.1 Machine commissioning

The commissioning of an MR-Linac system includes commissioning of the MRI imaging subsystem and Linac radiation delivery subsystem. Guidelines on the commissioning of this new type of MR-guided RT technology are not yet fully defined. Based on a multi-institutional study, Tijssen *et al* [14] suggested a set of commissioning protocols for MRI imaging system which included system configuration and connectivity (SCC) check, quality control (QC) test, and quality assurance (QA) measurements. The SCC check includes a general inventory (software configuration, receiver coils, and other peripherals) check and a test of the basic functionality of the MR scanner (e.g., DICOM image storing and transferring, scanner peripherals functionality). QC tests include assessments of B0 field, gradient field, and radiofrequency field and an interaction test between Linac and MRI subsystem. QA includes the measurements of signal-to-noise, low contrast detectability, and gradient fidelity. The complete list of the commissioning tests is given in table I of the reference [13].

In general, the guideline for testing, commissioning, and routine QA of the Linac radiation delivery subsystem follows the American Association of Physicists in Medicine (AAPM) code of practice for radiotherapy accelerators (TG45) [15], comprehensive QA for radiation oncology (TG40) [16], and QA of medical accelerators (TG142) [17]. The calibration of the photon beam is performed based on AAPM protocol for clinical reference dosimetry of high-energy photon and electron beams (TG51) [18] and with magnetic field correction [19–21]. However, since the geometry of the MR-Linac is different than the conventional radiation therapy Linac (e.g., bore-type machine, nonstandard source axis distance, beam line passing through the cryostat, MR gradient coil, anterior and posterior coil), it needs alternate methods of equipment alignment with radiation isocenter and additional tests accounting for MRI imaging system inclusion.

For the Unity system, the commissioning task includes a mechanical system check, dosimetry system check, and TPS beam model data collecting and TPS testing. The mechanical system check includes safety, gantry position, MV image quality, MLC accuracy, couch accuracy, MV isocenter radius, and MR-to-MV isocenter coincidence. The dosimetry system check includes beam output calibration, beam energy, profile flatness and symmetry, monitor unit (MU) linearity, constancy of output versus dose rate, constancy of the off-axis factor versus gantry angle, and cryostat characteristic output dependence of the gantry angle. The TPS beam model test includes the beam model data collecting and building, simple field percent depth doses and profiles comparison, CT to ED check, patient orientation on MVI and in adaptive plan matching test, heterogeneous test, clinical case delivery performance test, and workflow end-to-end test.

21.3.2 Machine QA

Major components of QA for MR-Linac and MRgART include QA for the MRI system, Linac, alignment of MRI and MV imaging systems, end-to-end workflow, and patient-specific checks. Details of these QA tasks have been published [14, 22]. A summary of these QA components is provided here.

21.3.2.1 MRI routine QA

The currently chosen devices for MRI QA include the Modus 3D geometry distortion phantom, Philips gradient nonlinearity (GNL) distortion phantom, Philips 40 cm disc

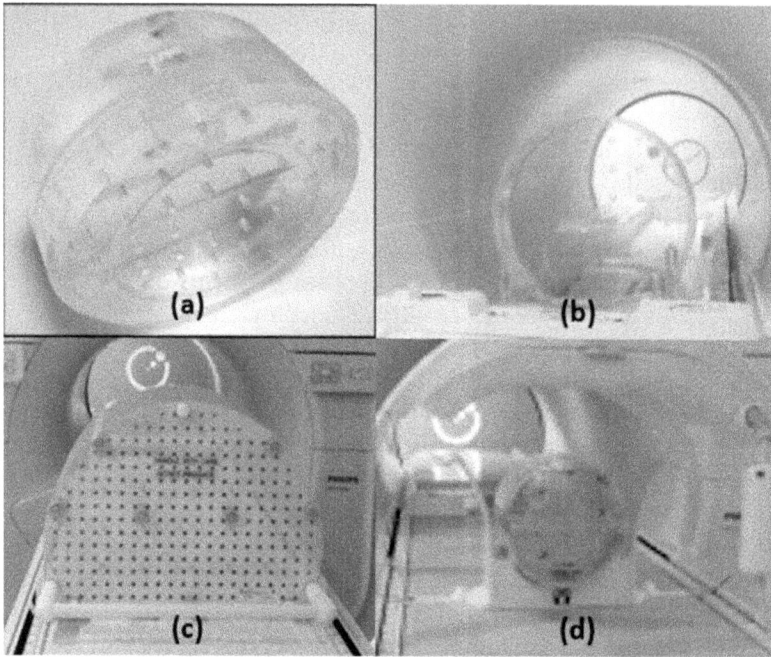

Figure 21.3. QA devices for MRI (a) Philips PIQT phantom, (b) Philips uniformity phantom, (c) Philips GNL distortion phantom, and (d) ACR phantom.

Uniformity phantom, Philips PIQT phantom, and ACR phantom. Figure 21.3 shows the QA devices for the MRI subsystem. Receiving coil element signal-to-noise ratio, scaling test of transverse, and coronal planes are tested daily. Weekly MR QA includes tests of flood field uniformity, spatial linearity, slice profile, and spatial resolution. MR geometric distortion and B0 uniformity are checked monthly. The details of the QA procedures and frequency are given in reference [22].

21.3.2.2 Linac routine QA

The QA devices used for Linac QA must be MR safe or MR conditional. The single detector for dosimetry QA may include but is not limited to PTW microdiamond detector, semiflex 3D ion chamber, farmer chamber, and standard imaging MR comparable A1 series chamber. The detector arrays for beam profile check may include MR compatible PTW Starcheck maxi, Octavius 1500, and Sun Nuclear IC profiler. The Sun Nuclear daily QA3 MR is under validation and should be commercially available soon for daily dosimetry checks. Currently available water tanks include the PTW manual MP1 MR 1D water tank and 3D beamscan MR.

The routine dosimetry QA includes checks on output, beam quality, profile constancy, backup monitor constancy, output factor, MU linearity, output constancy versus dose rate, and output constancy versus gantry angle. Mechanic QA includes checks on gantry angle accuracy, MLC and diaphragm position test, couch top position accuracy, beam limiting device field size, radiation isocenter radius, and MV imaging constancy. Frequency of the QA will follow the guidance of Roberts *et al* [22].

21.3.2.3 MR-to-MV alignment QA

The MRI isocenter can be different from the radiation isocenter. Even though the difference is subtle, it needs to be accounted for during the adaptive planning in TPS with daily MR images. For Elekta Unity, this difference was measured during machine installation and stored in the database for TPS. The purpose of the QA is to check whether the difference of the imaging isocenter and MV radiation center has changed. A phantom visible with both MR and MV should be used for this purpose. An MR-to-MV alignment QA phantom used for Unity is shown in figure 21.4. The phantom consists of several zirconium ceramic balls fixed in MR-visible fluid. The positions of the balls are compared in MR and MV images. The translation and rotation between two systems can be extracted from the difference of the ball positions. The frequency of the test is currently weekly and could be reduced based on the consistency of the QA data we have so far.

21.3.3 End-to-end workflow QA

The phantoms for end-to-end workflow QA should be MR visible. Currently available phantoms may include but are not limited to the computerized imaging reference systems thorax phantom, imaging and radiation oncology core MRgRT head and neck phantom, Sun Nuclear stereophan, standard imaging Lucy, RTsafe gels, and MR-to-MV alignment phantom. The purpose of the end-to-end workflow QA is to detect functionality and communications failure of all the major components of the MR-Linac in advance to ensure the accuracy and efficiency of online MgART. One example of the daily workflow QA for Elekta version

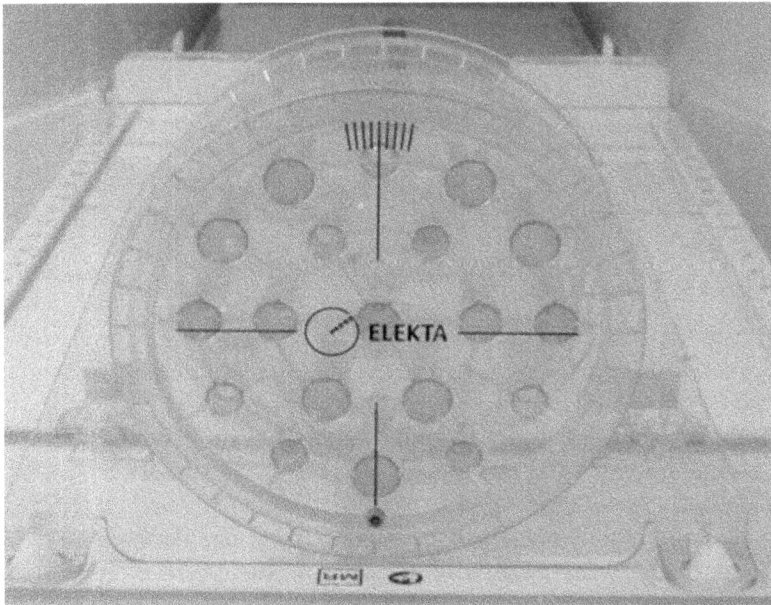

Figure 21.4. MR-to-MV alignment phantom.

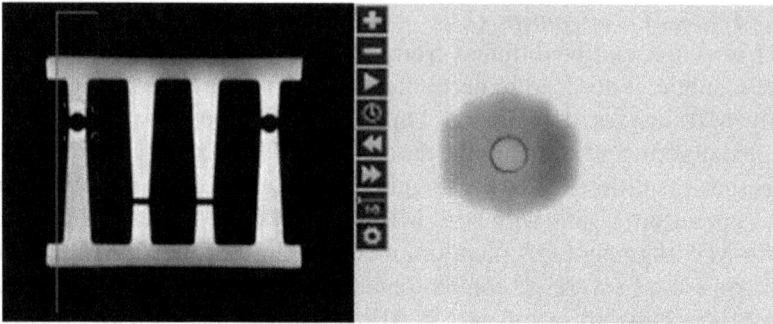

Figure 21.5. The comparison of the MLC shape in TPS and portal image. From [13], Figure 5, John Wiley & Sons.

MR-Linac was given in [13]. The workflow includes tests of MR daily image acquiring and transferring, image registration, online adaptive planning, adaptive plan QA and delivering, and motion monitoring during treatment plan delivery. The MR-to-MV alignment phantom is used as a test patient in the workflow. A reference plan with ceramic balls as targets was generated based on reference images. During the workflow, the daily images are rigidly registered with the reference images, and the adaptive plan is generated based on the isocenter shift. As shown in figure 21.2, the plan data integrity, dosimetry quality, and secondary dose verification are done with a software QA tool. In figure 21.5, the MLC shape in portal image is visually compared with that in TPS to verify the segment aperture morphing algorithm [23] in the adaptive plan to correctly drive the MLC leaves during the treatment plan delivery.

21.3.4 Patient-specific QA

The QA devices used for patient IMRT QA include MR compatible PTW Octavius, ScandiDos Delta4, and Sun Nuclear ArcCheck. Elekta provides a QA platform to help with positioning the QA device. The IMRT QA passing criterion is to have more than 95% of diodes within the gamma criterion of 3% to 3 mm.

21.4 MRgART for abdomen

The daily high soft tissue contrast and real-time imaging during radiation delivery offered by MR-Linac are highly desirable for the RT of abdominal targets where the high imaging contrast can improve the tumor and OAR delineation and the real-time imaging can help manage abdominal organ motion [24]. Radiation managements of tumors in the abdomen, such as pancreatic tumors, often have poor outcomes [25]. The accurate delineation and motion management in RT lead to the use of small margins, allowing dose escalation to the tumor and/or dose reduction to OARs [26], particularly to those sensitive gastrointestinal (GI) structures with serial functional subunits. The use of MR-Linac to safely deliver ablative doses in abdominal targets is being tested [27].

21.4.1 Motion of abdominal organs

Motion of abdominal targets and OARs consists of breathing motion, which is mostly in the superior-inferior direction with amplitudes that may exceed 2 cm [28], and the peristaltic motion, which is random and irregular [29–31]. Peristaltic motion may cause the drift of the target and OARs during the plan generation and delivery. To account for such drift, MRI may be acquired after the completion of plan generation and before beam-on.

The motion for abdominal targets and OARs needs to be accounted for during imaging, planning, and treatment delivery [28, 32]. Options to manage the motion include respiratory gating, breath hold, and dynamic tracking [33]. The images used to generate reference and daily plans should be acquired according to the motion management methods of delivery. Respiratory gating is a way to minimize the motion and margins; however, it increases the delivery times, which can be a problem for patients who cannot tolerate a long time on the couch, since the time for MRgART is generally long. Respiratory gating has been successfully performed using MRIdian for free breathing [3, 34] and breath hold inspiration [35] with four frames per second planar sagittal imaging, with gating boundaries set to 3–5 mm. At the time of writing, automatic respiratory gating is not available on the Unity system. Options for motion management include free breathing, shallow breathing, and breath hold with or without exception (manual) gating. The mid-position target, created from a time-weighted synthetic image based on 4D MRI [36], may be used to reduce the planning target volume (PTV) [37, 38] relative to full internal target volume (ITV) targets, especially for small targets and large motion amplitudes, but would typically require longer image reconstruction times. Instead of 4D MR, respiration-triggered MR images can be acquired if a gated treatment is intended. More advanced techniques, such as tracking or trailing [39, 40] or real-time planning [41, 42], are being actively developed.

21.4.2 Pre-beam daily imaging

A major advantage of MRgART is its improved soft tissue contrast of MRI compared with CT-based imaging [43–45]. For abdominal organs, Noel *et al* found the delineation precision with MR image to be adequate but also to depend on type of sequence and specific organs; e.g. duodenum and pancreas had worst precision (4–6 mm Hausdorff distance) [46]. Also, on-board MRI with MRIdian's low Tesla system was found to be superior to on-board cone beam CT [24] for many sites including abdomen. For online adaptive radiotherapy, there is timing pressure to acquire and reconstruct MRI quickly, which typically compromises MRI quality, and the times for 4D MRI are inherently long. Contrast agents typically are not permissible for online replanning of multiple fractions. Other challenging factors for achieving good image quality in abdomen include the organ motion and the requirement of acquiring imaging in accordance with the subsequent radiation delivery methods. Breath hold acquisition normally yields the best image quality but only if the treatment is delivered with breath hold. Respiration-triggered acquisition, although it can reduce motion artifacts, should be used only if the respiration-gated

Table 21.1. Reconstruction times for motion-averaged and mid-position 4D MR images.

Image size (#voxels)	Scanning time (s)		References
	Motion-averaged	Mid-position	
256	391	535	[48]
160	245	312	[48]
128	197	249	[48]

radiation delivered is used. For free breathing delivery, 4D MRI would be the choice for pre-beam imaging, and the resultant mid-position or motion-averaged images may be used for daily plan adaptation. In our implementation of 4D MRI, the three-dimensional (3D) golden angle radial stack of stars sequence is utilized due to its advantages of oversampling the center of k-space, which provides an inherent respiratory waveform for respiratory binning, and enabling highly parallelized reconstruction, which is performed on a dedicated high-speed, highly parallel (96 core) workstation [10, 47, 48]. The image acquisition and reconstruction times are in listed in table 1 of Mickevicius *et al*; for example, the time for a motion-averaged image of 256×256 is 6 min 31 s, while it is 8 min 55 s for mid-position (table 21.1) [48].

An imaging study to determine the best possible image technique for a patient before fractionated RT may be performed. For a given patient, 4D MR images with different contrast (e.g., T1 and mixed T2/T1 with and without fat suppression) are acquired, and the one with the best contrast is determined to be used for the subsequent daily pre-beam imaging for the patient. Mid-position or motion-averaged images can be reconstructed and utilized, selected based on their suitability for a particular case. Also, a separate quick scan to image air volumes (due to abdominal gas) may be conducted for the delineation of air volumes.

21.4.3 Beam-on imaging

MR-Linac systems offer real-time two-dimensional imaging, cine-MR, to be used to monitor anatomy change while the beam is being delivered. The temporal frequency of these images is in the order of four to eight frames per second. There are several methods that may improve the temporal frequency by non-conventional k-space sampling or reconstruction [41]. Although a cine image plane can be acquired on any arbitrary plane, the Unity system currently only allows orthogonal planes with an update rate of total 4.5 frames per second. For respiratory motion, the most useful plane is typically the sagittal or coronal, allowing visualization of motion in the superior-inferior direction. The latency is reported to be in the order of 0.5 s, but non-conventional k-space sampling, such as golden angle radial sampling [49] (not currently available on Unity) could achieve shorter times. Users can select which orthogonal planes will be used for each treatment, and the update frequency is shared among those planes. A single contour delineated online which is designated before-hand is overlaid with the images, and the planes are placed at the center of this

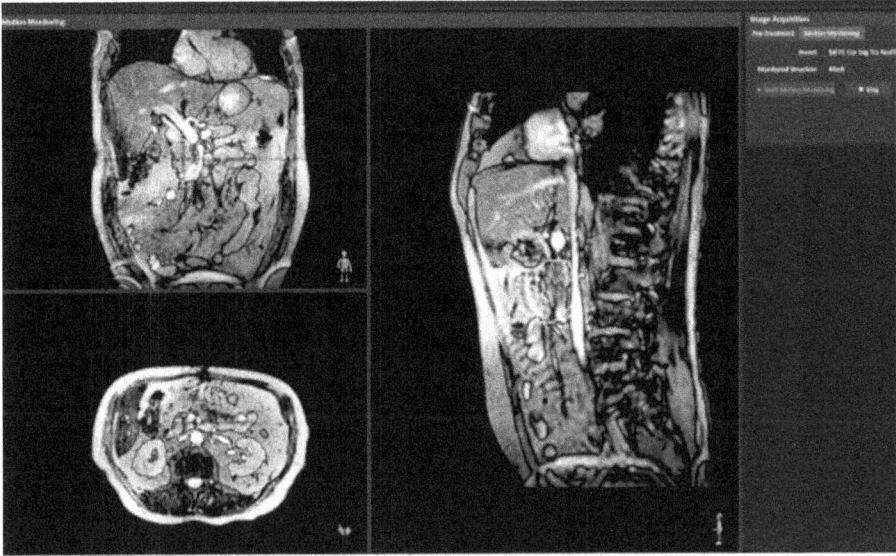

Figure 21.6. Motion monitoring screen and the cine-MR images in three orthogonal planes for a patient with pancreatic cancer. The PTV (motion-averaged full ITV) overlaid on the images.

contour. If the contour is too small, visualization may be obscured by saturation bands from other planes. The beam can be switched on or off based on the motion monitoring. Currently the Unity real-time imaging allows for only a single type of MR sequence, a balanced turbo field echo sequence (resolution: $1\times1\times5$ mm^3, echo time: 2 ms, repetition time: 4 ms, flip angle: 30°). In figure 21.6, the motion monitoring of the Unity system is shown with the PTV contour overlay for the treatment of a patient with pancreatic cancer. More flexibility in the MR sequences in the cine imaging would allow better visualization for a variety of target sites.

21.4.4 Biological imaging

Another advantage of high-field MR-Linac systems is the possibility of acquiring advanced quantitative MRIs, such as diffusion weighted imaging (DWI) or T1 or T2 mapping, at every fraction during the course of treatment without any additional burden to the patient. These longitudinal data can be analyzed to evaluate radiation responses of the tumor and/or OARs and thus may be used as treatment biomarkers. For example, DWI offers the possibility determining the tumor response map, to be used for adaptive update of dose delivery to radio-resistant regions such as hypoxic subvolumes [50], which was found to correlate with treatment response for pancreatic tumors as well [51]. With the help of longitudinal studies, it may be possible to use the advanced MR images for targeting subvolumes of the tumor with higher doses to increase the therapeutic ratio. In our clinic, the additional MR scans for longitudinal studies are acquired during the plan preparation time, with no increase to the patient's time on the treatment table [52]. Kooreman *et al* investigated the reliability of the

Unity MR-Linac system for quantitative imaging and concluded that it provides reliable data similar to diagnostic MR [53].

21.5 Online adaptive plan generation

For abdominal targets, daily plan adaptation offers improved plan quality by addressing interfractional changes of the target and OARs. As explained by Henke *et al* [27], this may mean either to recover the deterioration in the dose due to daily variation or to exploit the daily anatomy—when it is favorable—to further improve the dosimetry, e.g. by increasing the target dose. The approach is also known as isotoxic delivery, which was successfully tested by Henke *et al* on the MRIdian MR-Linac system [27]. Online replanning on top of already delivered dose to critical structures could also allow for higher doses in the remaining fractions; however, dose accumulation accuracy would be a critical component. Another advantage of online planning may be to have treatment margins tailored daily based on the daily target motion amplitudes [54]. Major steps in the current online adaptive planning workflow are described below.

21.5.1 Contour generation

Currently, a major time-consuming component of the replanning is contour generation, which has also been our experience, as can be seen from figure 21.7. For the abdominal targets, auto-segmentation is especially problematic with the currently available technology due to the GI structures such as bowels and colon that are complex in shape and are large in volume, requiring a long time for delineation. In addition, abdominal structures often experience large daily

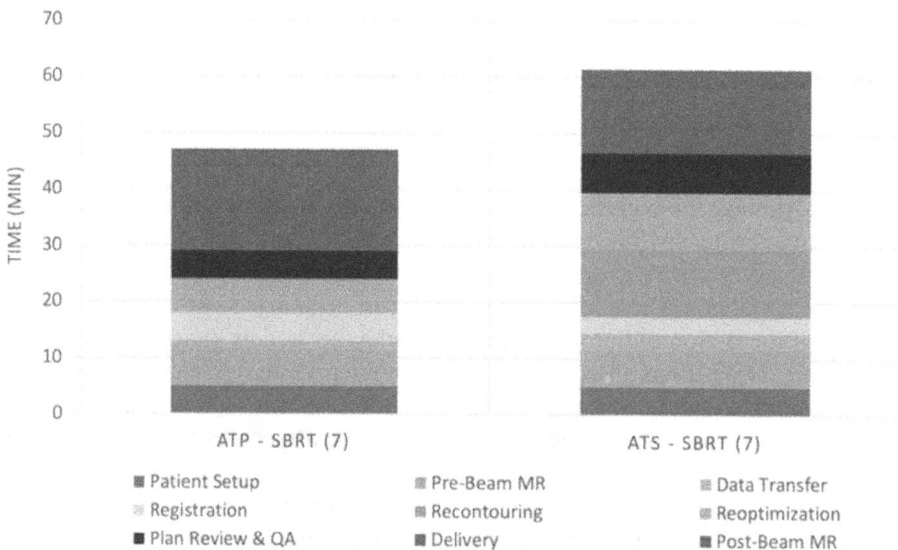

Figure 21.7. Median ATP and ATS session breakdowns for MR-guided online adaptive abdominal SBRT on the Elekta Unity MR-Linac. Reprinted from [10], Copyright (2020), with permission from Elsevier.

deformations due to the random changes in their contents. Consequently, atlas-based auto-segmentation methods work poorly. Auto-segmentation based on deep learning may be a solution and is currently under active development [55].

To reduce contouring times, ATS workflow may include a parallel contouring process where three separate operators simultaneously delineate the daily contours. The contours are transferred from a reference image set to daily MR and are distributed among a team of therapist, physicist, and a physician as follows: The physician delineates the targets, such as CTV, ITV, and PTV, as well as critical structures (bowels, colon, stomach, duodenum), while the physicist delineates the external patient body, air, and bony structures that are mainly necessary for density overrides [10]. The therapist edits several structures that are clearly identifiable such as liver, kidneys, and cord.

To further reduce the contouring times, GI structures may be edited only within close (e.g., 2 or 3 cm) proximity of the target. One feature of these structures is that they are mostly serial in their sensitivity to radiation and therefore only need to be constrained in maximum doses in the online optimization. This is a time-saving method, but to ascertain the underlying premise, it is necessary to maintain that these high doses only happen in the close proximity of the target by imposing adequate conformality constraints in the plan optimization. Also, this method is not applicable when/if these organs have dosimetric criteria with lower doses with isodose lines outside the 2 cm distance from the target.

21.5.2 Density overrides

In the current implementation, the two major clinical MR-Linac systems employ the following approaches for ED override of daily MRIs: MRIdian uses the American College of Radiology (ACR) deformable image registration (DIR) for mapping the CT densities from a reference (e.g. planning) CT, and the Unity system allows the user to override the MR by bulk densities for volumes of interest obtained from the reference CT. In both systems, the user needs to delineate the air regions and override their ED values.

Several solutions are being developed for generating synthetic CT images from the daily MRI [56], which can be roughly grouped into three categories: (1) voxel-based methods, where only the voxel intensity (or its proximity) of the MR image itself is used to determine the ED of a voxel; (2) atlas-based, or DIR-based, where the ED values are mapped from a reference CT image onto the target MR image; and (3) bulk density–based, where the contours on the daily MR image are delineated and overridden by predetermined ED values. Also, deep learning–based methods have been developed [57, 58] which can also be roughly categorized as voxel-based since only the daily MR image information is used to generate the ED values. The atlas-based methods offer the advantage of using the patient's own ED values, and since a reference CT is available for almost all patients undergoing RT, this option seems the most straightforward and computationally inexpensive solution. The regular DIR-related errors cause inaccuracies; however, the dosimetric effect of these is expected to be small [59]. One region where the DIR method does not work very well is the air/gas pockets in the abdominal structures [60], which occur randomly and therefore cannot be accurately located by an atlas-based approach.

The air/gas regions in the abdomen can pose a dosimetric issue, acerbated by the ERE, that may cause hot or cold spots at the vicinity of air–tissue interfaces and is unique to the high-field magnetic field of the Unity system.

The effect of the ERE can be successfully accounted for by IMRT optimization [61] once the air regions are delineated and overridden by air densities. However, intrafraction changes of the gas regions in the abdomen can result in dosimetric consequences and may necessitate updating the plan, which is challenging with the current imaging and planning technology. Currently the air/gas region delineation may be performed manually before the treatment starts by means of thresholding [60]; however, an automated method would be beneficial to increase speed and enable more frequent plan updating.

Thapa *et al* provided a solution to automate the air delineation by transferring the air-containing organs from a reference image by DIR, expanding by a margin of 1–2 cm to account for DIR inaccuracies, and applying a threshold to generate air volumes, which yielded dosimetrically similar results to air volumes manually delineated [60].

21.5.3 Online plan optimization

The current online optimization methods can be time consuming and often require human interaction and thus may be difficult to automate. The success of online optimization is mainly the ability to use the correct objective function (defined under 'IMRT constraints' (IC) in Monaco) for the optimization that would result in an acceptable quality daily dose distribution. For the ATS workflow, the IC are imported from the reference plan, and if necessary, they can be adjusted before or somewhat during the optimization. However, making big changes to IC is not very practical during optimization, both because it takes a lot of time and because it bears some risk of making bad adjustments. This approach assumes that daily anatomy and reference anatomy should be sufficiently similar so that IC working for reference anatomy would also work for the daily anatomy [62]. However, this assumption has limits, as the daily anatomy is always somewhat different which makes the difficulty of individual goals of the IC vary relative to each other and can cause undesirable results in daily optimization.

During our initial experience with our first couple of cases, we encountered several situations where the reference plan IC resulted in unacceptable dosimetry for the adaptive plan of the day. To prevent this happening, we adopted a practice to test the IC not only on the reference planning image but also on one or more daily anatomy by a dry run (generating a daily plan but not treating). Obviously, this still cannot also fully guarantee that the IC would perform well with future daily anatomies. Some considerations can be applied, such as keeping individual constraints from deviating too much from the desired dosimetric goals and especially from overshooting to force optimizer to work harder on opposing goals.

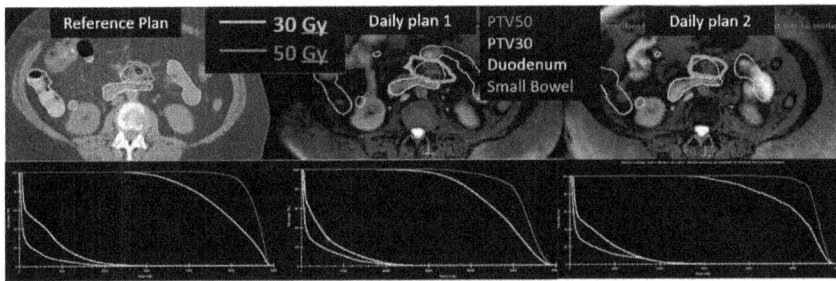

Figure 21.8. The variation in daily anatomy and the achievable dose to the target.

If multiple opposing goals are overshooting, these become rather unstable when the difficulty of those goals changes in a daily anatomy. However, even with these, fully automating the optimization is not yet achievable, and small adjustments to the IC are typically needed to improve the plan quality. More advanced IC definition, such as by scripting instead of weighted sum of individual goals, could help achieve the full automation of online optimization in the future implementation

Figure 21.8 depicts the daily dose distributions and dose–volume histograms (DVHs) of a patient treated for abdominal tumor. The target is between the small bowel and duodenum, which are limited by 32 Gy maximum dose, while the goal dose for gross target volume is 50 Gy and the PTV is 30 Gy. The daily anatomy variation was rather extensive; as can be seen on daily plan 1, the small bowel and duodenum almost fully surround the target, whereas on daily plan 2, the small bowel is much further from the target. The DVHs are also shown to be very different on these days. This illustrates the difficulty of having an IC that can still provide the best dose distribution under this variation and the necessity of human decision-making which is rather hard to automate.

21.5.4 Timing for online replanning process

With the currently available technologies, the time required to perform online replanning is generally long. Strategy and workaround are necessary to reduce the time to be practically acceptable and not to exceed the patient's tolerance of discomfort. For example, our parallel workflow for ATS treatments considerably reduce the contouring time [10]. The median time for daily online replanning for abdominal tumors treated with ATS workflow has been around 60 min, based on our own practice and reports from others. Although such a time frame is mostly tolerable, patients certainly might experience discomfort, such as when arm position is above the head. Henke *et al* [27] reported an online replanning time of about 80 min for respiratory-gated abdominal SBRT with MRIdian. This time span included a preliminary dose calculation with the original plan and a decision process whether to adapt or not. Although 45–60 min is tolerable for most patients, it is still desirable to reduce these times to relieve some patient discomfort, increase throughput, and eventually to be able to increase frequency of plan adaptations. Further improvements in the speed of plan preparation without compromising safety can be achieved by more automation. Currently several processes of the plan preparation process are not fully automatable.

As can be seen in figure 21.7 the processes with longest times are recontouring, beam delivery, optimization, pre-beam imaging and reconstruction, and plan review and QA. Contouring could be most affected by automation, and certain technologies report promising results as will be explained more in the section below. Optimization speed can be expected to improve further with more computing power and robust algorithms; however, there is also necessary manual involvement of adjusting the objective function that is currently difficult to automate. The delivery times for abdominal ATS are long, mainly since these are high-dose SBRT deliveries, and because Unity currently only allows step and shoot treatments. Once dynamic delivery (e.g. volumetric modulated arc therapy) becomes available, times should be reduced. The flattening filter free beam allows for high dose rate but also may require more beam modulation if the target is large, although this is typically not the case for abdominal targets. For image acquisition and reconstruction times, the main factor is the 4D MRI's longer image acquisition and reconstruction times necessary for moving abdominal targets. Acquisition times could be reduced with higher density phased array receive coils. Reconstruction times could always be improved with more computing power, as was done in our department with a dedicated 96 core workstation. There are also possible improvements to the speed of image acquisition in the development, mainly by reduced k-space acquisition and using deep learning to make up the missing data [41]. Plan review and QA times may be hardest to reduce via automation, and we are not aware of more efforts in that direction. If the anatomical change is expected to be small relative to a reference plan, the review process may be less crucial and may possibly be simplified.

21.6 Future considerations

MRgART is expected to result in significant improvement in RT. As an example, the so-called isotoxic treatments, enabled with MRgART, are reported to have a potential benefit for pancreatic tumors, which may lead to improvement in prognosis. Another improvement is the residual boost of sub-tumor volumes or radiation-resistant regions detected with daily biological MRI.

It is anticipated that technologic advances in the near future will focus on increasing automation in MRgART, allowing more effective and robust online replanning. For that purpose, deep learning–based technologies are being researched to automate contouring [55], MR image reconstruction [41], and dose calculation [42]. If these processes can succeed in reducing the plan generation to the order of a few minutes, the drift in the target position can be eliminated, which would further reduce the treatment margins.

21.7 Conclusion

MRgART with MR-Linac, offering online adaptation based on daily high soft-tissue contrast and functional images and motion management based on real-time imaging, is highly desirable for abdominal targets. Online adaptation and motion management will allow dose escalation to the tumor and/or dose reduction to OARs, improving treatment outcome for abdominal tumors.

Acknowledgements

We would like to thank our physician colleagues, Drs Bill Hall, Michael Straza, Chris Schultz, Awan Musaddiq, and Beth Erickson, for their significant contributions to our clinical MRgART program and to the content of this chapter.

References

[1] Hall W A et al 2019 The transformation of radiation oncology using real-time magnetic resonance guidance: a review Eur. J Cancer 122 42–52

[2] Liney G P, Whelan B, Oborn B, Barton M and Keall P 2018 MRI-linear accelerator radiotherapy systems Clin. Oncol. (R Coll. Radiol.) 30 686–91

[3] Kluter S 2019 Technical design and concept of a 0.35 T MR-Linac Clin. Transl. Radiat. Oncol 18 98–101

[4] Lagendijk J J, Raaymakers B W and van Vulpen M 2014 The magnetic resonance imaging-Linac system Semin. Radiat. Oncol. 24 207–9

[5] Keall P J, Barton M, Crozier S and Australian MRI-Linac Program, including contributors from Ingham Institute, Illawarra Cancer Care Centre, Liverpool Hospital, Stanford University, Universities of Newcastle, Queensland, Sydney, Western Sydney, and Wollongong 2014 The Australian magnetic resonance imaging-Linac program Semin. Radiat. Oncol. 24 203–6

[6] Fallone B G 2014 The rotating biplanar Linac-magnetic resonance imaging system Semin. Radiat. Oncol. 24 200–2

[7] Wachowicz K, De Zanche N, Yip E, Volotovskyy V and Fallone B G 2016 CNR considerations for rapid real-time MRI tumor tracking in radiotherapy hybrid devices: effects of B0 field strength Med. Phys. 43 4903

[8] Raaijmakers A J, Raaymakers B W and Lagendijk J J 2005 Integrating a MRI scanner with a 6 MV radiotherapy accelerator: dose increase at tissue-air interfaces in a lateral magnetic field due to returning electrons Phys. Med. Biol. 50 1363–76

[9] Winkel D et al 2019 Adaptive radiotherapy: the Elekta Unity MR-Linac concept Clin. Transl. Radiat. Oncol. 18 54 9

[10] Paulson E S, Ahunbay E, Chen X, Mickevicius N J, Chen G P, Schultz C, Erickson B, Straza M, Hall W A and Li X A 2020 4D-MRI driven MR-guided online adaptive radiotherapy for abdominal stereotactic body radiation therapy on a high field MR-Linac: implementation and initial clinical experience Clin. Transl. Radiat. Oncol. 23 72–9

[11] Zhang J, Ahunbay E and Li X A 2018 Technical Note: Acceleration of online adaptive replanning with automation and parallel operations Med. Phys. 45 4370–6

[12] Chen G P, Ahunbay E and Li X A 2016 Technical note: development and performance of a software tool for quality assurance of online replanning with a conventional Linac or MR-Linac Med. Phys. 43 1713

[13] Chen X, Ahunbay E, Paulson E S, Chen G and Li X A 2020 A daily end-to-end quality assurance workflow for MR-guided online adaptive radiation therapy on MR-Linac J. Appl. Clin. Med. Phys. 21 205–12

[14] Tijssen R H N, Philippens M E P, Paulson E S, Glitzner M, Chugh B, Wetscherek A, Dubec M, Wang J and van der Heide U A 2019 MRI commissioning of 1.5T MR-Linac systems – a multi-institutional study *Radiother. Oncol.* **132** 114–20

[15] Nath R, Biggs P J, Bova F J, Ling C C, Purdy J A, van de Geijn J and Weinhous M S 1994 AAPM code of practice for radiotherapy accelerators: report of AAPM Radiation Therapy Task Group No. 45 *Med. Phys.* **21** 1093–121

[16] Kutcher G J, Coia L, Gillin M, Hanson W F, Leibel S, Morton R J, Palta J R, Purdy J A, Reinstein L E and Svensson G K *et al* 1994 Comprehensive QA for radiation oncology: report of AAPM Radiation Therapy Committee Task Group 40 *Med. Phys.* **21** 581–618

[17] Klein E E *et al* 2009 Task Group 142 report: quality assurance of medical accelerators *Med. Phys.* **36** 4197–212

[18] Almond P R, Biggs P J, Coursey B M, Hanson W F, Huq M S, Nath R and Rogers D W 1999 AAPM's TG-51 protocol for clinical reference dosimetry of high-energy photon and electron beams *Med. Phys.* **26** 1847–70

[19] O'Brien D J, Roberts D A, Ibbott G S and Sawakuchi G O 2016 Reference dosimetry in magnetic fields: formalism and ionization chamber correction factors *Med. Phys.* **43** 4915

[20] Malkov V N and Rogers D W O 2018 Monte Carlo study of ionization chamber magnetic field correction factors as a function of angle and beam quality *Med. Phys.* **45** 908–25

[21] Malkov V N and Rogers D W O 2019 Erratum: Monte Carlo study of ionization chamber magnetic field correction factors as a function of angle and beam quality [*Med. Phys.* **45** 908–25 (2018)] *Med. Phys.* **46** 5367–70

[22] Roberts D A *et al* 2021 Machine QA for a 1.5 T MR-Linac: a report from the Elekta MR-Linac consortium *Med. Phys.* **48** e67–85

[23] Ahunbay E E, Peng C, Chen G P, Narayanan S, Yu C, Lawton C and Li X A 2008 An on-line replanning scheme for interfractional variations *Med. Phys.* **35** 3607–15

[24] Noel C E, Parikh P J, Spencer C R, Green O L, Hu Y, Mutic S and Olsen J R 2015 Comparison of onboard low-field magnetic resonance imaging versus onboard computed tomography for anatomy visualization in radiotherapy *Acta Oncol.* **54** 1474–82

[25] Vincent A, Herman J, Schulick R, Hruban R H and Goggins M 2011 Pancreatic cancer *Lancet* **378** 607–20

[26] Reyngold M, Parikh P and Crane C H 2019 Ablative radiation therapy for locally advanced pancreatic cancer: techniques and results *Radiat. Oncol.* **14** 95

[27] Henke L *et al* 2018 Phase I trial of stereotactic MR-guided online adaptive radiation therapy (SMART) for the treatment of oligometastatic or unresectable primary malignancies of the abdomen *Radiother. Oncol.* **126** 519–26

[28] Brandner E D, Wu A, Chen H, Heron D, Kalnicki S, Komanduri K, Gerszten K, Burton S, Ahmed I and Shou Z 2006 Abdominal organ motion measured using 4D CT *Int. J. Radiat. Oncol. Biol. Phys.* **65** 554–60

[29] Mostafaei F, Tai A, Omari E, Song Y, Christian J, Paulson E, Hall W, Erickson B and Li X A 2018 Variations of MRI-assessed peristaltic motions during radiation therapy *PLoS One* **13** e0205917

[30] Kumagai M, Hara R, Mori S, Yanagi T, Asakura H, Kishimoto R, Kato H, Yamada S, Kandatsu S and Kamada T 2009 Impact of intrafractional bowel gas movement on carbon ion beam dose distribution in pancreatic radiotherapy *Int. J. Radiat. Oncol. Biol. Phys.* **73** 1276–81

[31] Nakamoto Y, Chin B B, Cohade C, Osman M, Tatsumi M and Wahl R L 2004 PET/CT: artifacts caused by bowel motion *Nucl. Med. Commun.* **25** 221–5

[32] Case R B, Sonke J J, Moseley D J, Kim J, Brock K K and Dawson L A 2009 Inter- and intrafraction variability in liver position in non-breath-hold stereotactic body radiotherapy *Int. J. Radiat. Oncol. Biol. Phys.* **75** 302–8

[33] Wolthaus J W, Sonke J J, van Herk M, Belderbos J S, Rossi M M, Lebesque J V and Damen E M 2008 Comparison of different strategies to use four-dimensional computed tomography in treatment planning for lung cancer patients *Int. J. Radiat. Oncol. Biol. Phys.* **70** 1229–38

[34] Feldman A M, Modh A, Glide-Hurst C, Chetty I J and Movsas B 2019 Real-time magnetic resonance-guided liver stereotactic body radiation therapy: an institutional report using a magnetic resonance-Linac system *Cureus* **11** e5774

[35] Wojcieszynski A P *et al* 2016 Gadoxetate for direct tumor therapy and tracking with real-time MRI-guided stereotactic body radiation therapy of the liver *Radiother. Oncol.* **118** 416–8

[36] van de Lindt T N, Fast M F, van Kranen S R, Nowee M E, Jansen E P M, van der Heide U A and Sonke J J 2019 MRI-guided mid-position liver radiotherapy: validation of image processing and registration steps *Radiother. Oncol.* **138** 132–40

[37] Tai A, Chen X, Mickevicius N, Paulson E, Ahunbay E and Li X 2019 PTV reduction with time-weighted mid-position images for 4D-MRI guided online adaptive radiation therapy *Proc. 61st. Annual Meeting & Exhibition of the American Association of Physicists in Medicine* E392

[38] Wolthaus J, Sonke J J, Van Herk M and Damen E 2008 Reconstruction of a time-averaged midposition CT scan for radiotherapy planning of lung cancer patients using deformable registration a *Med. Phys.* **35** 3998–4011

[39] Keall P, Kini V, Vedam S and Mohan R 2001 Motion adaptive X-ray therapy: a feasibility study *Phys. Med. Biol.* **46** 1

[40] Trofimov A, Vrancic C, Chan T C, Sharp G C and Bortfeld T 2008 Tumor trailing strategy for intensity-modulated radiation therapy of moving targets *Med. Phys.* **35** 1718–33

[41] Terpstra M L, Maspero M, D'Agata F, Stemkens B, Intven M P, Lagendijk J J, Van den Berg C A and Tijssen R H 2020 Deep learning-based image reconstruction and motion estimation from undersampled radial k-space for real-time MRI-guided radiotherapy *Phys. Med. Biol.* **65** 155015

[42] Kontaxis C, Bol G, Lagendijk J and Raaymakers B 2020 DeepDose: towards a fast dose calculation engine for radiation therapy using deep learning *Phys. Med. Biol.* **65** 075013

[43] Khoo V S, Dearnaley D P, Finnigan D J, Padhani A, Tanner S F and Leach M O 1997 Magnetic resonance imaging (MRI): considerations and applications in radiotherapy treatment planning *Radiother. Oncol.* **42** 1–15

[44] Xi T, Guoqing H, Hong Q and Wei C 2005 Comparing gross tumor volume of delineation between CT and MRI for nasopharyngeal carcinoma *Chin.-German J. Clin. Oncol.* **4** 141–5

[45] Devic S 2012 MRI simulation for radiotherapy treatment planning *Med. Phys.* **39** 6701–11

[46] Noel C E, Zhu F, Lee A Y, Yanle H and Parikh P J 2014 Segmentation precision of abdominal anatomy for MRI-based radiotherapy *Med. Dosim.* **39** 212–7

[47] Mickevicius N J and Paulson E S 2017 Investigation of undersampling and reconstruction algorithm dependence on respiratory correlated 4D-MRI for online MR-guided radiation therapy *Phys. Med. Biol.* **62** 2910

[48] Mickevicius N, Straza M, Hall W and Paulson E 2019 First clinical use of 4D-MRI for online adaptive MR-gRT on a high field MR-Linac *Int. J. Radiat. Oncol. Biol. Phys.* **105** S30–1

[49] Borman P, Tijssen R, Bos C, Moonen C, Raaymakers B and Glitzner M 2018 Characterization of imaging latency for real-time MRI-guided radiotherapy *Phys. Med. Biol.* **63** 155023

[50] Thoeny H C and Ross B D 2010 Predicting and monitoring cancer treatment response with diffusion-weighted MRI *J. Magn. Reson. Imaging* **32** 2–16

[51] Dalah E, Erickson B, Oshima K, Schott D, Hall W A, Paulson E, Tai A, Knechtges P and Li X A 2018 Correlation of ADC with pathological treatment response for radiation therapy of pancreatic cancer *Transl. Oncol.* **11** 391–8

[52] Hal W A, Straza M W, Chen X, Mickevicius N, Erickson B, Schultz C, Awan M, Ahunbay E, Li X A and Paulson E S 2020 Initial clinical experience of stereotactic body radiation therapy (SBRT) for liver metastases, primary liver malignancy, and pancreatic cancer with 4D-MRI based online adaptation and real-time MRI monitoring using a 1.5 Tesla MR-Linac *PLoS One* **15** e0236570

[53] Kooreman E S, van Houdt P J, Nowee M E, van Pelt V W, Tijssen R H, Paulson E S, Gurney-Champion O J, Wang J, Koetsveld F and van Buuren L D 2019 Feasibility and accuracy of quantitative imaging on a 1.5 T MR-linear accelerator *Radiother. Oncol.* **133** 156–62

[54] Sarkar V, Lloyd S, Paxton A, Huang L, Su F-C, Tao R, Tward J, Zhao H and Salter B 2018 Daily breathing inconsistency in pancreas SBRT: a 4DCT study *J. Gastrointest. Oncol.* **9** 989

[55] Eppenhof K A, Maspero M, Savenije M, de Boer J, van der Voort van Zyp J, Raaymakers B W, Raaijmakers A, Veta M, van den Berg C and Pluim J P 2020 Fast contour propagation for MR-guided prostate radiotherapy using convolutional neural networks *Med. Phys.* **47** 1238–48

[56] Edmund J M and Nyholm T 2017 A review of substitute CT generation for MRI-only radiation therapy *Radiat. Oncol.* **12** 28

[57] Liu Y, Lei Y, Wang T, Kayode O, Tian S, Liu T, Patel P, Curran W J, Ren L and Yang X 2019 MRI-based treatment planning for liver stereotactic body radiotherapy: validation of a deep learning-based synthetic CT generation method *Br. J. Radiol.* **92** 20190067

[58] Han X 2017 MR-based synthetic CT generation using a deep convolutional neural network method *Med. Phys.* **44** 1408–19

[59] Ahunbay E E, Thapa R, Chen X, Paulson E and Li X A 2019 A technique to rapidly generate synthetic computed tomography for magnetic resonance imaging–guided online adaptive replanning: an exploratory study *Int. J. Rad. Oncol. Biol. Phys.* **103** 1261–70

[60] Thapa R, Ahunbay E, Nasief H, Chen X and Li X A 2020 Automated air region delineation on MRI for synthetic CT creation *Phys. Med. Biol.* **65** 025009

[61] Raaijmakers A, Hårdemark B, Raaymakers B W, Raaijmakers C and Lagendijk J J 2007 Dose optimization for the MRI-accelerator: IMRT in the presence of a magnetic field *Phys. Med. Biol.* **52** 7045

[62] Olberg S, Green O, Cai B, Yang D, Rodriguez V, Zhang H, Kim J S, Parikh P J, Mutic S and Park J C 2018 Optimization of treatment planning workflow and tumor coverage during daily adaptive magnetic resonance image guided radiation therapy (MR-IGRT) of pancreatic cancer *Radiat. Oncol.* **13** 1–8

IOP Publishing

Principles and Practice of Image-Guided Abdominal
Radiation Therapy

Yu Kuang

Chapter 22

Ultrasound guidance for abdominal radiation therapy

Amy S Yu and Dimitre H Hristov

Image-guided radiation therapy (IGRT) has become a routine radiotherapy technique for treating abdominal tumors such as those of the pancreas, liver, and prostate. However, real-time three-dimensional visualization of soft-tissue motion and deformation during the treatment when accurate targeting is most critical remains an unmet challenge with the most common radiotherapy platform. Various methods have been proposed to monitor target motion intrafractionally with linear accelerator add-on imaging devices, but these methods have several limitations when applied to the challenging problem of real-time soft-tissue visualization. Ultrasound imaging systems, on the other hand, offer nonionizing, real-time volumetric imaging with excellent soft-tissue contrast and can be used with existing radiation delivery systems. In this chapter, we will discuss the available IGRT ultrasound systems, their quality assurance, and future applications. The current research in the field of ultrasound imaging guidance for radiation therapy will also be discussed.

22.1 Introduction

External beam radiation therapy (EBRT) is used to treat >60% of all patients with cancer. Potent radiation doses with minimal treatment margins must be delivered in EBRT to maximize local tumor control and minimize toxicity to surrounding healthy tissue [1–3], but internal anatomy motion and deformation pose a fundamental threat to realizing these objectives. With the proliferation of hypofractionated radiotherapy treatment regimens such as stereotactic body radiotherapy (SBRT), interfractional and intrafractional imaging technologies are becoming increasingly critical to ensure safe and effective treatment delivery [4–10].

In the last decade, image-guided radiation therapy (IGRT) has become a routine radiotherapy technique for treating abdominal tumors such as those of the pancreas, the liver, and the prostate [11]. Intrafractional imaging and the ability to visualize soft tissue are important for IGRT, especially in the context of emerging SBRT approaches. Various methods have been proposed to monitor target motion intrafractionally with imaging devices added onto the most common C-arm linear accelerator delivery platform, but these methods have several limitations when applied to the challenging problem of real-time soft-tissue visualization. On-board three-dimensional (3D) kV MV^{-1} imaging is capable of producing soft-tissue images with excellent spatial resolution, but it does not provide adequate temporal sampling for real-time motion monitoring during beam delivery. Planar kV MV^{-1} imaging can be used for tracking in combination with implanted fiducials. However, this method is invasive, and the fiducials could migrate over the time [12, 13]. Furthermore, this approach requires frequent exposures delivering additional radiation doses that may become hazardous if frequency and technique are not properly controlled [14]. Bony anatomy may also be used for positioning purposes; however, this may not be accurate for prostate target positioning due to the differential filling of the bladder and rectum [15]. The Calypso real-time electromagnetic (EM) tracking system is another alternative method used for monitoring prostate motion during radiation treatment with high time resolution and precise localization [16]; however, image artifacts caused by transponders can preclude the use of magnetic resonance imaging (MRI) in post-treatment assessment [17], making its usefulness less appealing.

Ultrasound offers an attractive alternative to the previously described approaches for imaging radiotherapy targets outside the skull and lung. Ultrasound imaging systems offer nonionizing, real-time volumetric imaging with excellent soft-tissue contrast and can be used with existing radiation delivery systems. There are some systems which are currently used and evaluated in clinical settings. In this chapter, we will discuss the available IGRT ultrasound systems, their quality assurance, and future applications. We will also discuss current research in the field of ultrasound imaging guidance for radiation therapy.

22.2 Clinical interfractional ultrasound guidance systems

22.2.1 B-mode acquisition and targeting system

The first widely adopted ultrasound system for interfractional imaging was the B-mode acquisition and targeting (BAT) system (currently marketed as BatCam by Best NOMOS, Pittsburgh, PA; figure 22.1(a)) introduced in ultrasound markets in the late 1990s. In the BAT system, transabdominal ultrasound is used to image the target region in two near-orthogonal planes [18]. When the BAT system arrived on the market, ultrasound was the only method for obtaining soft-tissue interfractional images [cone-beam computed tomography (CBCT) was not introduced until 1996 in Europe and 2001 in the USA] [19]. Little *et al* [20] found that without orthogonal plane ultrasound imaging, prostate motion would have caused the target to move outside the planning target volume (PTV) in 23.3% to 41.8% of cases, concluding also that prostate organ motion dominates setup error. Studies of patients with prostate cancer by Lattanzi *et al* [18] demonstrated that alignment errors between a baseline computed tomography (CT) and ultrasound after initial positioning based

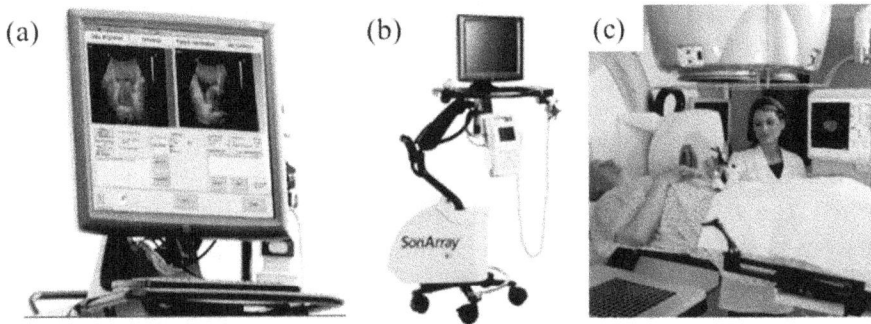

Figure 22.1. Interfractional ultrasound guidance systems. (a) Best NOMOS BAT. (b) Varian SonArray. (c) Elekta Clarity. [22], reproduced with permission © *Cureus*.

on skin marks often exceeded 5 mm using the BAT system, with maximum ranges of −26.8 to 33.8 mm in the anterior/posterior (AP) direction, −10.2 to 30.9 mm in the mediolateral (ML) direction, and −24.6 to 9.0 mm in the superior/inferior (SI) dimension. A study found that the BAT system improved positioning accuracy when compared with positioning using bony anatomy in a CBCT scan and significantly improved positioning accuracy over skin marks (with residual error magnitudes between 2.1 and 5.2 mm) [21].

Several studies have compared BAT positioning accuracy with x-ray fiducial marker positioning. Langen *et al* [23] showed that prostate alignment using the BAT system systematically differs from alignment using radiographic fiducial marker imaging in the SI (2.7 ± 3.9 mm) and ML (1.6 ± 3.1 mm) directions, but differences were minimal in the AP direction (0.2 ± 3.7 mm). Van den Heuvel *et al* [24] found that position shifts suggested by the BAT system were similar to fiducial marker shifts in the ML and AP directions but differed significantly in the SI direction; the researchers suspected a systematic error in their BAT positioning system.

22.2.2 SonArray systems

The SonArray system (Varian Medical Systems, Palo Alto, CA; figure 22.1(b)) was introduced a few years after the BAT system [25]. In contrast to the BAT system, SonArray uses 3D ultrasound images for image guidance. To this end, a sonographer sweeps the ultrasound probe across the target anatomy, capturing a series of two-dimensional (2D) slices that are reconstructed into a 3D volume based on the relative slice positions reported by an optical tracking system. Several studies have compared SonArray positioning accuracy with x-ray fiducial marker positioning. Peignaux *et al* [26] found significant differences between SonArray and x-ray fiducial marker positioning along the SI and AP axes, and Scarbrough *et al* [27] found a 3D distance discrepancy between the two methods of 8.8 mm (significantly >5 mm), concluding that larger target volume margins of ~9 mm are necessary for SonArray as compared with ~3 mm with kV x-ray in intrafractional imaging applications.

Although initial studies of the BAT and SonArray systems demonstrated promise of improved soft-tissue–based prostate positioning, cross-examination with x-ray fiducial imaging revealed systematic biases. Reliance on intermodality matching of ultrasound and CT image information is partially responsible for these systematic errors. For example, prostate volumes derived from CT scans are consistently larger than those derived from ultrasound images because of differing physical image contrast mechanisms and the inability of CT to differentiate structures well at low contrast [25]. Molloy et al [28] estimated displacement differences to be up to 9 mm in the ML direction and 3 mm in the AP direction. Cury et al [29] showed systematic differences in intermodality and intramodality positioning, finding significant mean differences of 0.9 ± 3.3 mm in the lateral and 6.0 ± 5.1 mm in the SI directions between the two techniques. Because comparison of intramodality ultrasound imaging and CT scans showed no significant differences in any direction, it was concluded that intramodality-based positioning is more accurate than intermodality positioning. Therefore, intramodality image matching has been recommended to minimize patient positioning error in ultrasound-guided radiotherapy [25].

22.2.3 Clarity system

While using 3D ultrasound images for guidance, SonArray still compares ultrasound with reference CT images and contours. On the other hand, the Clarity system (Elekta, Stockholm, Sweden; figure 22.1(c)) leverages intramodality image matching instead of the intermodality matching technique used by BAT and SonArray. Intramodality matching is achieved by capturing an ultrasound volume directly before or after the planning CT scan. By tracking the precise position and orientation of the ultrasound probe with respect to the CT scanner, the 3D ultrasound volume can be reconstructed in the reference frame of the CT. Directly prior to treatment, a new 3D ultrasound volume is captured in the linear accelerator room and reconstructed with respect to the linear accelerator reference frame. To position the patient, the ultrasound volumes collected during planning and prior to treatment are matched and used to determine patient anatomy offset relative to the desired planning position [30]. Since intramodality image matching has been shown to improve accuracy versus intermodality matching, the Clarity system should theoretically position patients more accurately than the BAT or SonArray system [31–36]. The Clarity system is not only a noninvasive but also a non-ionizing system allowing real-time soft-tissue imaging without the need for fiducial marker implantation. Research showed that the Clarity system is able to detect the prostate motion clinically. Richardson et al showed that Clarity system detected 3, 7, and 10 mm of target motion for 52%, 8%, and 2% of all 526 fractions, respectively. Moreover, 100% of patients experience at least one displacement at the 3 mm threshold across the treatment course, 60% at 7 mm and 35% at 10 mm [36].

22.3 Clinical intrafractional ultrasound guidance system

Intrafractional ultrasound guidance systems are the next step in the evolution of ultrasound imaging for radiotherapy, providing real-time, volumetric, markerless target tracking concurrent with beam delivery in packages that integrate with existing linear accelerators. In order to image and track patient anatomy with respect to the treatment beam, ultrasound image information must be transformed to the reference frame of the linear accelerator. As part of this process, the position of the ultrasound probe with respect to the linear accelerator must be known at all times. Potential techniques available to determine and track probe position in the frame of the linear accelerator include optical tracking, EM tracking, mechanical tracking, and x-ray tracking. The foundation of an intrafractional ultrasound guidance system is a hardware device to maintain the ultrasound probe in imaging position during therapy while the sonographer is outside the treatment room. The device must hold the probe in a way that maintains the therapy target within the ultrasound imaging field of view throughout treatment while minimizing possible interference with the linear accelerator, patient's body, and treatment beams. In this section a several key components of intrafractional ultrasound image acquisition devices for clinical use will be described.

22.3.1 AutoScan

The first ultrasound guidance system capable of intrafractional imaging was the Clarity AutoScan System (Elekta, AB, Stockholm, Sweden). The Clarity AutoScan builds upon the original Clarity system by replacing the 2D ultrasound imaging system with a 3D/four-dimensional (4D) ultrasound imaging system and adding a hardware fixture for hands-free transperineal prostate imaging. The hardware device is a simple, manually operated five-degrees-of-freedom fixture that is mounted to a plate on the treatment couch between the patient's legs (figure 22.2). The sonographer uses the fixture to lock the ultrasound probe into position after the initial transperineal imaging position is found, thus freeing the sonographer to exit the treatment room and deliver radiotherapy while the ultrasound probe remains in the imaging position. The 3D/4D probe enables volumetric ultrasound images to be automatically captured without physically moving the ultrasound probe head. The transperineal imaging setup is advantageous because the geometry of the fixture and probe keeps all system hardware out of the normal delivery plane for C-arm linear accelerators, thereby avoiding guidance system hardware interference with the radiation treatment process and enabling simultaneous ultrasound/CT imaging during the radiotherapy planning phase. Clinical studies of the system are underway to characterize the performance of the Clarity AutoScan for intrafractional

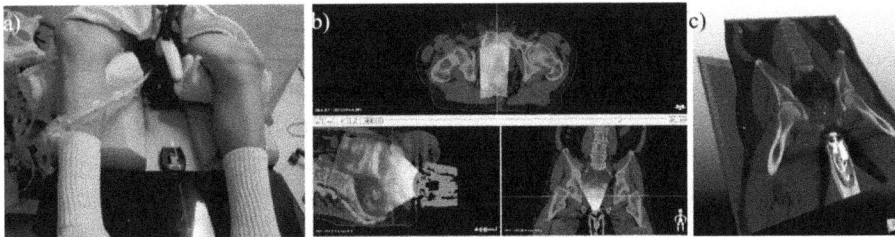

Figure 22.2. Acquiring ultrasound images during the CT simulation. (a) A patient is simulated with the AutoScan probe attachment board with frog-like leg position. (b) The CT simulation image with associated transperineal ultrasound image. (c) CT rendering of the patient and the ultrasound probe.

monitoring, and several results regarding its performance have already been reported [35–37].

22.3.2 Reference frame

AutoScan uses optical tracking comprising a camera fixed in the linear accelerator frame and a set of tracked markers attached to the ultrasound probe and the treatment couch. Spatial and temporal resolution of optical tracking is high, and for this reason most ultrasound guidance systems use this tracking modality [30, 38, 39], but optical tracking systems are limited by line-of-sight requirements between the camera and markers. EM tracking uses an EM emitter on the probe and a detector fixed in the linear accelerator room. EM tracking does not necessarily require a clear optical line of sight between emitter and detector but is very susceptible to noise caused by metallic objects in the operating vicinity.

Other approaches have also been investigated. Devices used to hold the ultrasound probe in place during intrafractional tracking can also be leveraged for tracking the probe. If the static or robotic device has sensors to measure each joint angle, the angles in combination with knowledge of the device geometry (forward kinematics) can be used to locate the probe in space. This tracking method is independent of the surrounding environment, but high accuracy can only be achieved by using very precise device manufacturing techniques and rigid materials.

Bruder *et al* [40] have investigated another approach to probe tracking utilizing the stereo x-ray cameras of the CyberKnife (Accuray, Inc., Sunnyvale, CA). After a non-orthogonal stereo camera calibration using x-ray phantoms, various marker geometries were positioned using a six-axis robotic arm, localized in six dimensions using algorithms developed for marker localization, and compared with results from the CyberKnife onboard system. Mean translational error for the newly developed software package was 0.218 mm, and rotational error was 0.076°. Results demonstrate high spatial accuracy; however, this method of tracking exposes the patient to x-ray radiation, and temporal resolution is limited by x-ray frequency, preventing capture of respiratory-induced probe motion.

22.3.3 Automatic tissue tracking

To enable intrafractional treatment intervention, ultrasound images collected during beam delivery must be processed in real time to extract soft-tissue motion information. 3D/4D ultrasound imaging for intrafractional guidance maximizes information content in images and enables true 3D motion tracking but suffers from slower frame rates and longer image processing times than 2D ultrasound imaging.

For this reason, the Clarity AutoScan System monitors intrafractional prostate motion by continuously gathering swept 2D slices of the region of interest in a cylindrical coordinate frame [30]. Rather than wait for a full 3D ultrasound sweep, 2D slices are registered in succession on each partially updated image using normalized cross-correlation. The registration algorithm calculates a correlation score for each iteration, and if the correlation is below a threshold chosen based on training data, a displacement is suspected, and the user is alerted to verify the registration.

Intensity-based image-to-image registration is used for prostate tracking. The tracking is limited to voxels within a 2 cm boundary surrounding the prostate contours [30]. The algorithm uses normalized cross-correlation as the cost-function calculated within 2 cm from the contours of prostate. The world coordinates X_w of the prostate (contours) center of mass (COM) is calculated, and the difference between the currently calculated COM and the reference COM position on the world determines the current prostate displacement. The registration is constrained to six degrees of freedom (translations and rotations with no deformations). The new sampled image is compared with the pretreatment ultrasound reference image acquired at beginning in cylindrical coordinates. The relationships between voxel coordinate $(X_p = i, j, k)$ and world coordinates X_w are defined in equations (22.1)–(22.3):

$$X_w = X_0 + D \left(\begin{bmatrix} i \cdot s_i \\ (j \cdot s_j + r)\cos(\alpha k) - r \\ (j \cdot s_j + r)\sin(\alpha k) \end{bmatrix} \right) \tag{22.1}$$

$$X_p = \begin{bmatrix} u \\ \sqrt{v^2 + w^2} \\ \tan^{-1}(v, w) \end{bmatrix} \tag{22.2}$$

$$[u \quad v \quad w] = \mathrm{diag}(s_i, s_j, s_j)^{-1} \cdot D^{-1}(X_w - X_0) \tag{22.3}$$

where s_i and s_j are the pixel scaling, α is the rotational step, r is the radial offset from the voxel origin (X_0) to the axis of rotation, and D is the matrix of direction cosines [41].

The field of view can be adjusted by the sweep angle to cover the desired field of view. The maximum scan angle is 75° in 0.5 s. The probe uses motorized control for the sweeping motion. The ultrasound frame rate (F) depends on the imaging parameters defined in equation (22.4):

$$F = \frac{\Delta\theta_{\text{sweep}}}{T_{\text{sweep}}\Delta\theta} \tag{22.4}$$

where $\Delta\theta_{\text{sweep}}$ is the sweep angle, $\Delta\theta$ is the angular spacing between frames, and T_{sweep} is the total sweep time.

There is an in-room tracker calibrated to the room coordinates and mounted on the wall to locate the reflective markers on the ultrasound probe. With the reflective markers, the coordinates of the ultrasound image are linked to room coordinates in the treatment room. The coordinate transformation between the ultrasound image and position in the treatment room is defined by the following equation:

$$r_R = {}^R T_T \times {}^T T_P \times {}^P T_F \times r_F \tag{22.5}$$

where r_R is the room coordinates, ^RT_T is the tracker-to-room transformation matrix, ^TT_P is the probe-to-tracker transformation matrix, ^PT_F is the 4×4 frame-to-probe transformation matrix, and r_F is a pixel in ultrasound 'frame' coordinates.

Lachaine and Falco show that the accuracy of the Clarity system tracking of a target in a water phantom is -0.2 ± 0.2 mm, 0.2 ± 0.4 mm, and 0.0 ± 0.2 mm in the AP, LR, and SI directions, respectively, by comparing Clarity tracking of a phantom with motion patterns fed to a robotic stage [30]. In a similar study, Abramowitz et al show 95% of maximum distance variation between Clarity measurements and optical control measurements of a robotically controlled probe to be less than 1.3 mm, with the majority less than 1 mm [42]. In both studies, target geometry and motion were informed by prostate motion profiles.

22.4 Quality assurance of clinical systems

Comprehensive performance evaluation and periodic quality control tests need to be designed and conducted for routine clinical use of ultrasound guidance. An American Association of Physicists in Medicine Task Group report comprehensively reviews quality assurance for interfractional guidance systems and recommends tests and their frequencies [25]. Since no such report exists for intrafractional ultrasound guidance, we discuss some of tests below using the Clarity AutoScan system as an example with a summary of the observed performance of the system under relevant treatment conditions [43].

22.4.1 Assumptions for speed of sound in tissue

Uncorrected speed of sound discrepancies in soft tissue can result in ultrasound target localization errors up to a few millimeters. Salter et al [44] found speed artifact errors of 0.7 mm cm^{-1} of fat traversed using experiments in a phantom. To correct for such discrepancies, Fontanarosa et al [45] apply a density-based speed of sound correction algorithm to a set of prostate, liver, and breast images collected from volunteers. Algorithms applied corrections to the speed of sound based on tissue densities drawn from CT scans registered to the simulation ultrasound scans. Results indicated prostate, liver, and breast centroid shifts of 3.6, 6, and 1.3 mm, respectively, and liver volume changes of up to 9% when compared with uncorrected data [45]. In later studies, a correction algorithm was applied to the scanning of a multimodality ultrasound phantom through three different layers of liquid, with results showing differences between ultrasound and CT images smaller than the resolution of the CT scan (around 0.7 mm in the image plane) [45]. Additional studies by this group estimated errors in prostate location based on speed-of-sound aberration that agreed well with previously published discrepancies between ultrasound and CT scans, with corrections to prostate centroids averaging 3.1 mm [46]. With these discrepancies exceeding typical tissue tracking accuracies of ultrasound guidance systems previously discussed, the results indicate a need to consider speed-of-sound aberrations in estimating tumor location.

22.4.2 Image quality

Tracking accuracy depends on how well the structure can be seen (i.e., good contrast will be tracked well) on the ultrasound image since the registration is intensity-based [30]. The live image is compared with the reference image, and the normalized cross-correlation is used as the cost-function. Therefore, if the image quality drops (e.g., acoustic shadows are induced), the registration algorithm might not be able to give optimal results, which likely will introduce errors into the tracking results. Yu *et al* [43] investigated this by introducing acoustic shadows with an air gap between the ultrasound probe and the surface of the phantom in order to simulate deterioration in ultrasound image quality as shown in figure 22.3. In the case of a phantom, the anatomy is well defined and the contrast of the prostate is easy to see, but for patients the prostate outline and contrast can vary. Structures including the rectum and pelvic bones within a patient also affect the ability to accurately image the prostate using ultrasound.

22.4.3 Motion frequency and the directional effect

In order to evaluate the impact of the motion frequency on the tracking accuracy and the precision in both ML and SI directions, in a phantom study motion was set to either a 20 s period or a 10 s period [43]. The effect of phantom motion frequency on the ultrasound tracking error and its standard deviation was determined using optical tracking for ground truth measurement. The precision of the ultrasound tracking performance in the SI direction was better than that in the ML direction (figure 22.4(b)). The root-mean-square (RMS) errors for SI and ML are 0.18 and 0.25 mm, respectively. The accuracy of ultrasound tracking performance in the ML direction was better than that in the SI direction (the mean position errors are 0.23 and 0.45 mm, respectively). The tracking accuracy and precision were better with a longer period (figure 22.4(c)). The RMS errors for low and high frequency were 0.25

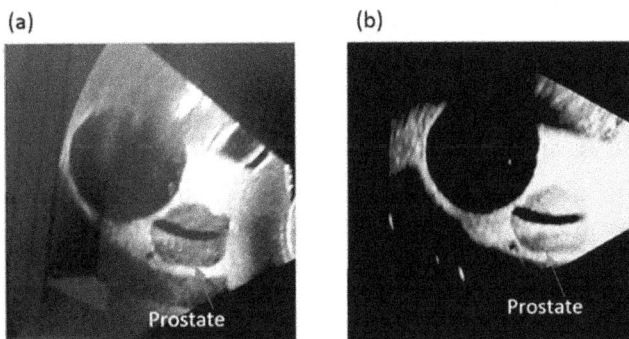

Figure 22.3. (a) A typical good quality ultrasound image of a male pelvic phantom. (b) A poor-quality ultrasound image. The contrast for the prostate is degraded by air gaps between the probe and phantom (arrows point the boundary of the prostate). The algorithm uses normalized cross-correlation as the cost-function calculated within 2 cm from the contours of prostate. Theoretically, if the contrast at boundary of prostate is compromised, the ultrasound tracking ability will decrease. [43], reproduced with permission © *Technology in Cancer Research & Treatment*.

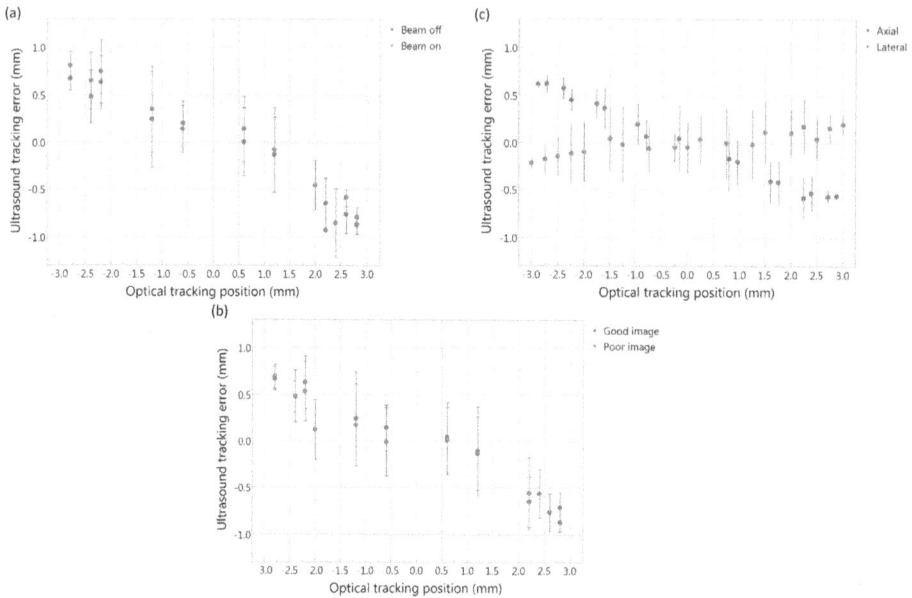

Figure 22.4. The average of the position difference recorded between the ultrasound system and the optical tracking system (y axis, ultrasound tracking error) at different motion phases (x axis). The error bar is the deviation of the position difference between two systems when the phantom goes through different phases of the sine wave and when the phantom has different velocity. (a) Beam off vs. beam on. (b) Axial (SI) vs. lateral (left-right). (c) High frequency (10 s period) vs. low frequency (20 s period). [43], reproduced with permission © *Technology in Cancer Research & Treatment*.

and 0.45 mm, respectively. The results from the phantom study indicate that the tracking accuracy in the ML direction is better than that in the SI direction. The tracking precision in the SI direction is better than that in the ML direction. In both directions tracking accuracy is within a millimeter when target motion is less than 3 mm.

22.4.4 Temporal accuracy (latency evaluation)

It is important to know if the intrafractional monitoring system can report the shift in real-time or with minimum time latency in order to determine the dosimetric effect accurately. Yu *et al* evaluated the temporal accuracy of the AutoScan system in tracking mode with the use of motion phantom and compared motion monitoring results between ultrasound and optical tracking systems [43]. Since the two tracking systems do not share a common synchronized time stamp, in order to determine the delay between actual phantom motion and the motion reported by the ultrasound tracking system, a video system was used to monitor both the optical and the ultrasound system, i.e. to put the two systems on the same clock with a frame rate of 30 frames per second (fps). The recorded video was reviewed frame by frame to examine the temporal signals displayed by both systems and determine the offset between the two signals. The first frame that reported the phantom in motion was located for each of the optical and the ultrasound systems. Since the frame rate of the video is known, the time delay between the two systems was calculated by the difference in frame number. The average difference in number of frames noting the initial phantom motion between the two systems was 5.2 for the Clarity AutoScan system (Elekta, AB, Stockholm, Sweden). With the known frame rate of 30 fps, the average time delay is 173 ms, and the standard deviation is 45.2 ms (n=10). If the finite video (30 fps) and optical tracking system (60 Hz) frame rates are included in the latency evaluation time delays, the upper limit latency is estimated to be 223 ms.

22.4.5 Imaging performance during beam delivery

In order to determine if radiation would affect the electronic processing of the ultrasound system, before using the ultrasound image for the intrafractional monitoring for radiation therapy, medical physicists need to examine the impact of radiation on the electronic processing of the ultrasound system which is present during actual patient treatments. The ultrasound tracking accuracy was evaluated with an independent optical tracking system under typical treatment conditions (beam on) [43]. A volumetric modulated arc therapy treatment plan using 15 MV beams was delivered while the ultrasound system was acquiring data in order to determine if radiation with some neutron component would affect performance [43]. In this study, the optical tracking patterns served as the ground truth. The mean position difference recorded between the ultrasound system and the optical tracking system (ultrasound tracking error) at different motion phases was used to evaluate the accuracy of the ultrasound system. The mean position value is the position difference detected by the ultrasound and optical systems calculated by the following equation:

$$\text{mean position errors} = \frac{1}{n}\sum_{i=0}^{i=n}(x_{i,\text{ultrasound}} - x_{i,\text{optical}}) \tag{22.6}$$

where $x_{(i,\text{ultrasound})}$ and $x_{(i,\text{optical})}$ are the phantom position detected by the ultrasound and the optical systems at certain phase. The phantom motion was set to ± 3 mm amplitude and 10 or 20 s period.

Linear regression fits of ultrasound-measured displacement versus optically measured displacements in the presence and lack of radiation and neutrons are used to evaluate the effect of the radiation. The fitting results (R2) for beam off and beam on are comparable, R2 = 0.958 and 0.956, respectively. The RMS errors are 0.31 and 0.32 mm for beam off and beam on. Moreover, the ultrasound tracking error and its standard deviation at different motion phases (when the phantom goes through different phases of the sine wave, the phantom has different velocity) as shown in figure 22.5(a) are comparable. Therefore, the study concluded that influence of radiation and neutrons on the tracking ability is negligible [43].

Other studies examining the effect of radiation on ultrasound image quality have produced dissimilar conclusions. Hsu *et al* used an Acuson 128/XP ultrasound scanner to image stationary and moving phantoms while operating an Elekta SL25

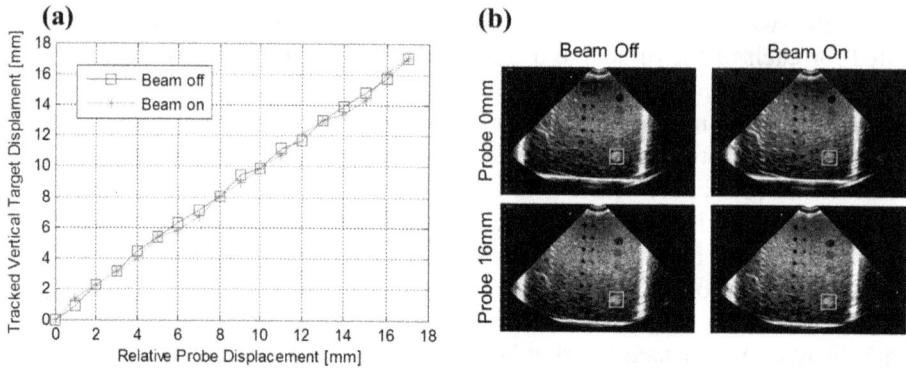

Figure 22.5. Comparison of ultrasound images and tracking performance with beam off and on. (a) True object displacement versus computed displacement using cross-correlation. (b) Tracked circular object indicated within ultrasound images at two displacement values. The square indicates the location of the tracked object based on the manually selected target template. [38], reproduced with permission © *Medical Physics*.

linear accelerator [47]. The experiment found the ultrasound images to be affected by a periodic noise at a frequency identical to the pulsing frequency of the treatment machine; however, this noise was found to have minimal effects on the precision of the tracking algorithm. Schlosser *et al* [38] found no spatial or temporal interference patterns in ultrasound images of a phantom acquired with a 3.5 MHz Interson 2D single-element probe during linear accelerator beam operation. Additionally, no significant difference in tracking accuracy was found between beam-off and beam-on cases (figure 22.5). The ultrasound imaging interference from a linear accelerator likely depends on the specifics of the ultrasound imaging system being used; for example, modern systems with improved shielding may better reject radio frequency interference from the linear accelerator.

22.5 Clinical applications

Ultrasound based image guidance systems offer real-time, markerless, volumetric imaging with excellent soft-tissue contrast, overcoming limitations of traditional x-ray or CT-based guidance for abdominal and pelvic cancer sites such as the liver and prostate.

A main challenge with ultrasound guidance is the interuser variability in both image acquisition and interpretation [23, 48]. Robinson *et al* [33] found prostate positioning error between Clarity scans and CT scans compared with a reference scan to be significant, observing a discrepancy of 5 mm or more between CT and ultrasound localizations in >80% of cases despite expert re-analysis of data.

The following needs to be considered when an ultrasound system is used for the interfractional and intrafractional image guidance:

1. Pressure applied by the ultrasound transducer on the patient's body can cause anatomy deformations and displacements of varying magnitude, depending on the properties and depth of the treatment target. Since pressure

applied during pretreatment imaging is not present during beam delivery, systematic positioning errors are common [49–51].

2. The quality and consistency of freehand ultrasound-based patient alignment is significantly dependent on the operator and level of training. Hence training on common, reproducible image acquisition techniques is of paramount importance.

3. The Clarity and BAT systems rely on freehand probe manipulation, which would pose an unacceptable hazard to the probe operator if performed during treatment delivery. Thus, no imaging is available during beam delivery, when accurate target tracking is most critical.

Due to the advantages of the ultrasound images, a concept for an add-on ultrasound soft-tissue image guidance system as a possible alternative to fiducial-based intrafractional imaging was introduced [38] and demonstrated that with operator-free soft-tissue real-time 2D or 3D imaging the system could be used for intrafractional monitoring of target displacements [52, 53].

22.5.1 Prostate

Because of accessibility to ultrasound imaging and the need for soft-tissue visualization, the prostate is the treatment site that has been the focus of clinical and research ultrasound guidance effort. Hypofractionated prostate treatment is contingent on having the technical means to deliver treatments with minimum risk of mis-targeting. Transabdominal ultrasound (TAUS) showed systemic errors compared with CT as well as gold seed verification [23]. The reason for the systemic error is that pressure impacts prostate localization, causing prostate movement during the TAUS acquisition [49] but not during the treatment. On the other hand, for transperineal ultrasound (TPUS), there is pressure applied to the prostate with the use of TPUS, but the pressure is always applied during the continuous ultrasound image acquisitions and the treatment. Therefore, TPUS prostate tracking is more attractive than the transabdominal approach for prostate localization, which faces challenges related to prostate displacement/deformation induced by probe pressure and interference of the imaging probe with beam paths [54]. Previous data show the feasibility of remotely controlled robotic ultrasound imaging for IGRT [38] to circumvent these challenges; however, limitations of the TAUS technique were prevalent. Foster *et al* showed that the isocenter determined by TAUS-based localization has a systematic shift compared with Calypso [55], bringing into question the usefulness of TAUS for prostate localization and monitoring. Another recent study compared TAUS with CBCT for prostate positioning and showed strong agreement between ultrasound and bone-matched CBCTs; however, the researchers noted limited feasibility due to bladder filling and user-dependent technique issues [56]. On the other hand, clinically, Yu *et al* did not notice consistent correlation between image quality and patient size and bladder filling with the use of TPUS [43]. This might be due to the smaller variation of the depth with the use of a transperineal probe compared with an abdominal probe. While bladder filling is indeed required with TAUS to provide acoustic imaging window, for the TPUS it appears to be less significant since the bladder is placed distally to the prostate along the ultrasound beam. This, along with the possible acoustic shadows from the pubic symphysis, limits the usefulness of TAUS for prostate localization.

Furthermore, the Clarity AutoScan system has the advantage of continuously taking images throughout the treatment without a manual operator present, thus moving from an ultrasound system which utilizes a manual probe for pretreatment setup to an automated scanning system capable of monitoring prostate motion both before and throughout treatment. The Clarity AutoScan system thus holds promise to move ultrasound IGRT into intrafractional continuous prostate monitoring. Knowing the accuracy, the precision and clinical feasibility of the Clarity AutoScan ultrasound system is crucial for implementation. Prostate motion during radiation treatment is expected; however, with current common on-board IGRT capabilities, it is impossible to determine the extent of this motion and its clinical impact. A recent study evaluated the potential for using TPUS as a means of pretreatment patient positioning for prostate radiotherapy [57]. Their data showed a strong agreement between CBCT and TPUS pretreatment patient positioning. An interesting clinical case was reported by Yu *et al*, as they used the Clarity AutoScan ultrasound system to acquire continuous TPUS images for monitoring of intrafractional prostate motion. A patient was positioned using CBCT, and once the patient was shifted to treatment position based on this CBCT, the ultrasound monitoring was started with a 3 mm threshold [43]. The zero position of the ultrasound monitoring is thus determined by the standard of practice CBCT. Any deviation beyond 3 mm from this position is signaled by the ultrasound system. Less than 1 min from initiating the monitoring and immediately prior to the commencement of treatment radiation delivery, the Clarity system signaled that the prostate had shifted more than 3 mm in the anterior direction. Beam initiation was halted, and the patient's position was observed. The patient's anatomy continued to shift in the anterior direction over the next several seconds and settled at a displacement of about 1 cm, as noted by the ultrasound monitoring. The anterior shift in prostate position was stable over tens of seconds and thus not indicative of a movement that would return to the treatment position on its own. A second CBCT was taken and verified the anterior shift noted by the Clarity system. The CBCT determined the shifts to be 1.1 cm in the anterior direction, which was in agreement with the ultrasound system. This example demonstrates the potential for intrafractional motion detection by the Clarity AutoScan ultrasound system. The prostate motion was undetected by standard CBCT IGRT protocol and only discovered with ultrasound prostate monitoring. Clearly, small random movements of the prostate during radiation treatment may not have large clinical impact. However, a stable large shift in prostate position, as described in this clinical case example, could result in missing the clinical target during radiation treatment. In this clinical example the authors showed that a positional deviation of 1 cm is possible and could potentially go undetected by standard IGRT protocols. This can potentially result in missing the treatment area without even recognizing the error. As presented in this clinical case example, the ultrasound tracking system would be able to recognize this shift and corrective actions could be taken prior to treatment initiation or even during beam on.

Han *et al* systematically compared and correlated the Clarity estimated prostate position with the prostate position visualized by 'pseudo-cine' on-treatment MV images of fiducials implanted in the prostate. The patients enrolled in this study underwent 195 treatment fractions including 70 boost fractions [37]. The ultrasound monitoring was available for 167 fractions. MV fiducial tracking and ultrasound monitoring were both available for 39 boost fractions. Continuous ultrasound motion tracking was done for 39.7 h. In a phantom the maximum tracking error was 1.1 mm, whereas in patients the maximum tracking error varied between 1.3 and 3.3 mm depending on the patient. The ultrasound measured mean prostate displacements across all patients and fractions were 0.3 ± 0.7 mm, -0.1 ± 0.7 mm, and -0.6 ± 1.1 mm in the SI, ML, and AP directions. The RMS was 0.7 ± 1.5 mm. The average motion of the prostate during kV image acquisition was 0.3 mm with a maximum motion of 0.8 mm. The average motion of the prostate between kV/CBCT acquisition and treatment delivery was 0.6 mm with a maximum motion of 1.3 mm over time periods ranging from 3.7 to 8.6 min. The use of 2D MV portal imaging for ground truth measurements is a limitation of the study. First, some of the discrepancies observed are caused by the uncertainty in localizing the prostate fiducials within the MV images. Second, evaluation of the ultrasound tracking accuracy is inherently performed in 2D, as it is limited to the plane of the MV images.

22.5.2 Liver

Liver is another soft tissue which can be visualized by ultrasound. Image-guided liver stereotactic ablative body radiotherapy (SABR) has emerged as a safe and effective noninvasive treatment option for patients with inoperable liver tumors [58]. Realizing the full potential of liver SABR by minimizing normal tissue irradiation and maximizing tumor dose is contingent on having the clinical capabilities to adequately manage respiratory-induced quasi-periodic liver motion [59, 60]. Abdominal compression techniques have been investigated to reduce respiratory liver motion, but in the majority of patients (60%) the magnitude of motion reduction is clinically not significant (<3 mm) [61]. In a different approach, respiratory gating [59, 62] or tracking [63] is being used to minimize the effect of liver motion on dosimetric coverage while keeping the volume of irradiated normal liver low. Since real-time soft-tissue imaging is lacking on current linear accelerator systems, fluoroscopy imaging or EM tracking of fiducials implanted in the liver has been used to guide respiratory gating [62,64–67]. However, the targeting accuracy with fiducials depends on marker–tumor distance [68], and thus substantial errors may occur if the fiducials are not placed in proximity to the target lesion [69]. Other concerns with this approach include unaccounted respiratory-induced liver deformation, the need for fiducial implantation in patients with inoperable disease, imaging dose for continuous intrafractional imaging, and artifacts on follow-up MRI with implanted EM markers.

Bloemen-van Gurp *et al* [70] demonstrated the feasibility of ultrasound-based image guidance in SBRT of liver lesions using the Clarity system, finding that the combined intra- and interobserver variability of image segmentation in scanned images was 4 mm (1 standard deviation) and could be reduced by 1.7 mm in the SI direction using active breathing control.

Initial pilot data were reported by Schlosser *et al* [52] from a prospective patient study performed with a robotic system in the treatment simulation setting and evaluating retrospectively the proportion of patients undergoing liver SABR for whom robotic ultrasound imaging can be deployed without interference in clinical volumetric modulated arc therapy (VMAT) plans. In this study, the authors presented the current iteration of their robotic ultrasound guidance system and the first examples of clinical imaging with the system in the setting of CT simulation of patients undergoing liver SABR. The investigators found that unattended robotic clinical liver imaging is possible. The study further indicated that for VMAT liver SABR, robotic ultrasound imaging of a relevant internal target would be possible in 85% of cases while using treatment plans currently deployed in the clinic [52]. With beam re-planning to account for the presence of robotic ultrasound guidance, intrafractional ultrasound was judged to be a potential option for 95% of the liver SABR cases.

22.5.3 Breast

Besides intrafractional monitoring during the radiation therapy, ultrasound image is also used to define the lumpectomy for a breast cancer treatment plan. It is difficult to precisely define the target (tumor bed) for partial-breast irradiation and the boost treatment followed by the whole breast irradiation to ensure that the PTV encompasses the tumor bed [71, 72] and to avoid irradiating surrounding normal tissues [73]. Wong *et al* [74] compared CT scans of patients with breast cancer taken just before radiation with Clarity ultrasound images taken at corresponding times to evaluate error in the Clarity system applied to breast imaging. The difference between CT and ultrasound images of the tumor bed centroid averaged 0.1 ± 2.8 mm, -0.2 ± 4.0 mm, and 0.4 ± 3.7 mm in the AP, LR, and SI directions, respectively, which was deemed clinically insignificant.

22.6 Ultrasound guidance research

Several research groups have introduced concepts for intrafractional ultrasound guidance systems leveraging robotic probe placement technology and real-time soft-tissue tracking to extend ultrasound imaging for other anatomical sites that require intrafractional motion monitoring during radiation delivery. In this section, we discuss a few robotic systems which are undergoing research and have potential for the future use.

22.6.1 Stanford imaging robot

Schlosser *et al* were the first to demonstrate the feasibility of robotic intrafractional ultrasound imaging in the context of radiotherapy guidance [38]. As a proof of concept, they developed and evaluated a robotic device designed to control 2D TAUS imaging of the prostate, shown in figure 22.6(a). The device has two active degrees of freedom for controlling probe pitch (tilt in the SI direction) and pressure against the abdomen and three passive, mechanically locking degrees of freedom for manually setting the probe position prior to the procedure. The active degrees of freedom can be remotely controlled during beam delivery using a joystick with force feedback known as a haptic device. With the telerobotic system, the authors demonstrated that gantry collisions are avoidable, stable remotely controlled prostate imaging is achievable in healthy human subjects over 10 min time periods, and robotic performance is not degraded during operation of a 15 MV radiation beam. Moreover, the research group at Stanford, in collaboration with SoniTrack Systems, Inc., has developed a second-generation custom-designed robotic device based on lessons learned from the first proof-of-concept prototype, shown in figure 22.6(b), to extend the application to abdomen. The goal of the research effort is to produce a simple, compact, human-safe robotic design that enables 3D/4D ultrasound imaging of any abdominal radiotherapy target, actively controls probe force, allows rapid and repeatable positioning of the ultrasound probe, and eliminates metal in areas exposed to CT/therapy radiation. The resulting design has a single active degree of freedom to control probe pressure against the patient; five passive, electronically actuated probe positioning degrees of freedom; and three passive, manually actuated positioning degrees of freedom [75]. The active degrees

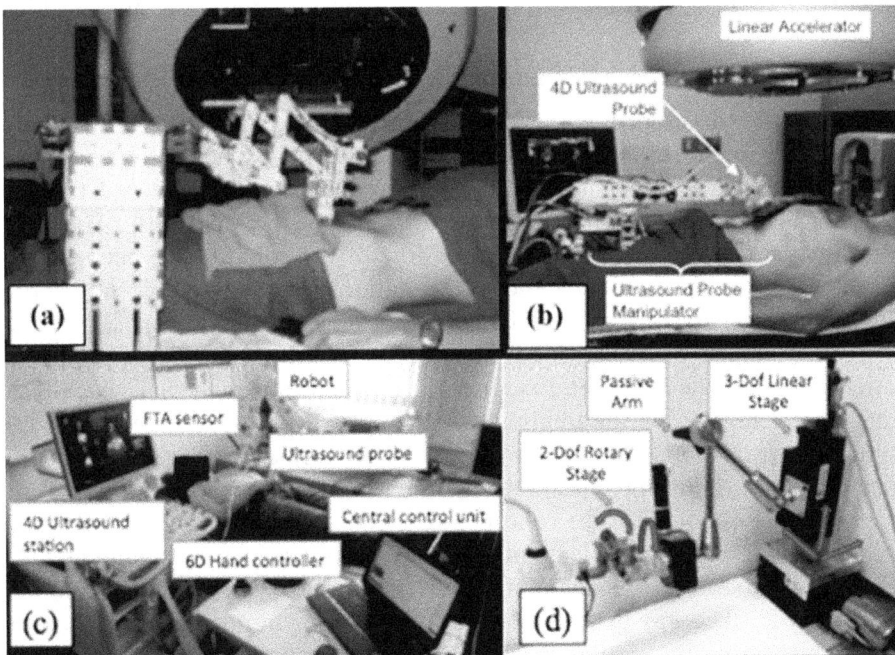

Figure 22.6. Devices to support intrafractional ultrasound imaging. (a) Stanford prostate imaging robot. (b) Stanford abdominal imaging robot. (c) University of Lubeck robot. (d) Johns Hopkins robot. [22], reproduced with permission © *Cureus*; [39], reproduced with permission © *IEEE Proceedings*.

of freedom are controlled with elastic force controllers that have the following advantages: (1) they provide easy back-drivability in case of power failure, (2) they eliminate need for a metallic force sensor near the ultrasound probe, and (3) they are inherently safe when used to control contact with human subjects. Metallic components in the CT/therapy field are eliminated by coupling the three-degrees-of-freedom robot wrist with remotely located sensors and actuators via a mechanical cable drive system. The nine-degrees-of-freedom robot design enables easy access to any abdominal target on the patient's body while accommodating a wide range of patient body shapes and avoiding potential collisions with the rotating linear accelerator. The robot has demonstrated successful imaging over extended time periods on the prostate, pancreas, liver, and kidneys of healthy volunteers.

22.6.2 University of Lubeck robotic platform

The University of Lubeck has developed a robotic ultrasound probe positioning system for CyberKnife radiotherapy using an off-the-shelf robotic arm as shown in figure 22.6(c) (Viper S850, Adept Technology, Inc.) [76]. The robot has six actively controlled degrees of freedom, enabling probe placement on nearly any part of the patient's body. User commands controlling robot pose are input using a six-degrees-of-freedom hand controller and sent via TCP/IP protocol to the control software [77]. Continuous high-quality imaging was confirmed by showing that image quality stayed above the threshold for continuous target tracking of heart volumes at least 95% of the time in three healthy human subjects over 30 min time intervals [76].

22.6.3 Johns Hopkins abdominal imaging robot

Sen *et al* have developed a custom-designed robotic manipulator, shown in figure 22.6(d), for ultrasound-guided radiotherapy with particular focus on overcoming inconsistencies in tumor localization between the planning and treatment phases due to tissue deformation induced by probe pressure [39]. The design incorporates a six-axis force/torque sensor for force feedback, five active degrees of freedom (three translational and two rotational—probe spin is not actively regulated), seven passive degrees of freedom, and optical robotic position tracking. In the CT planning phase, a model probe containing no metallic components is positioned against the abdomen to cause deformation similar to that which occurs during treatment. During the treatment setup phase, a virtual spring system helps the ultrasonographer manually move the probe to a position similar to that recorded during planning, inducing repeatable deformation and repeatable force. Virtual springs consist of motors in the robotic joints that exert force toward the position recorded during planning. Flexibility in the virtual springs accommodates any changes in position that the sonographer deems necessary. Treatment planning is carried out with the x-ray penetrable model probe being in position such that a CT scan can be conducted. In *ex vivo* experiments using a bovine liver fixed in gelatin with implanted fiducials, the system demonstrated repeatable arm placement with minimal effect on displacement of the fiducials, yielding after six repeated arm placements a mean absolute difference between fiducial displacements of

0.4 ± 0.4 mm in ultrasound images and 0.3 ± 0.2 mm in CT images acquired with the model probe [78]. Later *in vivo* experiments conducted in a canine model demonstrated mean 3D reproducibility of 0.6 to 0.7 mm, 0.3 to 0.6 mm, and 1.1 to 1.6 mm for the prostate, liver, and pancreas, respectively, under position control and controlled ventilation [51]. Force control proved less reproducible, however, indicating that position control rather than force control should be used for robotic substitution of real and model probes. Results indicated that the system shows promise for monitoring real-time organ motion, particularly under conditions of minimal probe pressure.

22.6.4 Treatment planning considerations with robotic ultrasound guidance

Because ultrasound probe placement is constrained by internal anatomy, and because intrafractional ultrasound imaging hardware (ultrasound probe, robot, probe tracking sensor) may absorb radiation and change the actual dose delivered to the patient, potential beam interferences must be taken into account during the radiotherapy treatment planning process. In general, this can be accomplished in two ways:

1. Avoid treatment beam positions that interfere with hardware.
2. Deliver radiation through the ultrasound image guidance hardware, and account for hardware in the treatment planning process.

Several groups have investigated the feasibility of strategy by studying whether constraints on beam angles imposed by intrafractional ultrasound image guidance hardware affect the quality of treatment. Wu *et al* found that avoiding the AP beam in radiation planning, which would pass through an ultrasound probe placed in transabdominal imaging position, resulted in a negligible effect of the transducer on delivered dose distributions [79]. Schlosser *et al* conducted a second feasibility study that compared a seven-beam clinical plan for a patient undergoing prostate intensity-modulated radiation therapy with a seven-beam plan for the same patient that excluded a 90° anterior sector in order to avoid ultrasound guidance hardware in the transabdominal imaging position. No impactful difference between the plans was found (figure 22.7) [38]. These two studies show that beam avoidance of ultrasound image guidance hardware is a feasible option for delivering prostate radiotherapy guided by TAUS imaging. Zhong *et al* examined the effect of probe orientation on liver SBRT plans that avoid an intrafractional ultrasound imaging probe. The study compared clinically accepted SBRT plans for ten patients with liver cancer with two new plans generated for each patient that avoided irradiating the ultrasound probe. One of these two plans positioned the probe on the surface of the abdomen parallel to the patient's longitudinal axis, and the other positioned the probe vertical to the longitudinal axis. Treatment plans could not be generated for two patients with superficially located tumors for either probe orientation. For the remaining eight patients, plans were successfully generated that did not show significant differences in dosage delivery metrics. With a treatment goal of delivering 37.5 Gy to the PTV in three fractions, average dose delivered to the 95% of the PTV

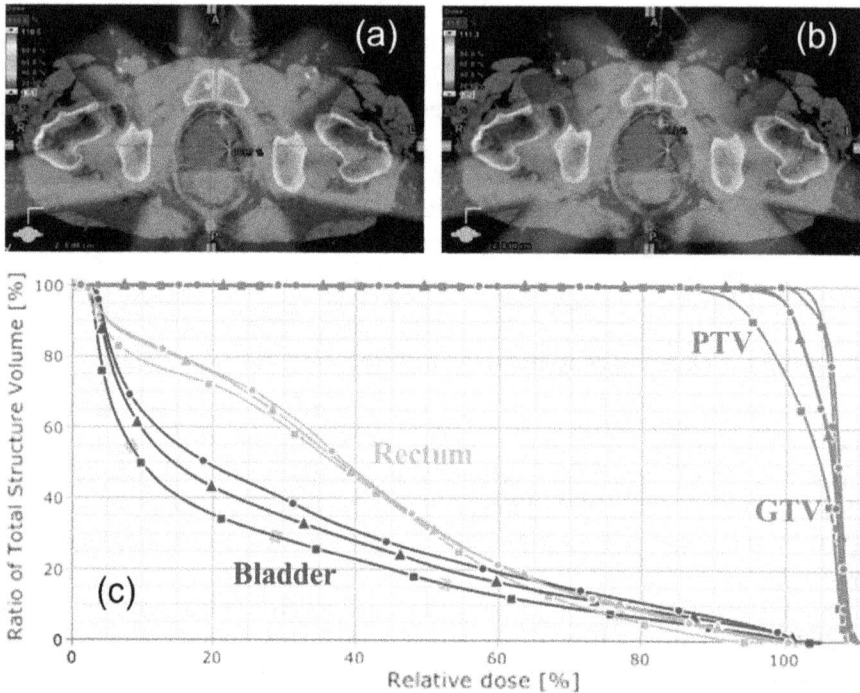

Figure 22.7. Seven-beam treatment plan comparison. (a) Axial dose distribution for a clinical prostate intensity-modulated radiation therapy (IMRT) plan. (b) Axial dose distribution for a reoptimized IMRT plan with restricted beam angles to avoid ultrasound probe and robot links. (c) Dose–volume histograms for the clinical IMRT plan (circles), reoptimized plan (triangles), and reoptimized plan with reduced margin (squares). Note that the IMRT plan with reduced margins underdoses original PTV but maintains groos tumor volume coverage and improves healthy tissue sparing. [22], reproduced with permission © *Cureus*.

was evaluated to be 38.63 ± 0.14 Gy for probe orientations parallel to the longitudinal axis, 38.48 ± 0.31 Gy for probe orientations vertical to the longitudinal axis, and 38.72 ± 0.14 Gy for clinical SBRT plans. The authors concluded that except for superficial lesions, real-time ultrasound monitoring during liver SBRT was clinically feasible [80].

22.6.5 Tissue tracking for intrafractional guidance

Combining operator-free imaging with soft-tissue tracking is essential for intrafractional ultrasound motion monitoring. This section reviews soft-tissue motion tracking methods implemented on 2D and 3D/4D ultrasound image streams for the specific purpose of intrafractional radiotherapy guidance.

To demonstrate early feasibility of 2D ultrasound in monitoring intrafractional soft-tissue displacements of the prostate, Schlosser *et al* developed a method using two tissue displacement parameters (TDPs) derived from the normalized cross-correlation similarity measure that characterized in-plane and out-of-plane displacement of the target volume in real time relative to a reference position [53]. The method successfully detected prostate displacements in healthy human subjects before they exceeded 2.3, 2.5, and 2.8 mm in the AP, SI, and ML directions at the 95% confidence level, with a total system lag averaging only 173 ms. False positives did not exceed 1.5 events over 10 min of continuous imaging. The authors performed an online demonstration of the system where a healthy human subject was asked to physically move hips at certain time intervals, causing a displacement of the prostate relative to a 'world' reference frame. Hip displacements were monitored using an external marker on the volunteer's hip and with the 2D ultrasound-based TDPs. The TDPs detected 10 out of 10 prostate displacements and registered zero false positives over the 12 min online test.

A similar demonstration of feasibility was carried out by Schlosser *et al* using 2D ultrasound for monitoring motion of the liver in human subjects [81]. Ultrasound image streams were acquired remotely in volunteers for 60–120 s. Concurrently, the position of an external infrared skin marker fixed to the subjects' abdomen was tracked. Within each image stream, the displacement of two separate liver features was monitored using normalized cross-correlation, one serving as a baseline 'target' and the other one as an internal target surrogate. Two models were fitted and used to predict target motion. The first one used the displacement of the external marker surrogate as an input signal. The second one used the displacement of the internal ultrasound surrogate feature as an input. Discrepancies between the measured target positions and those predicted by the models were quantified. In a separate analysis, the Pearson correlation coefficient and phase difference between the surrogate signals and the target signal were examined as a function of time. Error based on the external surrogate model was larger than 2.0 mm on average, at times exceeding 4.0 mm, while mean error was less than 1.0 mm using the internal ultrasound surrogate model. Pearson correlation coefficient averaged 0.83 between external surrogate motion and target motion, in contrast to 0.97 between internal surrogate and target. The study thus demonstrated superior tracking of target motion using ultrasound to monitor displacement of an internal feature when compared with tracking of an external surrogate.

For 3D and 4D tissue tracking, Harris *et al* assessed tracking accuracy and precision of real-time 4D ultrasound using a mechanically swept probe on both a tissue-mimicking phantom and on liver motion of healthy volunteers *in vivo* [82]. Using a 3D cross-correlation–based tracking algorithm, they found non-incremental tracking (comparing each volume with the first volume) to be superior to incremental tracking (comparing each volume to the next) and good agreement *in vivo* between cases tracked automatically and manually, with 1.7 mm mean error between the two measurements. Subsequent studies [83] conducted on liver motion *in vivo* examined the effect of volumetric imaging rates using a 4D matrix array ultrasound probe, finding that lower volume rates (2–12 Hz) resulted in RMS deviation values of 2–6 mm relative to the highest rates (24 Hz). In a third study, the accuracy of a mechanically swept 4D ultrasound probe for transperineal monitoring of prostate motion was evaluated using an ultrasound phantom undergoing prescribed motion [84]. The system tracked SI and AP motion to $\leqslant 0.81$ mm RMS error at a 1.7 Hz volume rate as compared with 0.74 mm for the CyberKnife system with which volumetric ultrasound was compared. Error was higher in the ML direction (elevational sweep direction for the ultrasound probe) but could be reduced to ≤2.0 mm using a correlational threshold.

In the robotic system developed by Bruder *et al* [85], template matching algorithms using sum of squared differences were employed for the purpose of tissue tracking. Similar to previously described methods, the process involved comparison between the current ultrasound volume and a template—a volume previously captured at a specific position. Unlike previously described rigid tracking algorithms, Bruder accounted for target deformation by developing a multi-template matching algorithm in which the current ultrasound volume was compared with a number of templates that represented snapshots of the patient's anatomy across a range of deformations. The algorithm was successfully employed in the University of Lubeck robotic system with 15 ms of processing time in practice.

Molloy *et al* conducted a study evaluating the spatial and temporal accuracy of ultrasound in tracking respiratory motion [86]. The system used an ultrasound transducer attached to a robotic arm registered in the frame of reference of the room to acquire an image stream. Spatially registered images gathered using the probe were then compared in real-time to a reference 3D data set, which was acquired beforehand using CT. Accuracy of the system was studied by measuring position of test objects within a phantom and comparing with 3D CT scan data. Dynamic properties were characterized using phantoms following programmed motion trajectories and comparing the system's performance to a 4D CT scan. Positional accuracy of the system was found to be better than 2 mm and typically 1 mm, and dynamic response produced a mean relative positional error of 1 mm if a latency of three video frames was incorporated, demonstrating that the technique could effectively track respiratory motion.

Kubota *et al* proposed a 3D tissue tracking algorithm for monitoring organs affected by respiratory motion [87]. The algorithm involved first manually identifying a region of interest that enveloped the target. A direction of maximum displacement due to respiratory motion was then identified prior to tracking by averaging images acquired at maximum inspiration and expiration. In tracking target motion in real-time, a pyramidal Lucas Kanade method was used to associate a large number of feature points in each frame with points in the previous frame, then to move the region of interest in a direction associated with the motion of these points. In order to adjust for deformations and stacked error over time, error correction was applied by comparing images at maximum inspiration/expiration through over multiple cycles and updating if necessary, under the assumption that target position in the image of maximum inspiration was constant between cycles. The method was validated by tracking the gallbladder in one subject and a liver vein in a second subject, comparing the proposed algorithm with a template matching algorithm and a second algorithm involving feature point tracking without error correction. The proposed algorithm outperformed both alternatives, allowing longer tracking times (up to 5 min) and consistent tracking through organ deformation and changes in cross-sectional position. Average tracking accuracy was 1.54 ± 0.9 mm as defined as deviation from the center of the region of interest and the center of the target organ designated by an experienced medical doctor. Computation time was 8 ms per frame for 2000 frames processed.

22.7 Conclusion

Recent years have seen increasing research and commercial activity toward ultrasound-based image guidance systems for radiation therapy. The BAT, SonArray, and Clarity systems for interfractional imaging demonstrated early success in deploying ultrasound for image guidance, but technical limitations stymied widespread adoption. The commercially available Clarity AutoScan system has moved ultrasound into the realm of intrafractional imaging for a specific application—treatment of prostate lesions. Current research aims to build out the components necessary for an intrafractional ultrasound image guidance system that integrates with existing linear accelerators and has the capability to image a variety of abdominal and pelvic cancers. A survey of emerging research shows great progress towards generalized intrafractional ultrasound guidance with robotically controlled ultrasound probe manipulators, robust 3D tissue tracking algorithms, and techniques for incorporating image guidance hardware into the radiotherapy treatment plan.

References

[1] Puck T T and Marcus P I 1956 Action of x-rays on mammalian cells *J. Exp. Med.* **103** 653–66

[2] Fertil B, Reydellet I and Deschavanne P J 1994 A benchmark of cell survival models using survival curves for human cells after completion of repair of potentially lethal damage *Radiat. Res.* **138** 61

[3] Timmerman R D 2008 An overview of hypofractionation and introduction to this issue of seminars in radiation oncology *Semin. Radiat. Oncol.* **18** 215–22

[4] Lo S S *et al* 2010 Stereotactic body radiation therapy: a novel treatment modality *Nat. Rev. Clin. Oncol.* **7** 44–54

[5] Papiez L and Timmerman R 2007 Hypofractionation in radiation therapy and its impact *Med. Phys.* **35** 112–8

[6] Minn A Y *et al* 2009 Pancreatic tumor motion on a single planning 4D-CT does not correlate with intrafraction tumor motion during treatment *Am. J. Clin. Oncol. Cancer Clin. Trials* **32** 364–8

[7] Feng M, Balter J M, Normolle D, Adusumilli S, Cao Y, Chenevert T L and Ben-Josef E *et al* 2009 Characterization of pancreatic tumor motion using cine MRI: surrogates for tumor position should be used with caution *Int. J. Radiat. Oncol.* **74** 884–91

[8] Pawlowski J M, Yang E S, Malcolm A W, Coffey C W and Ding G X 2010 Reduction of dose delivered to organs at risk in prostate cancer patients via image-guided radiation therapy *Int. J. Radiat. Oncol.* **76** 924–34

[9] Gierga D P *et al* 2004 Quantification of respiration-induced abdominal tumor motion and its impact on IMRT dose distributions *Int. J. Radiat. Oncol.* **58** 1584–95

[10] Wu Q J, Thongphiew D, Wang Z, Chankong V and Yin F F 2008 The impact of respiratory motion and treatment technique on stereotactic body radiation therapy for liver cancer *Med. Phys.* **35** 1440–51

[11] Das S *et al* 2014 Comparison of image-guided radiotherapy technologies for prostate cancer *Am. J. Clin. Oncol.: Cancer Clin. Trials* **vol. 37** 616–23

[12] Kitamura K *et al* 2002 Registration accuracy and possible migration of internal fiducial gold marker implanted in prostate and liver treated with real-time tumor-tracking radiation therapy (RTRT) *Radiother. Oncol.* **62** 275–81

[13] Ng J A *et al* 2012 Kilovoltage intrafraction monitoring for prostate intensity modulated arc therapy: first clinical results *Int. J. Radiat. Oncol. Biol. Phys.* **84** e655–61

[14] Simon J *et al* 2010 Epinal #2: 409 patients overexposed during radiotherapy for prostate cancer after daily use of portal imaging controls *Int. J. Radiat. Oncol.* **78** S361

[15] Ten Haken R K *et al* 1991 Treatment planning issues related to prostate movement in response to differential filling of the rectum and bladder *Int. J. Radiat. Oncol. Biol. Phys.* **20** 1317–24

[16] Murphy M J, Eidens R, Vertatschitsch E and Wright J N 2008 The effect of transponder motion on the accuracy of the Calypso electromagnetic localization system *Int. J. Radiat. Oncol. Biol. Phys.* **72** 295–9

[17] Zhu X, Bourland J D, Yuan Y, Zhuang T, O'Daniel J, Thongphiew D, Wu Q J, Das S K, Yoo S and Yin F F 2009 Tradeoffs of integrating real-time tracking into IGRT for prostate cancer treatment *Phys. Med. Biol.* **54** N393

[18] Lattanzi J *et al* 2000 Ultrasound-based stereotactic guidance in prostate cancer—quantification of organ motion and set-up errors in external beam radiation therapy *Comput. Aided Surg.* **5** 289–95

[19] Hatcher D C and Dugoni A A 2010 Operational principles for cone-beam computed tomography *J. Am. Dent. Assoc.* **141** 3S–6S

[20] Little D J, Dong L, Levy L B, Chandra A and Kuban D A 2003 Use of portal images and BAT ultrasonography to measure setup error and organ motion for prostate IMRT: implications for treatment margins *Int. J. Radiat. Oncol. Biol. Phys.* **56** 1218–24

[21] Trichter F and Ennis R D 2003 Prostate localization using transabdominal ultrasound imaging *Int. J. Radiat. Oncol. Biol. Phys.* **56** 1225–33

[22] Western C, Hristov D and Schlosser J 2015 Ultrasound imaging in radiation therapy: from interfractional to intrafractional guidance *Cureus* **7** e280

[23] Langen K M *et al* 2003 Evaluation of ultrasound-based prostate localization for image-guided radiotherapy *Int. J. Radiat. Oncol. Biol. Phys.* **57** 635–44

[24] Van Den Heuvel F *et al* 2003 Independent verification of ultrasound based image-guided radiation treatment, using electronic portal imaging and implanted gold markers *Med. Phys.* **30** 2878–87

[25] Molloy J A *et al* 2011 Quality assurance of U.S.-guided external beam radiotherapy for prostate cancer: report of AAPM Task Group 154 *Med. Phys.* **38** 857–71

[26] Peignaux K *et al* 2006 Clinical assessment of the use of the SonArray system for daily prostate localization *Radiother. Oncol.* **81** 176–8

[27] Scarbrough T J *et al* 2006 Comparison of ultrasound and implanted seed marker prostate localization methods: implications for image-guided radiotherapy *Int. J. Radiat. Oncol. Biol. Phys.* **65** 378–87

[28] Molloy J A, Srivastava S and Schneider B F 2004 A method to compare supra-pubic ultrasound and CT images of the prostate: technique and early clinical results *Med. Phys.* **31** 433–42

[29] Cury F L B *et al* 2006 Ultrasound-based image guided radiotherapy for prostate cancer-comparison of cross-modality and intramodality methods for daily localization during external beam radiotherapy *Int. J. Radiat. Oncol. Biol. Phys.* **66** 1562–7

[30] Lachaine M and Falco T 2013 Intrafractional prostate motion management with the Clarity AutoScan system *Med. Phys. Int. J.* **1** 72–80

[31] Fiandra C *et al* 2014 Impact of the observers' experience on daily prostate localization accuracy in ultrasound-based IGRT with the Clarity platform *J. Appl. Clin. Med. Phys.* **15** 168–73

[32] Mantel F *et al* 2016 Changes in penile bulb dose when using the Clarity transperineal ultrasound probe: a planning study *Pract. Radiat. Oncol* **6** e337–44

[33] Robinson D, Liu D, Steciw S, Field C, Daly H, Saibishkumar E P, Fallone G, Parliament M and Amanie J 2012 An evaluation of the Clarity 3D ultrasound system for prostate localization *J. Appl. Clin. Med. Phys.* **13** 100–12

[34] O'Shea T P, Bamber J C and Harris E J 2016 Temporal regularization of ultrasound-based liver motion estimation for image-guided radiation therapy *Med. Phys.* **43** 455–64

[35] Grimwood A, McNair H A, O'Shea T P, Gilroy S, Thomas K, Bamber J C, Tree A C and Harris E J 2018 In vivo validation of Elekta's Clarity AutoScan for ultrasound-based intrafraction motion estimation of the prostate during radiation therapy *Int. J. Radiat. Oncol. Biol. Phys.* **102** 912–21

[36] Richardson A K and Jacobs P 2017 Intrafraction monitoring of prostate motion during radiotherapy using the Clarity® Autoscan Transperineal Ultrasound (TPUS) system *Radiography* **23** 310–3

[37] Han B *et al* 2018 Evaluation of transperineal ultrasound imaging as a potential solution for target tracking during hypofractionated radiotherapy for prostate cancer *Radiat. Oncol* **13** 151

[38] Schlosser J, Salisbury K and Hristov D 2010 Telerobotic system concept for real-time soft-tissue imaging during radiotherapy beam delivery *Med. Phys.* **37** 6357–67

[39] Sen H T, Bell M A L, Iordachita I, Wong J and Kazanzides P 2013 A cooperatively controlled robot for ultrasound monitoring of radiation therapy *IEEE Int. Conf. on Intelligent Robots and Systems* **2013** 3071–6

[40] Bruder R, Ipsen S, Jauer P, Ernst F, Blanck O and Schweikard A 2013 MO-D-144-02: ultrasound transducer localization using the CyberKnife's x-ray system *Med. Phys.* **40** 405

[41] Brooks R 2011 Intrafraction prostate motion correction using a non-rectilinear image frame *Prostate Cancer Imaging. Image Analysis and Image-Guided Interventions: International Workshop, Held in Conjunction with MICCAI 2011, Toronto, Canada, September 22, 2011, Proceedings Lecture Notes in Computer Science* vol 6963 (Berlin: Springer), 57–9 LNCS

[42] Abramowitz M C *et al* 2012 Noninvasive real-time prostate tracking using a transperineal ultrasound approach *Int. J. Radiat. Oncol.* **84** S133

[43] Yu A S A S, Najafi M, Hristov D H D H and Phillips T 2017 Intrafractional tracking accuracy of a transperineal ultrasound image guidance system for prostate radiotherapy *Technol. Cancer Res. Treat.* **16** 1067–78

[44] Salter B J *et al* 2008 Evaluation of alignment error due to a speed artifact in stereotactic ultrasound image guidance *Phys. Med. Biol.* **53** N437–45

[45] Fontanarosa D, Van Der Meer S, Bloemen-Van Gurp E, Stroian G and Verhaegen F 2012 Magnitude of speed of sound aberration corrections for ultrasound image guided radiotherapy for prostate and other anatomical sites *Med. Phys.* **39** 5286–92

[46] Fontanarosa D *et al* 2013 A speed of sound aberration correction algorithm for curvilinear ultrasound transducers in ultrasound-based image-guided radiotherapy *Phys. Med. Biol.* **58** 1341–60

[47] Hsu A, Miller N R, Evans P M, Bamber J C and Webb S 2005 Feasibility of using ultrasound for real-time tracking during radiotherapy *Med. Phys.* **32** 1500–12

[48] Orton N P, Jaradat H A and Tomé W A 2006 Clinical assessment of three-dimensional ultrasound prostate localization for external beam radiotherapy *Med. Phys.* **33** 4710–7

[49] Fargier-Voiron M *et al* 2014 Impact of probe pressure variability on prostate localization for ultrasound-based image-guided radiotherapy *Radiother. Oncol.* **111** 132–7

[50] Serago C F *et al* 2002 Initial experience with ultrasound localization for positioning prostate cancer patients for external beam radiotherapy *Int. J. Radiat. Oncol. Biol. Phys.* **53** 1130–8

[51] Lediju Bell M A, Sen H T, Iordachita I, Kazanzides P and Wong J 2014 In vivo reproducibility of robotic probe placement for a novel ultrasound-guided radiation therapy system *J. Med. Imaging (Bellingham, Wash.)* **1** 025001

[52] Schlosser J *et al* 2016 Robotic intrafractional US guidance for liver SABR: system design, beam avoidance, and clinical imaging *Med. Phys.* **43** 5951–63

[53] Schlosser J, Salisbury K and Hristov D 2012 Online image-based monitoring of soft-tissue displacements for radiation therapy of the prostate *Int. J. Radiat. Oncol. Biol. Phys.* **83** 1633–40

[54] O'Shea T *et al* 2016 Review of ultrasound image guidance in external beam radiotherapy part II: intra-fraction motion management and novel applications *Phys. Med. Biol.* **61** R90–137

[55] Foster R D, Solberg T D, Li H S, Kerkhoff A, Enke C A, Willoughby T R and Kupelian P A *et al* 2010 Comparison of transabdominal ultrasound and electromagnetic transponders for prostate localization *J. Appl. Clin. Med. Phys.* **11** 57–67

[56] Li M, Ballhausen H, Hegemann N-S, Ganswindt U, Manapov F, Tritschler S, Roosen A, Gratzke C, Reiner M and Belka C 2015 A comparative assessment of prostate positioning guided by three-dimensional ultrasound and cone beam CT *Radiat. Oncol* **10** 82

[57] Fargier-Voiron M *et al* 2016 Evaluation of a new transperineal ultrasound probe for inter-fraction image-guidance for definitive and post-operative prostate cancer radiotherapy *Phys. Medica* **32** 499–505

[58] Van De Voorde L *et al* 2015 Image-guided stereotactic ablative radiotherapy for the liver: a safe and effective treatment *Eur. J. Surg. Oncol.* **41** 249–56

[59] Taguchi H *et al* 2007 Intercepting radiotherapy using a real-time tumor-tracking radiotherapy system for highly selected patients with hepatocellular carcinoma unresectable with other modalities *Int. J. Radiat. Oncol. Biol. Phys.* **69** 376–80

[60] Van Den Begin R *et al* 2014 Impact of inadequate respiratory motion management in SBRT for oligometastatic colorectal cancer *Radiother. Oncol.* **113** 235–9

[61] Eccles C L *et al* 2011 Comparison of liver tumor motion with and without abdominal compression using cine-magnetic resonance imaging *Int. J. Radiat. Oncol. Biol. Phys.* **79** 602–8

[62] Poulsen P R *et al* 2015 Respiratory gating based on internal electromagnetic motion monitoring during stereotactic liver radiation therapy: first results *Acta Oncol. (Madr)* **54** 1445–52

[63] Schweikard A, Shiomi H and Adler J 2004 Respiration tracking in radiosurgery *Med. Phys.* **31** 2738–41

[64] Habermehl D *et al* 2015 Evaluation of inter- and intrafractional motion of liver tumors using interstitial markers and implantable electromagnetic radiotransmitters in the context of image-guided radiotherapy (IGRT)—the ESMERALDA trial *Radiat. Oncol* **10** 143

[65] Poulsen P R *et al* 2014 Kilovoltage intrafraction motion monitoring and target dose reconstruction for stereotactic volumetric modulated arc therapy of tumors in the liver *Radiother. Oncol.* **111** 424–30

[66] Worm E S, Høyer M, Fledelius W and Poulsen P R 2013 Three-dimensional, time-resolved, intrafraction motion monitoring throughout stereotactic liver radiation therapy on a conventional linear accelerator *Int. J. Radiat. Oncol. Biol. Phys.* **86** 190–7

[67] Kitamura K *et al* 2003 Tumor location, cirrhosis, and surgical history contribute to tumor movement in the liver, as measured during stereotactic irradiation using a real-time tumor-tracking radiotherapy system *Int. J. Radiat. Oncol. Biol. Phys.* **56** 221–8

[68] Seppenwoolde Y *et al* 2011 Treatment precision of image-guided liver SBRT using implanted fiducial markers depends on marker-tumour distance *Phys. Med. Biol.* **56** 5445–68

[69] Wunderink W *et al* 2010 Potentials and limitations of guiding liver stereotactic body radiation therapy set-up on liver-implanted fiducial markers *Int. J. Radiat. Oncol. Biol. Phys.* **77** 1573–83

[70] Bloemen-van Gurp E, van der Meer S, Hendry J, Buijsen J, Visser P, Fontanarosa D, Lachaine M, Lammering G and Verhaegen F *et al* 2013 Active breathing control in combination with ultrasound imaging: a feasibility study of image guidance in stereotactic body radiation therapy of liver lesions *Int. J. Radiat. Oncol. Biol. Phys.* **85** 1096–102

[71] Romestaing P *et al* 1997 Role of a 10-Gy boost in the conservative treatment of early breast cancer: results of a randomized clinical trial in Lyon, France *J. Clin. Oncol.* **15** 963–8

[72] Bartelink H *et al* 2007 Impact of a higher radiation dose on local control and survival in breast-conserving therapy of early breast cancer: 10-year results of the randomized boost versus no boost EORTC 22881-10882 trial *J. Clin. Oncol.* **25** 3259–65

[73] Antonini N *et al* 2007 Effect of age and radiation dose on local control after breast conserving treatment: EORTC trial 22881-10882 *Radiother. Oncol.* **82** 265–71

[74] Wong P *et al* 2011 Use of three-dimensional ultrasound in the detection of breast tumor bed displacement during radiotherapy *Int. J. Radiat. Oncol. Biol. Phys.* **79** 39–45

[75] Schlosser J S 2013 Robotic Ultrasound Image Guidance for Radiation Therapy *PhD thesis* Stanford University

[76] Kuhlemann I, Bruder R, Ernst F and Schweikard A 2014 WE-G-BRF-09: force- and image-adaptive strategies for robotised placement of 4D ultrasound probes *Med. Phys.* **41** 523

[77] Ammann N 2012 Robotized 4D Ultrasound for Cardiac Image-Guided Radiation Therapy *Masters thesis* University of Lübeck

[78] Bell M L, Tutkun Sen H, Kazanzides P, Iordachita I, Teboh Forbang R, Boctor E, Lachaine M and Wong J *et al* 2013 SU-E-U-13: repeatability of robotic placement of ultrasound probes for an integrated US-CT approach to image-guided radiotherapy *Med. Phys.* **40** 376

[79] Wu J, Dandekar O, Nazareth D, Lei P, D'Souza W and Shekhar R *et al* 2006 Effect of ultrasound probe on dose delivery during real-time ultrasound-guided tumor tracking *Annual Int. Conf. of the IEEE Engineering in Medicine and Biology—Proc.* 3799–802

[80] Zhong Y *et al* 2013 Assessing feasibility of real-time ultrasound monitoring in stereotactic body radiotherapy of liver tumors *Technol. Cancer Res. Treat.* **12** 243–50

[81] Schlosser J, Salisbury K and Hristov D 2011 Image-based approach to respiratory gating for liver radiotherapy using a telerobotic ultrasound system *Int. J. Radiat. Oncol.* **81** S122

[82] Harris E J, Miller N R, Bamber J C, Symonds-Tayler J R N and Evans P M 2010 Speckle tracking in a phantom and feature-based tracking in liver in the presence of respiratory motion using 4D ultrasound *Phys. Med. Biol.* **55** 3363–80

[83] Bell M A L, Byram B C, Harris E J, Evans P M and Bamber J C 2012 In vivo liver tracking with a high volume rate 4D ultrasound scanner and a 2D matrix array probe *Phys. Med. Biol.* **57** 1359–74

[84] O'Shea T P *et al* 2014 4D ultrasound speckle tracking of intra-fraction prostate motion: a phantom-based comparison with X-ray fiducial tracking using CyberKnife *Phys. Med. Biol.* **59** 1701–20

[85] Bruder R, Ernst F, Schlaefer A and Schweikard A 2009 TH-C-304A-07: real-time tracking of the pulmonary veins in 3D ultrasound of the beating heart *Med. Phys.* **36** 2804

[86] Molloy J A and Oldham S A 2008 Benchmarking a novel ultrasound-CT fusion system for respiratory motion management in radiotherapy: assessment of spatio-temporal characteristics and comparison to 4DCT *Med. Phys.* **35** 291–300

[87] Kubota Y *et al* 2014 A new method for tracking organ motion on diagnostic ultrasound images *Med. Phys.* **41** 092901

IOP Publishing

Principles and Practice of Image-Guided Abdominal Radiation Therapy

Yu Kuang

Chapter 23

Synthetic computed tomography for abdominal image-guided radiation therapy

Richard L J Qiu, Yang Lei, Tonghe Wang, Walter J Curran, Tian Liu and Xiaofeng Yang

This chapter summarizes recent progress in synthetic computed tomography (CT) generation for abdominal image-guided radiation therapy (IGRT). It covers two main research topics: magnetic resonance imaging (MRI)–based synthetic CT generation and cone-beam CT (CBCT)-based synthetic CT generation. Within each topic, related literature is discussed. Emphasis is given to the methods adapting the frameworks of deep learning (DL). Specifically, comparisons are provided among DL-based methods for abdominal MRIs and CBCTs. Discussions are focused on the highlights and challenges of those synthetic CT generation methods. Finally, future research efforts are suggested to realize the full clinical implementation of synthetic CT for abdominal IGRT.

23.1 Introduction

Previous chapters review the current practice of image-guided radiation therapy (IGRT) in abdominal radiation therapy treatment as well as recent advances in adapting new imaging modalities for abdominal IGRT. The current standard of using cone-beam computed tomography (CBCT) as IGRT [1] provides better target alignment than using traditional 2D orthogonal radiograph pairs. However, CBCT has limited soft tissue contrast, which gives poor target visibility for abdominal cancer. Besides, it does not address the target motion induced by internal organ motion, which prevents the reduction of the planning target volume (PTV) margin that could further expand the therapeutic ratio of the radiation therapy. Recent work of ultrasound (US)-guided IGRT [2] and magnetic resonance imaging (MRI)–guided IGRT [3, 4] offers better alternatives than CBCT-guided IGRT. Both US and MRI have superior soft tissue contrast to computed tomography (CT) and CBCT. They not only help identify the abdominal lesions better and subsequently provide better patient alignment but also improve the delineations of target and organs at risk (OARs). Real-time intrafraction motion tracking is also feasible with custom-made US probes and MRI-guided teletherapy units such as the MRI–linear accelerator [5], which makes reducing the PTV margin possible. Therefore, it is ideal to treat abdominal cancer with US-guided IGRT or MRI-guided IGRT.

However, US and MRI do not carry the electron density information that is essential for the dose calculation of photon RT and proton RT in modern treatment planning systems. As such, in order to obtain the electron density information, a simulation CT is still required for treatment planning. To enable US-guided or MRI-guided IGRT, US-CT or MRI-CT image registrations are needed to transfer contours from planning CT to daily IGRT images. However, this additional step of image registration brings extra complexity and uncertainty with a systematic error of ~2–5 mm depending on the treatment sites [6], which could lead to geometric misalignment and additional PTV margin. Therefore, the single image modality treatment workflow has been proposed to circumvent the registration-induced error [7–11]. It would simplify the entire treatment, reduce overall cost, and avoid a CT radiation dose to patients. Additionally, with the increased availability of commercial MRI simulators [12] and the development of new MRI–IGRT techniques [13], high precision in target definition is the future for proton therapy. It means a solution to calculate dose with MRI only is greatly demanded.

To eliminate the need for CT, the idea of generating synthetic CT (sCT) from either MRI or US was proposed. Once achieved, treatment planning can be done on the sCT generated from the simulation MRI or US. Moreover, sCT can be created from daily IGRT imaging. Subsequently, accurate daily RT dose estimation can give quantitative assessments on PTV coverage and OAR doses. It would be a substantial upgrade from the current way of visual inspection of daily patient alignment that is prone to subjective bias and uncertainties. It would also lay the foundation for further treatment optimization such as adaptive radiation therapy (ART) [14]. For those reasons, research in sCT generation has drawn great attention in recent years.

In this chapter, we focus on studies of MRI-based sCT generation at the site of the abdomen. Compared with other sites, the abdomen is particularly challenging given the fact that significant artifacts can present in IGRT images when intrinsic organ motion is not properly managed. It is demanding for algorithms to make accurate sCT predictions if the adverse effect of those artifacts is not mitigated. Additionally, sCT generation algorithms confront difficulties because of the wide range of electron density/CT Hounsfield unit (HU) values attributed from the air cavities and small bony structures in the abdominal regime. In the following subsections, different methods of sCT generation are reviewed. Comparisons of the pros and cons are summarized in the discussion section.

As a side note, US-based sCT generation is intentionally left out. US-guided IGRT images have limited fields of view and inherent image artifacts. Its image quality is greatly affected by device setup condition and patient anatomy variations (patient size, air cavities, bone, etc). All those factors impose a huge challenge in deriving sCT from those images. Deformable registration methods [15] were introduced but limited to only prostate sites. It is difficult to use the same approach for abdominal sites.

CBCT-based sCT generation has been covered in chapter 18 of this book. Here a brief review of the recent papers done for the abdomen site is offered.

23.2 MRI-based sCT generation

Currently, MRI is routinely performed for patients with abdominal cancer who undergo RT because of its superior soft tissue contrast, which helps the tumor visualization and soft tissue delineation. Furthermore, MRI is capable of capturing a patient's breathing motion with great quality [16, 17]. For instance, MRI-guided RT can achieve gated RT treatment by checking the real-time tumor and/or close by surrogate locations [18]. Those technical advances enable stereotactic radiation therapy (SBRT), which delivers more conformal dose on targets. As such, those potential applications drive the need to have sCT for MRI-only treatment planning. In the literatures, the methods of generating sCT from MRIs can be divided into three categories [19]: segmentation-based methods, atlas-based methods, and machine learning-based methods.

23.2.1 Segmentation-based methods

The principle of segmentation-based methods is assigning bulk densities to different tissue groups classified either from CT or MRI images [20–24]. The typical workflow is shown in figure 23.1 [21].

23.2.1.1 Methods
MRI images are acquired at exhale or inhale breath hold for the patients. Breath hold CT scans are performed as well to get the electron density values for different tissue types. First, image preprocessing is typically done to address the residual intensity nonuniformity and reduce image noises. MRI difference images are created to enhance the differentiations between different tissue types. Then fuzzy C-means algorithms, which can bypass manual delineation, are applied to automatically segment MRI images into predetermined classes such as air, lungs, fat, high-density tissue, and spine. Additional efforts of localizing air, lung, and spine could be done to improve the results. Finally, predetermined tissue densities are assigned to each voxel by using the weighed probability of each tissue class. The sCT is eventually generated.

23.2.1.2 Results
Figure 23.2 shows the generated sCT (middle column), side by side with the water MRI and CT from Reference [24]. Because the HU numbers on the sCT were piecewise constant, details were lost in the sCT when compared with the standard CT.

The mean absolute error (MAE) between the sCT and standard CT HU values of various tissue types was in the range of around 20–50 HUs. However, large deviations were seen in the lung and spine. Dose of testing volumetric-modulated arc therapy (VMAT) photon RT plans were calculated on the sCT and standard CT.

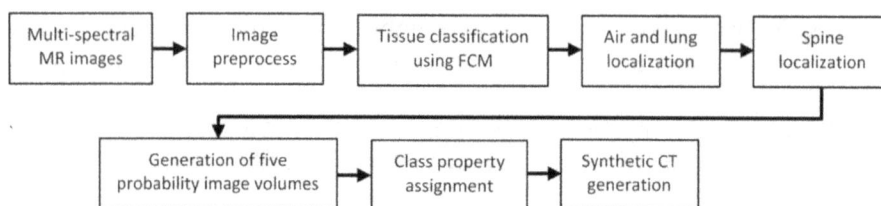

Figure 23.1. A typical workflow of the tissue classification process for sCT generation. FCM, fuzzy C-means algorithms; sCT, synthetic CT. [21] John Wiley & Sons.

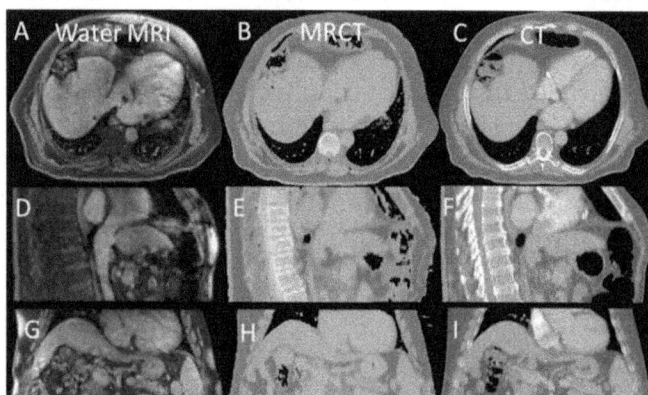

Figure 23.2. Co-registered axial, sagittal, and coronal views of the water MRI [(A), (D), and (G)] shown side by side with similarly windowed sCTs (MRI-derived synthetic CT volumes) [(B), (E), and (H)] and standard CTs [(C), (F), and (I)]. Reprinted from [24]. © IOP Publishing. Reproduced with permission. All rights reserved.

It was found that the difference was clinically insignificant [21, 24]. An example is illustrated in figure 23.3, from reference [24].

This approach is limited by the requirement to predetermine segmentation classes. Its clinical feasibility for online ART has not been demonstrated. Besides, air and bone segmentation from MRI images are quite challenging when misclassifications are seen. It may be able to produce reliable dosimetrical information for full arc VMAT plans. However, it is unknown whether this method would give accurate enough dosimetry in proton therapy, in which the range uncertainties matter a lot.

23.2.2 Atlas-based methods

Atlas-based methods use either a single atlas or a database of co-registered MRI and CT pairs to obtain an sCT [25–28]. The schematic workflow is shown in figure 23.4, from reference [25]. First, MRI and CT pairs are collected and registered. The subject MRI images are deformably registered with the atlas MRI images. Then, the resulting deformation vector fields are applied to the paired atlas CT images. In the end, the HU values from the deformed atlas CT image are then assigned to the voxels of the subject MRI images, which produces the sCT in the subject MRI coordinates. For multi-atlas approaches, the CT numbers from different atlases are weighted averaged via computing the voxel-wise median using a probabilistic

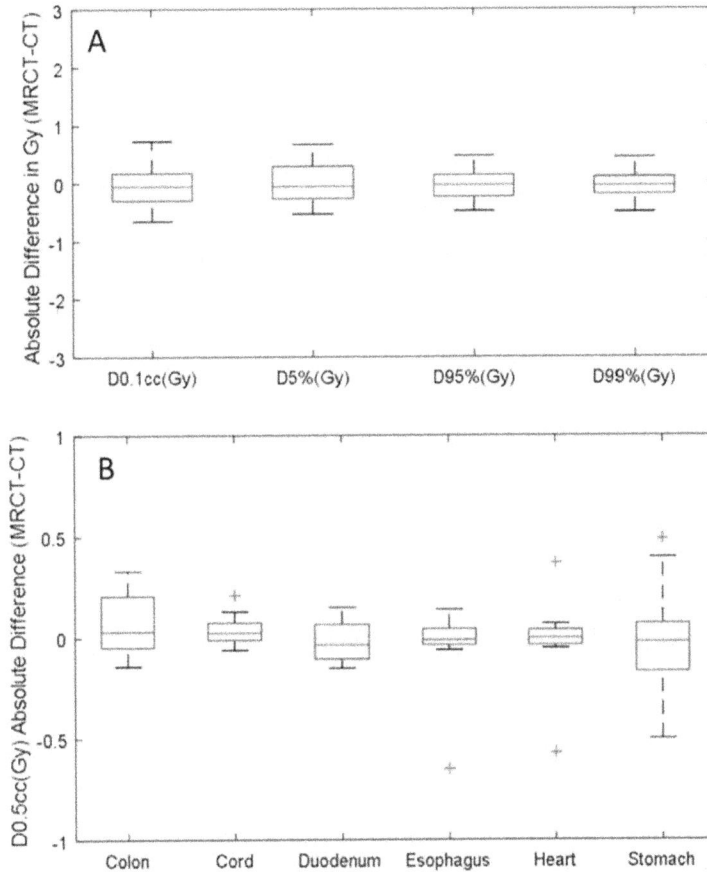

Figure 23.3. Clinically relevant dose–volume histogram (DVH) metric comparison for the PTVs (A) and OARs (B) for all 16 patients in the study. Outliers of the distributions are shown as plus signs. OAR, organs at risk; PTV, planning target volume. Reprinted from [24]. © IOP Publishing. Reproduced with permission. All rights reserved.

Bayesian framework, arithmetic mean process, or pattern recognition with Gaussian process or a local image similarity measure to generate the final sCT. It is worth pointing out that other, more complex approaches that are combined with machine learning techniques can be found in reference [27, 28]. The schematic flow chart of one example is shown in figure 23.5, which is from reference [28].

The atlas-based methods have been used mainly in the brain, head and neck (H&N), and prostate regime [25–28], in which reasonable image registrations are achievable because the anatomical variations are limited. In contrast, because of the organ motion-related large anatomy changes and image artifacts, robust image registrations in the abdominal area are difficult to get. Until now, there has not been any literature report in generating sCT using atlas-based methods for the abdominal site. Other than those obstacles, another drawback of this method is that it is computationally expensive and time consuming. Therefore, it is not suitable to be integrated into online ART process.

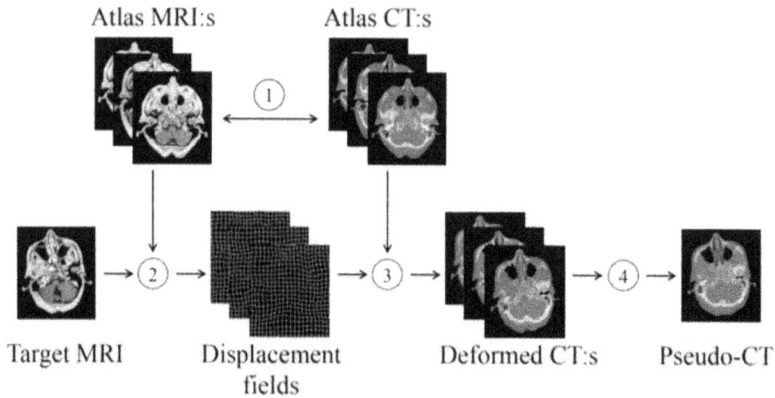

Atlas MRI:s   Atlas CT:s

Target MRI   Displacement   Deformed CT:s   Pseudo-CT
                       fields

Figure 23.4. An outline of the atlas-based regression algorithm for sCT/pseudo-CT generation: (1) MRI and CT pairs are collected and registered, (2) each atlas MRI is registered to the target MRI, (3) the resulting transformation is applied to the corresponding CTs, and (4) the set of deformed CTs are fused to a single sCT/pseudo-CT. Reprinted from [25]. © IOP Publishing. Reproduced with permission. All rights reserved.

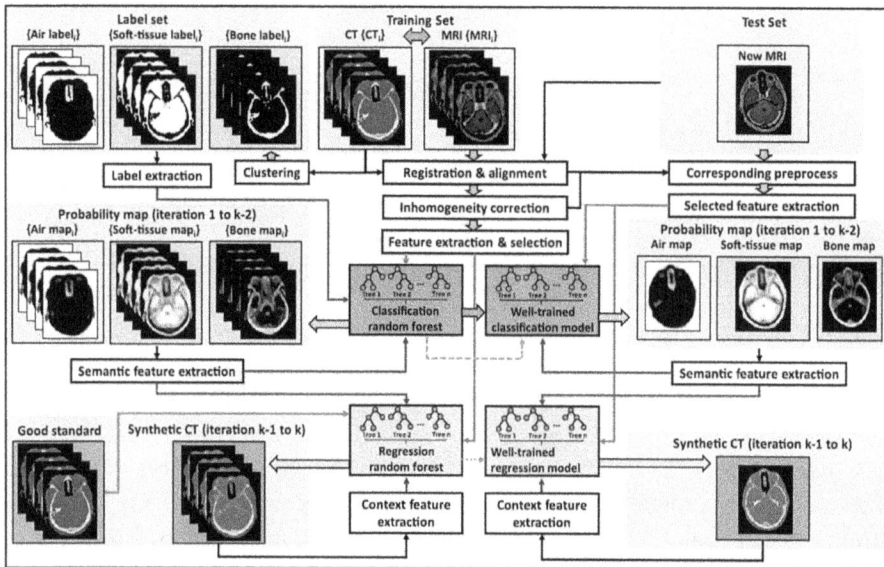

Figure 23.5. Schematic flow chart of the proposed algorithm for MRI-based sCT generation. The left part of this figure shows the training stage of the proposed method, which consists of classification random forest training, semantic feature extraction, context feature extraction, and regression random forest training. The right part of this figure shows the synthesizing stage in which a new MRI image follows a similar sequence to the training stage to generate a sCT image. Reprinted from [28]. © IOP Publishing. Reproduced with permission. All rights reserved.

23.2.3 Machine learning–based methods

Machine learning–based methods have gained increasing popularity in generating sCT. Sophisticated algorithms such as random forest and deep learning (DL) have shown promising results in generating brain, H&N, pelvic, and abdominal sCT [11,29–35]. Particularly, DL algorithms stand out because they can automatically extract underlying features from datasets to produce superior results. The overall idea of DL-based methods is to train a model by database of co-registered MRI and CT pairs. Once trained, it can produce an sCT in a few minutes when a new MRI is fed into the model. sCT images have the same structural information as MRIs, and the intensities are in HU values close to standard CT. Here we focus on papers specifically for abdominal sCT generation.

23.2.3.1 Methods

A sketch of the algorithm from reference [32, 33, 35] is illustrated in figure 23.6. The authors did not use the traditional convolution neural networks because it would give suboptimal results due to large organ motion and imperfect CT-MRI image registrations [36]. In the abdomen, CT-MR registrations tend to have substantial errors for places such as bone, bowel, and body surface. Another issue is that MRI has more structural information and contrast in soft tissue but less information at bone and air interfaces compared with CT. MR-to-CT generative adversarial networks (GANs) [37, 38] tend to generate erroneous prediction when there are many-to-one or one-to-many mappings. Therefore, the authors in reference [32, 33, 35] proposed a modified 3D dense-cycle GAN [39] that contains several dense blocks in the generator, which combine low and high frequency information to effectively represent image patches between MRI and CT. A 3D image patch was used as the input of the model to address the problem of possible cross-slice discontinuity [9]. A compound loss function was also employed to effectively differentiate the structure boundaries with significant HU variations and to retain the sharpness of the sCT image:

$$G = \arg_G^{\min} \{\lambda_{\text{adv}} L_{\text{adv}}(Z) + \lambda_{\text{distance}} L_{\text{distance}}(Z, Y)\} \tag{23.1}$$

where λ_{adv} and $\lambda_{\text{distance}}$ are balancing parameters. The adversarial loss function is defined as

$$L_{\text{adv}}(Z) = \text{MAD}(D(Z), 1) \tag{23.2}$$

The distance loss $L_{\text{distance}}(Z, Y)$ [40], an l_p-norm ($p = 1.5$) distance, termed mean p distance (MPD), was introduced. The generators are optimized as follows:

$$
(G_{\text{CT–MR}}, G_{\text{MR–CT}}) = \underset{G_{\text{CT–MR}}, G_{\text{MR–CT}}}{\arg\ \min}
$$

$$
\left\{
\begin{aligned}
&\lambda_{\text{adv}} MAD(D_{\text{MR}}(G_{\text{CT–MR}}(I_{\text{CT}})), 1) + \lambda_{\text{MPD}}^{\text{cycle}} \| G_{\text{MR–CT}}(G_{\text{CT–MR}}(I_{\text{CT}})), I_{\text{CT}} \|_p^p \\
&+ \lambda_{\text{GDL}}^{\text{cycle}} GDL(G_{\text{MR–CT}}(G_{\text{CT–MR}}(I_{\text{CT}})), I_{\text{CT}}) \\
&+ \lambda_{\text{MPD}}^{\text{fake}} \| G_{\text{CT–MR}}(I_{\text{CT}}), I_{\text{MR}} \|_p^p + \lambda_{\text{MPD}}^{\text{fake}} GDL(G_{\text{CT–MR}}(I_{\text{CT}}), I_{\text{MR}}) \\
&+ \lambda_{\text{adv}} MAD(D_{\text{CT}}(G_{\text{MR–CT}}(I_{\text{MR}})), 1) \\
&\quad + \lambda_{\text{MPD}}^{\text{cycle}} \| G_{\text{CT–MR}}(G_{\text{MR–CT}}(I_{\text{MR}})), I_{\text{MR}} \|_p^p \\
&+ \lambda_{\text{GDL}}^{\text{cycle}} GDL(G_{\text{CT–MR}}(G_{\text{MR–CT}}(I_{\text{MR}})), I_{\text{MR}}) \\
&+ \lambda_{\text{MPD}}^{\text{fake}} \| G_{\text{MR–CT}}(I_{\text{MR}}), I_{\text{CT}} \|_p^p + \lambda_{\text{MPD}}^{\text{fake}} GDL(G_{\text{MR–CT}}(I_{\text{MR}}), I_{\text{CT}})
\end{aligned}
\right\} \tag{23.3}
$$

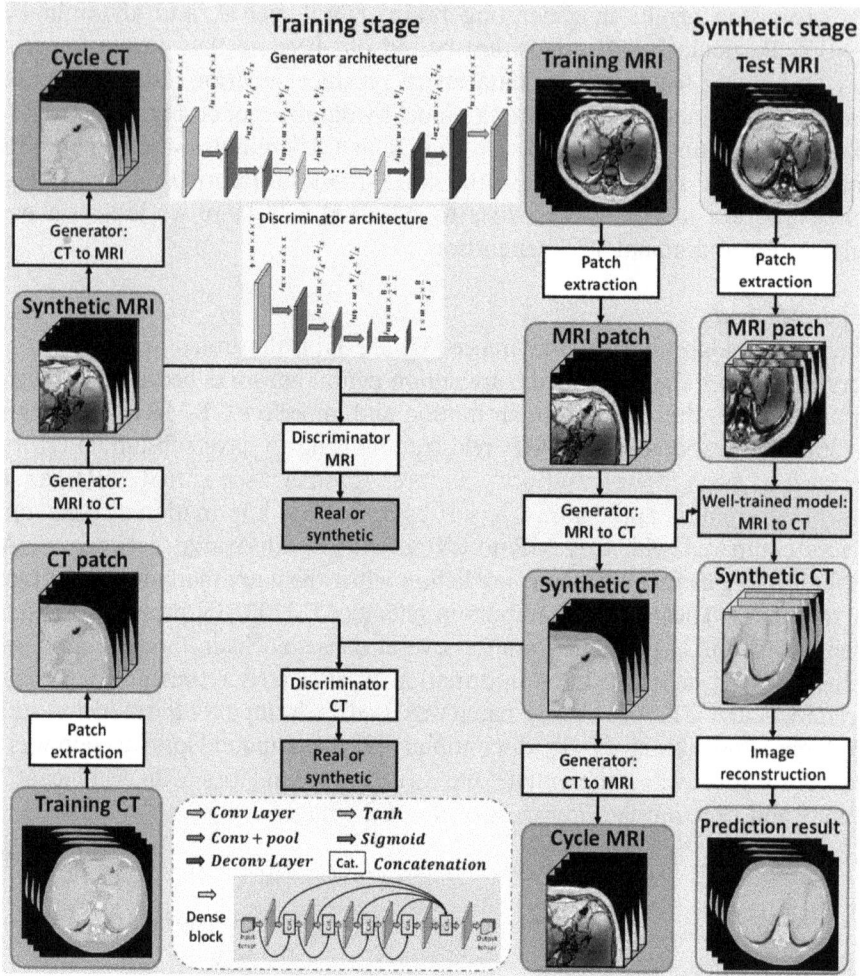

Figure 23.6. Schematic flow chart of the proposed algorithm for MRI-based sCT generation. The training stage consists of four generators and two discriminators. Each generator includes several dense blocks. The right side is the synthesizing stage, in which a new MRI image is fed into the trained model to generate the sCT image. Reprinted from [33]. © IOP Publishing. Reproduced with permission. All rights reserved.

$$GDL(Z, Y) = \sum_{i,j,k} \left\{ \begin{array}{l} \left\| \left| Z_{i,j,k} - Z_{i-1,j,k} \right| - \left| Y_{i,j,k} - Y_{i-1,j,k} \right| \right\|_2^2 + \\ \left\| \left| Z_{i,j,k} - Z_{i,j-1,k} \right| - \left| Y_{i,j,k} - Y_{i,j-1,k} \right| \right\|_2^2 \\ + \left\| \left| Z_{i,j,k} - Z_{i,j,k-1} \right| - \left| Y_{i,j,k} - Y_{i,j,k-1} \right| \right\|_2^2 \end{array} \right\} \quad (23.4)$$

where $\|\cdot\|_p^p$ denotes the l_p-norm and $GDL(\cdot)$ denotes the gradient descent loss function [41]. $\lambda_{\text{MPD}}^{\text{cycle}}$, $\lambda_{\text{GDL}}^{\text{cycle}}$, $\lambda_{\text{MPD}}^{\text{fake}}$, $\lambda_{\text{GDL}}^{\text{fake}}$, $\lambda_{\text{MPD}}^{\text{cycle}}$, $\lambda_{\text{GDL}}^{\text{cycle}}$, $\lambda_{\text{MPD}}^{\text{fake}}$, $\lambda_{\text{GDL}}^{\text{fake}}$ are regularization parameters for different regularizations. The discriminator loss is computed by MAD between the discriminator results of input synthetic and real images. To update all the hidden layers' kernels, the Adam gradient descent method is applied to minimize both generator loss and discriminator loss. In the final synthesizing stage, patches of the new arrival image are fed into the MRI-to-CT generator to obtain the sCT patch end-to-end mapping. Then, the final sCT image is obtained by patch fusion.

23.2.3.2 Results

Table 23.1 lists the statistics for the MAE, peak signal-to-noise ratio (PSNR), and normalized cross correlation (NCC) for the total of twenty-one patients. MAE represents the discrepancies between the predictions and the reference HU numbers. PSNR is the ratio between the maximum possible power of a signal and the power of corrupting noise that affects the fidelity of its representation. NCC is a measure of similarity between two series as a function of the displacement of one relative to the other. The three metrics are defined as

$$\mathrm{MAE} = \mid I_{\mathrm{CT}} - I_{\mathrm{sCT}} \mid / C \tag{23.5}$$

$$\mathrm{PSNR} = 10\log_{10}\left(\frac{Q^2}{\parallel I_{\mathrm{CT}} - I_{\mathrm{sCT}} \parallel_2^2 / C}\right) \tag{23.6}$$

$$\mathrm{NCC} = \frac{1}{C}\sum_{x,\,y,\,z}\frac{\left(I_{\mathrm{CT}}(x,\,y,\,z) - \mu_{\mathrm{CT}}\right)\left(I_{\mathrm{sCT}}(x,\,y,\,z) - \mu_{\mathrm{sCT}}\right)}{\sigma_{\mathrm{CT}}\sigma_{\mathrm{sCT}}} \tag{23.7}$$

where I_{CT} is the HU value of the ground truth CT image, I_{sCT} is the HU value of corresponding sCT image, Q is the maximal HU value between I_{CT} and I_{sCT}, and C is the number of voxels in the image. μ_{CT} and μ_{sCT} are the mean of the CT and PCT image, respectively. σ_{CT} and σ_{sCT} are the standard deviation of the CT and PCT image, respectively.

Figure 23.7 gives transversal, sagittal, and coronal views of an sCT example image next to the corresponding MR and CT. Except for the discrepancy at the spinous processes and body surface, the absolute HU difference between CT and sCT is small. A large discrepancy could be seen at the aorta site because of the presence of contrast during CT acquisition. Overall, the DL-based algorithm is able to predict accurate HU values.

Table 23.1. Statistics for the MAE, PSNR, and NCC values. Reprinted from [33]. © IOP Publishing. Reproduced with permission. All rights reserved.

	Mean (±SD)	Median	Min	Max
MAE (HU)	72.87±18.16	66.46	43.74	126.53
PSNR (dB)	22.65±3.63	23.35	13.46	28.25
NCC	0.92±0.04	0.93	0.81	0.97

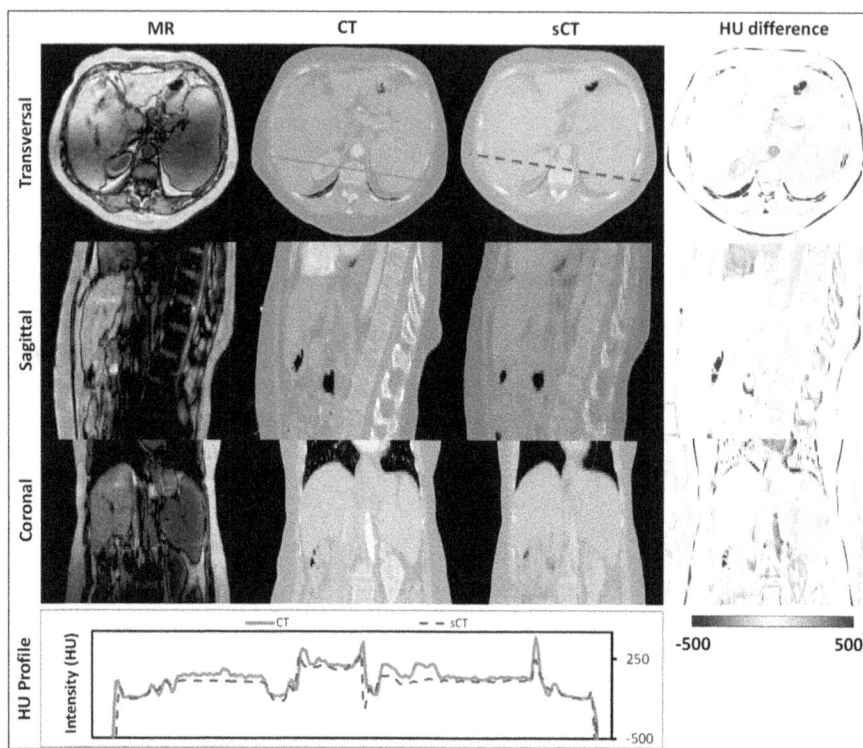

Figure 23.7. The transverse, sagittal, and coronal images of a representative patient. MR, CT, and sCT images and the HU difference between CT and sCT are presented. The CT and sCT voxel-based HU profiles are shown at the bottom to demonstrate the HU values highlighted in the solid and dash line in the transverse images. Figure from [32]. With permission from British Institute of Radiology.

23.2.3.3 Photon RT dose calculation on sCT

To have enough axial slices, which is necessary for dose calculation, original CT slices were added to the sCT set to create a data set of equal size. Two full arc coplanar beams were used for treatment planning. The prescription is 45 Gy in three fractions and normalized to 95% of the PTV receiving the prescription dose. Figure 23.8 gives an example of dose calculation on CT and sCT. Gamma analyses with 1%/1 mm criteria were carried out on the coronal plane at the level of the treatment isocenter, which were all above 99%. The differences between sCT and CT for liver V_{15} and stomach and bowel V_{20} and differences in PTV, OAR D_{min}, D_{10}, D_{50}, D_{95}, and D_{max} were evaluated. The maximum absolute mean dose difference among all the DVH metrics of PTV was 0.16 Gy and 0.14 Gy for the OARs. The absolute dose differences of other DVH metrics were all less than 0.3 Gy. Therefore, the dose calculation differences are negligible from the clinical perspective. It demonstrated that reliable dose calculation could be derived from sCT generated using MRI.

The authors further compared their method to two other methods including a 3D fully convolutional neural network (FCN) by Nie *et al* [38] and a GAN network

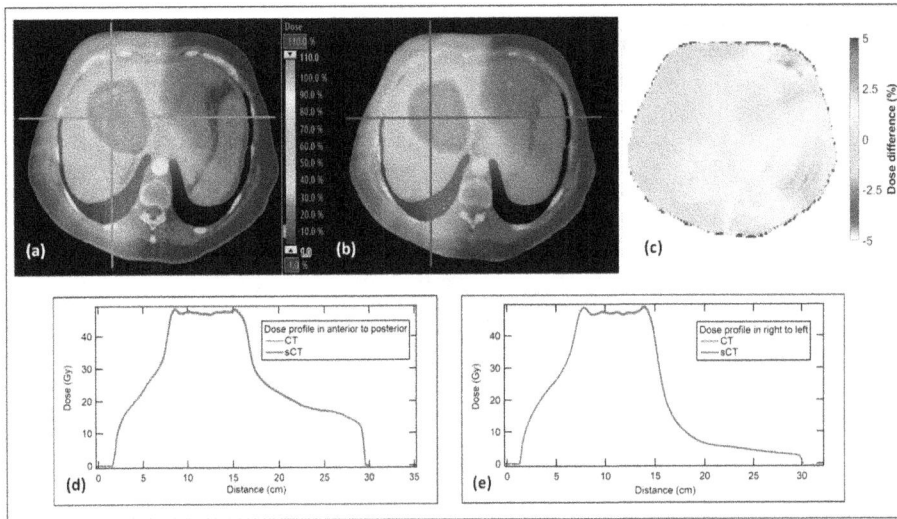

Figure 23.8. (a) Dose distribution calculated from the original CT. (b) Dose distribution calculated from sCT. (c) Dose difference (%) distribution. (d) Dose profiles of CT and sCT in anterior to posterior direction. (e) Dose profiles of CT and sCT in right to left direction. Figure from [32]. With permission from British Institute of Radiology.

done by the same group [42] on the same patient cohort. Through extracting patch-based deep features, FCN preserved the spatial information in the predicted sCT images. One step further, to produce more realistic sCT, Nie *et al* [39] used 3D FCN as a generator in a GAN framework by introducing an additional adversarial loss. It turned out the authors' method has better results, of which the generated sCT had more definitive boundaries and more accurate HU values. The MAE of the authors' methods was 72.87 ± 18.16 HU compared with 94.34 ± 21.06 and 86.40 ± 13.95 for the FCN and GAN based methods. In dose evaluation, they had smaller mean absolute dose differences for most of the DVH endpoints. However, the difference among the three methods were clinically insignificant, and all three were feasible for photon SBRT plan dose calculation.

23.2.3.4 Proton RT dose calculation on sCT
The dose calculation for proton therapy imposes higher demand on the quality of sCT because the required stopping power information is much more sensitive to local mismatch and HU accuracy. Liu *et al* [33] tested the proton RT dose calculation on their sCT generated by the proposed modified 3D dense-cycle-GAN. All proton plans were prescribed to 45 Gy in 25 fractions and normalized to 98% of the PTV receiving the full prescription dose. Two proton beams were used for treatment planning, and the beam angles were chosen to have minimal tissues in the beam path, avoiding hitting bone, bowel gas, and areas with large motion.

Overall, the generated sCT had shown decent accuracy in dose calculation. Mean gamma analysis pass rates of 1 mm/1%, 2 mm/2%, and 3 mm/3% criteria with 10%

dose threshold were 90.76±5.94%, 96.98±2.93%, and 99.37±0.99%, respectively, among 20 patients. Those numbers were similar to that of the pelvis study by Maspero *et al* [43], with a 98.4% pass rate of 2 mm/2% criteria, and the brain and prostate study by Koivula *et al* [44], with a 91% pass rate of 1 mm/1% criteria.

However, larger discrepancies were seen compared with the photon plan results. Figure 23.9 gives examples of dose calculations from plans of two patients. Clearly, it is easy to see that the voxel dose differences were more than 5% at the distal edge and tissue-air interface, such as the rib/lung cavity boundary. It could be due to HU prediction inaccuracy or the rib displacement between the CT and MRI scans. Figure 23.10 shows dosimetrical metrics of PTV, liver, and bowel. As shown, one outlier, plotted as the black "O" marker, has relatively large dose differences in PTV D_{max} and D_{95}. The DVH differences in liver were higher because of HU prediction inaccuracy in bone and liver. The major reason was imperfect CT/MRI registration

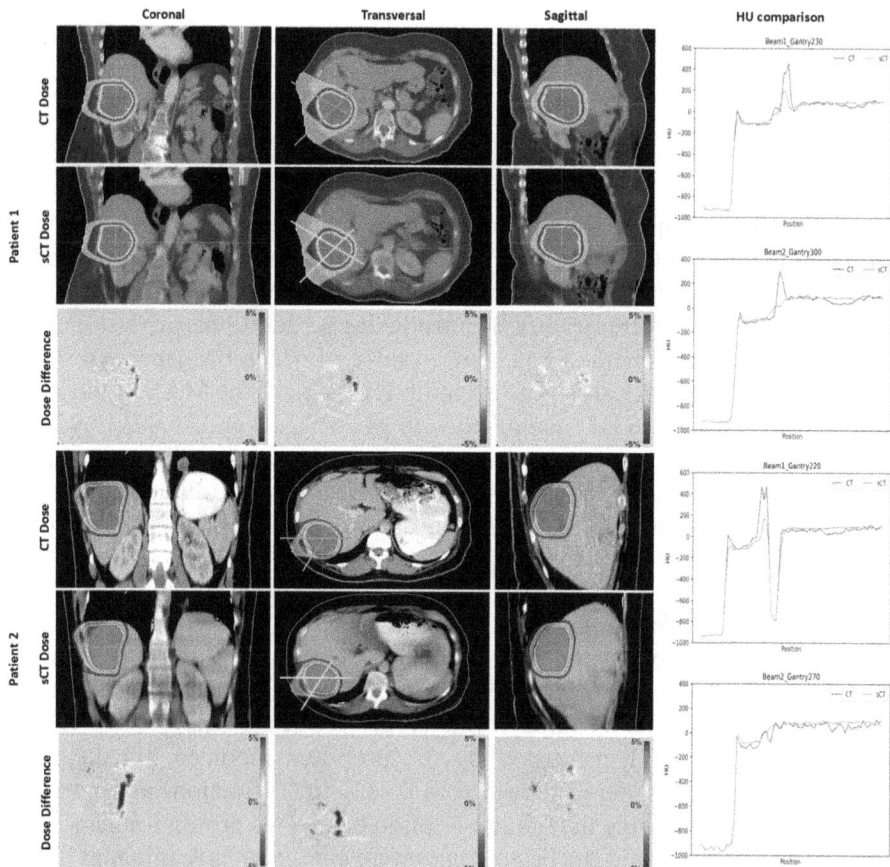

Figure 23.9. From left to right: coronal, transversal, and sagittal view. The dose differences between plans from two patients were demonstrated on original CT and sCT. Reprinted from [33]. © IOP Publishing. Reproduced with permission. All rights reserved.

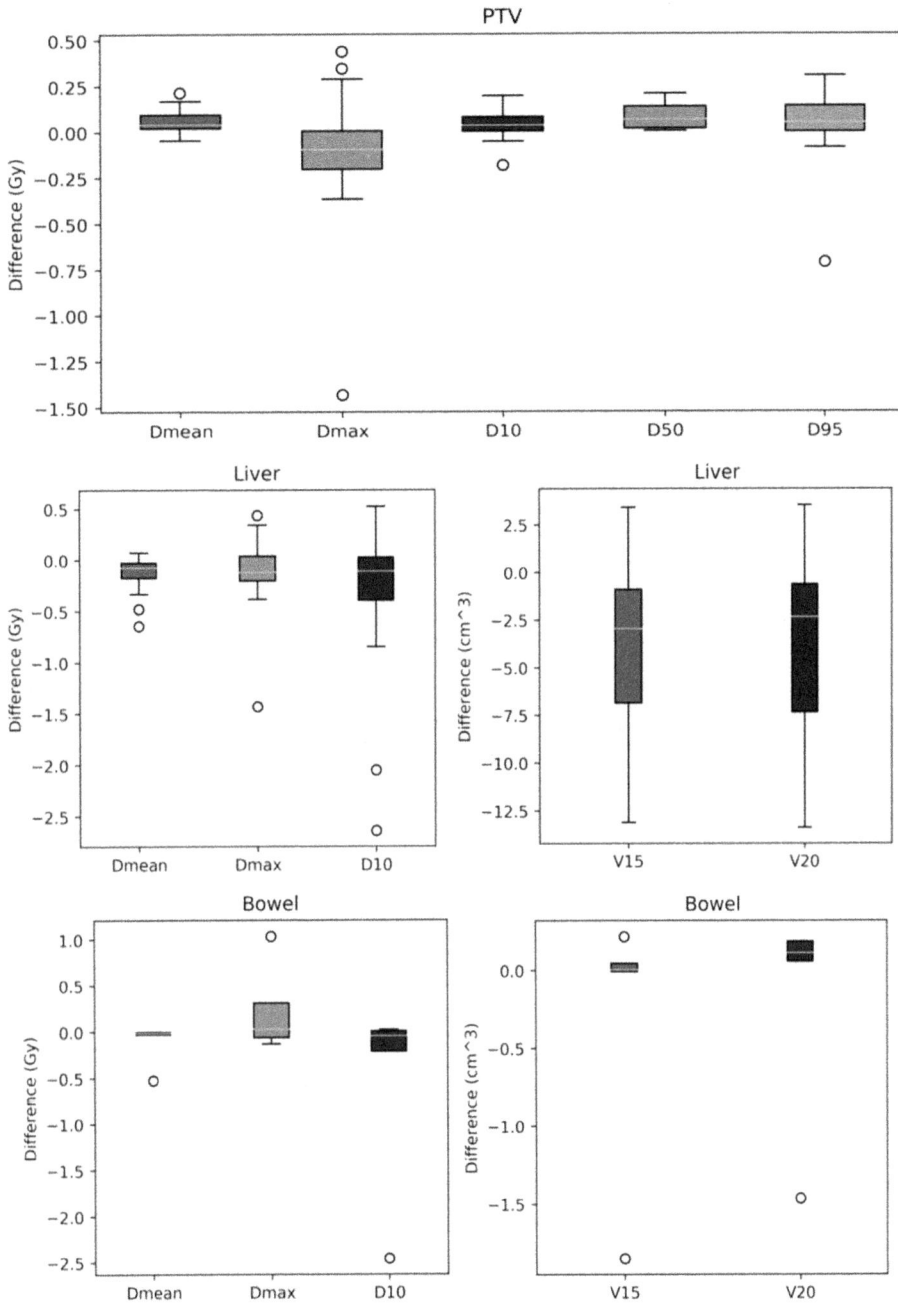

Figure 23.10. Box plots of DVH differences between sCT and CT for the PTV and OARs. The central orange line indicates the median value, and the borders of the box represent the 25th and 75th percentiles. The outliers are plotted by the black O marker. Reprinted from [33]. © IOP Publishing. Reproduced with permission. All rights reserved.

in the ground truth data set. Another factor is the proton beam overshooting, which is caused by the overall underestimation of HU values in sCT. Therefore, the accuracy of the dose calculation significantly depended upon the beam range accuracy.

The authors assessed the range uncertainty stemmed from the uncertainty in HU values of the sCT. As such, range evaluation was performed by deriving the line dose along the beam-line direction along the isocenter. The proton beam range was defined at the 80% of the spread-out Bragg peak plateau dose at the distal range. The range difference and relative range difference between planning CT and sCT were calculated by

$$\text{Range difference } = R_{80}\text{sCT} - R_{80}\text{CT} \tag{23.8}$$

$$\text{Relative range difference } = (R_{80}\text{sCT} - R_{80}\text{CT})/R_{80}\text{CT} \times 100\% \tag{23.9}$$

and compared with the Harvard Massachusetts General Hospital (MGH) [45] uncertainty criteria:

$$\text{Uncertainty } < 3.5\% R_{80}\text{CT} + 1 \text{ mm} \tag{23.10}$$

Other institutions such as MD Anderson and University of Pennsylvania apply looser criteria (3.5%+3 mm), whereas the University of Florida has tighter criteria (2.5%+1.5 mm). The results showed that most of the range displacements were within the MGH range uncertainty criteria acceptance level except two outlier beams, as shown in figure 23.11. The median absolute range difference was 0.17 cm. The maximum difference was found to be 0.56 cm. They were both higher than that in the brain study by Pileggi *et al* [46] (median of 0.05 cm and maximum of 0.44 cm) and that in the pelvis study [43] (average median of 0.01 cm). It was thought to be caused by the mismatch between CT and sCT because of the organ motion. Therefore, it is still not conclusive that one can use abdominal sCT for proton therapy dose calculation. More research is warranted.

23.3 CBCT-based synthetic CT generation

CBCT-based synthetic CT generation has been discussed in Chapter 18 of this book. Most of the early studies were done on brain and pelvis CBCT [47–56]. Overall, learning-based methods offered better sCT than conventional CBCT correction methods [57, 58]. Most recently, Liu *et al* [59] proposed a DL-based method to produce sCT for pancreatic ART. A modified cycle-GAN algorithm was used. The schematic flowchart can be found in figure 18.2. It was found that sCTs gave significant improvement over original CBCTs in dosimetry accuracy and were close to planning CT for photon dose calculation. As reported by Liu *et al* [59], with the same SBRT plan for each testing patient, no significant differences ($P > 0.05$) were observed in the PTV and OAR DVH metrics between the CT- and sCT-based plans, whereas significant differences ($P < 0.05$) were found between the CT- and the CBCT-based plans. Thus, it is promising to use sCT for accurate daily dose calculation, which makes ART possible.

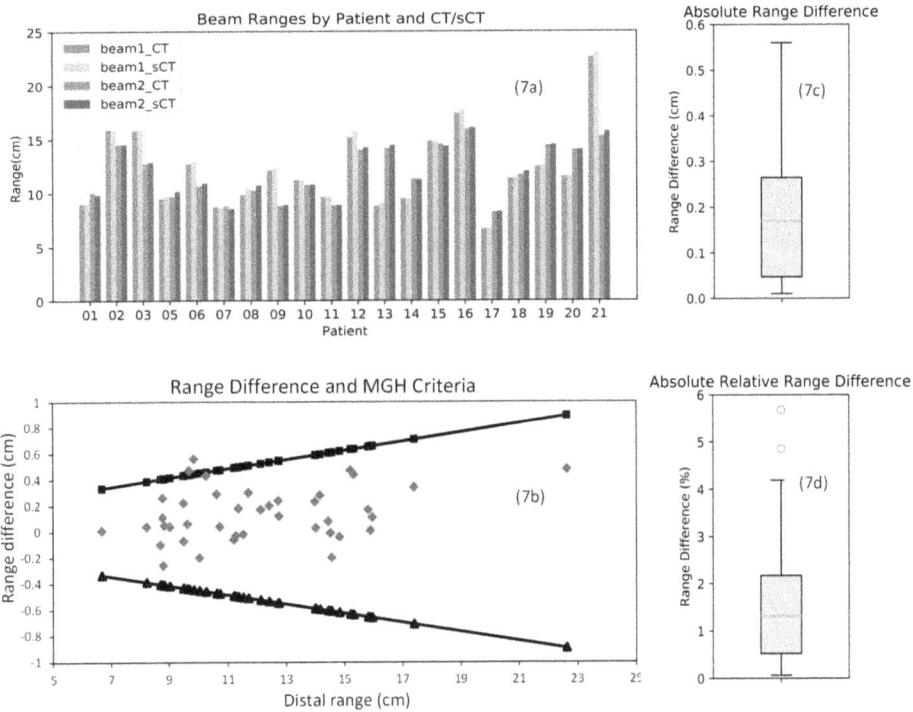

Figure 23.11. Range comparison between the plans created on CT and sCT. (a) Beam ranges of each beam of the entire patient cohort. (b) The red rhombus marker shows the distribution of the range differences as a function of the actual range value from the plan calculated on the original CT. The black square and triangle markers and the black lines represent the upper and lower limit for the MGH range uncertainty criteria. (c) Box plot of absolute range difference. (d) Box plot of absolute relative range difference. Reprinted from [33]. © IOP Publishing. Reproduced with permission. All rights reserved.

23.4 Discussion

The atlas-based methods are unfit for the task of generating abdominal sCT because of organ motion-related large anatomy changes and image artifacts. It is also computationally expensive and time consuming. There is not any existing literature for abdominal sCT using atlas-based methods.

Segmentation-based methods, although shown to be able to produce reliable dosimetrical information for full arc VMAT plans, have several inherent limitations. Predetermined segmentation classes are needed as input beforehand. Misclassifications are seen in air and bone. It is uncertain if it would give accurate enough dosimetry in proton therapy in which the range uncertainties matter greatly.

That leaves the machine learning-based methods to be the most promising solution. The sCT generated by DL algorithms gave small uncertainties. Once trained, those methods can generate sCT within minutes or less. When compared with FCN and GAN network, it was shown that the modified 3D dense-cycle-GAN gave better results. The sCT produced by all three methods seems to be suitable for photon RT plan dose calculation because the differences in dosimetrical metrics were clinically insignificant between CT and sCT.

However, several shortfalls were identified. There were large HU and dose discrepancies near the body/air interface and at bone/tissue boundaries. The results were less satisfying for proton therapy, in which the resulted range differences were larger than other anatomical sites and can lead to bigger differences in dosimetrical metrics. Several factors may be the root causes, such as the registration errors between CT and MRI images, MRI inhomogeneity, and distortion. To combat the issue of CT/MRI registration errors, better deformable image registration algorithms are needed. More advanced motion management and MRI techniques could help as well. Regarding MRI image artifacts, their effects on highly conformal RT such as liver proton therapy can be dramatic [60, 61]. Currently, nonstandard solutions to correct image distortion have been supplied by many manufacturers [62–64], and a standard guideline is under development (American Association of Physicists in Medicine Task Group No. 117). There are still limited studies on this topic. More tailored DL-based algorithms should be invented and tested to fill the gap in effort. Besides, most of the literature was done by a few groups. Therefore, more validation and investigation are needed before it can be implemented for clinical use.

In conclusion, sCT generation for abdominal IGRT is of great research and clinical interest to the medical physics and radiation oncology community. The quest has just started, and more adventures are on the way.

References

[1] Dawson L A and Jaffray D A 2007 Advances in image-guided radiation therapy *J. Clin. Oncol.* **25** 938–46
[2] Western C, Hristov D and Schlosser J 2015 Ultrasound imaging in radiation therapy: from interfractional to intrafractional guidance *Cureus* **7** e280
[3] Mittauer K *et al* 2018 A new era of image guidance with magnetic resonance-guided radiation therapy for abdominal and thoracic malignancies *Cureus* **10** e2422
[4] Mutic S and Dempsey J F 2014 The ViewRay system: magnetic resonance-guided and controlled radiotherapy *Semin. Radiat. Oncol.* **24** 196–9
[5] Lagendijk J J W, Raaymakers B W and van Vulpen M 2014 The magnetic resonance imaging-Linac system *Semin. Radiat. Oncol.* **24** 207–9
[6] Edmund J M and Nyholm T 2017 A review of substitute CT generation for MRI-only radiation therapy *Radiat. Oncol. (London, England)* **12** 28
[7] Arabi H *et al* 2018 Comparative study of algorithms for synthetic CT generation from MRI: consequences for MRI-guided radiation planning in the pelvic region *Med. Phys.* **45** 5218–33
[8] Chen S *et al* 2018 Technical note: U-net-generated synthetic CT images for magnetic resonance imaging-only prostate intensity-modulated radiation therapy treatment planning *Med. Phys.* **45** 5659–65
[9] Largent A *et al* 2018 Pseudo-CT generation for MRI-only radiotherapy treatment planning: comparison between patch-based, atlas-based, and bulk density methods *Int. J. Radiat. Oncol. Biol. Phys* **56** 17–8

[10] Koivula L *et al* 2017 Intensity-based dual model method for generation of synthetic CT images from standard T2-weighted MR images—generalized technique for four different MR scanners *Radiother. Oncol.* **125** 411–9

[11] Maspero M *et al* 2018 Dose evaluation of fast synthetic-CT generation using a generative adversarial network for general pelvis MR-only radiotherapy *Phys. Med. Biol.* **63** 185001

[12] Devic S 2012 MRI simulation for radiotherapy treatment planning *Med. Phys.* **39** 6701–11

[13] Oborn B M *et al* 2017 Future of medical physics: real-time MRI-guided proton therapy *Med. Phys.* **44** e77–90

[14] Yan D *et al* 1997 Adaptive radiation therapy *Phys. Med. Biol.* **42** 123–32

[15] van der Meer S *et al* 2016 Simulation of pseudo-CT images based on deformable image registration of ultrasound images: a proof of concept for transabdominal ultrasound imaging of the prostate during radiotherapy *Med. Phys.* **43** 1913-20

[16] Johansson A, Balter J and Cao Y 2018 Rigid-body motion correction of the liver in image reconstruction for golden-angle stack-of-stars DCE MRI *Magn. Reson. Med.* **79** 1345–53

[17] Stemkens B, Paulson E S and Tijssen R H N 2018 Nuts and bolts of 4D-MRI for radiotherapy *Phys. Med. Biol.* **63** 21tr01

[18] Wojcieszynski A P *et al* 2016 Gadoxetate for direct tumor therapy and tracking with real-time MRI-guided stereotactic body radiation therapy of the liver *Radiother. Oncol.* **118** 416–8

[19] Owrangi A M, Greer P B and Glide-Hurst C K 2018 MRI-only treatment planning: benefits and challenges *Phys. Med. Biol.* **63** 05tr01

[20] Hsu S H *et al* 2013 Investigation of a method for generating synthetic CT models from MRI scans of the head and neck for radiation therapy *Phys. Med. Biol.* **58** 8419

[21] Hsu S H *et al* 2020 A technique to generate synthetic CT from MRI for abdominal radiotherapy *J. Appl. Clin. Med. Phys.* **21** 136–43

[22] Chin A L *et al* 2014 Feasibility and limitations of bulk density assignment in MRI for head and neck IMRT treatment planning *J. Appl. Clin. Med. Phys.* **15** 100–11

[23] Korhonen J *et al* 2014 A dual model HU conversion from MRI intensity values within and outside of bone segment for MRI-based radiotherapy treatment planning of prostate cancer *Med. Phys.* **41** 011704

[24] Bredfeldt J S *et al* 2017 Synthetic CT for MRI-based liver stereotactic body radiotherapy treatment planning *Phys. Med. Biol.* **62** 2922–34

[25] Sjölund J *et al* 2015 Generating patient specific pseudo-CT of the head from MR using atlas-based regression *Phys. Med. Biol.* **60** 825

[26] Guerreiro F *et al* 2017 Evaluation of a multi-atlas CT synthesis approach for MRI-only radiotherapy treatment planning *Med. Phys.* **35** 7–17

[27] Lei Y *et al* 2018 MRI-based pseudo CT synthesis using anatomical signature and alternating random forest with iterative refinement model *J. Med. Imaging (Bellingham)* **5** 043504

[28] Lei Y *et al* 2019 MRI-based synthetic CT generation using semantic random forest with iterative refinement *Phys. Med. Biol.* **64** 085001

[29] Han X 2017 MR-based synthetic CT generation using a deep convolutional neural network method *Med. Phys.* **44** 1408–19

[30] Lei Y *et al* 2018 Pseudo CT estimation using patch-based joint dictionary learning *2018 40th Annual Int. Conf. of the IEEE Engineering in Medicine and Biology Society (EMBC) (Honolulu, HI)* (Piscataway, NJ: IEEE)

[31] Yang X *et al* 2019 MRI-based attenuation correction for brain PET/MRI based on anatomic signature and machine learning *Phys. Med. Biol.* **64** 025001

[32] Liu Y *et al* 2019 MRI-based treatment planning for liver stereotactic body radiotherapy: validation of a deep learning-based synthetic CT generation method *Br. J. Radiol.* **92** 20190067

[33] Liu Y *et al* 2019 MRI-based treatment planning for proton radiotherapy: dosimetric validation of a deep learning-based liver synthetic CT generation method *Phys. Med. Biol.* **64** 145015

[34] Liu L *et al* 2020 Abdominal synthetic CT generation from MR Dixon images using a U-net trained with 'semi-synthetic' CT data *Phys. Med. Biol.* **65** 125001

[35] Liu Y *et al* 2020 Liver synthetic CT generation based on a dense-CycleGAN for MRI-only treatment planning *SPIE Medical Imaging* 11313 *(Houston, TX)*

[36] Wolterink J M *et al* 2017 Deep MR to CT synthesis using unpaired data *Simulation and Synthesis in Medical Imaging (Québec City, Canada)* ed S A Tsaftaris *et al* (Cham: Springer International Publishing) pp 14–23

[37] Goodfellow I J *et al* 2014 Generative adversarial nets *Proc. 27th Int. Conf. on Neural Information Processing Systems—Volume 2 (Montreal, Canada)* (Cambridge, MA: MIT Press) pp 2672–80

[38] Emami H *et al* 2018 Generating synthetic CTs from magnetic resonance images using generative adversarial networks *Med. Phys.* **45** 3627–36

[39] Zhu J Y *et al* 2017 Unpaired image-to-image translation using cycle-consistent adversarial networks arXiv:1703.10593

[40] Nie D *et al* 2016 Estimating CT image from MRI data using 3D fully convolutional networks *Deep Learning and Data Labeling for Medical Applications. DLMIA LABELS 2016* (Lecture Notes in Computer Science vol 10008) (Cham: Springer)

[41] Nie D *et al* 2018 Medical image synthesis with deep convolutional adversarial networks *IEEE Trans. Biomed. Eng.* **65** 2720–30

[42] Nie D *et al* 2017 Medical image synthesis with context-aware generative adversarial networks *Int. Conf. on Medical Image Computing and Computer-Assisted Intervention 65* (Berlin: Springer) pp 2720–30

[43] Maspero M *et al* 2017 Feasibility of MR-only proton dose calculations for prostate cancer radiotherapy using a commercial pseudo-CT generation method *Phys. Med. Biol.* **62** 9159

[44] Koivula L, Wee L and Korhonen J J M P 2016 Feasibility of MRI-only treatment planning for proton therapy in brain and prostate cancers: dose calculation accuracy in substitute CT images *Med. Phys.* **43** 4634–42

[45] Paganetti H 2012 Range uncertainties in proton therapy and the role of Monte Carlo simulations *Phys. Med. Biol.* **57** R99–117

[46] Pileggi G *et al* 2018 Proton range shift analysis on brain pseudo-CT generated from T1 and T2 MR *Acta Oncol.* **57** 1521–31

[47] Hansen D C *et al* 2018 ScatterNet: a convolutional neural network for cone-beam CT intensity correction *Med. Phys.* **45** 4916–26

[48] Kida S *et al* 2018 Cone beam computed tomography image quality improvement using a deep convolutional neural network *Cureus* **10** e2548

[49] Xie S *et al* 2018 Scatter artifacts removal using learning-based method for CBCT in IGRT System *IEEE Access* **6** 78031–7

[50] Chen L *et al* 2020 Synthetic CT generation from CBCT images via deep learning *Med. Phys.* **47** 1115–25

[51] Harms J *et al* 2019 Paired cycle-GAN-based image correction for quantitative cone-beam computed tomography *Med. Phys.* **46** 3998–4009

[52] Kurz C *et al* 2019 CBCT correction using a cycle-consistent generative adversarial network and unpaired training to enable photon and proton dose calculation *Phys. Med. Biol.* **64** 225004

[53] Landry G *et al* 2019 Comparing Unet training with three different datasets to correct CBCT images for prostate radiotherapy dose calculations *Phys. Med. Biol.* **64** 035011

[54] Kida S *et al* 2020 Visual enhancement of Cone-beam CT by use of CycleGAN *Med. Phys.* **47** 998–1010

[55] Lei Y *et al* 2019 Learning-based CBCT correction using alternating random forest based on auto-context model *Med. Phys.* **46** 601–18

[56] Wang T *et al* 2019 Dosimetric study on learning-based cone-beam CT correction in adaptive radiation therapy *Med. Dosim.* **44** e71–9

[57] Thummerer A *et al* 2020 Comparison of CBCT based synthetic CT methods suitable for proton dose calculations in adaptive proton therapy *Phys. Med. Biol.* **65** 095002

[58] Nomura Y *et al* 2019 Projection-domain scatter correction for cone beam computed tomography using a residual convolutional neural network *Med. Phys.* **46** 3142–55

[59] Liu Y *et al* 2020 CBCT-based synthetic CT generation using deep-attention cycleGAN for pancreatic adaptive radiotherapy *Med. Phys.* **47** 2472-83

[60] Seibert T M *et al* 2016 Distortion inherent to magnetic resonance imaging can lead to geometric miss in radiosurgery planning *Pract. Radiat. Oncol.* **6** e319–28

[61] Wang H, Balter J and Cao Y 2013 Patient-induced susceptibility effect on geometric distortion of clinical brain MRI for radiation treatment planning on a 3T scanner *Phys. Med. Biol.* **58** 465

[62] Jovicich J *et al* 2006 Reliability in multi-site structural MRI studies: effects of gradient non-linearity correction on phantom and human data *Neuroimage* **30** 436–43

[63] Doran S J *et al* 2005 A complete distortion correction for MR images: I. gradient warp correction *Phys. Med. Biol.* **50** 1343

[64] Baldwin L N *et al* 2007 Characterization, prediction, and correction of geometric distortion in 3T MR images *Med. Phys.* **34** 388–99

IOP Publishing

Principles and Practice of Image-Guided Abdominal Radiation Therapy

Yu Kuang

Chapter 24

Virtual imaging for abdominal image-guided radiotherapy

You Zhang, Wendy Harris, Jing Wang and Lei Ren

This chapter is dedicated to virtual imaging for abdominal image-guided radio-therapy (IGRT). We first discuss pitfalls of traditional three-dimensional/four-dimensional cone-beam computed tomography for abdominal imaging and tumor localization, and introduce the concept of deformable registration-driven virtual computed tomography (CT), which morphs a high-quality prior CT image to on-board CT using the information from acquired cone-beam projections. With a focus on the site of liver, we show how virtual CT can effectively preserve high image quality while substantially reducing the image sampling requirement. Based on the concept of virtual CT, we further introduce biomechanical modeling–guided virtual CT generation, which addresses the low-contrast issues of abdominal imaging by incorporating tissue biomechanical properties to improve deformable registration accuracy. The concept, methodology, and results of virtual magnetic resonance imaging are also presented in the chapter, which aims to further enhance the image contrast to guide accurate abdominal IGRT.

24.1 Pitfalls of traditional cone-beam computed tomography imaging

On-board cone-beam computed tomography (CBCT) imaging has become a stand-ard of care in today's image-guided radiation therapy [1]. It has been routinely applied towards patient setup and tumor localization for a variety of treatment sites, including, but not limited to, head and neck, lung, abdomen, and pelvis [2–5]. For the abdominal site, however, the quality and accuracy of CBCT imaging are adversely affected by *respiration-induced tumor/organ motion* [6]. Respiratory motion introduces motion blurriness into the traditional three-dimensional (3D)

CBCT images, rendering it difficult to localize a moving target accurately [7]. To account for the respiratory motion, the concept of four-dimensional (4D) CBCT was introduced to generate respiratory phase-resolved CBCT volumes that capture the motion trajectory of anatomical structures [8–10]. Each phase of the 4D CBCT set is a 3D CBCT, representing a semi-static image of the anatomy along a nominal respiratory motion trajectory. Stacking all phases together, 4D CBCT reveals the full motion trajectory of the imaged anatomy to better guide tumor localization and motion management [11]. To generate a 4D CBCT set, we need to phase-sort cone-beam projections into different bins for phase-specific CBCT reconstruction [9]. However, the phase-sorting process reduces the projections available for CBCT reconstruction at each phase, leading to severe aliasing artifacts (figure 24.1(a)). To preserve the image quality of 4D CBCT at each phase, more projections need to be acquired than traditional 3D CBCT, leading to significantly increased imaging time and dose [12]. To reduce the imaging time and dose while preserving the image quality, different algorithms have been investigated for sparse-view 3D/4D CBCT reconstruction and enhancement [10,13–19]. Advanced 4D CBCT reconstruction algorithms usually explore the idea of image sparsity, a feature inherent in anatomical images, to combine data fidelity constraints with image-domain regularizations to improve the reconstruction accuracy from limited sampling [16, 20]. These algorithms, however, may over-smooth or fail to reconstruct detailed image textures and features, especially under substantial under-sampling scenarios.

In addition to the respiratory motion and the related 4D CBCT sampling and reconstruction issues, another major source of uncertainty for abdominal CBCT imaging is its *low soft-tissue contrast* [21]. The lack of soft-tissue contrast undermines a clinician's ability to accurately localize the tumor from the surrounding structures for motion tracking and beam targeting, even with respiratory motion-resolved 4D CBCT images (figure 24.1(b)). Although iodinated agents can be injected intravenously to enhance the contrast of abdominal tumors during the treatment simulation stage, they are rarely used for on-board image guidance [22]. Contraindications include the potential toxicity of the contrast agents and the fast wash-in and wash-out of the agents that lead to difficulties in injection and imaging

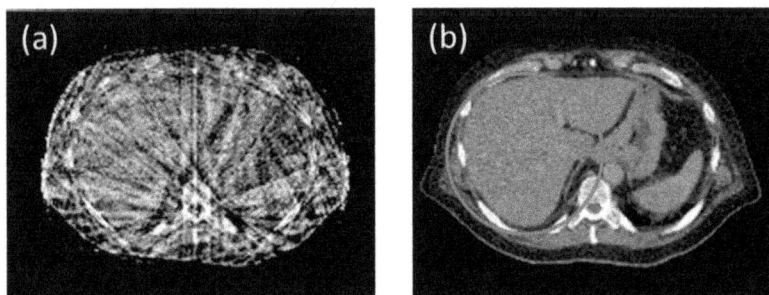

Figure 24.1. (a) One phase image of a liver 4D-CBCT set with severe under-sampling artifacts. (b) Low soft-tissue contrast of x-ray–based imaging affects the tumor localization accuracy for image guidance, even without motion/under-sampling artifacts.

timing management. Instead of the contrast agent, radiopaque fiducial markers may be implanted as motion surrogates to help localize the tumor and track its motion [23, 24]. The fiducial markers, however, require invasive implantation procedures that can further induce complications to organs already fragile from cancer. They may also migrate from place to place, leading to reduced motion correlation with the tumor. The fiducial markers cannot be used to track tumor deformation during respiratory motion, either.

24.2 The concept of virtual computed tomography

To address the issues of traditional CBCT imaging and reconstruction, a new approach to generate on-board CBCT-equivalent 'virtual' computed tomography (CT) images has gained research momentum recently [10, 14, 19, 25, 26]. The virtual CT approach proposes to use deformation vector fields (DVFs) to transfer an already-available prior CT with high image quality (for instance the CT image acquired during treatment simulation) to an on-board virtual CT that reflects the most up-to-date patient geometric and anatomical information (equation (24.1)).

$$CT_{virtual}(x) = CT_{prior}(x + \mathbf{DVF}) \tag{24.1}$$

CT_{prior} denotes the high-quality prior CT image, and $CT_{virtual}$ denotes the new virtual CT image as a deformed version of CT_{prior}. x denotes the CT image coordinates. $\mathbf{DVF}$ denotes the DVF that drives the deformation, usually through trilinear interpolations [14]. Instead of using the cone-beam projections to directly reconstruct CBCT images, the virtual CT approach uses them to guide the solution of the DVFs through optimizing an objective function as shown in equation (24.2).

$$\mathbf{DVF} = \text{argmin}_{\mathbf{DVF}} \left\{ \left\| \left(a \times \mathcal{A} CT_{prior}(x + \mathbf{DVF}) + b \right) - P \right\|_2^2 + \omega * E(\mathbf{DVF}) \right\} \tag{24.2}$$

$$E(\mathbf{DVF}) = \sum_{i=1}^{n_i} \sum_{j=1}^{n_j} \sum_{k=1}^{n_k} \sum_{m=1}^{3} \left(\left(\frac{\partial DVF_m(i, j, k)}{\partial x} \right)^2 + \left(\frac{\partial DVF_m(i, j, k)}{\partial y} \right)^2 + \left(\frac{\partial DVF_m(i, j, k)}{\partial z} \right)^2 \right) \tag{24.3}$$

As shown in equation (24.2), the objective function is formulated with two terms: a data fidelity term (first) and a regularization term (second). For the data fidelity term, $\mathcal{A}$ denotes the system matrix to generate digitally reconstructed radiographs (DRRs) using the imaging geometry matched to that of the acquired cone-beam projections P. a and b are intensity scaling and shift coefficients introduced to account for the inherent, non-deformation-induced intensity mismatches between the DRRs and the on-board projections. a and b can be explicitly computed using least-squares–based fitting [27] or implicitly by using normalized cross-correlation as the similarity metric instead of the sum of squared errors shown in equation (24.2) [28]. For the second regularization term, E denotes a deformation bending-energy function that is detailed in equation (24.3) [10]. The symbol m in equation (24.3) denotes the three Cartesian directions: x, y, and z. Symbols n_i, n_j, and n_k denote the

DVF dimensions along the three Cartesian directions, correspondingly. The data fidelity term and the energy regularization term are balanced by an energy weighting factor ω, which could be manually set or automatically updated [10, 14, 26]. The minimization of the data fidelity term in equation (24.2) matches the DRRs of the DVF-morphed virtual CT to the on-board cone-beam projections in the intensity domain, a key driving mechanism for the DVF optimization. The minimization of the deformation energy regularization term, on the other hand, works to enforce the DVF smoothness, which is an observed feature of naturally occurring deformation. It also reduces the DVF solution space for faster and stable convergence of optimization [10, 26, 29]. The objective function of equation (24.2) can be conveniently optimized through the non-linear conjugate gradient algorithm, since its gradient can be explicitly calculated [10]. As the original virtual CT technique solves 3D DVFs via matching two-dimensional (2D) projections, it is also often referred as the '2D–3D deformation' technique instead.

Through a deformation-driven approach, the virtual CT technique can effectively pass along the high-quality information from the prior CT image to the new virtual CT image. Detailed textures and features from the prior CT can be preserved. With the existing high-quality prior information, fewer on-board projections need to be acquired for virtual CT generation compared with the conventional CBCT imaging, leading to substantial imaging time and dose reduction [10, 14, 26]. These advantages position the virtual CT technique well for on-board 4D imaging, alleviating the trade-off between imaging time/dose and image quality. Note that for 4D imaging, the cone-beam projection set P of equation (24.2) is phase-specific. Each 4D phase is to be reconstructed independently, which can be parallelized for efficiency. Multiple studies on virtual CT have confirmed the benefits of this technique in generating high-quality, accurate, on-board 3D/4D images using sparse or even limited-angle projections to save imaging time and dose [10, 14, 19, 25]. The accurate Hounsfield unit (HU) information of virtual CT, inherited from the prior CT image, also allows accurate dose calculation and accumulation, paving the way for online adaptive radiotherapy [30, 31].

In addition to 4D imaging, the DVF-driven virtual CT approach can also potentially solve the low-contrast issue faced by traditional CBCT imaging for the abdominal site. Because the virtual CT can be generated by deforming the CT used for simulation and treatment planning, the tumor/normal organ segmentations on the simulation CT can be simultaneously propagated onto the new virtual CT image via DVF-driven contour propagation [32, 33]. The DVF-propagated contours will allow automatic tumor localization, even if manual visualization and localization of the tumors remain challenging on the virtual CT. The potential of automatic low-contrast tumor localization offers the virtual CT technique a unique advantage as compared to the other methods, which could allow a paradigm shift towards today's abdominal tumor localization conventions.

However, although the virtual CT technique theoretically allows accurate on-board tumor localization for all anatomical sites, its accuracy is ultimately dependent on the accuracy of the solved DVFs for different anatomical sites. As shown in equation (24.2), the 2D–3D deformation algorithm employed by the virtual CT technique solves the DVFs through a pure intensity-driven approach, whose

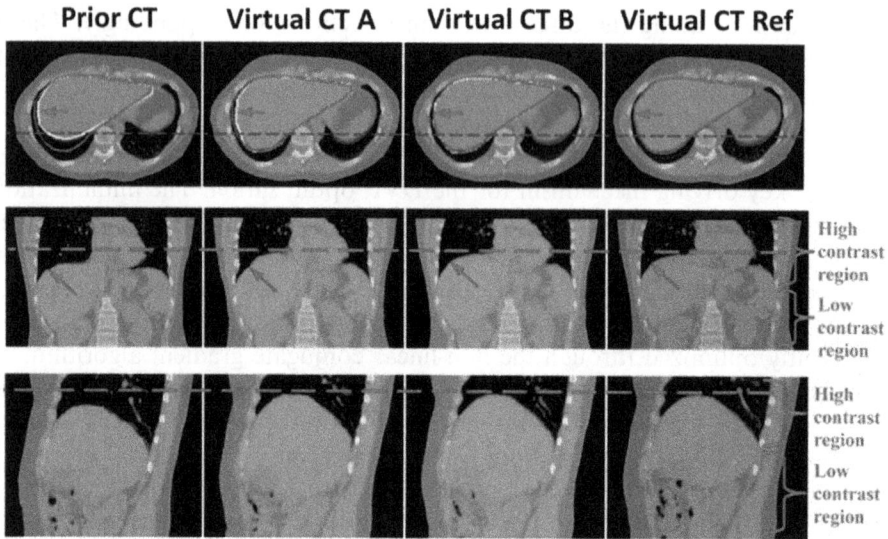

Figure 24.2. Comparison between different CT images at three views. The prior CT and the reference virtual CT (virtual CT Ref) were extracted from phase images of a 4D-CT set. Virtual CT A was generated via the 2D–3D deformation technique using 57 cone-beam projections simulated from the virtual CT Ref. The projections were evenly distributed within a 60° angular span with the angle central axis along the patient left lateral direction. Virtual CT B was similarly generated using 57 simulated projections. The projections covered a full 360° scan angle. Although both virtual CTs (A and B) were accurately generated in the high-contrast region when compared to the virtual CT Ref, they were both limited in accuracy for the low-contrast regions because of limited DVF accuracy there, which affects the accuracy of tumor localization in these regions. Reproduced from [25] with permission from Elsevier.

accuracy is highly dependent on the intensity variations induced by the deformation between the prior and virtual CT images. Due to the lack of soft-tissue contrast in the abdominal region, the deformation there leads to limited intensity variations on the CT image and on the corresponding cone-beam projections, making accurate DVF solution very challenging in these low-contrast areas. A study by Ren *et al* investigated the virtual CT approach for liver imaging and came to a conclusion that although the DVF at the liver boundary (which is of high contrast) can be accurately solved, the DVF within the liver remains inaccurate due to the limited image contrast there (figure 24.2) [25]. As the liver tumors are mostly within the low-contrast regions, the pure intensity-based virtual CT technique may not be able to localize these tumors accurately through DVF-driven contour propagation.

24.3 Biomechanical modeling–guided virtual CT generation

To further address the limited accuracy of the virtual CT technique when applied to low-contrast abdominal sites, a biomechanical modeling-guided virtual CT technique was recently developed [21, 34]. The biomechanical modeling–guided technique combines the 2D–3D deformation technique with biomechanical modeling–driven DVF optimization [35–38]. Being a pure physics-driven technique (compared to intensity-driven 2D–3D deformation), the biomechanical modeling technique can solve physically plausible and realistic DVFs. It models each organ of interest as a mesh composed of discrete tetrahedral finite elements. The collective movements of these finite elements, representing the intra-organ deformation, can be solved based on information of tissue elasticity properties (bio-material modeling) and organ boundary motion (displacement-based boundary condition). Previous studies have found it accurate in solving DVFs at low-contrast regions, which could be especially helpful for the abdominal sites [35].

In detail, for the biomechanical modeling–guided virtual CT technique, we used 2D–3D deformation to solve organ boundary DVFs and fed them into the biomechanical modeling step as the boundary condition of the organ tetrahedral mesh. Because the organ boundary is usually of higher contrast, the 2D–3D deformation-solved DVFs at the boundary are more accurate to drive biomechanical modeling [10, 25, 28]. Traditionally, the 2D–3D deformation technique only solves DVFs which by convention are defined on the voxel grid of the new $CT_{virtual}$ and point to CT_{prior}(equation (24.1)). We denoted such-defined deformation field $DVF_{forward}$. For boundary condition assignment of biomechanical modeling, however, the DVF needs to be defined on the voxel grid of CT_{prior} and point to $CT_{virtual}$, since the organ mesh is defined from the organ segmentations on CT_{prior} (figure 24.3). The corresponding DVF, $DVF_{inverse}$, is inverse to the $DVF_{forward}$ solved by the traditional 2D–3D deformation technique. To reduce the uncertainties of obtaining $DVF_{inverse}$ directly from inverting $DVF_{forward}$ after 2D–3D deformation, the traditional objective function (equation (24.2)) was modified to enforce a dual-directional approach to optimize $DVF_{forward}$ and $DVF_{inverse}$ simultaneously:

$$
\begin{aligned}
&DVF_{forward} \\
&= \mathrm{argmin}_{DVF_{forward}} \left\{
\begin{array}{l}
\left\| \left(a * \mathscr{A} CT_{prior}(x + DVF_{forward}) + b \right) - P \right\|_2^2 + \dfrac{\omega}{2} * E(DVF_{forward}) \\[2mm]
+ \left\| \mathscr{A}\left[\left(CT_{prior}(x + DVF_{forward}) \right)(x + DVF_{inverse}) \right] - P_{prior} \right\|_2^2 + \dfrac{\omega}{2} \\[2mm]
* E(DVF_{inverse})
\end{array}
\right\}
\end{aligned}
\tag{24.4}
$$

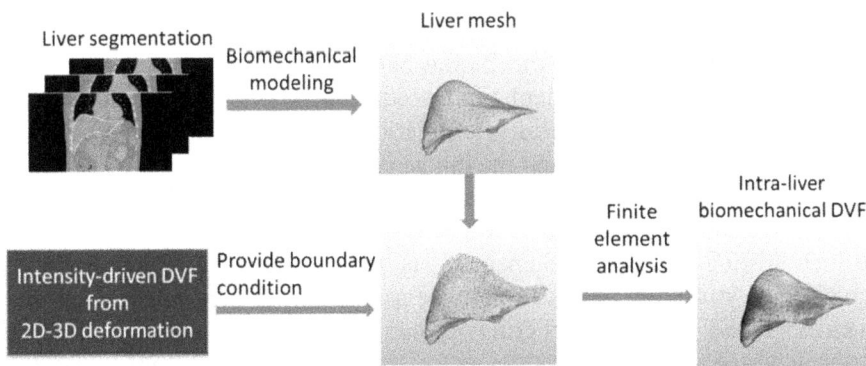

Figure 24.3. General workflow of the biomechanical-modeling-guided virtual CT generation technique. This workflow uses liver as an example, and it can be readily extended to other organs and anatomical sites.

Equation (24.4) shares the same denotations with equation (24.2) for most symbols. In equation (24.4), the new symbol P_{prior} denotes the projections generated from CT_{prior} to mirror the on-board projections P. P_{prior} and its hosting data fidelity term were designed to enforce the intensity matching of projections in the inverse direction, to maintain the accuracy of $\mathbf{DVF}_{\text{inverse}}$. For simplicity, P_{prior} was generated from CT_{prior} using the same system matrix as of P. To solve equation (24.4), a correspondence between $\mathbf{DVF}_{\text{inverse}}$ and $\mathbf{DVF}_{\text{forward}}$ also needs to be established. The $\mathbf{DVF}_{\text{inverse}}$ can be inverted from $\mathbf{DVF}_{\text{forward}}$ using multiple methods [18, 39, 40]. Here we simply set $\mathbf{DVF}_{\text{inverse}}$ as the negate of $\mathbf{DVF}_{\text{forward}}$, a simple but accuracy-validated strategy enabling fast optimization [18]. Similar to equation (24.2), the same conjugate gradient algorithm can be applied to solve equation (24.4) to derive the paired DVFs.

With the pair of DVFs solved, the $\mathbf{DVF}_{\text{inverse}}$ at the organ surface can then be applied as the boundary condition for finite element analysis to derive the intra-organ DVFs. To perform the finite element analysis [41], a model of the organ material is also needed. Many models have been proposed to represent soft tissues with adequate accuracy, and currently, there is no consensus over an 'optimal' model to work for all organs/scenarios. For the biomechanical modeling–guided virtual CT study on liver [21], the Mooney–Rivlin material model was adopted [34, 42, 43]. It is a hyper-elastic model that usually works well for large deformations, which are often observed in soft tissues. The strain energy density function (W) that characterizes the Mooney–Rivlin model is shown in equation (24.5):

$$W = c_1\left(\tilde{I}_1 - 3\right) + c_2\left(\tilde{I}_2 - 3\right) + 0.5K(J - 1)^2 \qquad (24.5)$$

The three terms in equation (24.5) characterize the Mooney–Rivlin material's responses to stress. The first two terms quantify the material's shape change towards stress, and the third term quantifies the material's volume change towards stress. $\tilde{I}_1$, $\tilde{I}_2$, and J are derivatives of DVFs. c_1, c_2, and K are the corresponding material coefficients to be provided. Through a parameter-sweeping strategy, the liver study provided the parameters as $c_1 = c_2 = 0.135$ kPa and $K = 27$ kPa [21]. Based on the organ tetrahedral mesh, the material model, as well as the boundary conditions provided by the intensity-driven 2D–3D deformation, the intra-organ DVFs can be solved via finite element analysis (figure 24.3).

The DVFs generated from the biomechanical modeling step, however, were defined on the tetrahedral elements and needed to be interpolated back onto the image voxel grid to allow trilinear-interpolation–based deformation. The barycentric coordinates were used to align the image voxel grids to the tetrahedral elements [45]. An image voxel resides within a tetrahedral element, only if all the (four) barycentric coordinates of the voxel respective to the tetrahedron are nonnegative. If none such hosting tetrahedral element exists, the one closest to the voxel grid was assigned. Each voxel grid will eventually be assigned one corresponding tetrahedral element. The DVF at each voxel grid was then calculated by weighting the DVFs at the four nodes of the tetrahedral element with the barycentric coordinates.

To further improve the DVF accuracy, the biomechanically corrected DVF solved above could be further fed back into 2D–3D deformation as a new initial

$DVF_{inverse}$ for further optimization and fine-tuning, until pre-defined convergence criteria are met. The liver study [21] iteratively optimized and corrected the DVFs back and forth between the 2D–3D deformation step and the biomechanical modeling step until it met the convergence criteria defined on projection data fidelity. After convergence, the final $DVF_{inverse}$ and $DVF_{forward}$ were obtained, and the virtual CT was generated by deforming the CT_{prior} using the final $DVF_{forward}$. Similar to the traditional virtual CT technique, the biomechanical modeling-guided virtual CT technique also generates each phase volume individually and independently. For a 4D virtual CT set, each phase volume can be generated concurrently in parallel.

A liver study evaluated the biomechanical modeling–guided virtual CT technique on a cohort of seven liver patients. For each patient, the study extracted the 0% phase volume of the simulation 4D-CT image set as CT_{prior} and the 50% phase volume as the 'gold-standard' reference $CT_{virtual}$. Twenty cone-beam projections evenly covering a 360° scan angle were simulated via a Monte Carlo package and used for CBCT reconstruction or virtual CT generation. As shown (figure 24.4), the CBCT image directly reconstructed by the Feldkamp–Davis–Kress (FDK) algorithm suffers from a

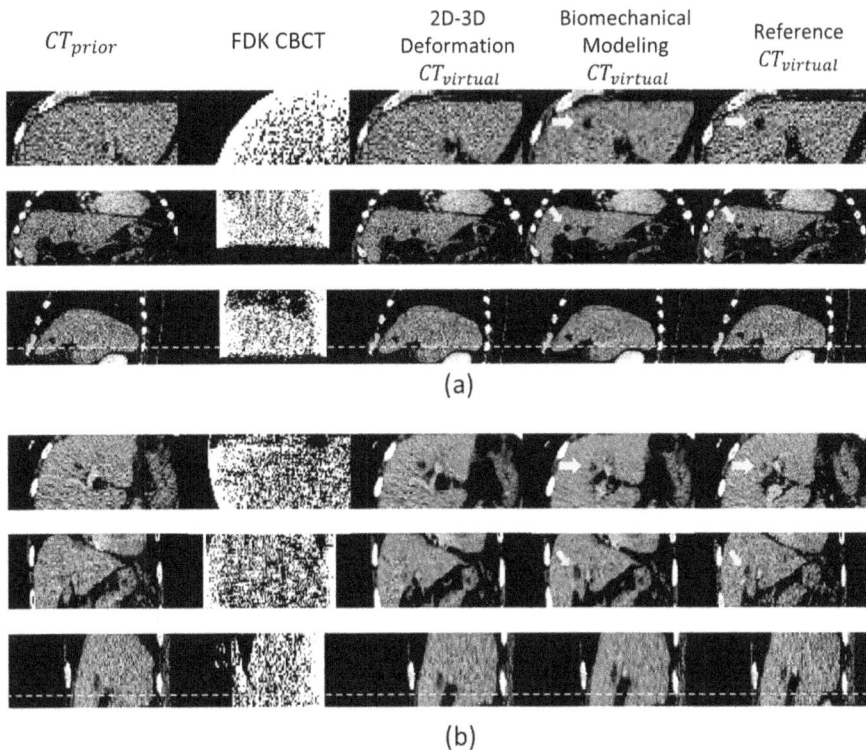

Figure 24.4. Comparison of images at three views for two different patient cases (a) and (b). From left to right: the prior CT image, the CBCT image directly reconstructed by the clinical FDK algorithm [47], the 2D–3D deformation-generated virtual CT image, the biomechanical modeling–guided virtual CT image, and the reference virtual CT image. The arrows point to the tumor regions. Reproduced from [21]. With permission from Elsevier.

substantial level of noise that prevents accurate tumor localization. The 2D–3D deformation-generated $CT_{virtual}$ inherits the high-quality information from the CT_{prior} but suffers from low-accuracy intra-liver DVFs as evidenced by the unmatched liver tumor region. With the accuracy-boost from finite element modeling, the biomechanical modeling-guided technique has generated a $CT_{virtual}$ with the tumor area well matched to the reference $CT_{virtual}$, enabling accurate tumor localization.

The accuracy of DVF-driven tumor localization was evaluated in figure 24.5. As shown, the 4D liver tumor/cyst motion tracked by the biomechanical modeling-guided virtual CT technique matched significantly better with that manually tracked through physicians' tumor segmentations. The biomechanical modeling–guided virtual CT technique promises a potential paradigm shift in abdominal target localization, by allowing markerless, contrast-agent–free abdominal target localization with limited projections (reduced imaging dose/time) [21]. Similar to the conventional intensity-driven virtual CT technique, the high-quality on-board images and the DVFs offered by the biomechanical modeling–guided virtual CT technique can also benefit online adaptive radiotherapy, which is especially relevant to the abdominal sites given the complex organ arrangements and respiratory motion there [46].

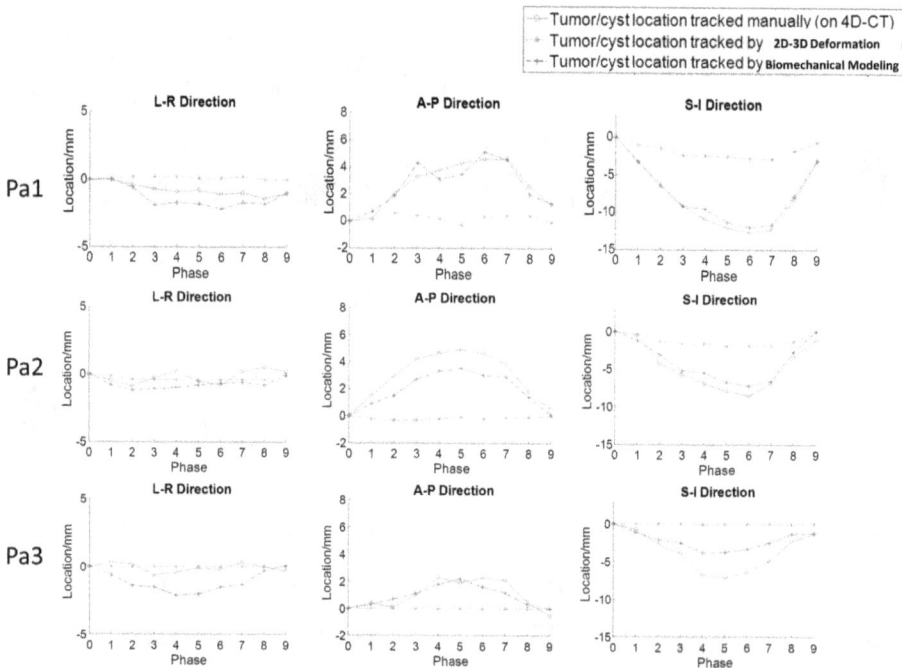

Figure 24.5. Comparison of 4D tumor localization accuracy between using 2D–3D deformation-based DVFs and using biomechanical modeling-guided DVFs, for three patient cases. For each patient case, phase 0% of a simulation 4D-CT set was used as the prior CT volume. All other 4D-CT phase volumes were used to generate cone-beam projections for DVF-based virtual CT generation and liver tumor contour propagation. The tumors on the original 4D-CT sets were manually contoured by radiation oncologists as the reference for comparison. Reprinted from [21] with permission from Elsevier.

24.4 From virtual CT to virtual magnetic resonance imaging

Compared with other types of treatments, for stereotactic body radiation therapy (SBRT), accurate tumor localization is especially critical due to the high fractional doses involved with SBRT treatments [19,48–50]. The virtual CT technique, especially the one with biomechanical modeling guidance, enables accurate tumor localization in low-contrast regions for safe and effective SBRT delivery. However, although the DVF-driven virtual CT technique allows localization without necessarily visualizing the tumors directly, it is always desired to have a crystal-clear image allowing us to see tumors as well as all organs at risk to minimize the localization and treatment uncertainties. On-board magnetic resonance imaging (MRI) has been introduced for target localization [51–53]. Using MRI for on-board target localization can substantially improve localization accuracy SBRT treatments in the abdominal region since MRI has better soft-tissue contrast with no ionizing radiation dose. At the moment, MRI-equipped radiotherapy machines are only available in a limited number of facilities because of its high cost [54]. Conventional linear accelerators (LINACs) with on-board kilovoltage (kV) imaging systems, however, are much more widely available in facilities. It would be advantageous to generate MRI-quality images with high soft-tissue contrast using prior MRIs and on-board kV projections from a LINAC for target localization. Based on the foundation of the virtual CT technique, a new method was developed that utilized deformable image registration, prior MRIs, and on-board limited kV cone-beam projections from a conventional LINAC to generate on-board virtual 4D-MRI to improve the soft-tissue localization accuracy for liver SBRT treatments without the need for a magnetic resonance (MR)–guided radiotherapy machine.

The overall workflow of the virtual 4D-MRI technique is shown in figure 24.6. A 4D-MRI is acquired during the simulation stage and MRI_{prior} is defined as the end of expiration (EOE) phase of the 4D-MRI. The on-board virtual 4D-MRI at each respiratory phase is considered a deformation of the MRI_{prior} and can be solved based on equation (24.6):

$$\text{Virtual } 4DMRI(i, j, k) = MRI_{prior}\big(i + \text{DVF}_x(i, j, k), \quad j + \text{DVF}_y(i, j, k),$$
$$k + \text{DVF}_z(i, j, k)\big) \tag{24.6}$$

DVF_x, DVF_y, and DVF_z are the DVFs along x, y, and z directions. A data fidelity constraint is used to solve the DVFs based on phase-sorted on-board kV projections. In order to generate DRRs to match with the kV projections to evaluate the data fidelity constraint, a synthetic CT prior, sCT_{prior}, is needed. The MRI_{prior} is used to generate this synthetic CT at the EOE phase. DRRs from the deformed sCT_{prior} are then calculated to match with the corresponding phase-sorted on-board kV projections (OBI). The data fidelity term is shown in equation (24.7).

$$DRR\big(\text{DVF}, sCT_{prior}\big) = OBI \tag{24.7}$$

Figure 24.7 shows a flowchart for generating the sCT_{prior} from the MRI_{prior} and a prior 3D CT volume, which can be the EOE phase of a simulation 4D-CT set.

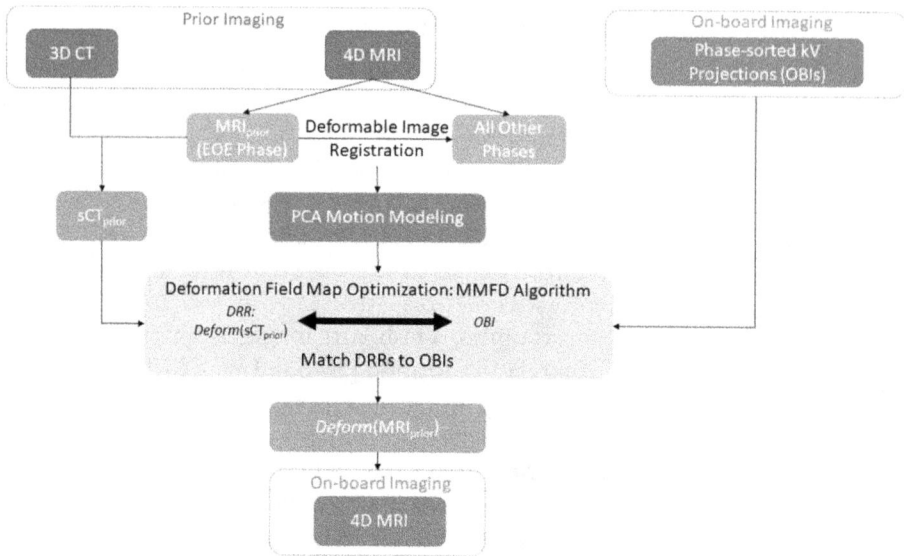

Figure 24.6. Overall workflow to generate on-board virtual 4D MRI from prior 4D MRI and on-board kV projections.

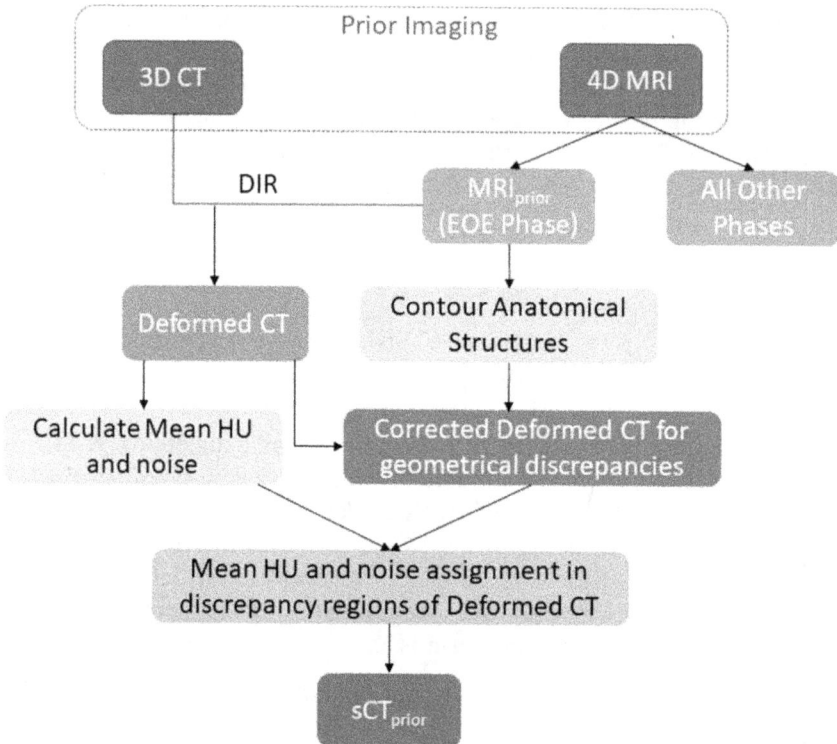

Figure 24.7. Workflow to generate sCT_{prior} from a 3D CT volume and MRI_{prior}. DIR stands for deformable image registration.

MRI_{prior} is registered to the 3D CT via deformable image registration. A clinician manually contours organs in both the MRI_{prior} and the deformed CT. The contours in the MRI_{prior} are considered as ground truth and the liver contour is overridden in the deformed CT to be that of the MRI_{prior} liver contour. The area of the overridden liver contour that does not overlap with the original deformed CT contour is called the discrepancy region. In the discrepancy region, mean HU and noise values from nearby areas in the deformed CT are used to fill in as heterogeneous HU values. The textures of the liver were generated in the deformed CT directly based on the deformable registration [56–58]. The sCT_{prior} is then used to optimize the DVFs using a motion modeling and free-form deformation (MMFD) algorithm.

The MMFD method is used to solve for the DVFs to generate the on-board virtual 4D-MRI. In the motion modeling part of the optimization, principal component analysis (PCA) is used to reduce the number of variables that need to be solved for in the DVF. MRI_{prior} is deformed to all other phases of the prior 4D-MRI to obtain DVFs. Then PCA is used to extract the first three principal motion modes $\left\{ \widetilde{D}_0^{\,j} \right\}$. The DVF to be solved can be expressed as a linear combination of the three principal motion modes, as shown in equation (24.8).

$$DVF = D_{0,\,ave} + \sum_{j=1}^{3} w_j \widetilde{D}_0^{\,j} \tag{24.8}$$

$D_{0,\,ave}$ is the average of the original DVFs from the prior 4D-MRI. The w_j are the weighting coefficients and are optimized by using the data fidelity term in equation (24.7). A gradient descent optimization is used to minimize the differences between the deformed sCT_{prior} with the on-board kV projections.

The DVF solved by the motion modeling step is a coarse estimation and used as an initial DVF for the following free-form deformation optimization. In the Free Form algorithm, the DVF is further fine tuned, and every voxel in the DVF can deform freely without any assumptions from prior motion modes. In the free-form optimization, the DVF is solved by minimizing the same bending-energy as equation (24.3) to preserve its smoothness, such that the data fidelity constraint is met. Once the final DVF is solved for after the MMFD optimization, the on-board virtual 4D-MRI at each respiratory phase is generated by deforming the MRI_{prior} based on the deformation field map solved for each phase, as shown by equation (24.6).

Phantom and preliminary patient studies have demonstrated excellent performance in generating on-board virtual 4D-MRI. For the phantom study, a digital anthropomorphic phantom, extended cardiac torso (XCAT), was used to simulate the prior 4D-MRI, ground-truth on-board 4D-MRI, and ground-truth on-board 4D-CT. Siddon's ray-tracing technique was used to generate on-board cone-beam projections of different phases using the full-fan geometry [59, 60]. Figure 24.8(a) shows MRI_{prior}, ground-truth on-board CT image in end of inspiration (EOI) phase, ground-truth on-board MRI in EOI phase, and estimated on-board MRI in EOI phase. Figure 24.8(b) shows the subtraction images for axial, coronal, and sagittal views, as in figure 24.8(a). Table 24.1 shows volume percent difference (VPD), volume dice coefficient (VDC),

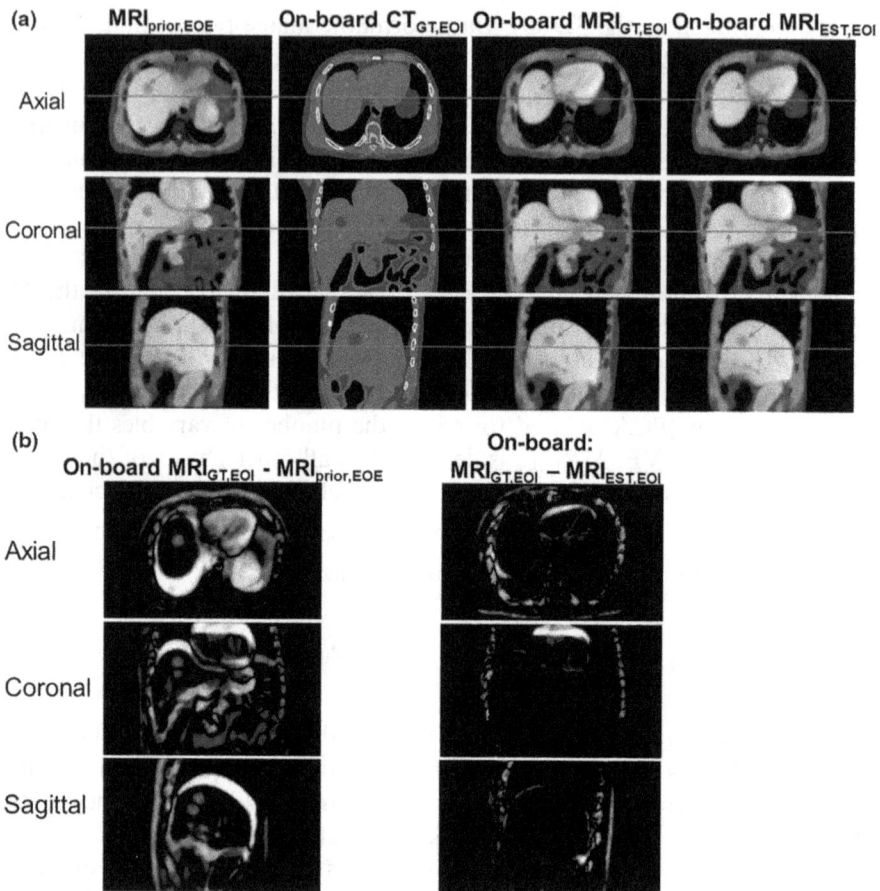

Figure 24.8. (a) The columns from left to right are the following: MRI_{prior} at EOE phase, ground-truth on-board CT at EOI phase ($EOI_{GT,EOI}$), ground-truth on-board MRI at EOI phase ($MRI_{GT,EOI}$), and estimated on-board MRI at EOI phase ($MRI_{EST,EOI}$) for XCAT scenario 2, simulating respiratory amplitude decrease from simulation to treatment. The rows represent axial, coronal, and sagittal views, respectively. The horizontal line and arrows indicate areas for comparison. (b) Subtraction images for axial, coronal, and sagittal images shown in (a). Reprinted from [55] with permission from Elsevier.

and center of mass shift (COMS) for the estimated and ground-truth tumor volumes in the XCAT scenarios for a variety of kV projection angles and numbers. VPD and COMS definitions can be found in [17], and VDC definition can be found in [19]. The virtual 4D-MRI was generated for three different on-board scenarios: scenario 1 has no respiratory change from simulation to treatment, scenario 2 has amplitude decrease in both superior–inferior (SI) and anterior–posterior (AP) directions from simulation to treatment, and scenario 3 has amplitude increase from simulation to treatment. Evaluation was performed at EOI phase, as to show the greatest deformation from MRI_{prior}, which is defined at EOE phase. Based on the XCAT data, on-board virtual 4D-MRI was successfully estimated for all XCAT scenarios using all scan angles and number of projections that were simulated.

Table 24.1. VPD (%), VDC, and COMS (mm) values for the on-board virtual MRI with the three XCAT scenarios. All results are shown for the EOI phase images. Reprinted from [55] with permission from Elsevier.

		Scenario 1	Scenario 2	Scenario 3
		VPD (%)		
Ortho-view 30°	102 projections	9.89	9.00	11.58
Ortho-view 15°	168 projections	10.92	8.87	11.80
Single-view 100°	167 projections	9.93	8.61	14.35
Single-view 100°	41 projections	11.87	8.61	13.01
Single-view 200°	81 projections	11.40	7.21	11.15
		VDC		
Ortho-view 30°	102 projections	0.95	0.96	0.94
Ortho-view 15°	168 projections	0.95	0.96	0.94
Single-view 100°	167 projections	0.95	0.96	0.93
Single-view 100°	41 projections	0.94	0.96	0.93
Single-view 200°	81 projections	0.94	0.96	0.94
		COMS (mm)		
Ortho-view 30°	102 projections	0.79	0.80	1.05
Ortho-view 15°	168 projections	0.93	0.76	1.07
Single-view 100°	167 projections	0.82	0.67	1.25
Single-view 100°	41 projections	0.94	0.69	1.04
Single-view 200°	81 projections	0.92	0.57	0.95

For the patient study, a retrospective analysis was performed using one liver cancer patient who had both 4D MRI and 4D CT data. The 4D MRI data was used as the 'prior' data, and the 4D CT data were used as the 'on-board' volumes for the study. Orthogonal-view 30° scan angle CBCT projections (102 total projections) were simulated from the 4D CT data and used as on-board kV projections. Figure 24.9 shows MRI_{prior} at EOE phase, on-board CT at EOI phase, reference on-board MRI at EOI phase, and estimated on-board MRI at EOI phase for the patient data. The reference on-board MRI was generated by deforming the EOI phase of the prior 4D-MRI to the EOI phase of the on-board 4D-CT. The reference on-board MRI visually matches well with the estimated on-board MRI.

The on-board virtual 4D-MRI technique is a novel concept to use prior images to enhance the on-board image contrast for liver SBRT target localization by using a conventional LINAC with on-board kV imaging capabilities. The method provides a potential solution to address the issue of poor image quality and soft-tissue contrast and utilize the current LINACS with x-ray capabilities to provide MR-guided radiotherapy to benefit abdominal cancers. The method can also be implemented by using other modalities for the prior images to generate on-board multimodality images. It can also be applied for proton or heavy ion therapy machines with x-ray imaging systems to generate on-board multimodality images. This method is not yet available

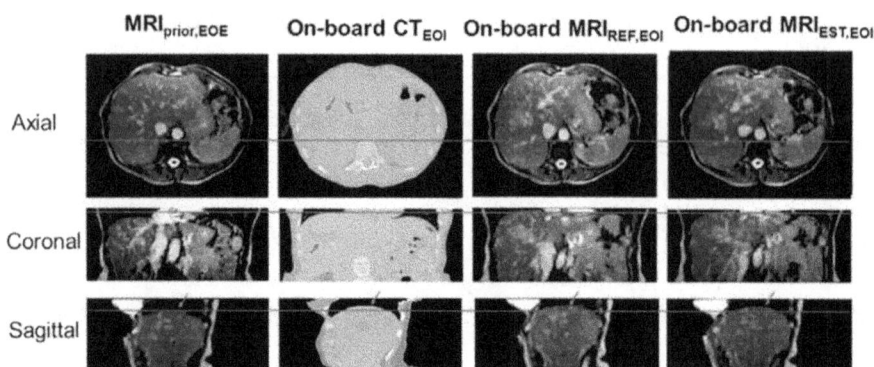

Figure 24.9. The columns from left to right show the following: MRI_{prior} at EOE phase ($MRI_{prior,EOE}$), on-board CT at EOI phase (CT_{EOI}), reference on-board MRI at EOI phase ($MRI_{REF,EOI}$), and estimated on-board MRI at EOI phase ($MRI_{EST,EOI}$) for the patient study. The rows show the axial, coronal, and sagittal views, respectively. The horizontal line and arrows indicate areas for comparison. Reprinted from [55] with permission from Elsevier.

clinically, but preliminary results from XCAT phantom studies and a single liver patient demonstrated the potential for virtual MRI-based image guidance using a conventional LINAC. More details about the method can be found in [55].

References

[1] Xing L *et al* 2006 Overview of image-guided radiation therapy *Med. Dosim.* **31** 91–112
[2] Borst G R *et al* 2007 Kilo-voltage cone-beam computed tomography setup measurements for lung cancer patients; first clinical results and comparison with electronic portal-imaging device *Int. J. Radiat. Oncol. Biol. Phys.* **68** 555–61
[3] Wen N *et al* 2007 Dose delivered from Varian's CBCT to patients receiving IMRT for prostate cancer *Phys. Med. Biol.* **52** 2267
[4] Case R B *et al* 2009 Inter-and intrafraction variability in liver position in non–breath-hold stereotactic body radiotherapy *Int. J. Radiat. Oncol. Biol. Phys.* **75** 302–8
[5] Den R B *et al* 2010 Daily image guidance with cone-beam computed tomography for head-and-neck cancer intensity-modulated radiotherapy: a prospective study *Int. J. Radiat. Oncol. Biol. Phys.* **76** 1353–59
[6] Lewis J H *et al* 2011 Mitigation of motion artifacts in CBCT of lung tumors based on tracked tumor motion during CBCT acquisition *Phys. Med. Biol.* **56** 5485
[7] Vergalasova I, Maurer J and Yin F F 2011 Potential underestimation of the internal target volume (ITV) from free-breathing CBCT *Med. Phys.* **38** 4689–99
[8] Sonke J J *et al* 2005 Respiratory correlated cone beam CT *Med. Phys.* **32** 1176–86
[9] Vergalasova I, Cai J and Yin F F 2012 A novel technique for markerless, self-sorted 4D-CBCT: feasibility study *Med. Phys.* **39** 1442–51
[10] Zhang Y *et al* 2013 A technique for estimating 4D-CBCT using prior knowledge and limited-angle projections *Med. Phys.* **40** 121701
[11] Sweeney R A *et al* 2012 Accuracy and inter-observer variability of 3D versus 4D cone-beam CT based image-guidance in SBRT for lung tumors *Radiat. Oncol.* **7** 81

[12] Lu J *et al* 2007 Four-dimensional cone beam CT with adaptive gantry rotation and adaptive data sampling *Med. Phys.* **34** 3520–9

[13] Sidky E Y and Pan X 2008 Image reconstruction in circular cone-beam computed tomography by constrained, total-variation minimization *Phys. Med. Biol.* **53** 4777–807

[14] Ren L *et al* 2008 A novel digital tomosynthesis (DTS) reconstruction method using a deformation field map *Med. Phys.* **35** 3110–5

[15] Rit S *et al* 2009 On-the-fly motion-compensated cone-beam CT using an *a priori* model of the respiratory motion *Med. Phys.* **36** 2283–96

[16] Jia X *et al* 2010 GPU-based fast cone beam CT reconstruction from undersampled and noisy projection data via total variation *Med. Phys.* **37** 1757–60

[17] Jia X *et al* 2011 GPU-based iterative cone-beam CT reconstruction using tight frame regularization *Phys. Med. Biol.* **56** 3787–807

[18] Wang J and Gu X 2013 Simultaneous motion estimation and image reconstruction (SMEIR) for 4D cone-beam CT *Med. Phys.* **40** 101912

[19] Zhang Y *et al* 2015 Preliminary clinical evaluation of a 4D-CBCT estimation technique using prior information and limited-angle projections *Radiother. Oncol.* **115** 22–9

[20] Hansen D C and Sørensen T S 2018 Fast 4D cone-beam CT from 60 s acquisitions *Phys. Imag. Radiat. Oncol.* **5** 69–75

[21] Zhang Y *et al* 2019 4D liver tumor localization using cone-beam projections and a biomechanical model *Radiother. Oncol.* **133** 183–92

[22] Beddar A S *et al* 2008 4D-CT imaging with synchronized intravenous contrast injection to improve delineation of liver tumors for treatment planning *Radiother. Oncol.* **87** 445–8

[23] Park J C *et al* 2012 Liver motion during cone beam computed tomography guided stereotactic body radiation therapy *Med. Phys.* **39** 6431–42

[24] Zeng C *et al* 2019 Intrafraction tumor motion during deep inspiration breath hold pancreatic cancer treatment *J. Appl. Clin. Med. Phys.* **20** 37–43

[25] Ren L *et al* 2012 Development and clinical evaluation of a three-dimensional cone-beam computed tomography estimation method using a deformation field map *Int. J. Radiat. Oncol. Biol. Phys.* **82** 1584–93

[26] Wang J and Gu X 2013 High-quality four-dimensional cone-beam CT by deforming prior images *Phys. Med. Biol.* **58** 231–46

[27] Li R *et al* 2010 Real-time volumetric image reconstruction and 3D tumor localization based on a single x-ray projection image for lung cancer radiotherapy *Med. Phys.* **37** 2822–6

[28] Zhang Y *et al* 2015 Preliminary clinical evaluation of a 4D-CBCT estimation technique using prior information and limited-angle projections *Radiother. Oncol.* **115** 22 9

[29] Lu W *et al* 2004 Fast free-form deformable registration via calculus of variations *Phys. Med. Biol.* **49** 3067–87

[30] Wu Q J *et al* 2011 Adaptive radiation therapy: technical components and clinical applications *Cancer J.* **17** 182–9

[31] Zhang Y, Yin F F and Ren L 2015 Dosimetric verification of lung cancer treatment using the CBCTs estimated from limited-angle on-board projections *Med. Phys.* **42** 4783–95

[32] Hardcastle N *et al* 2013 Accuracy of deformable image registration for contour propagation in adaptive lung radiotherapy *Radiat. Oncol.* **8** 243

[33] Simon A *et al* 2015 Roles of deformable image registration in adaptive RT: from contour propagation to dose monitoring *2015 37th Annual Int. Conf. of the IEEE Engineering in Medicine and Biology Society (EMBC)* (Piscataway, NJ: IEEE)

[34] Zhang Y, Tehrani J N and Wang J 2017 A Biomechanical Modeling Guided CBCT Estimation Technique *IEEE Trans. Med. Imaging* **36** 641–52

[35] Brock K K *et al* 2005 Accuracy of finite element model-based multi-organ deformable image registration *Med. Phys.* **32** 1647–59

[36] Werner R *et al* 2009 Patient-specific finite element modeling of respiratory lung motion using 4D CT image data *Med. Phys.* **36** 1500–11

[37] Al-Mayah A *et al* 2010 Biomechanical-based image registration for head and neck radiation treatment *Phys. Med. Biol.* **55** 6491–500

[38] Al-Mayah A *et al* 2011 Toward efficient biomechanical-based deformable image registration of lungs for image-guided radiotherapy *Phys. Med. Biol.* **56** 4701

[39] Gobbi D G and Peters T M 2003 Generalized 3D nonlinear transformations for medical imaging: an object-oriented implementation in VTK *Comput. Med. Imaging Graph.* **27** 255–65

[40] Chen M *et al* 2008 A simple fixed-point approach to invert a deformation field *Med. Phys.* **35** 81–8

[41] Maas S A *et al* 2012 FEBio: finite elements for biomechanics *J. Biomech. Eng.* **134** 011005

[42] Nasehi Tehrani J and Wang J 2015 mooney-Rivlin biomechanical modeling of lung with Inhomogeneous material *Conf. Proc. IEEE Eng. Med. Biol. Soc.* **2015** 7897–900

[43] Tehrani J N *et al* 2015 Sensitivity of tumor motion simulation accuracy to lung biomechanical modeling approaches and parameters *Phys. Med. Biol.* **60** 8833–49

[44] Zhang Y *et al* 2019 Enhancing liver tumor localization accuracy by prior-knowledge-guided motion modeling and a biomechanical model *Quant. Imag. Med Surg* **9** 1337–49

[45] Zhong Z *et al* 2016 4D cone-beam CT reconstruction using multi-organ meshes for sliding motion modeling *Phys. Med. Biol.* **61** 996

[46] Boldrini L *et al* 2019 Online adaptive magnetic resonance guided radiotherapy for pancreatic cancer: state of the art, pearls and pitfalls *Radiat. Oncol.* **14** 71

[47] Feldkamp L A, Davis L C and Kress J W 1984 Practical cone-beam algorithm *J. Opt. Soc. Am. A* **1** 612–19

[48] Ren L, Zhang Y and Yin F F 2014 A limited-angle intrafraction verification (LIVE) system for radiation therapy *Med. Phys.* **41** 020701

[49] Harris W *et al* 2016 A technique for generating volumetric cine-magnetic resonance imaging *Int. J. Radiat. Oncol. Biol. Phys.* **95** 844–53

[50] Harris W *et al* 2017 Estimating 4D-CBCT from prior information and extremely limited angle projections using structural PCA and weighted free-form deformation for lung radiotherapy *Med. Phys.* **44** 1089–104

[51] Fallone B G *et al* 2009 First MR images obtained during megavoltage photon irradiation from a prototype integrated linac-MR system *Med. Phys.* **36** 2084–8

[52] Raaymakers B W *et al* 2009 Integrating a 1.5 T MRI scanner with a 6 MV accelerator: proof of concept *Phys. Med. Biol.* **54** N229–37

[53] Mutic S and Dempsey J F 2014 The ViewRay system: magnetic resonance-guided and controlled radiotherapy *Semin. Radiat. Oncol.* **24** 196–9

[54] Winkel D *et al* 2019 Dosimetric benefit of the first clinical SBRT of lymph node oligometastases on the 1.5 T MR-linac *Radiother. Oncol.* **133** EP–2006

[55] Harris W *et al* 2018 A novel method to generate on-board 4D MRI using prior 4D MRI and on-board kV projections from a conventional LINAC for target localization in liver SBRT *Med. Phys.* **45** 3238–45

[56] Dowling J A *et al* 2012 An atlas-based electron density mapping method for magnetic resonance imaging (MRI)-alone treatment planning and adaptive MRI-based prostate radiation therapy *Int. J. Radiat. Oncol. Biol. Phys.* **83** e5–11

[57] Kraus K M *et al* 2017 Generation of synthetic CT data using patient specific daily MR image data and image registration *Phys. Med. Biol.* **62** 1358–77

[58] Guerreiro F *et al* 2017 Evaluation of a multi-atlas CT synthesis approach for MRI-only radiotherapy treatment planning *Phys. Med.* **35** 7–17

[59] Siddon R L 1985 Fast calculation of the exact radiological path for a three-dimensional CT array *Med. Phys.* **12** 252–5

[60] Yan H *et al* 2007 Accelerating reconstruction of reference digital tomosynthesis using graphics hardware *Med. Phys.* **34** 3768–76

IOP Publishing

Principles and Practice of Image-Guided Abdominal
Radiation Therapy

Yu Kuang

Chapter 25

Y-90 radioembolization for liver cancer

Tianjun Ma and Naichang Yu

Yttrium-90 embedded microsphere based radioembolization has become one of the major treatment modalities in the management of unresectable primary and metastatic liver tumors. Significant progress has been made in the physical aspects of the treatment, as well as the clinical aspects of efficacy and safety of the treatment. In this chapter, we review the normal liver and tumor vascular structures relevant for the treatment, geometric and radiological properties of the microspheres, imaging procedures for the planning and execution of the treatment, and imaging for the post treatment dose assessment.

25.1 Introduction

25.1.1 Rationale for radioembolization

Liver cancer is the seventh most common cancer and the third most common cause of cancer mortality worldwide. For patients with early stage cancer who have good liver reserves, surgery (liver resection and liver transplant) is the choice of treatment that may be curative. However, most patients with liver cancer, when diagnosed, are ineligible for surgery due to poor liver function. For these patients, locoregional therapies, including radiofrequency ablation, stereotactic body radiation therapy (SBRT), conventional transarterial chemoembolization (TACE), and radioembolization, are documented in the National Comprehensive Cancer Network guidelines as effective methods to limit tumor progression and bridge the patient to transplant as they allow the patient more time to wait for donor organs [1–3].

For normal liver parenchyma, the portal vein system supplies the majority of blood (around 80%), whereas for tumorous liver tissue, the hepatic arterial system supplies up to 80% of the blood. This contrast of the source of blood supply between the normal parenchyma and tumorous liver tissues has been exploited to supply cancer-fighting agents through the liver arteries, to be deposited preferentially in the tumors, as in TACE and radioembolization.

doi:10.1088/978-0-7503-2468-7ch25

Radioembolization has been gaining increasing traction as a modality of palliative treatment to limit tumor progression. Currently, in clinical practices, the agent is Y-90 embedded glass or resin microspheres. These microspheres are delivered through an appropriately placed catheter through the hepatic artery and are preferentially lodged permanently in the capillaries to occlude the capillaries, and the short-range beta radiation from the microspheres also delivers radiation doses to the tumor. In the last 20 years, much progress has been made in the understanding of the microscopic distribution of the microspheres in the tumor and normal liver, special dose distribution characteristics of radiation delivered by the microspheres, the imaging of the distribution of the microspheres, and the planning and evaluation of the treatments with refined dose distribution. These signs of progress and the understanding of the dose-dependency of clinical efficacy and toxicity of the treatment have made this treatment modality a choice in the palliation of liver tumors for both primary and metastatic liver tumors.

25.1.2 Liver tumors and radioembolization efficacy

Clinical benefits have been shown for hepatocellular carcinoma (HCC), intra-hepatic cholangiocarcinoma (ICC), and liver metastases. HCC accounts for the majority (75% to 85%) of the primary liver cancers. The portal vein is the major blood supply for the normal liver, whereas for HCC and hypervascular metastases, the hepatic artery becomes the dominant blood supplier [4]. For hypovascular metastases (such as colorectal cancer), the hepatic artery also has a greater blood supply compared with the liver parenchyma [5]. Compared with other treatment options, radioembolization has demonstrated durable local control and a good safety profile [6].

ICC accounts for 10% to 15% of primary liver cancers [7]. ICC lesions are often asymptomatic, resulting in patients often presenting with locally advanced tumors. Similar to HCC, surgery offers the highest curative potential; however, many tumors are deemed unresectable at the time of diagnosis [8]. Radioembolization offers an alternative option for primary and secondary ICC lesions [9, 10].

For liver metastasis, better tumor response, longer time to progression with similar survival outcome, and acceptable toxicity have also been reported [11–16].

25.1.3 Liver anatomy

Knowledge of the liver blood flow helps explain the hydraulic flow of the micro-sphere as well as its dose delivery. The liver is a wedge-shaped organ located mainly in the upper right quadrant of the abdominal cavity, inferior to the diaphragm, and to the right of the stomach and gallbladder [17]. The falciform ligament divides the liver into the left and right lobe, more specifically into four lobes (left, right, caudate, and quadrate). However, microsphere brachytherapy delivers dose according to the functionality of the liver, described by the widely used Couinaud classification system [18]. In the Couinaud system, the liver is divided into eight functional segments based on a transverse plane through the bifurcation of the main portal vein [19]. Each segment has its own vascular inflow, outflow, and biliary drainage. Each segment has a centrally located branch of the portal vein, hepatic artery and bile duct as inflow, and a peripherally located branch of hepatic veins as outflow. This autonomy enables the resection of a given segment independent of the remaining segments. Similarly, if the radioactive microspheres can be purposely delivered to the desired segment while maintaining the integrity of the remaining segments, normal liver function can be preserved with minimal invasion.

25.1.4 Patient eligibility

To determine whether radioembolization is appropriate for each patient, specific indications and contraindications are given. There are many clinical trials utilizing radioembolization considering unresectable primary and secondary liver malignancies.

General indications include, other than unresectable hepatic primary or secondary disease, a life expectancy greater than three months, age greater than 18 years, and Eastern Cooperative Oncology Group (ECOG) performance status less than or equal to 2.

General contraindications include previous external beam radiation therapy to the liver, limited liver reserve, extensive and untreated portal hypertension, uncorrected flow to the gastrointestinal tract, and greater than 20% of lung shunting or estimated single session dose to lung greater than 30 Gy [20, 21].

25.1.5 Radionuclide

Y-90 is a pure beta-emitting radioactive isotope without primary gamma-ray components. It emits maximum energy of 2.27 MeV electrons with mean energy at 0.93 MeV. The maximum range of electrons is 11 mm with a mean of 2.5 mm. Y-90 decays into Zr-90 with a half-life of approximately 64.1 h. It is widely used for liver radioembolization compared with other radioisotopes, such as I-131, Re-188, and Ho-166.

There are two commercially available Y-90 microsphere products available for clinical practice : SIR-Spheres® (Sirtex Medical Limited, St Leaonards, NSW, Australia) and TheraSphere™ (Boston Scientific Corporation, Ottawa, Canada). SIR-Spheres® are biocompatible Y-90 doped resins with a median diameter of 32.5 μm ranging from 20 to 60 μm, while TheraSphere™ is glass microspheres containing the radioactive isotope, with a mean diameter of 20–30 μm. One of the major differences between SIR-Spheres® and TheraSphere™ is that SIR-Spheres® have only one activity size in a single vial whereas TheraSphere™ has multiple activity sizes per vial. The other major difference is the assumed activity per particle: 50 Bq for resin microspheres compared with 2500 Bq for glass microspheres. Therefore, for a given activity, SIR-Spheres® will have more particles than TheraSphere™, which is why early stasis has been seen more in cases with SIR-Spheres® than with TheraSphere™ [22]. These differences between the two types of microspheres lead to differences in microscopic dose distribution characteristics, as well as to possible differences in embolization effect.

SIR-Spheres® resin microspheres are approved for the treatment of unresectable metastatic liver tumors from primary colorectal cancer with adjuvant intrahepatic artery chemotherapy of FUDR (floxuridine) by the USA Food and Drug Administration (FDA). TheraSphere™ is approved for radiation treatment or as a neoadjuvant to surgery or transplantation in patients with unresectable HCC who can have appropriately positioned hepatic arterial catheters placed under a Humanitarian Device Exemption (HDE) in the USA. The HDE means that even though the device demonstrates safety usage, the effectiveness is not assured, which means the institutional review board (IRB) must oversee the usage of the product.

25.2 Treatment preparation

Pretreatment assessment, including history, laboratory testing, and imaging, is essential for treatment planning as well as post-treatment evaluation. Usually, patients follow the pretreatment protocol approximately one to two weeks before the treatment.

25.2.1 Angiography

For a patient potentially undergoing microsphere brachytherapy, pretreatment angiograms will be performed. A catheter will be placed to the hepatic artery via a patient's femoral artery by the interventional radiologist. After verification of the catheter position through digital subtraction angiography (DSA), the catheter will be secured in place. The angiograms will be used to assess the blood flow and guide the catheter delivery positioning as indicated in figure 25.1. The angiograms should at least include abdominal aortogram, superior mesenteric and celiac arteriogram, and right and left hepatic arteriogram. Depending on the complicity of the aberrant vascular structure, further angiograms may be required since hepatic vessel variation is expected in 45% of the patient population [23].

As a safety precaution, prophylactic embolization is recommended for all extrahepatic vessels, including the gastroduodenal, right gastric, and potentially other extrahepatic vessels, to prevent the inadvertent deposition of the microsphere into the gastrointestinal tract. Angiograms should be performed immediately before the treatment to confirm the arterial anatomy has not deviated from the initial angiograms.

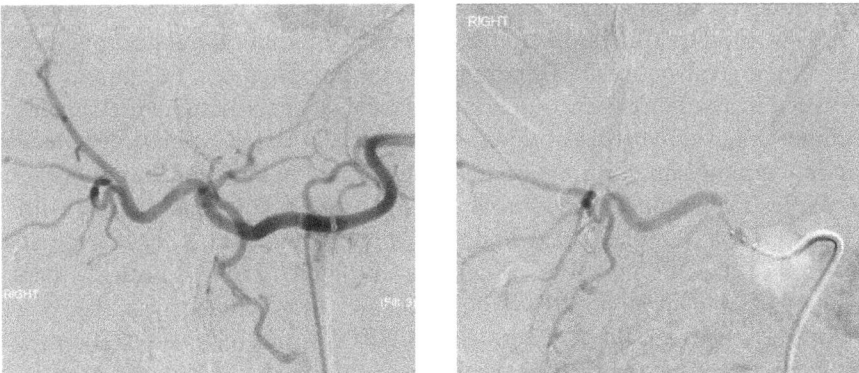

Figure 25.1. DSA images before (left) and after (right) the blockage of the blood vessel.

25.2.2 Lung shunt

Ideally, the Y-90 microspheres should form embolization at the hepatic terminal arteries inside the tumor. In reality, the microspheres can bypass the capillary bed of the liver to the lungs through direct arteriovenous shunting, resulting in radiation to the lungs [20]. If the patient has considerable shunting activity of the Y-90 microspheres to the lungs, the microsphere treatment will increase the risk of radiation-induced pneumonitis [24]. Evaluation procedures need to be done to accurately assess the ratio of the lung shunt or lung shunt fraction (LSF).

To evaluate whether the patient has a considerable amount of lung shunt, after the angiographic evaluation, successful deployment of the catheter, and the embolization of extrahepatic vessels, Technetium-99m–labeled macroaggregated albumin (^{99m}Tc-MAA) is infused through the delivery point (end of the catheter) into the liver. Depending on the manufacturer, ^{99m}Tc-MAA usually comes in a vial of 10 mL solution containing aggregated albumin particles; at least 90% of the particle diameters are between 10 and 90 μm, with maximum diameter not exceeding 150 μm. Due to the similarity in dimension, ^{99m}Tc-MAA is used as a surrogate to predict the distribution of Y-90 microspheres. However, the dimension and the shape of the ^{99m}Tc-MAA is not exactly the same as either of the commercially available Y-90 microspheres. There will be differences between the real-treatment and ^{99m}Tc-MAA simulation based on the similarity assumption.

After manufacture, the bounds between ^{99m}Tc-pertechnetate and albumin will remain for 6–12 h depending on the manufacturer. Due to the short effective half-life of ^{99m}Tc-MAA, it is recommended that imaging begin within at least 1 h of administration and not to exceed 4 h. Conventionally, a planar scintigraphy image will be taken to detect the two-dimensional (2D) distribution of ^{99m}Tc-MAA.

Once the 2D images are acquired, regions of interest (ROIs) need to be drawn on the acquired images, specifically lung and liver. The counts inside each ROI are used to calculate the lung shunt based on the equation below:

$$LSF = \frac{Count_{Lung}}{Count_{Lung} + Count_{Liver}}$$

Any patients with lung shunt larger than 20% will be disqualified from the Y-90 radioembolization treatment due to the high risk of radiation-induced pneumonitis.

Even though conventional 2D gamma images provide a way to estimate the lung shunt, it is inherently deficient in certain aspects, such as tissue heterogeneity and three-dimensional activity distribution. Moreover, 2D-based estimation suffers great uncertainty in lung shunt estimation [25]. Single-photon emission computed tomography (SPECT)/computed tomography (CT), a SPECT reconstruction with CT-based corrections, is implemented clinically to improve the accuracy in assessing the disposition of the ^{99m}Tc-MAA as shown in figure 25.2. However, there are challenges associated with SPECT/CT. One of the challenges is the potential misregistration between the ROIs in lengthy SPECT and relatively static lung CT, which is introduced by the breathing motion during the SPECT acquisition, as well as spillover due to the partial volume averaging effect. One way to avoid this

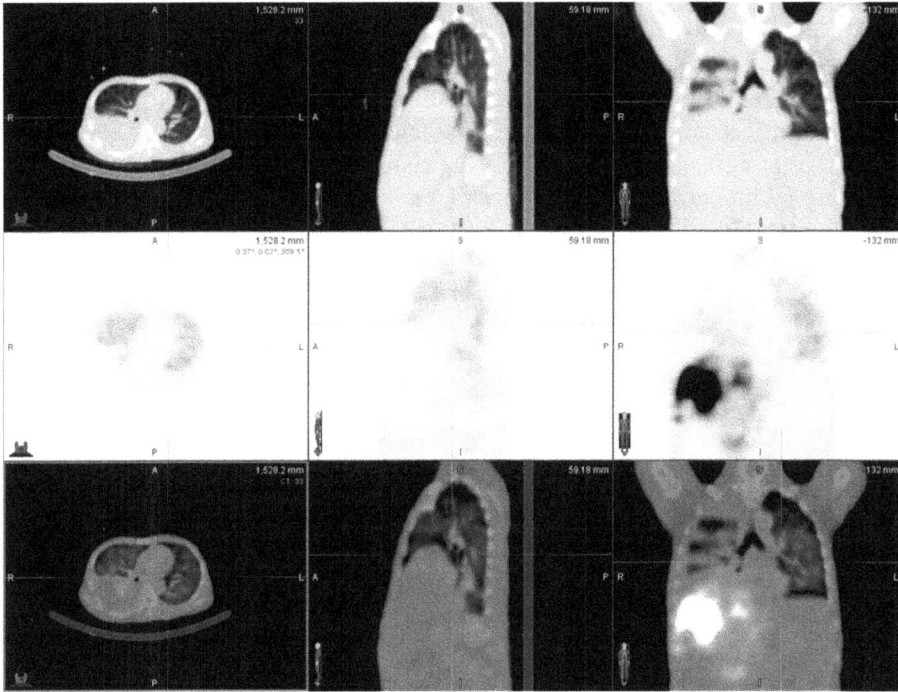

Figure 25.2. Images of SPECT/CT for lung shunt estimation.

inconsistency is to exclude the lung region within 2 cm of the diaphragm [25]. Another approach is to use a motion management technique, such as gating, to alleviate the motion effect [26].

Due to the variation in physical shape and dimension between MAA particles and microspheres, the LSF may not be accurately and reproducibly estimated. A recent study has investigated the usage of identical particles to perform a pretreatment procedure. Different imaging techniques, including planar images, SPECT/CT, and positron emission tomography (PET)/CT, were compared in a phantom study using a Y-90–based pretreatment procedure [27]. The researchers found that planar images overestimated the LSF, which is consistent with the finding of Yu *et al* [25]. The clinical SPECT reconstruction protocol also overestimated the LSF compared with SPECT with a Monte Carlo–based reconstruction protocol and PET reconstructions. There is also a clinical trial investigating whether low-dose Y-90 microspheres can be used as an alternative to MAA (NCT04172714).

25.2.3 Treatment planning dosimetry

25.2.3.1 Medical Internal Radiation Dose and body surface area dosimetry
The Medical Internal Radiation Dose (MIRD) Committee of the Society of Nuclear Medicine developed the current absorbed dose calculation standard for Y-90 microspheres [28]. A uniformed tissue dose is assumed in this dose schema; thus the dose rate inside the tissue can be viewed as

$$\dot{D} = k\frac{A}{m}\overline{E},$$

where k is a unit conversion constant, A is the activity, m is the mass, and E is the average emitted energy. Due to the characteristic of beta decay, the energy is assumed to deposit completely within the tissue itself. Additionally, Y-90 microspheres are permanently implanted, which means all the Y-90 will decay inside the tissue. Thus,

$$D = k\frac{A_0}{m}\overline{E}\frac{T_{1/2}}{\ln(2)} = 49.38\frac{A_0}{m}.$$

For SIR-Spheres®, body surface area (BSA) is recommended by the manufacturer as a general approach to estimate the prescribed activity. BSA value must be calculated first as

$$BSA\ (m^2) = 0.20247 \times \text{height(m)}^{0.725} \times \text{weight(kg)}^{0.425}.$$

Then the prescribed activity of SIR-Spheres® (GBq) for whole liver or bilobar treatment can be calculated as

$$A(GBq) = (BSA - 0.2) + \left(\frac{V_{\text{Tumor}}}{V_{\text{Tumor}} + V_{\text{Normal liver}}}\right),$$

where A is the prescribed activity, V_{Tumor} is the total tumor volume, and $V_{\text{Normal liver}}$ is the normal liver tissue.

The prescribed activity of TheraSphere™ treatment is calculated as

$$A\ (GBq) = \frac{D\ (Gy) \times\ m_{\text{Liver}}\ (kg)}{49.38 \times (1 - LSF)(1 - R)},$$

where A is the prescribed activity or injected activity as indicated by the manufacturer, D is the dose to the liver, m_{Liver} is the liver mass, LSF is the lung shunt (can be abbreviated as L or F as indicated by the manufacturer), and R is the residual activity after the treatment. The typical dose to the liver is between 80 and 150 Gy.

Unlike SIR-Spheres®, TheraSphere™ is available in six different dose sizes, from 3 to 20 GBq, provided in 0.6 mL of sterile, pyrogen-free water in a 1.0 mL v-bottom vial, secured with a clear acrylic vial shield. In addition, the microspheres vial can be shelved up to 12 days following the calibration time stamp.

25.2.3.2 Anatomic partition model

A refinement of the previous models takes advantage of pretreatment ⁹⁹ᵐTc-MAA SPECT images. The liver is divided into tumor and normal liver compartments, and the ratio of activities $R_{\text{T/N}}$ is defined as

$$R_{T/N} = \left[\frac{A_{\text{Tumor}}}{m_{\text{Tumor}}}\right] \bigg/ \left[\frac{A_{\text{Normal liver}}}{m_{\text{Normal liver}}}\right]$$

where A_{Tumor} and $A_{\text{Normal liver}}$ are the activity inside the tumor and normal liver volume, respectively, m_{Tumor} is the mass of tumor, and $m_{\text{Normal liver}}$ is the mass of normal liver in kilogram.

And the prescribed activity will be

$$A(GBq) = D(Gy) \times \frac{[(R_{T/N} \times m_{\text{Tumor}}) + m_{\text{ Normal liver}}]}{[49.38 \times (1 - LSF)(1 - R)]}$$

The partition model requires accurate segmentation and quantification of the preplan imaging and therefore is commonly used for well-defined tumors [29].

25.2.4 Radiation lobectomy

In contrast to multifocal bilobar (or whole liver) delivery, radiation lobectomy was initially described in 2009 [30, 31]. In addition to the general requirements for Y-90 treatment, patients who are eligible for radiation lobectomy usually have good performance status and liver function, as well as a unilobar disease site [32]. Even though radiation lobectomy is off-label usage of the Y-90 device, using radiation lobectomy with Y-90 as a bridge to surgical resection has been proven to be safe and effective [33]. It has a comparable effect in hypertrophy to portal vein embolization, which is a standard technique to prepare patients with surgically unresectable disease to increase future liver remnant. A higher target dose (200 Gy) compared with the conventional 120–140 Gy prescription is usually used with TheraSphere™ [34]. For dosimetry calculation, replace the total liver mass with the lobe mass, and the infusion location will change from proper hepatic artery to lobar artery.

25.2.5 Radiation segmentectomy

Shortly after radiation lobectomy, radiation segmentectomy was described in 2011 [35]. Instead of injecting the entire lobar with Y-90 microspheres, the investigators administered the radiation through segmental arteries, thereby sparing the remaining segments with a boosted median dose. Similar to radiation lobectomy, radiation segmentectomy is an off-label use for both Y-90 products. To be eligible, generally, the patient's disease is confined to one or two segments. In this case, the calculated mass for the radiation dose will be the segment mass instead of the liver mass. Recent data provided promising results for tumoricidal effect with minimal side effects; however, further investigation is still needed [36]. A study has also been done to combine radiation segmentectomy with liver SBRT for HCC [37]. Patients are often offered SBRT due to the incomplete treatment of radioembolization or chemoembolization, whereby the location of the untreated lesion may not be effectively treated through intravascular therapy. In addition, gross vascular invasion penetrating deeper into the venous structures would also benefit from combined therapy. However, a major difficulty is to generate a composite dose for evaluation purpose, since the dose calculation methods between the two techniques are vastly different.

25.2.6 Activity measurement methods

A standard method to measure the beta emitter is through liquid-scintillation counting to determine the activity of the radionuclides. Different methods can be applied to determine the efficiency: the Centro de Investigaciones Energéticas, Medioambientales y Tecnológicas (CIEMAT)/National Institute of Standards and Technology (NIST) efficiency tracing method and the triple-to-double coincidence ratio method. This type of calibration is typically performed at a country's national metrology institute, which is the NIST in the USA.

For local clinics, validating the Y-90 activity upon receiving the microspheres should be integrated into the microsphere radioembolization program. Depending upon the local clinic setup, the activity verification step can be done by a dose calibrator located in the radiation safety department, nuclear medicine department, or radiation oncology department. The variations of the source position, measuring geometry, and the materials of the container can affect the final reading from the local calibrator. Thus, the activity verification procedure of the microspheres should be done in a fixed geometrical technique, potentially with multiple measurements at symmetrical angles. Moreover, a local reference should be established and documented as a consistency measure.

25.2.7 Administrative and regulatory issues

The USA Nuclear Regulatory Commission (NRC) deems Y-90 microspheres as a medical device as they are not radiopharmaceuticals. The licensee must follow the requirements for brachytherapy sources listed in 10 CFR 35 unless otherwise specified for regulatory relief. If a licensee is starting a microsphere brachytherapy program in an agreement state, they should contact the local regulatory agency for any additional requirements.

A trained physician specialized in radiation therapy, nuclear medicine, or interventional radiology who has completed the vendor-specific training and met the supervised case number requirement can pursue becoming an authorized user (AU) of Y-90 microspheres.

25.3 Treatment

25.3.1 Treatment procedure

Depending on the tumor locations, the number of lesions, arterial flow, and the liver function, the microsphere treatment can be done within a single segment of the liver through a major artery or an arterial branch through catheterization or radiation segmentectomy (usually infuses Y-90 microspheres to one or two hepatic segments). This enables elevated dosage to the tumor and decreased radiation to healthy hepatic tissue. Radiation lobectomy is comparably less precise and sparing, treating either the entire left or right hepatic lobe. Radiation lobectomy is recommended when surgical resection is deemed not feasible or the remnant liver is not able to function adequately, which results in tumor control and contralateral lobe hypertrophy [33, 38]. The final option is whole liver radioembolization when numerous tumors are scattered in both lobes. The liver treatment can be done either sequentially (treating each lobe at a time with at least 30 day intervals) or singularly (treating both lobes at the same time). Fewer adverse side effects were observed in the sequential technique [39].

Both manufacturers have published their recommended approach to administering Y-90 microspheres. An administration toolkit is required along with multiple accessories. The AU should perform the treatment according to the recommended procedure.

A written directive should be recorded, including the patient name, treatment date, treatment site, radioactive material and its manufacturer, prescribed dose or activity, maximum dose or activity to tissues outside the primary site due to shunting, and the signature of the AU.

25.3.2 Medical event

A Y-90 microsphere medical event defined by the NRC in 10 CFR 35.3045(2) is described below:

1. The total source strength administered differing by 20% or more from the total source strength documented in the post-implantation portion of the written directive;
2. The total source strength administered outside of the treatment site exceeding 20% of the total source strength documented in the post-implantation portion of the written directive; or
3. An administration that includes any of the following:
   (a) Wrong radionuclide;
   (b) Wrong individual;
   (c) Sealed source(s) implanted directly into a location discontiguous from the treatment site, as documented in the post-implantation portion of the written directive; or
   (d) A leaking sealed source resulting in a dose that exceeds 0.5 Sv to an organ or tissue.

There are many reported medical events in the NRC archive. It is always a good practice to learn from the errors of the past. In a recent NRC report regarding Y-90 microsphere brachytherapy [40], there are 152 records during the fiscal years 2008–2017. The report mentions that a common theme among the majority of the nonsignificant events is the partial or total clogging of the catheter or device, resulting in no dose being delivered to the patient. Twenty-eight out of 152 events were deemed to be significant or abnormal occurrences, which were categorized into three groups: dose to unintended site (59%), too much dose to intended site (31%), and wrong patient (10%). One of the 28 cases fell into two categories (unintended site and wrong patient).

25.3.3 Radiation safety

All personnel involved in the procedure should wear proper personal protective equipment, including a lead apron, a whole-body dosimeter, and potentially an extremity dosimeter. A proper radiation detector, such as a handheld Geiger–Muller counter, should be used to survey all the personnel leaving the procedure room during and after the radioembolization treatment for contamination. The hands, shoes and body surface of each person should be surveyed. It is recommended to do the survey away from the patient and from radioactive waste to reduce the background level and increase the detectability of contamination.

25.3.4 Patient release

NRC 10 CFR 35.75 provides guidance about releasing patients with radioactive implants. Any individual with administered unsealed byproduct materials or implants containing byproduct material may be authorized for release if the total effective dose equivalent to any other individual from exposure to the released individual is not likely to exceed 5 mSv in 1 year. If the total effective dose equivalent is not likely to exceed 1 mSv, no written instructions are needed. Even though beta emissions are generally attenuated sufficiently by the patient's tissue, it is best practice to inform patients and their family about actions to keep doses to other individuals as low as reasonably achievable.

25.4 Post-treatment evaluation

25.4.1 Imaging evaluation

Post-treatment imaging and dose verification are essential to calculate the implanted microsphere dose distribution and evaluate the overall microsphere brachytherapy treatment quality [41]. Y-90 produces bremsstrahlung photons with a wide range of energy spectrum and no photopeak. Therefore, traditionally, 2D planar or SPECT images over the thoracic and abdominal region are acquired through the Y-90 bremsstrahlung radiation to quantitatively assess the plan quality due to low image quality. Nowadays, post-treatment imaging is performed mostly through SPECT/CT and PET/CT for quantitative images as demonstrated in figures 25.3 and 25.4, respectively. A trace of Y-90 will decay to an excited 0^+ state of Zr-90, which decays immediately, generating a positron and electron pair through pair production. Utilizing the annihilation of the positron, PET scans provide far better image quality when co-registered with CT to provide quantitative Y-90 images. The enforcement of the post-imaging procedures can help ensure no medical event has occurred.

Multiple studies have been done to evaluate and improve the SPECT/CT and PET/CT imaging techniques in Y-90 post-radioembolization evaluation [42–46]. With proper reconstruction protocol and calibration correction factors, either SPECT/CT or PET/CT can be used for post-radioembolization monitoring. Currently, there is no standardized protocol for post-radioembolization imaging in terms of acquisition and reconstruction settings; therefore, differences may exist between studies even within the same modality [47]. A multicenter phantom study has been performed to quantitatively evaluate the image accuracy and consistency for post-radioembolization dosimetry; the researchers found that current-generation time-of-flight scanners could consistently and accurately reconstruct Y-90 activity concentrations [44].

Additionally, PET/magnetic resonance imaging (MRI) has been recently implemented in assessing Y-90 post-radioembolization activity as illustrated in figure 25.5.

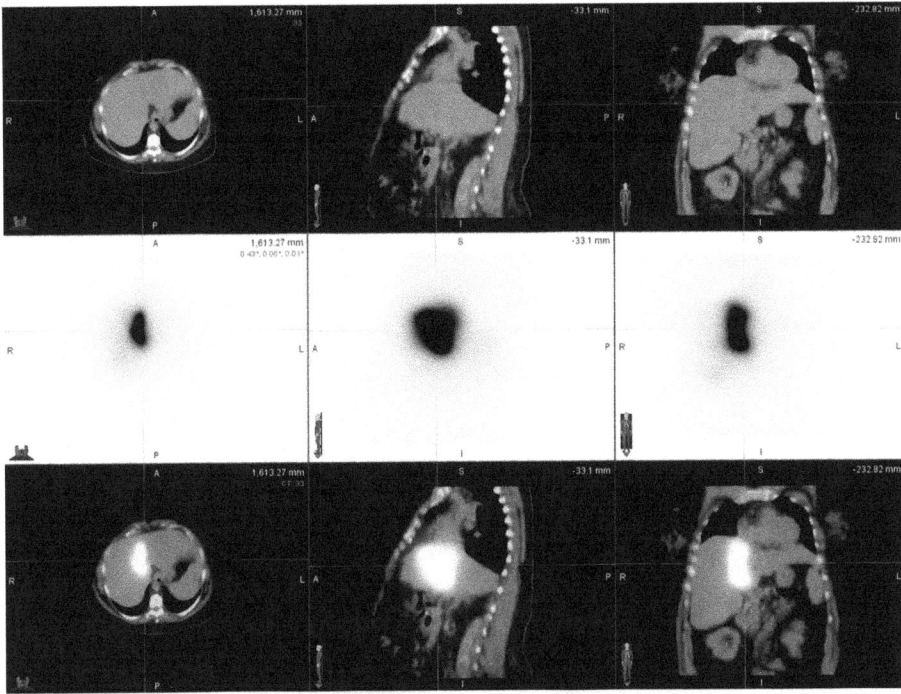

Figure 25.3. Images of SPECT/CT for post-radioembolization evaluation.

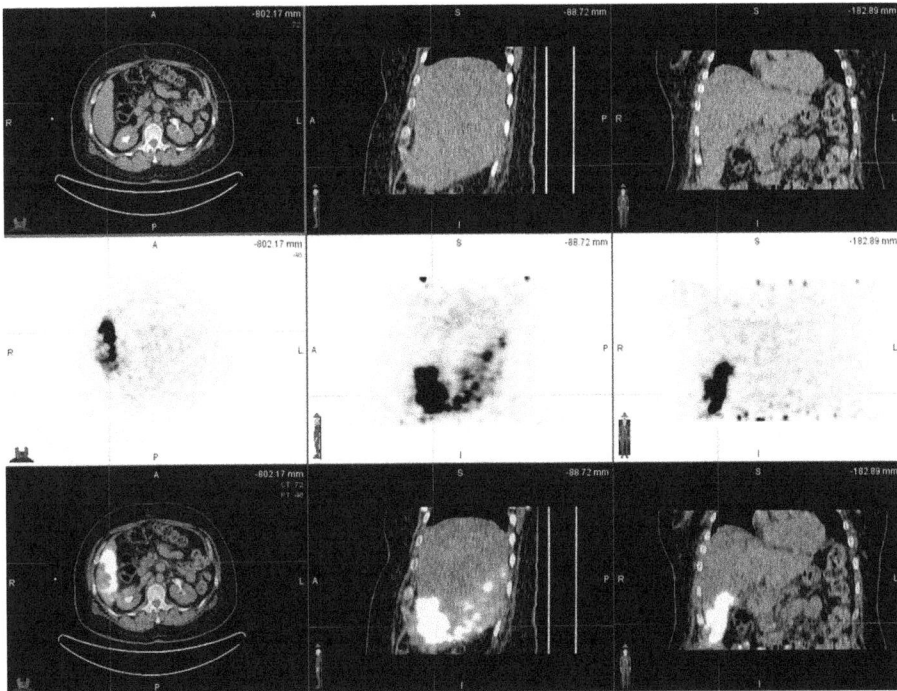

Figure 25.4. Images of PEC/CT for post-radioembolization evaluation.

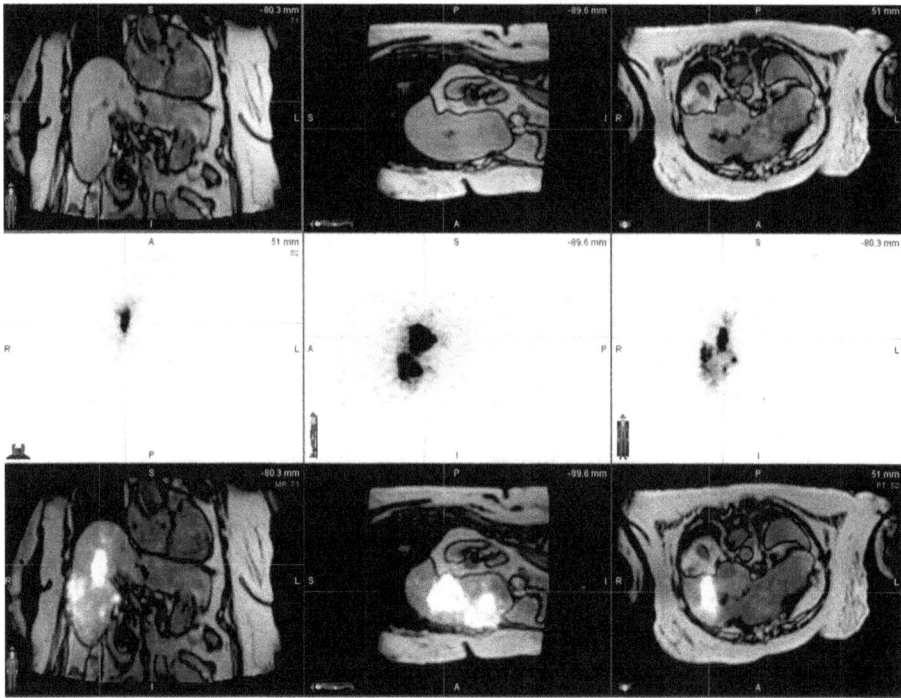

Figure 25.5. Images of PET/MRI for post-radioembolization evaluation.

Different from PET/CT's sequential data acquisition approach, PET/MRI offers simultaneous data acquisition and the opportunity to correct the liver motion due to respiration during PET reconstruction. In addition, high soft-tissue contrast MRI images can be used for partial volume correction in PET [48]. Studies have been done to determine the relationship between dose deposition measured by PET/MRI and the lesion response by Y-90 microsphere radioembolization [49]. And the PET/MRI technique has also been proven to be able to produce acceptable and reproducible measurements at a multi-institutional level [50].

With advanced imaging techniques, research can help accurately establish relationships between tumor dose and tumor response. A recent retrospective SPECT/CT study established a correlation of voxel-level absorbed dose and biological effective dose with tumor response [51]. Meanwhile, based on the accurately assessed tumor dose and tumor response status, a defined tumor dose threshold of 225 Gy was established for HCC [52]. A PET/CT post-treatment imaging study found that for liver metastases from colorectal cancer, lesions receiving an average dose of >50 Gy are likely to respond significantly; meanwhile, lesions receiving an average dose <20 Gy are unlikely to respond [53]. And potentially, accurate dose estimation could give confidence about treatment planning and even personalized treatments [41]. With the help of imaging guidance, simultaneous dose distribution might be possible [54].

25.4.2 Complications

The common post-radiation symptoms, such as fatigue, nausea, vomiting, anorexia, fever, diarrhea, etc, may be observed in the radioembolization patient population. Additionally, there are published data sets identifying a number of serious potential adverse events, including acute pancreatitis, acute gastritis, acute cholecystitis, radiation pneumonitis, radiation hepatitis, and radioembolization-induced liver disease [55].

25.5 Example workflow

As described in the *Journal of the American College of Radiology*, Y-90 is more complicated than most of the other radioisotopes in current practice [56]. Initial licensing and personnel training must be done before the clinical implementation of the radioembolization procedure per NRC regulation 10 CFR 35 or local agency if it is an agreement state [57]. As mentioned earlier, TheraSphere™ is under the provisions of the HDE approved by the FDA; thus an additional IRB approval is required.

Upon the success of initial licensing, the involved team members are recommended to form a committee to formally establish a general workflow for patients receiving Y-90 for liver disease, due to the complexity of the team assembly. Traditionally, the team will include the interventional radiologist(s), the radiation oncologist(s), the medical physicist(s), the radiation safety officer(s), and the nurse(s). An example workflow of Y-90 radioembolization is presented in figure 25.6.

After the patient is deemed a potential candidate for the Y-90 radioembolization procedure, they must go through multiple imaging procedures to determine whether Y-90 is a safe procedure for this patient. A lung shunt estimation imaging using ^{99m}Tc is required; depending on the imaging technique, the liver volume can be estimated through SPECT/CT, PET/CT, or a standalone CT if planar imaging is used for the lung shunt. If the lung shunt result satisfies the indication criteria, scheduling should be done to decide when to treat, as well as the location and volume of the lesion(s) that to be treated at the scheduled time in a timely manner. Scheduling and communication among team members, especially involving multiple personnel, need to be carefully arranged to avoid rescheduling the patient [58]. Once all the necessary information is in place, the activity for the desired dose can be calculated and ordered for the scheduled date. Attention needs to be paid to the different shelf times of SIR-Spheres® and TheraSphere™ when ordering them. Once the shipment is received, wipe tests should be done to make sure the package is intact.

On the treatment day, an in-house activity verification test should be done with a dose calibrator to ensure that the correct vial (for TheraSphere™) or correct amount of activity is drawn into the syringe (for SIR-Spheres®). In the meantime, the interventional radiology team will prepare the patient, re-evaluate the patient's condition, and get ready for the Y-90 administration. Immediately before the administration, a series of measurements need to be done to document the initial exposure rate at a fixed geometry; an example template made for TheraSphere™ is shown in figure 25.7. Then, the AU will administer the dose to the patient. The AU

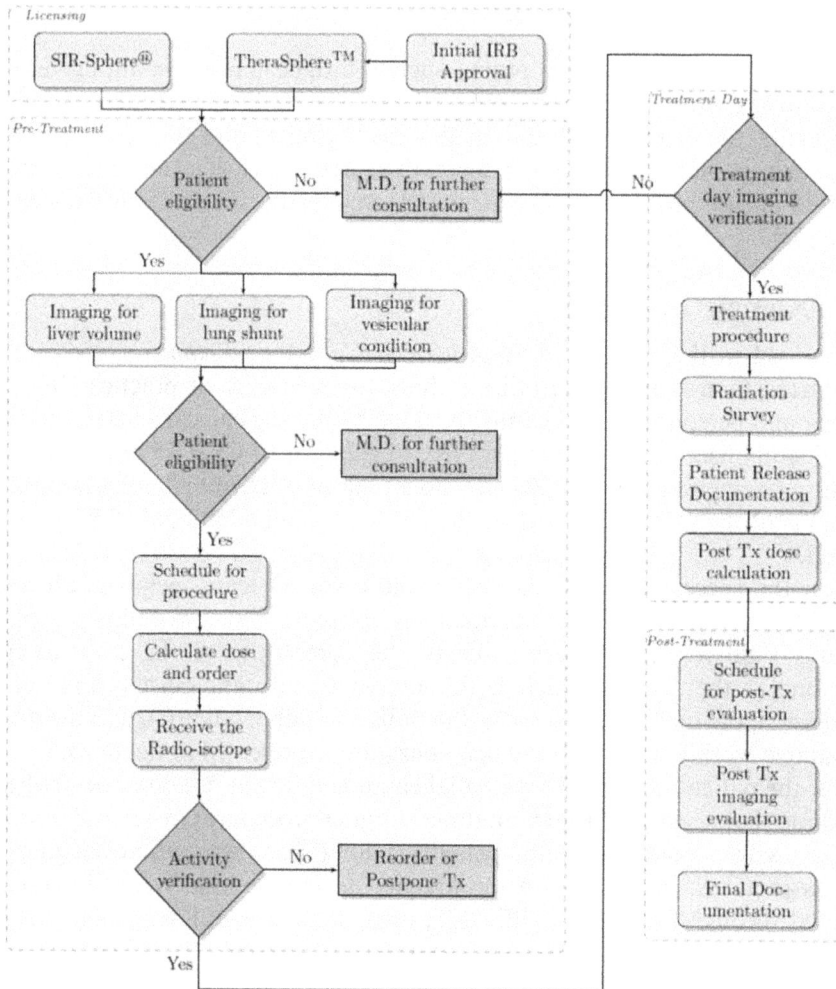

Figure 25.6. Example workflow for Y-90 radioembolization treatment.

can be a radiation oncologist, nuclear medicine physician, or interventional radiologist. Therefore, the easiest way is to train the interventional radiologist to be an AU to facilitate the entire procedure. Immediately after the procedure, collect all the contaminated material, and take another measurement with the same fixed geometry which would reflect the residual dose in the vial. The ratio of the exposure rate before and after is used to estimate the administered dose.

After the procedure is completed, carefully survey the interventional radiology suite to make sure there is no radioactive material contamination on physicians or staff or on any surface inside the suite. Since Y-90 is a pure beta emitter, external exposure to the public is not an issue. Hand the safety instructions to patients when they are ready to be released. A post-imaging evaluation should be performed immediately after the procedure to evaluate the accuracy of the dose administered.

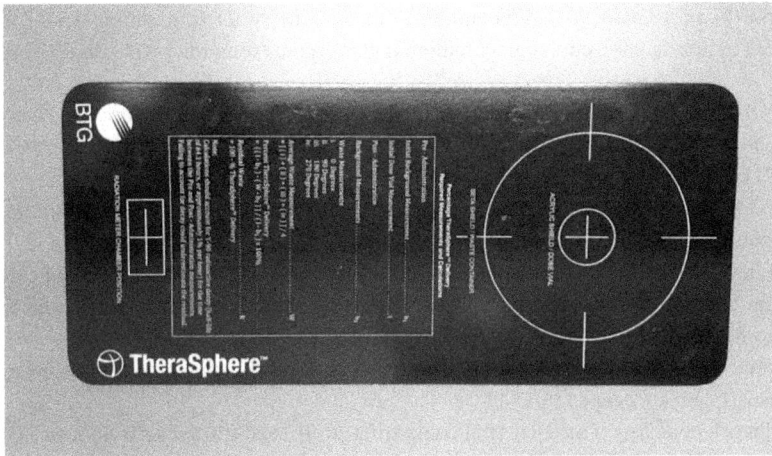

Figure 25.7. Fixed geometry template for TheraSphere™ measurement.

If everything looks normal in the post-imaging evaluation, then the patient should be scheduled for routine follow-ups.

References

[1] The University of Michigan Comprehensive Cancer Center 2006 Hepatobiliary cancers. Clinical practice guidelines in oncology *J. Natl. Compr. Canc. Netw.* **4** 728–50

[2] Kulik L M *et al* 2006 Yttrium-90 microspheres (TheraSphere®) treatment of unresectable hepatocellular carcinoma: downstaging to resection, RFA and bridge to transplantation *J. Surg. Oncol.* **94** 572–86

[3] Salem R *et al* 2016 Y90 radioembolization significantly prolongs time to progression compared with chemoembolization in patients with hepatocellular carcinoma *Gastroenterology* **151** 1155–63

[4] Boas F E *et al* 2015 Classification of hypervascular liver lesions based on hepatic artery and portal vein blood supply coefficients calculated from triphasic CT scans *J. Digit. Imaging* **28** 213–23

[5] Boas F E *et al* 2016 Quantitative measurements of enhancement on preprocedure triphasic CT can predict response of colorectal liver metastases to radioembolization *Am. J. Roentgenol.* **207** 671–5

[6] Saini A *et al* 2019 History and evolution of yttrium-90 radioembolization for hepatocellular carcinoma *J. Clin. Med.* **8** 55

[7] Bray F *et al* 2018 Global cancer statistics 2018: GLOBOCAN estimates of incidence and mortality worldwide for 36 cancers in 185 countries *CA. Cancer J. Clin.* **68** 394–424

[8] Tan J C C, Coburn N G, Baxter N N, Kiss A and Law C H L 2008 Surgical management of intrahepatic cholangiocarcinoma – a population-based study *Ann. Surg. Oncol.* **15** 600–8

[9] Salem R and Thurston K G 2006 Radioembolization with yttrium-90 microspheres: a state-of-the-art brachytherapy treatment for primary and secondary liver malignancies – part 3: comprehensive literature review and future direction *J. Vasc. Interv. Radiol* **17** 1571–93

[10] Nezami N *et al* 2018 90Y radioembolization dosimetry using a simple semi-quantitative method in intrahepatic cholangiocarcinoma: glass versus resin microspheres *Nucl. Med. Biol.* **59** 22–8

[11] Gray B *et al* 2001 Randomised trial of SIR-Spheres®plus chemotherapy vs. chemotherapy alone for treating patients with liver metastases from primary large bowel cancer *Ann. Oncol.* **12** 1711–20

[12] Goin J E *et al* 2003 Treatment of unresectable metastatic colorectal carcinoma to the liver with intrahepatic Y-90 microspheres: dose-ranging study *World J. Nucl. Med.* **2** 216–25

[13] Sharma R A *et al* 2007 Radioembolization of liver metastases from colorectal cancer using yttrium-90 microspheres with concomitant systemic oxaliplatin, fluorouracil, and leucovorin chemotherapy *J. Clin. Oncol.* **25** 1099–106

[14] Mulcahy M F *et al* 2009 Radioembolization of colorectal hepatic metastases using yttrium-90 microspheres *Cancer* **115** 1849–58

[15] Hendlisz A *et al* 2010 Phase III trial comparing protracted intravenous fluorouracil infusion alone or with yttrium-90 resin microspheres radioembolization for liver-limited metastatic colorectal cancer refractory to standard chemotherapy *J. Clin. Oncol.* **28** 3687–94

[16] Cosimelli M *et al* 2010 Multi-centre phase II clinical trial of yttrium-90 resin microspheres alone in unresectable, chemotherapy refractory colorectal liver metastases *Br. J. Cancer* **103** 324–31

[17] Tortora G and Derrickson B 2018 *Principles of Anatomy and Physiology* (New York: Wiley)

[18] Germain T, Favelier S, Cercueil J-P, Denys A, Krausé D and Guiu B 2014 Liver segmentation: practical tips *Diagn. Interv. Imaging.* **95** 1003–16

[19] Rutkauskas S, Gedrimas V, Pundzius J, Barauskas G and Basevicius A 2006 Clinical and anatomical basis for the classification of the structural parts of liver *Med* **42** 98–106

[20] Kennedy A *et al* 2007 Recommendations for radioembolization of hepatic malignancies using yttrium-90 microsphere brachytherapy: a consensus panel report from the radio-embolization brachytherapy oncology consortium *Int. J. Radiat. Oncol. Biol. Phys.* **68** 13–23

[21] Dezarn W A *et al* 2011 Recommendations of the American Association of Physicists in Medicine on dosimetry, imaging, and quality assurance procedures for 90Y microsphere brachytherapy in the treatment of hepatic malignancies *Med. Phys.* **38** 4824–45

[22] Piana P M *et al* 2014 Early arterial stasis during resin-based yttrium-90 radioembolization: incidence and preliminary outcomes *HPB* **16** 336–41

[23] van den Hoven A F *et al* 2014 Identifying aberrant hepatic arteries prior to intra-arterial radioembolization *Cardiovasc. Intervent. Radiol.* **37** 1482–93

[24] Leung T W T *et al* 1995 Radiation pneumonitis after selective internal radiation treatment with intraarterial 90yttrium-microspheres for inoperable hepatic tumors *Int. J. Radiat. Oncol. Biol. Phys.* **33** 919–24

[25] Yu N *et al* 2013 Lung dose calculation with SPECT/CT for 90Yittrium radioembolization of liver cancer *Int. J. Radiat. Oncol. Biol. Phys.* **85** 834–9

[26] Bastiaannet R, Viergever M A and De Jong H W A M 2017 Impact of respiratory motion and acquisition settings on SPECT liver dosimetry for radioembolization *Med. Phys.* **44** 5270–9

[27] Kunnen B *et al* 2018 Radioembolization lung shunt estimation based on a 90Y pretreatment procedure: a phantom study *Med. Phys.* **45** 4744–53

[28] Bolch W E, Eckerman K F, Sgouros G and Thomas S R 2009 MIRD pamphlet no. 21: a generalized schema for radiopharmaceutical dosimetry standardization of nomenclature *J. Nucl. Med.* **50** 477–84

[29] Tafti B A and Padia S A 2019 Dosimetry of Y-90 microspheres utilizing Tc-99m SPECT and Y-90 PET *Semin. Nucl. Med.* **49** 211–7

[30] Gaba R C *et al* 2009 Radiation lobectomy: preliminary findings of hepatic volumetric response to lobar yttrium-90 radioembolization *Ann. Surg. Oncol.* **16** 1587–96

[31] Siddiqi N H and Devlin P M 2009 Radiation lobectomy: a minimally invasive treatment model for liver cancer: case report *J. Vasc. Interv. Radiol* **20** 664–9

[32] Malhotra A, Liu D M and Talenfeld A D 2019 Radiation segmentectomy and radiation lobectomy: a practical review of techniques *Tech. Vasc. Interv. Radiol.* **22** 49–57

[33] Vouche M *et al* 2013 Radiation lobectomy: time-dependent analysis of future liver remnant volume in unresectable liver cancer as a bridge to resection *J. Hepatol.* **59** 1029–36

[34] Garin E *et al* 2013 Boosted selective internal radiation therapy with 90 Y-loaded glass microspheres (B-SIRT) for hepatocellular carcinoma patients: a new personalized promising concept *Eur. J. Nucl. Med. Mol. Imaging* **40** 1057–68

[35] Riaz A *et al* 2011 Radiation segmentectomy: a novel approach to increase safety and efficacy of radioembolization *Int. J. Radiat. Oncol. Biol. Phys.* **79** 163–71

[36] Jia Z, Wang C, Paz-Fumagalli R and Wang W 2019 Radiation segmentectomy for hepatic malignancies: indications, devices, dosimetry, procedure, clinical outcomes, and toxicity of yttrium-90 microspheres *J. Interv. Med* **2** 1–4

[37] Hardy-Abeloos C *et al* 2019 Safety and efficacy of liver stereotactic body radiation therapy for hepatocellular carcinoma after segmental transarterial radioembolization *Int. J. Radiat. Oncol. Biol. Phys.* **105** 968–76

[38] Lewandowski R J *et al* 2016 90Y radiation lobectomy: outcomes following surgical resection in patients with hepatic tumors and small future liver remnant volumes *J. Surg. Oncol.* **114** 99–105

[39] Rhee T K *et al* 2008 Tumor response after yttrium-90 radioembolization for hepatocellular carcinoma: comparison of diffusion-weighted functional MR imaging with anatomic MR imaging *J. Vasc. Interv. Radiol.* **19** 1180–86

[40] Smith T W, Huntsman D C and Sant R L 2018 *Nuclear Material Events Database: Medical Events Involving Y-90 Microsphere Brachytherapy* INL/LTD-18-44325 Idaho National Laboratory https://www.nrc.gov/docs/ML1805/ML18057A901.pdf

[41] Alsultan A A, Smits M L J, Barentsz M W, Braat A J A T and Lam M G E H 2019 The value of yttrium-90 PET/CT after hepatic radioembolization: a pictorial essay *Clin. Transl. imaging* **7** 303–12

[42] Rong X *et al* 2012 Development and evaluation of an improved quantitative 90Y bremsstrahlung SPECT method *Med. Phys.* **39** 2346–58

[43] Dewaraja Y K *et al* 2017 Improved quantitative 90Y bremsstrahlung SPECT/CT reconstruction with Monte Carlo scatter modeling *Med. Phys.* **44** 6364–76

[44] Willowson K P *et al* 2015 A multicentre comparison of quantitative 90Y PET/CT for dosimetric purposes after radioembolization with resin microspheres: the QUEST phantom study *Eur. J. Nucl. Med. Mol. Imaging* **42** 1202–22

[45] Yue J *et al* 2016 Comparison of quantitative Y-90 SPECT and non-time-of-flight PET imaging in post-therapy radioembolization of liver cancer *Med. Phys.* **43** 5779–90

[46] Siman W, Mikell J K and Kappadath S C 2016 Practical reconstruction protocol for quantitative 90Y bremsstrahlung SPECT/CT *Med. Phys.* **43** 5093–103

[47] Bastiaannet R *et al* 2018 The physics of radioembolization *EJNMMI Phys* **5** 22

[48] Eldib M *et al* 2016 Optimization of yttrium-90 PET for simultaneous PET/MR imaging: a phantom study *Med. Phys.* **43** 4768–74

[49] Fowler K J *et al* 2016 PET/MRI of hepatic 90Y microsphere deposition determines individual tumor response *Cardiovasc. Intervent. Radiol.* **39** 855–64

[50] Maughan N M *et al* 2018 Multi institutional quantitative phantom study of yttrium-90 PET in PET/MRI: the MR-QUEST study *EJNMMI Phys* **5** 7

[51] Kappadath S C *et al* 2018 Hepatocellular carcinoma tumor dose response after 90Y-radioembolization with glass microspheres using 90Y-SPECT/CT-based voxel dosimetry *Int. J. Radiat. Oncol. Biol. Phys.* **102** 451–61

[52] Chan K T *et al* 2018 Prospective trial using internal pair-production positron emission tomography to establish the yttrium-90 radioembolization dose required for response of hepatocellular carcinoma *Int. J. Radiat. Oncol. Biol. Phys.* **101** 358–65

[53] Willowson K P *et al* 2017 Clinical and imaging-based prognostic factors in radioembolisation of liver metastases from colorectal cancer: a retrospective exploratory analysis *EJNMMI Res* **7** 1–13

[54] Ting C-H, Lin K-H, Chiu C-H, Lee R-C and Huang W-S 2020 Simultaneous time-of-flight PET/MR identifies the hepatic 90Y-resin distribution after radioembolization *Clin. Nucl. Med.* **45** e92–3

[55] Riaz A, Awais R and Salem R 2014 Side effects of yttrium-90 radioembolization *Front. Oncol.* **4** 198

[56] Sturchio G M and Pooley R A 2018 The physics of radioisotope treatment is not always the same *J. Am. Coll. Radiol. JACR* **15** 1125

[57] 2016 *Yttrium-90 Microsphere Brachytherapy Sources and Devices TheraSphere® and SIR-Spheres®: Licensing Guidance* Nuclear Regulatory Commission https://www.nrc.gov/docs/ML1535/ML15350A099.pdf

[58] Cai B *et al* 2017 Process improvement for the safe delivery of multidisciplinary-executed treatments—a case in Y-90 microspheres therapy *Brachytherapy* **16** 236–44

IOP Publishing

Principles and Practice of Image-Guided Abdominal
Radiation Therapy

Yu Kuang

Chapter 26

Artificial intelligence in radiation therapy for abdominal cancer

Dan Nguyen and Mu-Han Lin

This chapter is dedicated to artificial intelligence (AI) technologies and applications in radiation therapy for abdominal cancer. Since the explosion of AI technologies in recent years, deep learning (DL) has been utilized in improving the radiation therapy workflow and demonstrated the potential to revolutionize the entire radiation therapy treatment workflow. We present the various AI methods in several key areas of radiation therapy for abdominal cancer, including imaging & diagnosis, treatment planning, treatment delivery, and follow up. We discuss the pros and cons of the current AI technology, and the future landscape and prospects for AI in radiation therapy for abdominal cancer.

26.1 Introduction

Abdominal cancers have some of the highest death rates among cancer types. The fact the majority of abdominal cancer cases are diagnosed at an advanced stage leads to limited options for treatment and poor prognosis [1]. For example, liver cancer is the third leading cause of cancer death globally in 2018, where 781 631 out of 841 080 people worldwide diagnosed with liver cancer died from the disease. Liver is one of the organs that are the most prone to develop metastases. Pancreatic cancer also had high incidence of death in 2018, when 432 232 out of 45 8918 people worldwide diagnosed with pancreatic cancer died from the disease [1]. It also has a poor five-year overall survival of 9% [2].

Radiation therapy alone and in conjunction with chemotherapy, surgery, and immunotherapy is often used for abdominal cancer treatment [3, 4]. While the effectiveness of radiation therapy has been proven, challenges remain in the early diagnosis, tumor and organ delineation, image guidance and adaptive therapy to

doi:10.1088/978-0-7503-2468-7ch26

account for the daily anatomy variation, and outcome prediction. The soft tissue surrounding the abdominal organs leads to poor contrast on computed tomography (CT) images due to the Hounsfield units (HU) being similar among different organs. In addition, the shapes of certain organs, such as the liver, differ wildly among different patients and change from day to day. This leads to difficulties during the treatment imaging and segmentation phases and leads to challenges for the remaining downstream aspects of radiation therapy, such as treatment planning and daily treatment delivery. Although magnetic resonance imaging (MRI)–guided and cone beam computed tomography (CBCT)–guided adaptive radiotherapies are clinically available to adjust the treatment based on the patient's anatomy on the treatment day, efficient real-time image segmentation and replanning are required to ensure a seamless treatment.

Over the last several years, artificial intelligence (AI) has made major advances in many areas of computer vision and decision-making. In particular, advances in the creation and development of convolutional neural networks (CNNs) [5], as well as the fully convolutional network (FCN) [6] and the U-net [7], have revolutionized tasks that require processing of image information. These deep learning models are capable of learning to extract their own features from the raw data, process them, and perform a specific task. Deep learning has the capability to unearth useful information from lower-contrast data that may be difficult to discern with the human eye. Deep learning advances have been applied in radiation therapy treatment of patients with abdominal cancer, particularly in the areas of diagnosis, segmentation, synthetic CT generation, treatment planning, and toxicity prediction. The abdominal cancer sites with deep learning applications described in the literature are the liver, pancreas, and kidney.

In this chapter, we will review the state-of-the-art AI that has been applied to patients with abdominal cancer treated with radiation therapy and discuss future directions for the field. Figure 26.1 summarizes the ways in which AI can contribute in each step of the radiation therapy workflow including early diagnosis, synthetic CT generation for pretreatment and online adaptive planning, image segmentation, treatment planning, daily image guidance, online adaptive planning, and outcome prediction.

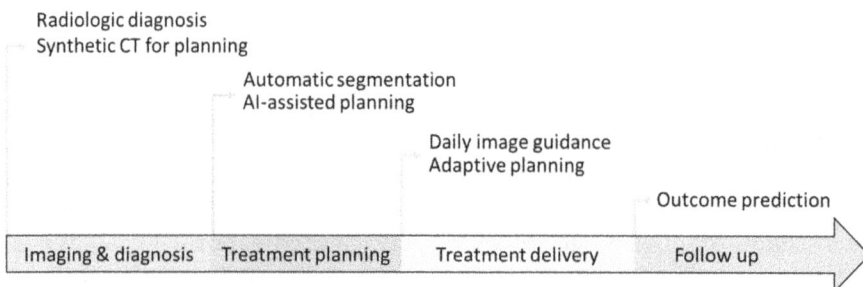

Figure 26.1. Applications of AI in the radiation treatment workflow for abdominal cancer.

26.2 Imaging and diagnosis

26.2.1 Radiologic diagnosis

Successful early screening and diagnosis of the disease may substantially improve the treatment effectiveness and, thus, the overall survival and quality of life of the patient. The addition of AI into the screening and diagnostic processes may make it possible to catch diseases sooner and with higher accuracy than before. A pilot study on pancreatic cancer found that an AI model was capable of accurately detecting pancreatic cancer up to 20 months prior to the patient's diagnosis [8]. These finding were presented by Malhotra *et al* on 2 July 2020 at the European Society for Medical Oncology World Congress on Gastrointestinal Cancer.

Many studies in the last several years have focused on pancreatitis (CP) and pancreatic cancer (PC). Zhu *et al* developed a support vector machine (SVM) method [9] for distinguishing CP and PC. They recruited 262 patients with PC and 126 patients with CP for the study; extracted 105 features from the data; selected the best 16 features for training the SVM; and obtained an average accuracy, sensitivity rate, and specificity rate of 94%, 96%, and 93%, respectively, for distinguishing PC and CP. Das *et al* developed a neural network–based classification model for identifying and diagnosing PC [10]. They first used principal component analysis to reduce the data dimension, and the neural network was trained on the resulting information. The developed model could correctly classify PC, with an area under the receiver operating characteristic (ROC) curve (AUC) of 0.93. Cazacu *et al* studied the use of an artificial neural network for differentiating CP and PC, and they found that their neural network was capable of differentiating PC and CP with a sensitivity of 94.64% and a specificity of 94.44% [11].

Wang *et al* developed a real-time framework for detection of colonoscopic polyp and adenoma and tested their framework on a prospective randomized control study [12]. The authors created a deep learning CNN architecture based on an existing framework called SegNet [13]. The model is trained to take in a colonoscopy image as input and is trained to output a pixel-level probability map that classifies each pixel as containing a polyp or not, as well as bounding boxes of the polyp's location. The prospective comparative study included 1058 patients, of whom roughly half were randomly assigned to either the routine colonoscopy procedure or the AI-assisted coloscopy procedure. The researchers found that the AI arm had statistically significantly increased adenoma detection rates of 29.1%, as compared with 20.3% for the routine clinical arm. Specifically, the AI framework was better at catching diminutive adenomas than the routine procedure. Both arms performed similarly for larger adenomas. Kiani *et al* proposed a deep learning–based assistant to aid physicians to distinguish between hepatocellular carcinoma and cholangiocarcinoma by using whole-slide images (WSIs) [14]. The authors used a densely connected CNN, also known as DenseNet [15], and trained the model to predict hepatocellular carcinoma and cholangiocarcinoma using the WSIs as input. Their model was capable of predicting with an accuracy of 0.885 on the validation data

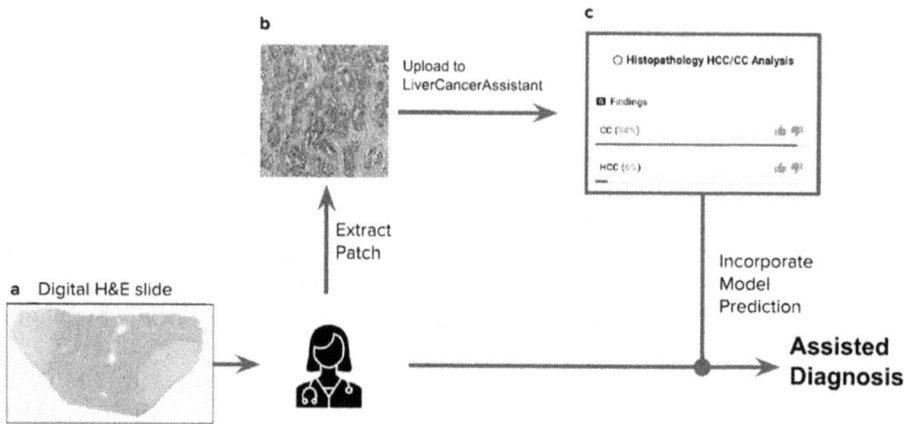

Figure 26.2. Proposed workflow for AI-assisted diagnosis from Kiani *et al* [14] (2020). With permission of Springer (CC BY 4.0).

and 0.842 on the test set. Furthermore, the authors performed a physician study and found that, through the workflow outlined in figure 26.2, accuracy significantly improved for a subset of pathologists who were gastrointestinal subspecialists, non-gastrointestinal subspecialists, and trainees.

26.2.2 Synthetic CT for planning

CT image is the primary image modality employed for the radiation treatment planning. However, with other image modalities offering superior soft tissue contrast or online three-dimensional (3D) patient anatomy, it is preferable to convert these images into CT format and use them for planning directly. Synthetic CT generation has become a popular field of research, and with the advent of deep learning, it may be possible to create realistic synthetic CTs from other imaging modalities. MRI is a widely used modality and has exceptional soft tissue contrast for consistent and accurate target delineation, and many centers are even migrating towards an MRI-only treatment process. In an adaptive radio-therapy setting, CBCT images are taken while the patient is on the treatment couch to determine if planning adaptation is necessary. However, during the planning process, accurate dose calculation requires having the accurate HU of CTs, which MRI images lack and CBCT images provide inaccurately. Deep learning has made it possible to perform style transfer to convert these MRI and CBCT images to synthetic CTs that have accurate HU.

Liu *et al* focused heavily on synthetic CT generation using cyclic generative adversarial network (CycleGAN) deep learning architectures over a series of studies. In one study, the authors investigated MRI to synthetic CT generation for the liver for the purposes of stereotactic body radiation therapy (SBRT) [16] and proton therapy [17]. The study used 21 patients with co-registered MRI and CT image pairs

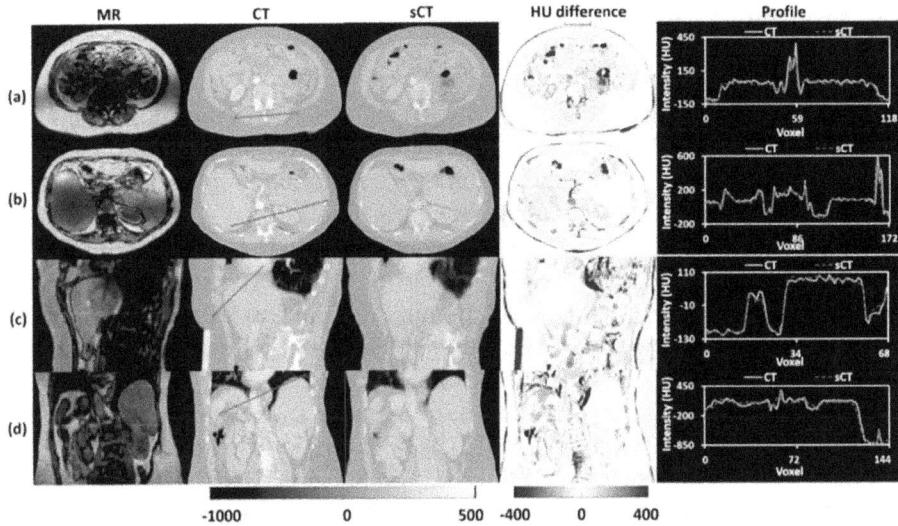

Figure 26.3. Magnetic resonance image, CT image, and synthetic CT image from Liu *et al* [17]. The line profiles comparing CT and synthetic CT show very good agreement. © IOP Publishing. Reproduced with permission. All rights reserved.

and, due to the low number of patients in the study, used leave-one-out cross validation for evaluating the framework. CycleGAN is known to be able to perform style transfer while maintaining macro anatomies and structures of the source (MRI) data [18]. The mean absolute error (MAE) in the synthetic CT, compared with the ground truth CT, was found to be 72.87 ± 18.16 HU. In order to assess the impact of the synthetic CT error on radiation therapy plans, the authors recalculated the dose from the plan onto both the synthetic CT and ground truth CT for comparison. For the SBRT plans, the authors evaluated the dose–volume histogram metrics, found no significant differences between the two doses, and found the 1%/1 mm gamma passing rate to be over 99%. For the proton plans, the authors found the mean 3D gamma analysis similarity passing rate to be 90.76% ± 5.94% (1%/1 mm), 96.98% ± 2.93% (2%/2 mm), and 99.37% ± 0.99% (3%/3 mm). Qualitative results comparing MRI, CT, and synthetic CT are shown in figure 26.3.

Liu *et al* also applied their CycleGAN method for CBCT to synthetic CT generation for pancreatic adaptive radiation therapy [19]. In addition, for this study, they added a self-attention property to CycleGAN. Using a leave-one-out cross validation procedure on 30 patients with PC with paired CBCT and CT images, the authors trained a model that was capable of achieving an MAE of 56.89 ± 13.84 HU, an improvement over the average absolute difference of 81.06 ± 15.86 HU between the unconverted CBCT and the CT. The authors showed that, while the two recalculated doses were significantly different in dose–volume histogram metrics, the difference between those of the synthetic CT and the CT were not significantly different.

26.3 Treatment planning

26.3.1 Automatic segmentation

The major challenge in abdominal organ segmentation is that the abdominal organs are surrounded by soft tissue and have similar HU in CT images, leading to poor contrast. This can lead to misclassification of the abdominal organs that are similar in shape and density. The advent of deep learning allows AI models to detect nuances in the images in order to accurately find the organ boundary and correctly segment the organs. The study by Hu *et al* proposed a fully automatic method for segmenting multiple organs from 3D abdominal CT images [20]. The organs in the study include liver, spleen, and both kidneys. They utilized deep CNNs for organ detection and segmentation, and an additional multi-phase evolution to refine the segmentation. The authors trained the deep learning framework using abdominal CT scans from a public data set, MICCAI-SLiver07, from the Medical Image Computing and Computer Assisted Intervention Society (MICCAI), and evaluated it on 140 patients. The study had found high Dice similarity coefficients (DSCs) of 0.96 (liver), 0.94 (spleen), and 0.95 (kidneys), with an average computation time of 125 s per CT volume.

For segmentation focused on patients with renal cancer, Jackson *et al* developed a segmentation model using 3D CNNs for right and left kidneys for non-contrast CT images [21]. The authors trained the deep learning model on 89 patients with metastatic prostate cancer and evaluated its performance on a separate test set of 24 patients. The neural network was integrated into the hospital database and was able to predict the kidney contours in 90 s. The authors found that the CNN was capable of achieving a mean DSC of 0.91 (right kidney) and 0.86 (left kidney), but, however, they noted low DSC scores for patients with cystic kidneys, as shown in patients 2 and 5 in figure 26.4. The authors also used the dose maps obtained from post-treatment single-photon emission computerized tomography (SPECT) imaging and found that there was no significant different in the absorbed dose to the predicted segmentation and the ground truth segmentation of the kidneys.

Several segmentation models have been studied specifically for liver cancer. The study by Yuan proposed the automatic segmentation of the liver and its tumors using a hierarchical deconvolutional neural network [22]. The study utilized MICCAI's 2017 Liver Tumor Segmentation Challenge (LiTS), which had multiple tasks on contrast-enhanced CT images including liver tumor segmentation, liver segmentation, and tumor burden estimation. The study's proposed framework, using a convolution-deconvolution neural network, was able to achieve a DSC of 0.963 for liver segmentation, a DSC of 0.657 for liver tumor segmentation, and root-mean-square error of 0.017 for tumor burden estimation, using 70 of the challenge patients as testing cases. Ben-Cohen *et al* utilized an FCN for liver segmentation and lesion detection [23]. Training an FCN on a small data set of 40 total patients and using threefold cross validation, the authors found that their framework was capable of achieving a true positive rate of 0.86. Lu *et al* proposed a framework using CNNs and graph cuts for automatic 3D liver localization and segmentation [24]. This

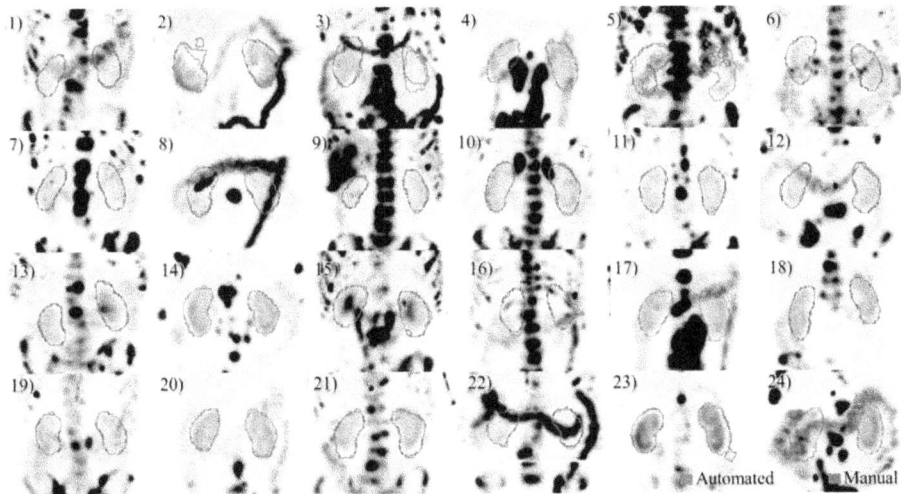

Figure 26.4. AI-predicted segmentation (red) versus manual segmentation (blue) from Jackson *et al* [21]. The larger mismatches in contours for patients 2 and 5 are in the kidney's cystic regions. Reprinted from [21] with permission of Frontiers.

particular method is a two-step process. First, there a liver detection and stochastic segmentation using a CNN. Then there is a refinement procedure using graph cuts to improve the segmentation from the first step. The framework was tested on two public databases, 3D-IRCADb-01 and MICCAI-SLiver07. The study calculated five metrics, including volumetric overlap error, relative volume difference, average symmetric surface distance, root-mean-square symmetric surface distance, and maximum symmetric surface distance. The calculated metrics were found to be 9.36, 0.97%, 1.89, 4.15 and 33.14 mm, respectively, for the 3D-IRCADb-01 data set and 5.9, 2.7%, 0.91, 1.88 and 18.94 mm, respectively, for the MICCAI-SLiver07 data set. Qin *et al* used a superpixel-based and boundary-sensitive method with their CNN for liver segmentation [25]. The framework used a multistep process, shown in figure 26.5, where the CT image regions with similar HU were first converted into superpixel regions. The pixel classification was broken down into three regions: inside the liver, outside the liver, and the liver boundary. A saliency map was calculated for each CT and was used to direct the patch-based sampling for training. Finally, the CNN was trained to predict the liver boundary. With the proposed framework, the authors achieved an extremely high DSC of 0.9731 ± 0.0036, using tenfold cross validation on a 100-patient data set. Christ *et al* utilized a cascaded FCN architecture for liver and tumor segmentation from CT and MRI volumes [26]. The framework consists of two FCNs. The first FCN performs a course segmentation to create the liver contour and a region of interest. The region of interest is then fed into a second FCN which performs the fine-tuned segmentation of its lesions. The authors showed that, for both CTs and MRIs, this method achieves means DSC scores of 0.94% and 0.61 for the CT liver and lesions and 0.87 and 0.70 for the MRI liver and lesions, respectively.

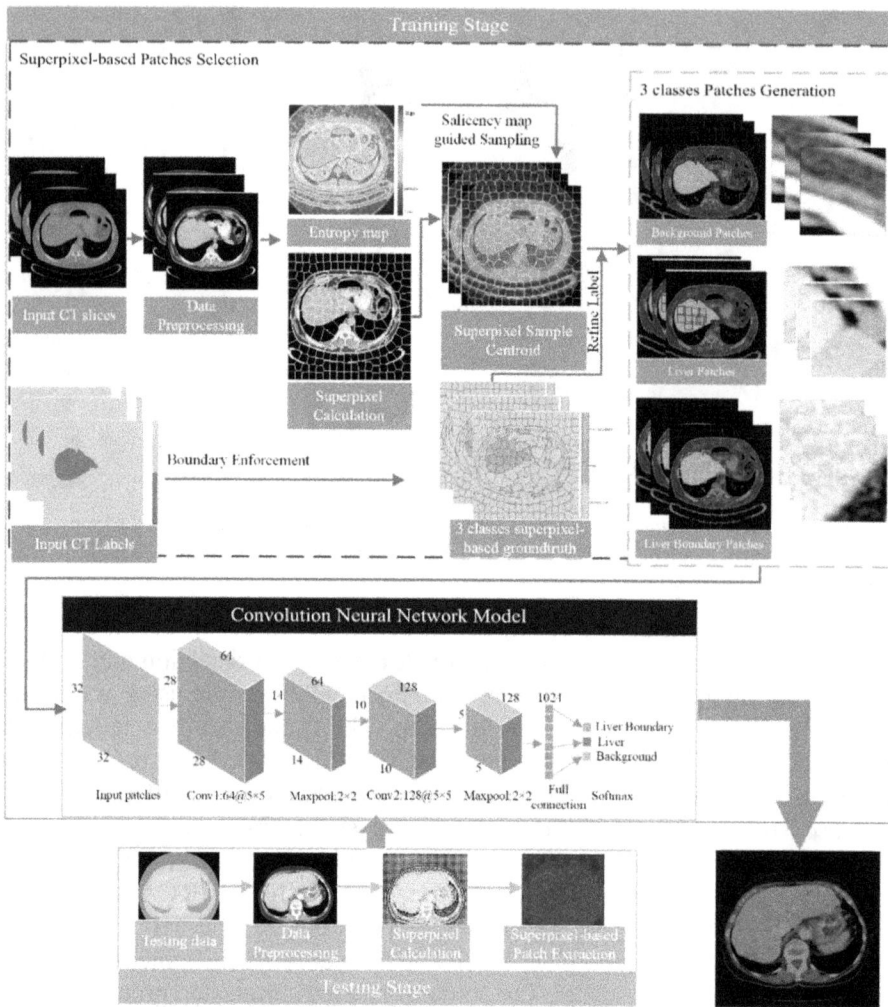

Figure 26.5. Schematic of superpixel-based and boundary-sensitive method with the CNN for liver segmentation from the study by Qin *et al* [25]. © IOP Publishing. Reproduced with permission. All rights reserved.

26.3.2 AI-assisted planning

Since the advent of intensity modulated radiation therapy (IMRT) [27] and, subsequently, volumetric modulated arc therapy (VMAT) [28], treatment planning has drastically evolved, substantially improving plan quality and patient survival and quality of life. However, these advanced optimization methods are complex and require an enormous amount of manual tuning on several hyperparameters—such as organ weights—from a treatment planner to achieve a reasonable clinical plan. Moreover, several rounds of feedback between the planner and physician are required for the plan to be finally approved by the physician for treatment delivery.

In order to remedy this, many researchers have started to look towards AI-based frameworks. Much of the early work to solve this problem focused on knowledge-based planning (KBP) [29–41], which developed models to predict a patient's dose–volume histogram and dose constraints using historical patient plans and information. As deep learning models quickly advanced over the last several years, it became possible to perform volumetric voxel-based dose predictions, and many studies have begun to apply neural networks to dose prediction for many cancer sites, including abdomen, prostate, head and neck, and lung [42–52]. These dose prediction models have shown to be extremely accurate at predicting the 3D dose will small error. While KBP, a machine learning–based tool, holds the promise of improving treatment planning quality and efficiency, deep learning dose prediction tools, predicting the dose without a real planning process, have the potential to (1) assist the physician to construct a more precise directive that is personalized to the patient without going through the planning process and (2) be used as a decision support tool for adaptive treatment. The dose predictor can be directly used in planning to improve the overall process and, ultimately, the plan quality.

Cozzi *et al* investigated the use of KBP for intensity modulated proton therapy on patients with hepatocellular cancer (HCC) [53]. A commercial KBP, RapidPlan, was tuned to a data set of 65 patients with HCC and validation on an additional 15 patients. The plans were compared with plans generated manually from the manual optimization method that is commonly utilized today in the clinic, as well as with plans generated using a photon-based KBP model for VMAT (RapidArc). They found that the RapidPlan proton plans outperformed RapidArc plans while maintaining equivalent quality against the manual proton plans. Zhang *et al* built a KBP model for patients with liver HCC treated with tomotherapy [54]. They tuned the model on 50 patients and tested the model on an additional 20 patients. For the test patients, both manual planning and the KBP-based planning were done for comparison. The found that the KBP plan yielded a lower mean dose and V15 of the normal liver, as well as mean doses to the left kidney and small bowel, as shown in figure 26.6. The total planning time was significantly reduced on the KBP arm.

For the abdomen, Campbell *et al* developed artificial neural networks to predict the voxel doses from patients with PC receiving an arc-based SBRT plan, using a set of geometric and plan parameters as input [55]. Using 29 patients for training and 14 patients for validation, the group was able to develop a model capable of predicting within a mean absolute dose error of less than 10%, with the largest errors in the low-dose regions of the patient. Thomas *et al* developed an artificial neural network that learned to predict the dose to each voxel on patients with abdominal cancer treated with adaptive MRI-guided radiation therapy, using 16 different hand-selected input variables [56]. The study used 310 plans from 53 patients and performed a cross validation procedure, splitting the data into 48 patients for training and five patients for validation in each fold. The researchers found that the model was capable of predicting the voxel-based dose within an absolute error of 3.5 ± 2.4 Gy, as shown in figure 26.7.

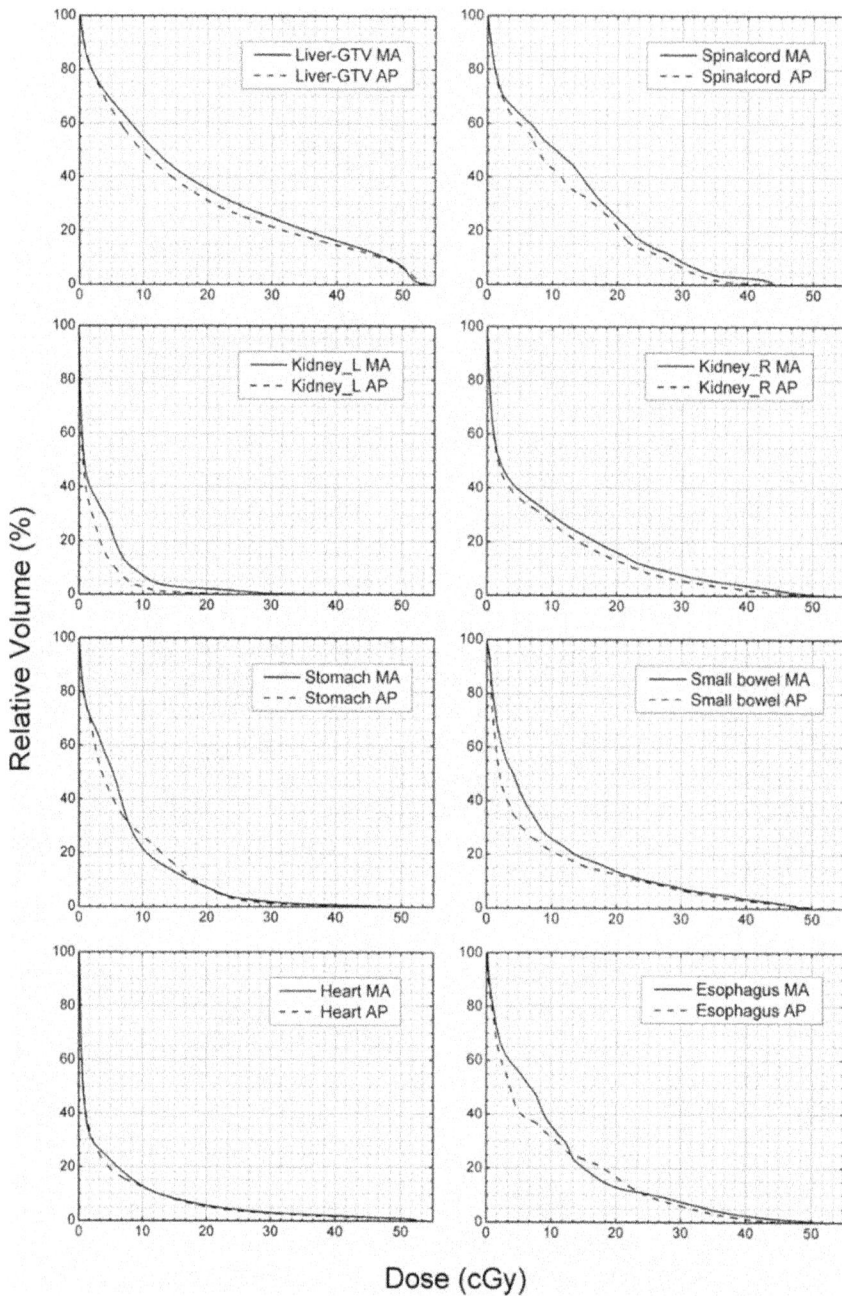

Figure 26.6. Dose–volume histogram comparison of the manual plans (solid) and the KBP-based plan (dashed), from Zhang *et al.* Reprinted from [54] with the permission of AIP Publishing (CC BY).

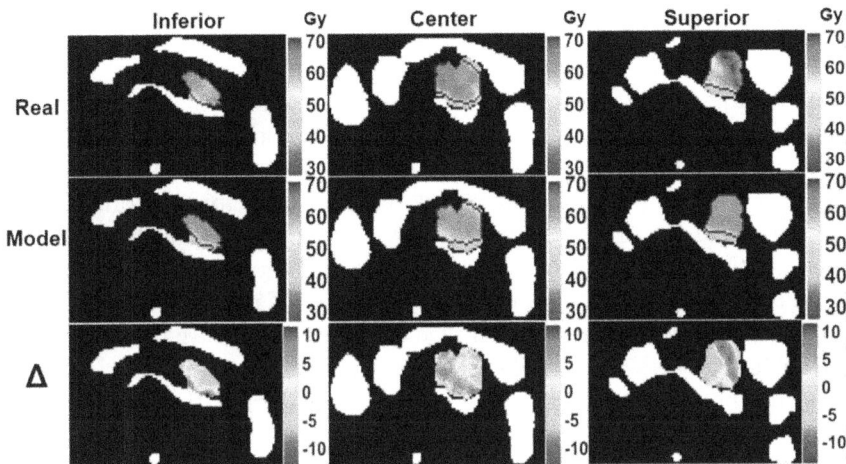

Figure 26.7. Comparison of real (top) and model (middle) doses, as well as their differences (real − model) (bottom) from the study by Thomas *et al.* Reprinted from [56] with the permission of AIP Publishing (CC BY).

26.4 Treatment delivery

26.4.1 Daily image guidance

Daily image guidance is imperative for each fraction with which the patient is treated. It can be used for patient alignment to the plan's coordinate system, ensuring that the treatment is being delivered to the correct location. In addition, it can be used to identify and match the patient to the correct treatment. Lamb *et al* developed a framework that uses a 2D-to-3D image registration technique to both identify and localize the patient [57]. Fixed in-room kilovoltage image systems take some daily setup x-ray images. Using these setup images, the 2D-to-3D registration algorithm aligns the patient's planning CT to the correct position. Digitally reconstructed radiographs are then computed from the CT to compare against the setup x-rays to ensure that the alignment is correct. Lamb *et al* found that 100% of incorrectly matched patients were identified by the system. Jani *et al* developed an imaging system that is able to take the daily 3D CBCT or megavoltage CT setup images and automatically aligns the patient's planning CT anatomy to the patient's current position on the treatment table [58]. This allows for the patient to be aligned to each of their fractions throughout their treatment course. In addition, the authors also developed their framework to be able to identify the patient using the daily CT, to make sure that the treatment is given to the correct patient. The framework utilizes feature selection via principal component analysis, and linear discriminant analysis was used to classify the patient as the correct or incorrect patient for the plan. For alignment, the framework utilized similarity metrics, including mutual information, cross-correlation, and structural similarity [59]. The authors found a low misclassification error of under 2% on tomotherapy images and under 3.5% on TrueBeam images, with all incorrectly matched patients having a correlation coefficient of less than 0.5.

26.4.2 Adaptive planning

Frameworks that focus on adaptive planning leverage information from the initial segmentation and treatment planning phase in order to accurately predict aspects of the plan for a later fraction, when the patient's anatomy has changed enough to warrant plan adaption. These types of AI frameworks are typically designed to be very fast due to the short time constraints planners have for adapting a treatment plan. Paulson *et al* implemented a magnetic resonance (MR)–guided online adaptive radiotherapy (MRgOART) workflow and evaluated it on patients with abdominal cancer treated with SBRT [60]. The specific MRgOART platform was created by the developers of the Elekta Unity MR-Linac [61]. The team developed a system that utilizes an 'adapt to shape' method, which generates a new plan based on the patient's current anatomy of the day. After contour editing, the adaptive plans are generated using frameworks that can find the Pareto optimal plan with the same tradeoffs as the reference plan. The authors found that the mean ArcCheck 3D and the daily adaptive plan's secondary dose calculation's gamma passing rate was 99.3% and 97.5%, respectively.

Liang *et al* proposed an auto-contouring method for abdominal MRgOART [62]. The authors use two types of information in their AI framework, which they call top-down and bottom-up information. Top-down data are calculated directly on the MR images, where it roughly transfers the initial planning contours onto the current fraction's image. The bottom-up data are calculated using an SVM on the pixel data. The two types of information are combined using AI fusion. The schematic is shown in figure 26.8. The framework Liang *et al* developed was able to achieve a DSC of over 0.82, which is much higher than the DSC of 0.7 that suggests suitable overlap [63].

For MRgOART on PC, Liang *et al* proposed a fast square-window CNN architecture for gross tumor volume segmentation [64]. From 37 MRI sets from

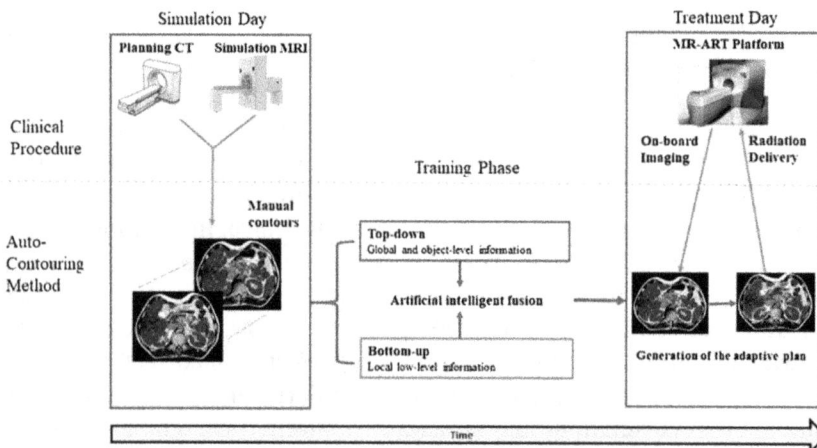

Figure 26.8. Schematic of the auto-contouring algorithm by Liang *et al.* Reprinted from [62], Copyright (2018), with permission from Elsevier.

Transverse   Sagittal   Coronal

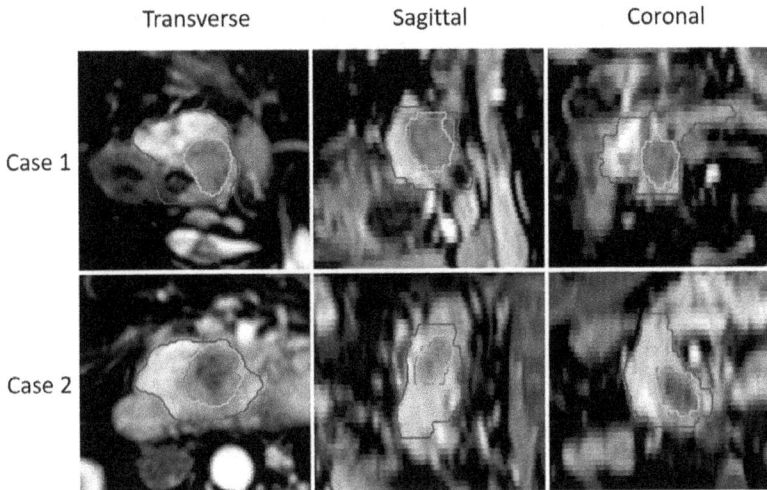

Case 1

Case 2

Figure 26.9. AI-predicted contours (cyan), manually segmented contours (red), and pancreas head contours (blue), as shown by the results from Liang *et al.* Reprinted from [64], Copyright (2020), with permission from Elsevier.

27 patients, the authors generated 245 000 normal and 230 000 tumor patches for training the deep learning model. On another 19 MRI test sets from 13 test patients, the model was able to achieve a DSC of 0.73 ± 0.09, a Hausdorff distance of 8.11 ± 4.09 mm, and a mean surface distance of 1.82 ± 0.84 mm. Visually, the deep learning model's predications had high agreement with the ground truth as shown in figure 26.9. In addition, the authors performed a human radiation oncologist interobserver study and found that the interobserver variations to be 0.71 ± 0.08, 7.36 ± 2.72 mm, and 1.78 ± 0.66, respectively, and found the difference between model's variability versus that of the human physicians to be statistically insignificant. Liu *et al* proposed a pancreatic multi-organ segmentation framework using an attention U-net framework [65]. Using 30 patients with pancreatic SBRT as the training set, the framework was trained to segment the liver, stomach, spinal cord, kidneys, bowels, and duodenum. The attention mechanism was utilized to more accurately locate the organ boundaries. The study found that, on the test data set, the attention U-net was capable of accurately segmenting each organ with a mean DSC of 0.86 ± 0.02 (liver), 0.88 ± 0.06 (stomach), 0.75 ± 0.04 (spinal cord), 0.86 ± 0.04 (left kidney), 0.87 ± 0.06 (right kidney), 0.89 ± 0.05 (large bowel), 0.86 ± 0.04 (small bowel), and 0.79 ± 0.04 (duodenum).

Fu *et al* also developed a segmentation method using a CNN-based correction network for MRgOART [66]. The treatment sites they investigated were the liver, stomach, pancreas, adrenal gland, and prostate. Their CNN first performs a voxelwise prediction on the image to segment it, and then the correction network finetunes the prediction, fixing incorrect classifications from the CNN. The model was trained, validated, and tested on 100, 10, and 10 data sets, respectively. The authors found that their model outperformed a technique known as conditional random field and had DSC scores of 95.3 ± 0.73 (liver), 93.1 ± 2.22 (kidneys), 85.0 ± 3.75 (stomach), 86.6 ± 2.69 (bowel), and 65.5 ± 8.90 (duodenum).

26.5 Follow-up

26.5.1 Outcome prediction

Toxicities induced by radiation on healthy tissue remain a major issue and limiting factor for radiation therapy. Due to the advent and improvements of deep learning, highly accurate toxicity risk assessment models can be achieved. These can greatly improve treatment planning quality. Ibragimov *et al* looked into the prediction of hepatobiliary toxicity from liver SBRT treatment [67] using a deep learning CNN. The authors first pretrained the CNN on CT images of 2644 human organs, and then a transfer learning technique was applied to fine-tune the model with a training set of 125 patients with liver cancer treated with SBRT. The authors additionally performed a study that inserted pretreatment features, including data such as patient demographics and underlying liver diseases, into the fully connected layers of the network. They estimated the toxicity to the organs at risk (OARs) due to radiation by using the saliency maps from the CNN. The authors found that the deep learning CNN model alone could accurately predict the hepatobiliary toxicity with an AUC of 0.79, and, with the pretreatment features, the AUC improved to 0.85. The authors later performed a related study for the identification of critical regions associated with the toxicities after liver SBRT [68]. They trained CNNs to take CT images and radiation therapy doses as inputs and to predict the post-treatment toxicities without any additional anatomical information such as the segmentation of the OARs. The authors found that, for 122 patients treated with liver SBRT with a median follow-up of 13 months, their model was able to accurately predict the toxicity with an AUC of 0.85, without the knowledge of any OAR segmentations.

26.6 Discussion and conclusion

Since the explosion of AI technologies in recent years, deep learning has been utilized in improving radiation therapy workflow and has demonstrated potential to revolutionize the entire radiation therapy treatment workflow for abdominal cancers. Currently, deep learning research for abdominal cancers mostly focuses on segmentation, CT synthesis, and toxicity prediction. Treatment planning, daily image guidance, motion management, and adaptive decision support are growing areas for AI-related projects. It may be feasible to have a model predict the treatment plan directly from the CT simulation. For daily image guidance and motion management, further improvement can be made with real-time internal motion prediction and correction. CBCT-based adaptive planning for abdominal cancer also has many research opportunities for improving workflow efficiency and plan quality. Despite the advantages of the AI technique, there are barriers to a widespread clinical use. Although the AI models are trained to perform 'human-like' tasks, the diversity of clinical operation can compromise AI's effectiveness in real-world operation. While there has been progress in the AI field on interpretability, deep learning models largely operate as a complex black box, and their performance is highly dependent on the data set that they train on. In addition, these models tend to be poor at extrapolation, yielding poor results when the patient data are outside of the domain that the model trained on. Along with the obstacle of data sharing among clinics, limited access to AI expertise and to large repository data for AI modeling makes it difficult for clinics to initiate AI application. However, as the AI landscape continues to grow, there are still many research prospects for improving model robustness and uncertainty estimations as a pathway towards safer AI. We expect more researchers to fully investigate these areas in abdominal cancer radiation therapy.

References

[1] Bray F *et al* 2018 Global cancer statistics 2018: GLOBOCAN estimates of incidence and mortality worldwide for 36 cancers in 185 countries *CA: A Cancer J. Clin.* **68** 394–424

[2] Siegel R L, Miller K D and Jemal A 2019 Cancer statistics, 2019 *CA: A Cancer J. Clin.* **69** 7–34

[3] McCloskey S A and Yang G Y 2009 Benefits and challenges of radiation therapy in gastric cancer: techniques for improving outcomes *Gastrointest. Cancer Res.: GCR* **3** 15

[4] Badiyan S *et al* 2020 Immunotherapy and radiation therapy for gastrointestinal malignancies: hope or hype? *Transl. Gastroenterol. Hepatol.* **5** 21

[5] LeCun Y *et al* 1989 Backpropagation applied to handwritten zip code recognition *Neural Comput.* **1** 541–51

[6] Long J, Shelhamer E and Darrell T 2015 Fully convolutional networks for semantic segmentation *Proc. IEEE Conf. on Computer Vision and Pattern Recognition* 3431–40

[7] Ronneberger O, Fischer P and Brox T 2015 U-Net: convolutional networks for biomedical image segmentation *Int. Conf. on Medical Image Computing and Computer-Assisted Intervention* 234–41

[8] Malhotra A, Rachet B, Bonaventure A, Pereira S and Woods L 2020 SO-13 Can we screen for pancreatic cancer? Identifying a sub-population of patients at high risk of subsequent diagnosis using machine learning techniques applied to primary care data *Ann. Oncol.* **31** S221–2

[9] Zhu M *et al* 2013 Differentiation of pancreatic cancer and chronic pancreatitis using computer-aided diagnosis of endoscopic ultrasound (EUS) images: a diagnostic test *PLoS One* **8** e63820

[10] Das A, Nguyen C C, Li F and Li B 2008 Digital image analysis of EUS images accurately differentiates pancreatic cancer from chronic pancreatitis and normal tissue *Gastrointest. Endosc.* **67** 861–7

[11] Cazacu I M *et al* 2019 Artificial intelligence in pancreatic cancer: toward precision diagnosis *Endosc. Ultrasound* **8** 357

[12] Wang P *et al* 2019 Real-time automatic detection system increases colonoscopic polyp and adenoma detection rates: a prospective randomised controlled study *Gut* **68** 1813–9

[13] Badrinarayanan V, Kendall A and Cipolla R 2017 SegNet: a deep convolutional encoder-decoder architecture for image segmentation *IEEE Trans. Pattern Anal. Mach. Intell.* **39** 2481–95

[14] Kiani A *et al* 2020 Impact of a deep learning assistant on the histopathologic classification of liver cancer *NPJ Digital Med.* **3** 1–8

[15] Huang G, Liu Z, van der Maaten L and Weinberger K Q 2017 Densely connected convolutional networks *30th IEEE Conf. on Computer Vision and Pattern Recognition (Cvpr 2017)* **1** 2261–9

[16] Liu Y *et al* 2019 MRI-based treatment planning for liver stereotactic body radiotherapy: validation of a deep learning-based synthetic CT generation method *Br. J. Radiol.* **92** 20190067

[17] Liu Y *et al* 2019 MRI-based treatment planning for proton radiotherapy: dosimetric validation of a deep learning-based liver synthetic CT generation method *Phys. Med. Biol.* **64** 145015

[18] Zhu J-Y, Park T, Isola P and Efros A A, *Proc. IEEE Int. Conf. on Computer Vision* 2223–32

[19] Liu Y *et al* 2020 CBCT-based synthetic CT generation using deep-attention cycleGAN for pancreatic adaptive radiotherapy *Med. Phys.* **47** 2472–83

[20] Hu P *et al* 2017 Automatic abdominal multi-organ segmentation using deep convolutional neural network and time-implicit level sets *Int. J. Comp. Assist. Radiol. Surg.* **12** 399–411

[21] Jackson P *et al* 2018 Deep learning renal segmentation for fully automated radiation dose estimation in unsealed source therapy *Front. Oncol.* **8** 215

[22] Yuan Y 2017 Hierarchical convolutional-deconvolutional neural networks for automatic liver and tumor segmentation arXiv:1710.04540

[23] Ben-Cohen A, Diamant I, Klang E, Amitai M and Greenspan H 2016 *Deep Learning and Data Labeling for Medical Applications* (Berlin: Springer) pp 77–85

[24] Lu F, Wu F, Hu P, Peng Z and Kong D 2017 Automatic 3D liver location and segmentation via convolutional neural network and graph cut *Int. J. Comp. Assist. Radiol. Surg.* **12** 171–82

[25] Qin W *et al* 2018 Superpixel-based and boundary-sensitive convolutional neural network for automated liver segmentation *Phys. Med. Biol.* **63** 095017

[26] Christ P F *et al* 2017 Automatic liver and tumor segmentation of CT and MRI volumes using cascaded fully convolutional neural networks arXiv:1702.05970

[27] Brahme A 1988 Optimization of stationary and moving beam radiation therapy techniques *Radiother. Oncol.* **12** 129–40

[28] Otto K 2008 Volumetric modulated arc therapy: IMRT in a single gantry arc *Med. Phys.* **35** 310–17

[29] Zhu X *et al* 2011 A planning quality evaluation tool for prostate adaptive IMRT based on machine learning *Med. Phys.* **38** 719–26

[30] Appenzoller L M, Michalski J M, Thorstad W L, Mutic S and Moore K L 2012 Predicting dose-volume histograms for organs-at-risk in IMRT planning *Med. Phys.* **39** 7446–61

[31] Wu B *et al* 2014 Improved robotic stereotactic body radiation therapy plan quality and planning efficacy for organ-confined prostate cancer utilizing overlap-volume histogram-driven planning methodology *Radiother. Oncol.* **112** 221–6

[32] Shiraishi S, Tan J, Olsen L A and Moore K L 2015 Knowledge-based prediction of plan quality metrics in intracranial stereotactic radiosurgery *Med. Phys.* **42** 908–17

[33] Moore K L, Brame R S, Low D A and Mutic S 2011 Experience-based quality control of clinical intensity-modulated radiotherapy planning *Int. J. Radiat. Oncol. Biol. Phys.* **81** 545–51

[34] Shiraishi S and Moore K L 2016 Knowledge-based prediction of three-dimensional dose distributions for external beam radiotherapy *Med. Phys.* **43** 378–87

[35] Wu B *et al* 2009 Patient geometry-driven information retrieval for IMRT treatment plan quality control *Med. Phys.* **36** 5497–505

[36] Wu B *et al* 2011 Data-driven approach to generating achievable dose–volume histogram objectives in intensity-modulated radiotherapy planning *Int. J. Radiat. Oncol. Biol. Phys.* **79** 1241–47

[37] Wu B *et al* 2013 Using overlap volume histogram and IMRT plan data to guide and automate VMAT planning: a head-and-neck case study *Med. Phys.* **40** 021714

[38] Tran A *et al* 2017 Predicting liver SBRT eligibility and plan quality for VMAT and 4π plans *Radiat. Oncol.* **12** 70

[39] Yuan L *et al* 2012 Quantitative analysis of the factors which affect the interpatient organ-at-risk dose sparing variation in IMRT plans *Med. Phys.* **39** 6868–78

[40] Lian J *et al* 2013 Modeling the dosimetry of organ-at-risk in head and neck IMRT planning: An intertechnique and interinstitutional study *Med. Phys.* **40** 121704

[41] Folkerts M M, Gu X, Lu W, Radke R J and Jiang S B 2016 SU-G-TeP1-09: modality-specific dose gradient modeling for prostate imrt using spherical distance maps of PTV and isodose contours *Med. Phys.* **43** 3653–54

[42] Barragán-Montero A M *et al* 2019 Three-dimensional dose prediction for lung IMRT patients with deep neural networks: robust learning from heterogeneous beam configurations *Med. Phys.* **46** 3679–91

[43] Kearney V, Chan J W, Haaf S, Descovich M and Solberg T D 2018 DoseNet: a volumetric dose prediction algorithm using 3D fully-convolutional neural networks *Phys. Med. Biol.* **63** 235022

[44] Ma J *et al* 2021 A feasibility study on deep learning-based individualized 3D dose distribution prediction *Med. Phys.* **48** 4438–47

[45] Bohara G, Barkousaraie A S, Jiang S and Nguyen D 2020 Using deep learning to predict beam-tunable Pareto optimal dose distribution for intensity modulated radiation therapy arXiv:2006.11236

[46] Nguyen D *et al* 2019 3D radiotherapy dose prediction on head and neck cancer patients with a hierarchically densely connected U-net deep learning architecture *Phys. Med. Biol.* **64** 065020

[47] Nguyen D *et al* 2019 A feasibility study for predicting optimal radiation therapy dose distributions of prostate cancer patients from patient anatomy using deep learning *Sci. Rep.* **9** 1076

[48] Nguyen D, Barkousaraie A S, Shen C, Jia X and Jiang S 2019 Generating pareto optimal dose distributions for radiation therapy treatment planning *Lect. Notes Comput. Sci.* **11769** 59–67

[49] Nguyen D *et al* 2019 Incorporating human and learned domain knowledge into training deep neural networks: a differentiable dose volume histogram and adversarial inspired framework for generating Pareto optimal dose distributions in radiation therapy *Med. Phys.* **47** 837–49

[50] Fan J *et al* 2019 Automatic treatment planning based on three-dimensional dose distribution predicted from deep learning technique *Med. Phys.* **46** 370–81

[51] Liu Z *et al* 2019 A deep learning method for prediction of three-dimensional dose distribution of helical tomotherapy *Med. Phys.* **46** 1972–83

[52] Chen X, Men K, Li Y, Yi J and Dai J 2019 A feasibility study on an automated method to generate patient-specific dose distributions for radiotherapy using deep learning *Med. Phys.* **46** 56–64

[53] Cozzi L, Vanderstraeten R, Fogliata A, Chang F-L and Wang P-M 2021 The role of a knowledge based dose–volume histogram predictive model in the optimisation of

intensity-modulated proton plans for hepatocellular carcinoma patients *Strahlenther. Onkol.* **197** 332–42

[54] Zhang Y *et al* 2018 A knowledge-based approach to automated planning for hepatocellular carcinoma *J. Appl. Clin. Med. Phys.* **19** 50–9

[55] Campbell W G *et al* 2017 Neural network dose models for knowledge-based planning in pancreatic SBRT *Med. Phys.* **44** 6148–58

[56] Thomas M A, Fu Y and Yang D 2020 Development and evaluation of machine learning models for voxel dose predictions in online adaptive magnetic resonance guided radiation therapy *J. Appl. Clin. Med. Phys.* **21** 60–9

[57] Lamb J M, Agazaryan N and Low D A 2013 Automated patient identification and localization error detection using 2-dimensional to 3-dimensional registration of kilovoltage x-ray setup images *Int. J. Radiat. Oncol. Biol. Phys.* **87** 390–3

[58] Jani S S, Low D A and Lamb J M 2015 Automatic detection of patient identification and positioning errors in radiation therapy treatment using 3-dimensional setup images *Pract. Radiat. Oncol.* **5** 304–11

[59] Wang Z, Bovik A C, Sheikh H R and Simoncelli E P 2004 Image quality assessment: from error visibility to structural similarity *IEEE Trans. Image Process.* **13** 600–12

[60] Paulson E S *et al* 2020 4D-MRI driven MR-guided online adaptive radiotherapy for abdominal stereotactic body radiation therapy on a high field MR-Linac: Implementation and initial clinical experience *Clin. Transl. Radiat. Oncol.* **23** P72–9

[61] Winkel D *et al* 2019 Adaptive radiotherapy: the Elekta Unity MR-Linac concept *Clin. Transl. Radiat. Oncol.* **18** 54–9

[62] Liang F *et al* 2018 Abdominal, multi-organ, auto-contouring method for online adaptive magnetic resonance guided radiotherapy: an intelligent, multi-level fusion approach *Artif. Intell. Med.* **90** 34–41

[63] Zou K H *et al* 2004 Statistical validation of image segmentation quality based on a spatial overlap index1: scientific reports *Acad. Radiol.* **11** 178–89

[64] Liang Y *et al* 2020 Auto-segmentation of pancreatic tumor in multi-parametric MRI using deep convolutional neural networks *Radiother. Oncol.* **145** 193–200

[65] Liu Y *et al* 2020 *Medical Imaging 2020: Imaging Informatics for Healthcare, Research, and Applications* vol 11318 (Bellingham, WA: International Society for Optics and Photonics)

[66] Fu Y *et al* 2018 A novel MRI segmentation method using CNN-based correction network for MRI-guided adaptive radiotherapy *Med. Phys.* **45** 5129–37

[67] Ibragimov B *et al* 2018 Development of deep neural network for individualized hepatobiliary toxicity prediction after liver SBRT *Med. Phys.* **45** 4763–74

[68] Ibragimov B *et al* 2020 Deep learning for identification of critical regions associated with toxicities after liver stereotactic body radiation therapy *Med. Phys.* **47** 3721–31

www.ingramcontent.com/pod-product-compliance
Lightning Source LLC
Chambersburg PA
CBHW082118210326
41599CB00031B/5803